))) Energy Humanities

⟩⟩⟩ Energy Humanities
An Anthology

Edited by Imre Szeman and Dominic Boyer

Johns Hopkins University Press *Baltimore*

© 2017 Johns Hopkins University Press
All rights reserved. Published 2017
Printed in the United States of America on acid-free paper
9 8 7 6 5 4 3 2 1

Johns Hopkins University Press
2715 North Charles Street
Baltimore, Maryland 21218-4363
www.press.jhu.edu

Library of Congress Cataloging-in-Publication Data

Names: Szeman, Imre, 1968– | Boyer, Dominic.
Title: Energy humanities : an anthology / edited by Imre Szeman and Domi-
    nic Boyer.
Description: Baltimore : Johns Hopkins University Press, 2017. | Includes
    bibliographical references.
Identifiers: LCCN 2016019648| ISBN 9781421421889 (hardcover : acid-free
    paper) | ISBN 9781421421896 (paperback : acid-free paper) | ISBN
    9781421421902 (electronic) | ISBN 1421421887 (hardcover : acid-free
    paper) | ISBN 1421421895 (paperback : acid-free paper) | ISBN 1421421909
    (electronic)
Subjects: LCSH: Humanities—Philosophy. | Power resources—Social aspects.
    | Power resources—Philosophy. | Power resources—Political aspects. |
    Power resources—Moral and ethical aspects. | Fossil fuels—Social aspects.
    | Nuclear energy—Social aspects. | Petroleum—Social aspects.
Classification: LCC T14 .E58 2017 | DDC 333.79—dc23
    LC record available at https://lccn.loc.gov/2016019648

A catalog record for this book is available from the British Library.

*Special discounts are available for bulk purchases of this book. For more informa-
tion, please contact Special Sales at 410-516-6936 or specialsales@press.jhu.edu.*

Johns Hopkins University Press uses environmentally friendly book materials, including recycled text paper that is composed of at least 30 percent
post-consumer waste, whenever possible.

Contents

Acknowledgments

Critical engagements with energy have taken many forms and have multiple origins. The beginning point of this project in the energy humanities dates back to March 2013, when two key sites of research on energy and culture—the Center for Energy and Environmental Research in the Human Sciences (CENHS) at Rice University and the Petrocultures Research Group at the University of Alberta—first had an opportunity to share ideas and resources. Since then, members of each group have collaborated on a number of research and writing projects, including "The Rise of the Energy Humanities," the 2014 op-ed in *University Affairs* that gave an emerging field of research the name it now bears. The present volume owes its life first of all to those researchers at CENHS and in Petrocultures who have contributed to development of this emergent field, including Bill Arnold, Lynn Badia, Brent Bellamy, Gwen Bradford, Joe Campana, Adam Carlson, Ann Chen, Cecily Devereux, Farès el-Dahdah, Randal Hall, Cymene Howe, Richard Johnson, David Kahane, Jordan Kinder, Jeff Kripal, Caroline Levander, Elizabeth Long, Carrie Masiello, Cyrus Mody, Tim Morton, Albert Pope, Alexander Regier, Matthew Schneider-Mayerson, Derek Woods, and Jack Zammito.

We thank the distinguished visiting scholars and guest lecturers at CENHS and Petrocultures, each of whom has also helped give form and substance to energy humanities, including Paolo Bacigalupi, Darin Barney, Ursula Biemann, Warren Cariou, Dipesh Chakrabarty, Tom Cohen, Paul Edwards, Jón Gnarr, Akhil Gupta, Graham Harman, Gabrielle Hecht, Brenda Hillman, Dale Jamieson, Eric Klinenberg, Stephanie LeMenager, Graeme Macdonald, Mika Minio-Paluello, Timothy Mitchell, Laura Nader, Naomi Oreskes, Karen Pinkus, Doug Rogers, Antti Salminen, Suzana Sawyer, Kristin Shrader-Frechette, Allan Stoekl, Bron Taylor, Anna Tsing, and Marina Zurkow. A special thanks to Sheena Wilson, cofounder and co-director of Petrocultures, and to Michael O'Driscoll and Mark Simpson, who have been at the center of the After Oil research initiative at the University of Alberta. Their energies on behalf of energy have been infectious, multiplying possibilities in ways that the first law of thermodynamics suggests should be impossible—which is true, too, of the work of Jennifer Wenzel, who has also played an essential role in bringing greater attention to the study of energy. Special thanks also to the Rice University Provost's Office for ongoing support of CENHS and energy humanities in general.

This book benefited from the great enthusiasm of Matthew McAdam, Senior Humanities Editor at Johns Hopkins University Press, who understood the aims and ambitions of this project from the very beginning. Thanks to David Janzen, who helped us with permissions—an essential task for a book project like this one— and Miriam Mabrouk, who assisted with copy-edits. A special shout-out goes to Jeff Diamanti for feedback on the introduction and section introductions.

Finally, a huge shout-out goes to Justin Sully, who performed the same magic here that he has on any number of projects. This volume would not have found its way into print without his efforts on our behalf.

))) Energy Humanities

Introduction: On the Energy Humanities

Imre Szeman and Dominic Boyer

Energy Humanities: An Anthology brings together research that attends to the social, cultural, and political challenges posed by global warming and environmental damage and destruction. As the title suggests, the pieces collected here focus on a specific issue in relation to today's environmental challenges: energy. The use and abuse of energy have had a significant impact—perhaps *the* most significant impact—on the shape in which we find the planet today. This is especially the case when it comes to the use of fossil fuels—first coal, and then oil and natural gas. The pieces brought together here address the social as well as environmental consequences of energy once it gets industrialized across the globe. This volume makes a strong case for why it is essential to better understand the import and impact of energy when it comes to trying to puzzle out how we might address global warming. It does so not by pointing out that we remain dependent on forms of dirty energy that continue to increase the level of CO_2 in the atmosphere—or not only by doing so: for most of the planet's inhabitants, this is no longer a mystery.[1] Rather, *Energy Humanities* draws critical attention to the fact that energy is absolutely necessary for modern societies. To be modern is to depend on the capacities and abilities generated by energy. Without the forms of energy to which we've had access and which we've come to take for granted, we would never have been modern. We are citizens and subjects of fossil fuels through and through, whether we know it or not. And so any meaningful response to climate change will have to tarry with the world and the people that have been made from oil.

This strong equation of energy and modernity has two consequences. First, it necessitates a fundamental reconsideration of our understanding of the forces that have given shape to modernity. Our dominant narrative of the modern combines the expansion of rights and freedoms, the advent of scientific insights and technological innovations, and the ballooning of capitalist economies, holding these very different spheres of social life together under the sign of "progress" in a powerful way. The work of critical theory in the humanities and social sciences has been to pull apart the clunky (albeit effective) apparatus of an enlightened modernity, exposing the multiple fictions of this narrative and bringing to light the truths of the modern buried beneath the shiny drama of progress that proclaims that each year is better (richer, bigger, freer) than the one before it. That rights and freedoms—when and where they exist at all—have to take place through a process of Kantian maturation, rather than being enabled all at once, points to the limits of a liberalism born in the Industrial Revolution rather than speaking to its supposed self-evidence; and as critics of colonialism and postcolonialism have repeatedly shown, the progress and growth of the global North have been made possible only by centuries of exploitation of the people and resources of the global South.

As the contributions to this book highlight, these invaluable, important critiques of modernity have nevertheless left a key element out of our understanding of the modern—energy. Economic growth, as well as the expansion of access to the goods and services we have come to associate with the experience of modernity, is a *direct* consequence of the massive expansion of energy use by human communities, especially (though not only) in the global North;[2] the capacities and freedoms that are connected to the modern, from the opening up of leisure time to expectations of almost unfettered mobility, are similarly the consequence of a world awash in the kilocalories generated primarily by fossil fuels. While the story of modernity isn't reducible to the use of energy on an ever-greater scale, an account of its developments, transgressions, and contradictions that fails to address the role played by energy in shaping its infrastructures (cities designed around automobiles) and its subjectivities (mobile consumers with near-infinite powers—such as communicating with someone across the globe), and everything else in between, can't help but misrepresent the forces and processes shaping historical development, especially over the past two centuries. That access to and the struggle over energy have had a role in shaping modern geopolitics is evident; witness the protracted struggle over power in Africa and the Middle East and the role played by access to oil in shaping conflict in World War II.[3] What is less evident, however, is the degree to which the energy riches of the past two centuries have influenced our relationships to our bodies, molded human social relations, and impacted the imperatives of even those varied activities we group together under the term "culture."[4]

In the modern era, the rapid expansion of humans on the planet, from an estimated population of 1 billion in 1800 to 7.3 billion in 2016, has been facilitated by (perhaps even animated by) growth in the availability and accessibility of energy. And these increases in human population and energy consumption have had, in turn, a decisive impact on the state of the environment.[5] The second consequence of adding energy to our accounts of the modern experience is that it offers us a new vantage point on global warming and environmental crisis. One of the principle causes of global warming has been the emission of CO_2 produced by the burning of large quantities of fossil fuels. The problem of global warming is, at its core, an energy problem. The link between energy use and global warming may seem to be an obvious-enough point: the operations of industrial capitalism and the civilization it has brought into existence have had a deleterious impact on the global environment. It makes sense that there would be a focus in environmental studies on shifts in how we employ fossil fuels (e.g., switching from coal and oil to natural gas) or on the transitions away from fossil fuels to other forms of renewable energy. Too often, however, these changes are envisioned as narrowly technical ones. Much of the contemporary discussion about energy in relation to the environment imagines energy as an input into modern social and material processes that doesn't alter their character or nature very much, if at all: it's seen as little more than the gas that runs the engine of a society whose shape and form are largely independent of it.[6] But just as energy is essential to a fuller understanding of modernity, its critical role in shaping existing social structures, lived and material infrastructures, and even cultural practices points to those sites in which changes will *have* to take place if we are to address global warming. Even if it envisions difficult, large-scale shifts in the dominant source of energy, the existing language of energy transition is most often defensive, insisting on changes in input in order to preserve global capitalism and its systems of property and profit.[7] The texts in *Energy Humanities* move beyond the limits of such affirmations of the present state of things and speak instead to the widespread social, cultural, and political changes that are necessary if we are to truly address global warming and its multiple consequences.

As an increasing number of researchers have insisted, the challenge of addressing global warming isn't fundamentally a scientific or technological one.[8] Environmental scientists have played a crucial role in identifying the causes and consequences of global warming, including projections of what might occur if we fail to keep increases in global temperature to less than 2.0°C, as it appears we are poised to do.[9] However, the next steps in addressing environmental crisis will have to come from the humanities and social sciences—from those disciplines that have long attended to the intricacies of social processes, the nature and capacity of political change, and the circulation and organization of symbolic meaning through culture. This constitutes an enormous challenge and is one that we have barely begun to take up. What we need to do is, first, grasp the full intricacies of our imbrication with energy systems (and with fossil fuels in particular), and second, map out other ways of being, behaving, and belonging in relation to both old and new forms of energy. The task is nothing less than to reimagine modernity, and in the process to figure ourselves as different kinds of beings than the ones who have built a civilization on the promises, intensities, and fantasies of a particularly dirty, destructive form of energy. It is a large enough challenge that many engaged in research in the energy humanities wonder whether we have the conceptual, affective, material, and collective capacities to take it on.

The refigurations to which the work of energy humanities draws attention go beyond changes to driving habits or the establishment of stricter policies on emissions and the energy efficiency of new homes. The more difficult changes are those that are harder to see, name, or grasp, those zones of experience and expectation generated by our energy systems that we take as equivalent to normal life—what might well be described as the energy dimension of the "spontaneous consent" of hegemony. The sharpest critics working today on the concatenation of oil and culture explore the depths of being-in-relation to our era's dominant form of energy. "Energy systems are shot through with largely unexamined cultural values, with ethical and ecological consequences," writes Stephanie LeMenager.[10] Frederick Buell argues that "it has become impossible not to feel that oil at least partially determines cultural production and reproduction on many levels." "Nowadays," he writes, "energy is more than a constraint; it (especially oil) remains an essential (and, to many, *the* essential) prop underneath humanity's material and symbolic cultures."[11] The degree to which energy has shaped modern forms of life and ways of being means that the energy humanities have to be seen as more than just a specialist field of study—a subset of environmental studies, for instance. The claim being made by this volume is a much stronger one. "The mansion of modern freedoms stands on an ever-expanding base of fossil fuel use," writes Dipesh Chakrabarty. "Most of our freedoms are energy intensive."[12] Anyone interested in understanding the material, social, and symbolic operations of an issue as important as (for instance) human freedoms *must* take into account the significance of energy in enabling the very possibility of these freedoms, and must certainly do so if they want to grapple with their continuation or extension in an era of environmental challenges and diminishing energy resources.[13] Every evocation of Rousseau or Jefferson today needs to be accompanied by information on per capita energy use and knowledge about the sources and implications of this energy configuration for the operations of politics at every scale, from personal politics to geopolitics.[14]

Energy humanities is a burgeoning field, with a huge amount of research developing over the past decade.[15] The work collected here emerges out of the specific coordinates of our contemporary environmental crises and struggles over the use and abuse of energy that have made the broad social significance of energy increasingly difficult to avoid. Like any new area of research, recent explorations of energy and society build on earlier studies that have addressed the social and cultural import of fossil fuels. Lewis Mumford's influential *Technics and Civilization* (1934) was among the first books to attend to the social impacts of shifts in energy, recognizing the broad changes produced by (for example) the movement from coal-fired steam power to the electric motors that were emerging in the 1930s.[16] In "Energy and the Evolution of Culture" (1943), anthropologist Leslie White linked cultural development directly to the amount of energy available to human communities; his attention to the link between the ever-greater use of fossil fuels and the expansion of social systems was repeated in anthropological studies following the 1973 oil crisis, and again in the past few years in the wake of the 2008 financial crisis.[17] In early environmental studies, E. J. Schumacher's influential *Small Is Beautiful: A Study of Economics As If People Mattered* (1973) begins by noting the short- and long-term consequences of the ever-expanding use of fossil fuels—a source of fuel that isn't renewable, generates pollution, and reinforces capitalism's insistence that "bigger is better."[18] The connections that have been repeatedly drawn between the growth in the size of human communities and the growth in their economies has a long tradition of critical analysis—another work we might mention in this vein is Jean-Claude Debeir, Jean-Paul Deléage, and Daniel Hémery's *In the Servitude of Power: Energy and Civilization through the Ages* (1986)[19]—although one whose force and effectivity have ebbed and waned along with the price of energy and the difficulty of keeping the social import of fossil fuels front and center for academics and publics alike.

What distinguishes contemporary critical attention to energy and fossil fuels is the growing recognition that we now fully inhabit the difficult circumstances of which Mumford, White, and other critics forewarned. As the energy source around which we have shaped our social and economic development, the fact that fossil fuels are in ever-greater demand at a moment when there are anxieties about their long-term availability, as well as growing environmental challenges to their necessity and legitimacy, means that energy is on our minds as never before. While there are fluctuations in the demand for oil at any given moment, even in the best-case scenario outlined by the World Energy Council, we can expect to use 27% more energy in 2050 than today.[20] At the same time, the most recent report from the Emission Database for Global Atmosphere Research suggests that annual global emissions of CO_2 have increased *significantly* since the UN Framework Convention on Climate Change in 1992, an agreement whose aim was to have had the opposite outcome.[21] The difficult coordinates of our own circumstances do not stop there. Access to energy is a key component of development. Those countries whose citizens currently use significantly less energy than the average European or North American have expectations of using more, so as to gain the capacities and opportunities that attend the expanded use of energy.[22] While some of this energy will come in the form of renewables, the infrastructures and mechanisms supporting global modernity demand the use of fossil fuels, which means that, in large part, the development of the global South requires the increased use of fossil fuels.[23] In the tension established between North and South, and between oil producers and consumers, the opening decades of the twenty-first century are unwittingly establishing the conditions for an expansion and intensification of the geopolitical conflict around energy—something about which a global political and economic elite seem aware, but about which they seem inclined to do relatively little.

This gap between knowledge and action is important in how we figure the next steps in environmental politics. Despite ample evidence to the contrary, there continues to be belief and expectation that scientific evidence will, of its own accord, communicate—and hence trigger—the social and political changes needed to address climate change. This is one of the hoped-for outcomes of such expansive collections of scientific expertise as the Intergovernmental Panel on Climate Change (IPCC), whose fifth iteration brought together the work of thousands of scientists and reported that it is "extremely likely" (95%–100% probability) that humans are the dominant cause of global warning. And yet, as more and more scholars are coming to recognize, quantification of global environmental threats through scientific research has, on its own, "failed to effect anything resembling the radical change likely to be required in order to avert environmental catastrophe."[24] The frustrating impasses that have appeared in naming environmental problems have characterized the communication and analysis of energy as well. In *Carbon Nation: Fossil Fuels in the Making of American Culture* (2014), historian Bob Johnson remarks that "we industrial peoples have preferred to keep our energy dependencies out of sight."[25] One of the issues explored by many of the contributions in *Energy Humanities* is the structure and function of what might be termed "energy epistemologies."[26] Not only energy in general but also fossil fuels in particular have been surprisingly hard to figure—narratively, visually, conceptually—as a central element of the modern. Petroleum firms have been among the biggest companies in the world since the modern advent of oil, and they remain so even in an era of computers and social media. An alarming array of everyday goods, without which we might find it hard to live, are made up of petroleum by-products.[27] And the geopolitics of the modern era—and especially of the period following World War II—have been shaped by the struggle over access to and control over fossil fuels. Despite this, recent critical scholarship has had to account for the ways in which fossil fuels have managed to hide in plain sight/site, evading inclusion in our economic calculations as much as in our literary fictions.[28]

Recent film, fiction, and visual arts have also explored the character of our energy epistemologies, with the aim of grasping the curious invisibility of such a powerful substance as oil, while also trying to render fuels nameable, readable, and visible. One of the takeaways of this volume is a broader understanding of the peculiar, if hitherto unremarked, philosophical characteristics of fossil fuels, and perhaps, too, of the dominant energy source of any given era.[29] If it has been so difficult to grasp and grapple with so important an element, it is in many respects because fossil fuels are saturated into every aspect of our social substance. The dark black, inky liquid that we sometimes encounter as oil is in fact a ruse: it gives away this obvious sign of itself, dead and harmless, so that it might all the more powerfully inhabit and shape the modern under the cover and with the force of its own darkness.

How might we use the critical insights provided by research in energy humanities to develop a different relationship to energy, to fossil fuels? One beginning point is to consider how we have imagined our relation to history.[30] We've tended to allow history to just happen to us. In the modern period, this is in part due to our faith in the forward and upward pull of technology, and in part because the calculus of progress insists that we will, by the forward march of time alone, of necessity be better off than our predecessors. This is not to say that history hasn't also been shaped and guided by those with a vested interest in retaining or attaining power, and equally by those who have wished to challenge and unnerve social, political, and economic privilege. What we haven't done—or perhaps haven't had to do before now—is take on the collective challenge of planning what comes next, and in the fullest way possible.

In the context of a now almost universally accepted faith in free markets, the suggestion of something akin to central planning can't help but invoke images—and fears—of failed, clunky, Soviet-era plans to increase collective prosperity and reshape subjectivities at the same time. Yet it is difficult to see how we might engage in the energy transition we need without plans that bring together scientific knowledge about the causes and consequences of global warming with social and cultural insights into the shape and character of our oil subjectivities. To date, the hope has been that market forces will, if managed properly, address the self-same problems they have generated. This has been, in large measure, the official response to climate change, as represented by the Kyoto Agreement and the follow-up series of international climate summits that resulted in the UN Climate Change Conference in Paris in 2015. Assigning a cost to CO_2 emissions might well help to slow down the increasing warmth of the atmosphere, at least somewhat. But placing one's faith in environmental change in a market system built around growth and profit, endless expansion, and the bottom line, and one, furthermore, premised in a fundamental way on disavowing or negating the value of natural systems, is questionable, to say the least. At the heart of the energy humanities is a political project unlike any we've encountered before. There may have been coal capitalism and oil capitalism; there cannot be solar or wind capitalism.[31] As we figure out how to no longer be oil subjects inhabiting destructive petrocultures (and it is worth remembering that the Soviet system was as much a petroculture as the capitalist variant of modernity), we will need to undertake a sociopolitical revolution that is both necessary and unavoidable.

But what will that revolution look like?[32] Energy provides us with a vector to newly imagine societies defined by an equality of opportunities and capacities—communities in which, for the first time in history, we are always already attuned to our relations to natural systems. For instance, what if our political freedoms were to now come with a material component—an equity of kilocalories or British thermal units (Btu) assigned to each individual, determined in part by how much energy the planet could bear? Are there ways in which newfound attention to energy might reinvigorate our politics, allowing us to position our material demands and impact on the planet at the core of social equity?

The revolution that energy could produce would need to attend to more than just the sharing of kilocalories. In "Nature and Revolution," Herbert Marcuse writes, "Our world emerges not only in the pure form of time and space, but also, and *simultaneously*, as a totality of sensuous qualities—object not only of the eye (synopsis) but of *all* human senses (hearing, smelling, touching, tasting). It is this qualitative, elementary, unconscious, or rather preconscious, constitution of the world of experience, it is this primary experience itself which must change radically if social change is to be radical, qualitative change."[33] Critical theory has sought to draw our attention to the multiple ways in which we are other than we imagine ourselves to be—for instance, as revealed by Marx's critique of capitalism, Freud's analysis of the liberal subject, and Nietzsche's assault on morality and philosophy. To this, the essays in this volume add an account of the energy unconscious. Our everyday practices and activities have been shaped by energy in a way that we have never fully understood. If we are to be able to address the environmental challenges we currently face, we need to understand that something like "primary experience" in Marcuse's account has been constituted by fossil fuels. If one aspect of our revolutionary transition will concern social uses of energy, another will refigure the coordinates of our primary experience, doing away with (for instance) the fundamental divide between human and nature on which the modern has been built.

To move forward, our critical work will also have to push past our inherited categories of analysis and action. Bruno Latour has noted, for example, that the critique of Enlightenment rationality that once fueled critical theory has inadvertently played into the hands of climate change deniers and racial essentialists.[34] Other scholars have noted how our epistemic tools for revolution and redemption are deeply entangled with the magnitudes of energy promised by fossil fuels.[35] Still more unsettling questions have been raised by materialist feminist scholars who argue that even terms like "Anthropocene" can reproduce the conditions of anthrocentrism they purport to analyze. Stacy Alaimo writes, for example, that we should consider how easily Anthropocene "becomes enlisted in all too familiar formulations, epistemologies, and defensive maneuvers—modes of knowing and being that are utterly incapable of adequately responding to the cataclysmic complexities of the anthropocene itself." "Anthropocene" even contains a "veneer of species pride" in its geo(onto)logical formulation, which is figured around an implicit sense that no other species could affect the lifeworld of all other species. And Claire Colebrook asks whether even the posthuman embrace of living systems might not be "a way of avoiding the extent to which man is a theoretical animal, a myopically and malevolently self-enclosed machine whose world he will always view as present for his own edification."[36]

One generative response to such concerns, as Donna Haraway has recently suggested, is to further diversify our critical conceptual resources for interrogating our current ecological condition—engaging our situation in the Plantationocene,

Capitalocene, and even Chthulucene as well—while also resolutely committing ourselves to "join forces to reconstitute refuges, to make possible partial and robust biological-cultural-political-technological recuperation and recomposition, which must include mourning irreversible losses."[37] We view the rise of energy humanities as part of this project of recuperation and recomposition. As fragile rather than omnipotent creatures, *Homo sapiens* have long sought to harness other forms of energy to magnify and extend their capacities. As that harnessing intensified with the mastery of the enormous energic potentiality of fossil fuels, human industry accelerated, creating more and more machines, institutions, expectations, and practices dependent on new energy magnitudes.[38] That acceleration has, as discussed above, led us to the brink of ecological catastrophe. Not all humans share equal culpability in this process, of course. We must interrogate the "we" that is the subject of climatological and ecological responsibility. Only certain populations in the world drove the globalization of fuel-intensive life, and they did so through centuries of colonizing violence. More than that, northern white masculinity continues to epitomize the apex species logic of entitlement that has brought us to our current situation; the Anthropocene has, in other words, always been the Andropocene.

Energy humanities thus retains a deep kinship and intimate conversation with environmental humanities, particularly with the pathbreaking efforts of materialist feminist thinkers to deliver new critical intellectual resources for understanding and remediating the biotic, social, cultural, and political dimensions of human and nonhuman life. The point of energy humanities is not to constitute a new explanatory causal monopoly (in the manner of Leslie White's argument that all life can now be reduced to energy) that can then be used to dominate other analytics into submission. The point is rather to turn phenomena such as global warming, species extinction, and environmental degradation inside out, so as to reveal how the use and abuse of energy have contributed to the making of what Anna Tsing terms the "damaged planet." We wish to shed light on the fuel apparatus of modernity, which is all too often invisible or subterranean, but which pumps and seeps into the groundwaters of politics, culture, institutions, and knowledge in unexpected ways. Moreover, energy humanities aspires to provide a speculative impulse as well as critical diagnostics. The works included here by artists and writers such as Margaret Atwood, Paolo Bacigalupi, and Marina Zurkow schematize the futures that beckon if our current trajectories remain uninflected. They also probe and surface the contradictions of our contemporary condition, materializing and communicating them in new and provocative ways. There is a place for sober criticism and discussion in the enterprise of energy humanities; there is also a place for surreal vision and wild imagination. It will take all the capacities of the arts and humanities to help transform this modernity. We hope only that this volume contributes a step in that direction, toward conversations and collaborations we've long waited to have with one another about what we want this century to become.

Notes

1. Even in the United States, which once remained an exception, a survey released in October 2015 indicated that 70% of Americans believe the science behind global warming—the highest since 2008. See Emma Howard, "Rising Numbers of Americans Believe Climate Science, Poll Shows," *Guardian*, October 13, 2015. Available at http://www.the guardian.com/environment/2015/oct/13/rising-numbers-of-american-believe-climate -science-poll-shows.

2. See, for instance, Edward Renshaw, "The Substitution of Inanimate Energy for Animal Power," *Journal of Political Economy* 71, no. 3 (1963): 284–92; William McNeill, *Something New under the Sun: An Environmental History of the Twentieth-Century World* (New York: Norton, 2000); and Astrid Kander, Paolo Malanima, and Paul Warde, *Power to the People: Energy in Europe over the Last Five Centuries* (Princeton, NJ: Princeton University Press, 2014). The last makes a strong case that, without the energy available from fossil fuels, modern economic growth would have been impossible.

3. On the latter, see Daniel Yergin, *The Prize: The Epic Quest for Oil, Money, and Power* (New York: Free Press, 1991).

4. See, for instance, Imre Szeman's claim that "instead of challenging the fiction of surplus—as we might have hoped or expected—literature participates in it just as surely as every other social narrative in the contemporary era. Ever more narrative, ever more signification, ever more grasping after social meaning: what literature shares with the Enlightenment and capitalism is the implicit longing for the plus beyond what is." Imre Szeman, "Literature and Energy Futures," *PMLA* 126, no. 2 (March 2011): 324. In a similar fashion, Frederick Buell has mapped how a dialectic of exuberance and catastrophe characteristic of modernity has found its way into culture: "In popular and also high cultural discourse, people's bodies and psyches are refigured as oil-electric-energized systems, and avant-garde artists become the experts who most aggressively convert these energetics into new styles, new aesthetics, new poetics." Frederick Buell, "A Short History of Oil Cultures: Or, the Marriage of Catastrophe and Exuberance," *Journal of American Studies* 46, no. 2 (2012): 286–87.

5. In its 2015 *Oil Market Report*, the International Energy Agency estimated global demand for oil for the fourth quarter of the year to be 95.47 million barrels per day. In 1980, by comparison, total world oil consumption was 59.93 million barrels per day—a 63% increase over this period. Over 1 year, this constitutes a difference of more than 13 billion barrels of oil.

6. While the work of Vaclav Smil on oil and fossil fuels is exemplary, for him, too, energy has relatively little impact on society and culture. For Smil, "the amount of energy at a society's disposal puts clear limits on the overall scope of action" and little more. "Timeless literature, painting, sculpture, architecture, and music," he writes, "show no correlation with advances in energy consumption." Vaclav Smil, *Energy in World History* (Boulder, CO: Westview, 1994), 252.

7. Jonathon Porritt's *The World We Made: Alex McKay's Story from 2050* (2013) offers a prime example of this defensive view of transition. Porritt narrates a prospective end to the age of oil via a carefully managed global retreat from the use of 76 million barrels per day in 2017 to only 4 million in 2048. In Porritt's story governments react (all of a sudden) to a 2016 Intergovernmental Panel on Climate Change (IPCC) report, as well as to the growing cost of oil (!), and begin to actively make use of alternative energy sources such as solar power and biomass, as well as substituting algae-based materials for the "plastics, pharmaceuticals, paints, lubricants and so on" (171). "Of all of the projections I've used in this book," Porritt writes, "I suspect it's this one [the drop in energy use] that may cause more eyebrows to be raised than any other" (176).

In Porritt's account, the global shift away from oil is driven by the increasing cost of getting it out of the ground—a fact that in the real world seems to have had relatively little impact on oil production, and certainly not to the degree he suggests. Porritt views the infrastructure of the "world that we made" as being able to sustain a transition to other forms of energy over a short period of time without any major disruptions in global capitalism, the size of populations, transportation systems, or other elements of the infrastructure of modernity. The aim of the view from the future he offers is to highlight energy input changes that need to be made in order to *sustain* existing politico-economic forms and the beliefs and practices that accompany them, rather than drawing attention to their implications for both environmental futures and social justice. Indeed, in Porritt's view of things, even if oil is an incredibly important fuel source, there is in many respects nothing special about it. It is a source of energy—one of many such sources—and while its impact on the form and character of contemporary life might well be large, it has not played an especially determinate role in shaping modern life. Rather, for Porritt (as for many others who try to outthink or think past the limits of oil availability), oil is a neutral substance that can be replaced in time by other forms of energy; the task of environmentalists is not to address the expectations and structures of modernity that are enabled by and also sustain oil cultures, but to work to generate new energy inputs so that modernity can continue along unabated. See Jonathon Porritt, *The World We Made: Alex McKay's Story from 2050* (London: Phaidon, 2013).

8. See, for instance, Andrew J. Hoffman, *How Culture Shapes the Climate Change Debate* (Stanford: Stanford University Press, 2015), one of an increasing number of books making this case.

9. See Robin McKie, "World Will Pass Crucial 2C Global Warming Limit, Experts Warn," *Observer*, October 10, 2015. Available at http://www.theguardian.com/environment/2015/oct/10/climate-2c-global-warming-target-fail.

10. Stephanie LeMenager, *Living Oil: Petroleum and Culture in the American Century* (Oxford: Oxford University Press, 2013), 4.

11. Buell, "Short History of Oil Cultures," 274.

12. Dipesh Chakrabarty, "The Climate of History: Four Theses," *Critical Inquiry* 35 (2009): 208.

13. See Ian Morris, *Foragers, Farmers and Fossil Fuels* (Princeton, NJ: Princeton University Press, 2015).

14. Stephanie LeMenager notes, "I became frustrated while writing *Living Oil* by how

much of what I think of as progressive modernity—feminism, environmentalism even, as it has been expressed in the U.S. in particular—is actually tied to assumptions, but also objects and paths, that have been created by fossil fuel energy." The politics of energy reaches across the political spectrum, as well as across scales of the political. See Brent Ryan Bellamy, Stephanie LeMenager, and Imre Szeman, "When Energy Is the Focus: Aesthetics, Politics, and Pedagogy: A Conversation," *Postmodern Culture*, forthcoming.

15. In addition to those in this volume, contributors to the field include Lynn Badia, Gretchen Bakke, Ross Barrett, Ericka Beckman, Brent Bellamy, Amanda Boetzkes, Frederick Buell, Cara Daggett, Mona Damluji, Jeff Diamanti, Danine Farquharson, Sarah Fredericks, John Bellamy Foster, David Haberman, Dan Hackbarth, Jacob Darwin Hamblin, Peter Hitchcock, Matthew Huber, Naomi Klein, Kairn Klieman, Toby Lee, Jenny Lin, Ernst Logar, Andreas Malm, Arthur Mason, Ellen McLarney, John-Andrew McNeish, Marty Melosi, James Nisbet, Wendy Parker, Claire Pentecost, Fiona Polack, Doug Rogers, Peter Shulman, Rebecca Slayton, Janet Stewart, Michael Truscello, Ilana Xinos, Eric Winsberg, Daniel Worden, and Natasha Zaretsky.

16. Lewis Mumford, *Technics and Civilization* (New York: Harcourt, Brace, 1934).

17. Leslie White, "Energy and the Evolution of Culture," *American Anthropologist* 45, no. 3 (1943): 335–56. For an overview of the history of energy in anthropology, see Dominic Boyer, "Energopower: An Introduction," *Anthropology Quarterly* 87, no. 2 (2014): 309–33.

18. E. J. Schumacher, *Small Is Beautiful: Economics As If People Mattered* (New York: Perennial, 2010).

19. Jean-Claude Debeir, Jean-Paul Deléage, and Daniel Hémery, *In the Servitude of Power: Energy and Civilization through the Ages*, trans. John Barzman (London: Zed Books, 1991).

20. World Energy Council, *World Energy Insight 2013* (World Energy Council, 2013), https://www.worldenergy.org/publications/2013/world-energy-insight-2013/.

21. Netherlands Environmental Assessment Agency, *Trends in Global CO_2 Emissions: 2014 Report* (The Hague, 2014).

22. The figures detailing per capita energy usage offer a stark reminder of the planet's discrepancies. The US Energy Information Administration reports that in 2012, a resident of the United States used 313 million Btu of energy per capita; in Haiti, the figure is 3.13—a 100-fold difference.

23. See, e.g., Akhil Gupta, "An Anthropology of Electricity from the Global South," *Cultural Anthropology* 30, no. 4 (2015); and John-Andrew McNeish, Axel Borchgrevink, and Owen Logan, eds., *Contested Powers: The Politics of Energy and Development in Latin America* (London: Zed Books, 2015).

24. Sverker Sörlin, "The Changing Nature of Environmental Expertise," *Eurozine*, November 19, 2013, n.p.

25. Bob Johnson, *Carbon Nation* (Lawrence: University Press of Kansas, 2014), xxix.

26. For an elaboration of the idea of energy epistemologies, see Imre Szeman, "How to Know about Oil: Energy Epistemologies and Political Futures," *Journal of Canadian Studies / Revue d'études canadiennes* 47, no. 3 (2013): 145–68.

27. The list of products made from petroleum includes ink, tires, vitamin capsules, eyeglasses, footballs, detergents, parachutes, fertilizers, panty hose, aspirin, dyes, yarns,

nail polish, plastics, dentures, bandages, linoleum, hair coloring, surfboards—in a word: everything.

28. For a discussion of energy and economics, see Philip Mirowski, *More Heat than Light: Economics as Social Physics, Physics as Nature's Economics* (Cambridge: Cambridge University Press, 1989); and Timothy Mitchell, *Carbon Democracy* (New York: Verso, 2011). For a discussion of energy and literature, see Amitav Ghosh, "Petrofiction," *New Republic*, March 2, 1992, 29–34; and Patricia Yaeger et al., "Literature in the Ages of Wood, Tallow, Coal, Whale-Oil, Gasoline, Atomic Power and Other Energy Sources," *PMLA* 126, no. 2 (2011): 305–26.

29. Antti Salminen and Tere Vadén, *Energy and Experience: An Essay in Nafthology* (Chicago: MCM Prime, 2015).

30. For an elaboration of the complex relation of energy to history, see Jennifer Wenzel, introduction to *Fueling Culture*, ed. Imre Szeman, Jennifer Wenzel, and Patricia Yaeger (New York: Fordham University Press, 2016).

31. As Daniel Tanuro points out, "generalized commodity production has brought humanity so close to the abyss that a new long wave of growth—whether 'green,' 'selective,' or 'left-wing'—would result in a dreadful climate shift" (100). Tanuro argues that the demands of energy transition and global warming necessitate that Marxists, too, have to abandon productivist accounts of development in imagining post–fossil fuel societies and economies. Daniel Tanuro, "Marxism, Energy, and Ecology: The Moment of Truth," *Capitalism Nature Socialism* 21, no. 4 (2010): 89–101.

32. Dominic Boyer, "Revolutionary Infrastructure," in *The Promise of Infrastructure*, ed. Hannah Appel, Nikhil Anand, and Akhil Gupta (Durham, NC: Duke University Press, forthcoming).

33. Herbert Marcuse, "Nature and Revolution," in *The Essential Marcuse: Selected Writings of Philosopher and Social Critic Herbert Marcuse*, ed. Andrew Feenberg and William Leiss (Boston: Beacon, 2007), 237.

34. Bruno Latour, "Why Has Critique Run Out of Steam?," *Critical Inquiry* 30, no. 2 (2004): 225–48.

35. Boyer, "Revolutionary Infrastructure"; Mitchell, *Carbon Democracy*.

36. Claire Colebrook, "Not Symbiosis, Not Now: Why Anthropogenic Climate Change Is Not Really Human," *Oxford Literary Review* 34, no. 2 (2012): 198–99.

37. Donna Haraway, "Anthropocene, Capitalocene, Chthulucene: Making Kin," *Environmental Humanities* 6 (2015): 159–65.

38. Karen Pinkus, *Fuel: A Speculative Dictionary* (Minneapolis: University of Minnesota Press, 2016).

Another Storm Is Coming
Judy Natal

In the face of Superstorm Sandy that blasted in on the heels of Hurricanes Katrina and Rita, artist Judy Natal had to admit that the optimistic beliefs that guided her previous future-focused, environmentally driven work, *Future Perfect*, had been rapidly evaporating. Her belief that we would do the right thing and make the right environmental choices had been replaced by a visceral, ever-present anxiety about the future. *Another Storm Is Coming* is a premonition of a much darker future. Full of ominous warnings and a steadily escalating sense of foreboding, it tracks the impending disaster that is unfolding as environmental issues become more and more pressing and the need to act more immediate.

The land and water between New Orleans and Houston, commonly referred to as the Energy Coast, is a watery mirage created by human intervention and land use and devastated by hurricanes, flooding, loss of wetlands, and oil spills. It is also the "eye of the needle," where oil production and transportation in the U.S. reach their zenith. The area literally floats—a "terra infirma" that is liquid and fluidly unstable as a result of hurricanes, flooding, loss of wetlands, and the rise of the oceans as well as disastrous oil spills that have devastated both animal and human ecosystems

Natal explores both the economic boom and tragic events that forever unite these two cities, while imagining the future potential for environmental disaster that continues to exist on the Energy Coast. Drawing lines between Houston's Astrodome and the Superdome in New Orleans to demarcate the Gulf Coast, Natal eventually focuses on Galveston Bay and Port Arthur, Texas, among the largest oil ports in the country, and Cameron Parish, Louisiana, where coastline erosion, loss of wetlands, saltwater intrusion, and offshore drilling activities continue to show the effects of multiple hurricanes that devastated these areas. She commemorates the tenth anniversary of Hurricane Katrina and reflects upon our stubborn insistence to ignore all the warning signs of climate change.

Commissioned by Rice University's Center for Energy and Environmental Research in the Human Sciences (CENHS), *Another Storm Is Coming* was exhibited in conjunction with FotoFest 2016 International Biennial, whose theme was *Changing Circumstances: Looking at the Future of the Planet*.

Judy Natal: www.judynatal.com; *Another Storm Is Coming*
CENHS: http://culturesofenergy.com/

Institutional Critique

Amy De'Ath

poem based on an essay by Clint Burnham & other sentiments

If you ever walked down a street and
saw some disaster porn I will tell you
it was
        Burtynsky
Never mind the sublime sugar-daddy
journalistic for the most part
yet still expressed crudely
as I want to show:

Huge   Californian romance
Monumental & manmade
the aesthetic & the repulsive
and three times gushing
I've totally liberated myself
using a self-conscious technique
and a release from gravity
                    release from   form
and with respect to this very
fantasy of liberation and
fetishistic disavowal

I will tell you:

That art journalism unwittingly
raises the same Achilles' heel of
the artwork in the act of praising it
        like Mama raises the bar;
That we are compelled to progress
        toward the illiteracy of the future
And you can enjoy a privileged gaze
        on the despoliation of Mama

free of didacticism,

                            unchained from politics,

no one will hold it against you
but against will hold it so

fuck you negative sublime
ah you toxic sublime
fuck you technological sublime
ah you inverted sublime
fuck you industrial sublime

Burtynsky I told you I'm not
trying to editorialize, this is not
an indictment of the industry, this is
                                  what it is?
we are compelled to progress
to a dry toxic wastebed
Burtynsky I'm one of the foot soldiers
in the war on sustainability

Evidently Burtynsky wants his cake
And eat it.

Burtynsky eat your cake.
Now: apologize.

I've totally librettoed myself
such a horrific fear of dismemberment
Ah, breakdown in Kantian categories
Hmm, terrifying snowslide
Beautiful objects, horrible scenes
where obsolescence is obsolete
a screen against the abyss of desire
a densified oil filter which as
            in a photograph imagines for you

well fuck you tar sands
ah you tar sands
as the plague of scale
warps around our feet
then a poem purports to be progressive

      political in any way
anyway it could be brute presence
infinitesimally small marks
stuplimity or
rubbish &

mere middlebrow kitsch

Note: Some of the language and ideas in this poem are drawn from Clint Burnham's essay "Photography from Benjamin to Žižek, via the Petrochemical Sublime
of Edward Burtynsky."

Energy and Modernity: Histories and Futures

One of the most generative insights emerging from the work of energy humanities is the extent to which the affordances of modern life depend on certain magnitudes, forms, infrastructures, and cultures of energy. There is, for example, scarcely a modern convenience that does not require electricity for its operation. Given that that electrical power is supplied throughout most of the world by a majority mix of fossil and nuclear fuels, our every experience and practice of modern living is enabled, in a fundamental way, by particular fuel infrastructures. Fuel and electrical dependency is so banal in its omnipresence that it receives strikingly little public attention and commentary. But in an era of fossil fuel–induced climate change, our energic conditions of possibility are becoming increasingly difficult to ignore. A kind of "energy unconscious" forms in which the desire to retain and/or increase our energy-intensive modernity grinds against a suppressed awareness of the impossibility of maintaining our energic status quo.

Part of the critical work of energy humanities is to sound the depths of this energy unconscious, probing the symptoms and effects of various modernities and their entanglements with fuel and electricity. At the same time, especially in the context of literature and the arts, we find a parallel speculative impulse—often narrated in the future perfect mode of "what will have been"—to imagine how modern life might be reinvented for a post-carbon era. This first section explores these two dimensions of energy humanities through three sets of texts. The first collection (Chakrabarty, Szeman, and Nye) reflects on the conjuncture of energy and modernity and explores its implications for our critical-analytical practice in the humanities. The second group of texts (Neruda, Calvino, and Collis) consists of three works of literature, each of which unpacks a distinct and pivotal moment in fossil fueled modernity. Finally, four works (Scheer, Oreskes/Conway, Bacigalupi, and Atwood) engage speculatively with energy futures reimagining utopian and dystopian trajectories extending forward from our present condition.

Dipesh Chakrabarty's already-classic essay "The Climate of History" explores how humanist practices of historiography might respond to the geological recognition of the Anthropocene, a period that challenges conventional distinctions between "human history" and "natural history" and the separation of the historical trajectory of *Homo sapiens* from the timelines of other species. Chakrabarty argues ultimately that "climate change poses for us a question of a human collectivity, an us, pointing to a figure of the universal that escapes our capacity to experience the world. It is more like a universal that arises from a shared sense of a catastrophe" and thus, for him, evidence of a contra-Hegelian "negative universal history" that can help to decenter the figure of "the human" in world history.

Imre Szeman's "System Failure" pursues the inextricable linkage between oil and capital and, in light of discourses on climate change and peak oil, assays dominant narratives of petrocapitalism to better understand public reckoning with oil (its use or its end) as a form of disaster. Strategic realist narratives, Szeman writes, hold at their core the "blunt need for nations to protect themselves from energy disruptions." Techno-utopian narratives meanwhile cathect scientific and technological forms of salvation. Finally, eco-apocalyptic narratives take the oil–capital nexus head-on but predict an inexorably grim future. Szeman concludes, provocatively, that the problem is not that the end of oil can't be imagined. Rather, "these discourses are unable to mobilize or produce any response to a disaster we know is a direct result of the law of capitalism—limitless accumulation—it is easy to see that nature will end before capital." He thus advocates a new practice of political economic criticism that takes humanity's planetary impact much more fully into account.

Finally, David Nye's "The Great White Way" takes us back to the dawn of the "electrical sublime," revealing through fine-grained social history how electrification transformed both the lifeworld of America's cities and the cultural imagination of its artists in the late nineteenth and early twentieth centuries. Spectacular lighting changed the temporality of urban life and made dark spaces seem drab, indeed gloomy. Artificial illumination became a crucial technology of national imagining, a "white magic" whose contrasting shadows became home to new depths of secrecy and depravity in the aesthetics of *noir*.[1]

Pablo Neruda's poem "Standard Oil Co." is a child of this time, documenting the internationalization of the oil industry and a new apparatus of global power. Its poetics play with the forces unleashed from "subterranean estates, / and dead years." From a world of "sinister depth" a new form of dominion leaks out from oil wells, spreading across social and natural landscapes. When the arteries of oil begin to flow, Standard Oil arrives, "with its checks and its guns, / with its governments and its prisoners." Its dark legions then reorganize political power, remaking the fates of persons and nations, uniting a world under one "petroliferous moon." Neruda's poem captures the "resource curse" of fossil fuel development decades before that language began to circulate in the human sciences.

Italo Calvino's 1974 short story "The Petrol Pump" captures the ethos of a different moment, the oil shocks of the 1970s, when cornucopian fantasies of endless oil first became unsettled. Calvino's narrator is invested in fossil fuel's engine of modern life—the accelerated mobility to go wherever one wishes, even when there seems no place to go. Low on fuel, desperately searching for an open petrol pump, he feels anxieties of dispossession and the compression of time: "experiencing simultaneously the rise, apex, and decline of the so-called opulent societies, the same way a rotating drill pushes in an instant from one millennium to the next." His discovery of a new self-service pump proves a momentary revelation until he realizes the banality of its automation. It is only when a young woman mistakes him for a pump attendant that he rediscovers his vitality in the gift of petrol. "I want her to understand that this is an extreme act of love on my part, I want to involve her in the last blast of heat the human race can make its own, an act of love that is an act of violence too, a rape, a mortal embrace of subterranean powers." The story ends with the pair in suspension on the cusp of divorce from the Earth— "that ruthless devourer of living substances"—poised to start over again, beyond oil and the humanity that has been made from it.

Flash forward to the twenty-first century, where Stephen Collis in "Reading Wordsworth in the Tar Sands" confronts us with the effects of "unconventional" oil production on landscapes, culture, and the imagination in the tar sands of Alberta. Like Wordsworth, Collis finds the walking of land essential to his creative process. Sadly, there is little uplifting or picturesque to be found in the lands around Fort McMurray and Fort McKay. It reeks of death, devastation, and false promises. But Collis walks it with eyes open:

> Perhaps I digress—the occasion
> Was a public walk on a
> Public road—but the aesthetics
> Of the place is pure negation—
> Open maw is no landscape
> Ripped wound no terrain
> There is no viewpoint despite
> The signs and picnic tables of
> Doom's treeless playground.

The poem's work is to excoriate, literally, to tear off the hide of petronormality that refuses to acknowledge the environmental costs of this particularly brutal form of fossil fuel extraction, all done in the name of creating an Albertan economic miracle.

> Wordsworth there are things
> That are fucked up
> That we live among.

Much writing on energy futures, even when not explicitly eco-apocalyptic in tone and content, seems strikingly petrofatalist in its stunted capacity to imagine a post–fossil fuel future. The final set of writings in this section shows why this need not be so. In an excerpt from *The Solar Economy*, Hermann Scheer makes his case for a turn from the invisible hand of the market to the visible hand of the sun. Scheer, the chief architect of Germany's renewable energy transition, offers an inspiring vision of solar economy and solar citizenship relieved of the congenital inefficiencies and inequalities of long-chain energy systems like fossil and nuclear fuels. By reemphasizing the primary economy over the market economy, by rejecting the centralized authoritarian systems empowered by fossil fuels, and by embracing the fact that all the earth's wealth is ultimately owed to the sun, Scheer envisions a future modernity that promises greater local political independence paired with a more just and equitable global division of labor.

Naomi Oreskes and Erik Conway offer a different, but equally provocative, experiment in an excerpted chapter from *The Collapse of Western Civilization*. Writing in the model of a textbook history from the year 2393, the authors chronicle the miserable end of petromodernity. From the political persecution of climate scientists in the 2030s through to catastrophic droughts, sea-level rise, widespread refugeeism, and plague in the remainder of the twenty-first century, the authors deploy tropes of tragedy and how-could-they-not-have-acted-in-time remonstration grooved by memories of global devastation that brought the human species to the brink of extinction. "There is no need to rehearse the details of the human tragedy that occurred; every schoolchild knows of the terrible suffering. Suffice it to say that total losses—social, cultural, economic, and demographic—were greater than any in recorded human history." In the end, there is even a techno-utopian twist: humanity was only saved by a rogue scientist and her bioengineered black lichenized fungus that spread across the world sucking down carbon dioxide at an unprecedented rate.

Paolo Bacigalupi's extrapolative science fiction is less sanguine about the potential for technoscientific salvation. In this excerpt from his novel *The Windup Girl*, we find instead a portrait of a world centuries deeper into the Anthropocene and with no end in sight. A planet largely devoid of fossil fuel use has renationalized to a great extent, with competition over fuel and resource use spawning conflicts and genocide across the world. Advances in kinetic energy capture have helped to plug some gaps in our energic infrastructures—computer terminals are powered by foot pedals, factories are powered by massive kink springs turned by genetically engineered behemoths known as Megadonts—but the overall mood is one of poverty and desperation. Global business and science still do their work, but only in a corporate-driven neo-imperial mode in which genetic engineering becomes a weapon for assimilating or destroying national food stocks and thus for forcing markets open to future Monsantos. Meanwhile, sea-level rise has continued unabated, with Bangkok only preserved by massive sea walls withstanding the weight of the blue ocean. Bacigalupi's is a *noir* future but one not devoid of resilient, imaginative creatures that survive and even occasionally thrive in the ruins of the late Anthropocene.

Margaret Atwood, another writer who has gifted us some of the most moving works of "climate fiction," offers one of the more powerful statements in recent years of the stakes of climate change in her "It's Not Climate Change—It's Everything Change." Atwood offers glimpses of three futures. The first is comforting and utopian, a future in which the post-carbon transition has been managed smoothly and we are all living a happy, modern, sustainable life. The second future is dystopian and fear inducing, a world that has bungled its future beyond oil and instead has devolved into chaos and violence. Finally, there is a world in which some countries have tried and succeeded to live beyond oil whereas others, like Atwood's native Canada, have seemingly doubled down on fossil fuels. This is a world familiar to us, but one in which, as Atwood notes, the conversation concerning oil and the imaginableness of a post-carbon future have changed dramatically in only the past half decade. In the end, Atwood underscores the importance of the work that writers like her and other energy humanists have been doing to create new synapses in human imaginations, perhaps even new values: "Could cli-fi be a way of educating young people about the dangers that face them, and helping them to think through the problems and divine solutions? Or will it become just another part of the 'entertainment business'? Time will tell. But . . . the outbreak of such fictions is in part a response to the transition now taking place—from the consumer values of oil to the stewardship values of renewables." Atwood reminds that "we [humans] don't always act in our own worst interests," but her question as to whether we have enough time left to do what needs to be done—to avoid the scenarios brilliantly and unsettlingly described by Oreskes and Conway and by Bacigalupi, among others—remains an open one. The moral being: the work of energy humanities is only just beginning.

Notes

1. See David Nye, *Electrifying America: Social Meanings of a New Technology, 1880–1940* (Cambridge, MA: MIT Press), 66.

))) The Climate of History: Four Theses

Dipesh Chakrabarty

The current planetary crisis of climate change or global warming elicits a variety
of responses in individuals, groups, and governments, ranging from denial, dis-
connect, and indifference to a spirit of engagement and activism of varying kinds
and degrees. These responses saturate our sense of the now. Alan Weisman's best-
selling book *The World without Us* suggests a thought experiment as a way of experiencing our present: "Suppose that the worst has happened. Human extinction is
a fait accompli. . . . Picture a world from which we all suddenly vanished. . . .
Might we have left some faint, enduring mark on the universe? . . . Is it possible
that, instead of heaving a huge biological sigh of relief, the world without us would
miss us?"[1] I am drawn to Weisman's experiment as it tellingly demonstrates how
the current crisis can precipitate a sense of the present that disconnects the future
from the past by putting such a future beyond the grasp of historical sensibility.
The discipline of history exists on the assumption that our past, present, and future are connected by a certain continuity of human experience. We normally
envisage the future with the help of the same faculty that allows us to picture the
past. Weisman's thought experiment illustrates the historicist paradox that inhab-
its contemporary moods of anxiety and concern about the finitude of humanity.
To go along with Weisman's experiment, we have to insert ourselves into a future
"without us" in order to be able to visualize it. Thus, our usual historical practices
for visualizing times, past and future, times inaccessible to us personally—the
exercise of historical understanding—are thrown into a deep contradiction and
confusion. Weisman's experiment indicates how such confusion follows from our
contemporary sense of the present insofar as that present gives rise to concerns
about our future. Our historical sense of the present, in Weisman's version, has
thus become deeply destructive of our general sense of history.

I will return to Weisman's experiment in the last part of this essay. There is
much in the debate on climate change that should be of interest to those involved
in contemporary discussions about history. For as the idea gains ground that the
grave environmental risks of global warming have to do with excessive accumulation in the atmosphere of greenhouse gases produced mainly through the burning
of fossil fuel and the industrialized use of animal stock by human beings, certain
scientific propositions have come into circulation in the public domain that have
profound, even transformative, implications for how we think about human history or about what the historian C. A. Bayly recently called "the birth of the mod-

Critical Inquiry 35 (Winter 2009): 197–222. Copyright 2009, University of Chicago Press.

ern world."[2] Indeed, what scientists have said about climate change challenges not only the ideas about the human that usually sustain the discipline of history but also the analytic strategies that postcolonial and postimperial historians have deployed in the last two decades in response to the postwar scenario of decolonization and globalization.

In what follows, I present some responses to the contemporary crisis from a historian's point of view. However, a word about my own relationship to the literature on climate change—and indeed to the crisis itself—may be in order. I am a practicing historian with a strong interest in the nature of history as a form of knowledge, and my relationship to the science of global warming is derived, at some remove, from what scientists and other informed writers have written for the education of the general public. Scientific studies of global warming are often said to have originated with the discoveries of the Swedish scientist Svante Arrhenius in the 1890s, but self-conscious discussions of global warming in the public realm began in the late 1980s and early 1990s, the same period in which social scientists and humanists began to discuss globalization.[3] However, these discussions have so far run parallel to each other. While globalization, once recognized, was of immediate interest to humanists and social scientists, global warming, in spite of a good number of books published in the 1990s, did not become a public concern until the 2000s. The reasons are not far to seek. As early as 1988 James Hansen, the director of NASA's Goddard Institute of Space Studies, told a Senate committee about global warming and later remarked to a group of reporters on the same day, "It's time to stop waffling . . . and say that the greenhouse effect is here and is affecting our climate."[4] But governments, beholden to special interests and wary of political costs, would not listen. George H. W. Bush, then the president of the United States, famously quipped that he was going to fight the greenhouse effect with the "White House effect."[5] The situation changed in the 2000s when the warnings became dire, and the signs of the crisis—such as the drought in Australia, frequent cyclones and brush fires, crop failures in many parts of the world, the melting of Himalayan and other mountain glaciers and of the polar ice caps, and the increasing acidity of the seas and the damage to the food chain—became politically and economically inescapable. Added to this were growing concerns, voiced by many, about the rapid destruction of other species and about the global footprint of a human population poised to pass the nine billion mark by 2050.[6]

As the crisis gathered momentum in the last few years, I realized that all my readings in theories of globalization, Marxist analysis of capital, subaltern studies, and postcolonial criticism over the last 25 years, while enormously useful in studying globalization, had not really prepared me for making sense of this planetary conjuncture within which humanity finds itself today. The change of mood in globalization analysis may be seen by comparing Giovanni Arrighi's masterful history of world capitalism, *The Long Twentieth Century* (1994), with his more recent *Adam Smith in Beijing* (2007), which, among other things, seeks to understand the implications of the economic rise of China. The first book, a long meditation on the chaos internal to capitalist economies, ends with the thought of capitalism burning up humanity "in the horrors (or glories) of the escalating violence that has accompanied the liquidation of the Cold War world order." It is clear that the heat that burns the world in Arrighi's narrative comes from the engine of capitalism and not from global warming. By the time Arrighi comes to write *Adam Smith in Beijing*, however, he is much more concerned with the question of ecological limits to capitalism. That theme provides the concluding note of the book, suggesting the distance that a critic such as Arrighi has traveled in the 13 years that separate the publication of the two books.[7] If, indeed, globalization and global warming are born of overlapping processes, the question is, how do we bring them together in our understanding of the world?

Not being a scientist myself, I also make a fundamental assumption about the science of climate change. I assume the science to be right in its broad outlines. I thus assume that the views expressed particularly in the 2007 Fourth Assessment Report of the Intergovernmental Panel on Climate Change of the United Nations, in the *Stern Review*, and in the many books that have been published recently by scientists and scholars seeking to explain the science of global warming leave me with enough rational ground for accepting, unless the scientific consensus shifts in a major way, that there is a large measure of truth to anthropogenic theories of climate change.[8] For this position, I depend on observations such as the following one reported by Naomi Oreskes, a historian of science at the University of California, San Diego. Upon examining the abstracts of 928 papers on global warming published in specialized peer-reviewed scientific journals between 1993 and 2003, Oreskes found that not a single one sought to refute the "consensus" among scientists "over the reality of human-induced climate change." There is disagreement over the amount and direction of change. But "virtually all professional climate scientists," writes Oreskes, "agree on the reality of human-induced climate change, but debate continues on tempo and mode."[9] Indeed, in what I have read so far, I have not seen any reason yet for remaining a global warming skeptic.

The scientific consensus around the proposition that the present crisis of climate change is man-made forms the basis of what I have to say here. In the interest of clarity and focus, I present my propositions in the form of four theses. The last three theses follow from the first one. I begin with the proposition that anthropogenic explanations of climate change spell the collapse of the age-old humanist distinction between natural history and human history and end by returning to the question I opened with: How does the crisis of climate change appeal to our sense of human universals while challenging at the same time our capacity for historical understanding?

Thesis 1: Anthropogenic Explanations of Climate Change Spell the Collapse of the Age-Old Humanist Distinction between Natural History and Human History

Philosophers and students of history have often displayed a conscious tendency to separate human history—or the story of human affairs, as R. G. Collingwood put it—from natural history, sometimes proceeding even to deny that nature could ever have history quite in the same way humans have it. This practice itself has a long and rich past of which, for reasons of space and personal limitations, I can only provide a very provisional, thumbnail, and somewhat arbitrary sketch.[10]

We could begin with the old Viconian–Hobbesian idea that we, humans, could have proper knowledge of only civil and political institutions because we made them, while nature remains God's work and ultimately inscrutable to man. "The true is identical with the created: verum ipsum factum" is how Croce summarized Vico's famous dictum.[11] Vico scholars have sometimes protested that Vico did not make such a drastic separation between the natural and the human sciences as Croce and others read into his writings, but even they admit that such a reading is widespread.[12]

This Viconian understanding was to become a part of the historian's common sense in the nineteenth and twentieth centuries. It made its way into Marx's famous utterance that "men make their own history, but they do not make it just as they please" and into the title of the Marxist archaeologist V. Gordon Childe's well-known book, *Man Makes Himself*.[13] Croce seems to have been a major source of this distinction in the second half of the twentieth century through his influence on "the lonely Oxford historicist" Collingwood, who, in turn, deeply influenced E. H. Carr's 1961 book, *What Is History?*, which is still perhaps one of the best-selling books on the historian's craft.[14] Croce's thoughts, one could say, unbeknown to his legatees and with unforeseeable modifications, have triumphed in our understanding of history in the postcolonial age. Behind Croce and his adaptations of Hegel and hidden in Croce's creative misreading of his predecessors stands the more distant and foundational figure of Vico.[15] The connections here, again, are many and complex. Suffice it to say for now that Croce's 1911 book, *La filosofia di Giambattista Vico*, dedicated, significantly, to Wilhelm Windelband, was translated into English in 1913 by none other than Collingwood, who was an admirer, if not a follower, of the Italian master.

However, Collingwood's own argument for separating natural history from human ones developed its own inflections, while running, one might say, still on broadly Viconian lines as interpreted by Croce. Nature, Collingwood remarked, has no "inside." "In the case of nature, this distinction between the outside and the inside of an event does not arise. The events of nature are mere events, not the acts of agents whose thought the scientist endeavours to trace." Hence, "all history properly so called is the history of human affairs." The historian's job is "to think himself into [an] action, to discern the thought of its agent." A distinction, therefore, has "to be made between historical and non-historical human actions. . . . So far as man's conduct is determined by what may be called his animal nature, his impulses and appetites, it is non-historical; the process of those activities is a natural process." Thus, says Collingwood, "the historian is not interested in the fact that men eat and sleep and make love and thus satisfy their natural appetites; but he is interested in the social customs which they create by their thought as a framework within which these appetites find satisfaction in ways sanctioned by convention and morality." Only the history of the social construction of the body, not the history of the body as such, can be studied. By splitting the human into the natural and the social or cultural, Collingwood saw no need to bring the two together.[16]

In discussing Croce's 1893 essay "History Subsumed under the Concept of Art," Collingwood wrote, "Croce, by denying [the German idea] that history was a science at all, cut himself at one blow loose from naturalism, and set his face towards an idea of history as something radically different from nature."[17] David Roberts gives a fuller account of the more mature position in Croce. Croce drew on the writings of Ernst Mach and Henri Poincaré to argue that "the concepts of the natural sciences are human constructs elaborated for human purposes." "When we peer into nature," he said, "we find only ourselves." We do not "understand ourselves best as part of the natural world." So, as Roberts puts it, "Croce proclaimed that there is no world but the human world, then took over the central doctrine of Vico that we can know the human world because we have made it." For Croce, then, all material objects were subsumed into human thought. No rocks, for example, existed in themselves. Croce's idealism, Roberts explains, "does not mean that rocks, for example, 'don't exist' without human beings to think them. Apart from human concern and language, they neither exist nor do not exist, since 'exist' is a human concept that has meaning only within a context of human concerns and purposes."[18] Both Croce and Collingwood would thus enfold human history and nature, to the extent that the latter could be said to have history, into purposive human action. What exists beyond that does not "exist" because it does not exist for humans in any meaningful sense.

In the twentieth century, however, other arguments, more sociological or materialist, have existed alongside the Viconian one. They too have continued to justify the separation of human from natural history. One influential though perhaps infamous example would be the booklet on the Marxist philosophy of history that Stalin published in 1938, *Dialectical and Historical Materialism*. This is how Stalin put the problem:

> Geographical environment is unquestionably one of the constant and indispensable
> conditions of development of society and, of course, . . . [it] accelerates or retards its

development. But its influence is not the *determining* influence, inasmuch as the changes and development of society proceed at an incomparably faster rate than the changes and development of geographical environment. In the space of 3000 years three different social systems have been successfully superseded in Europe: the primitive communal system, the slave system and the feudal system. . . . Yet during this period geographical conditions in Europe have either not changed at all, or have changed so slightly that geography takes no note of them. And that is quite natural. Changes in geographical environment of any importance require millions of years, whereas a few hundred or a couple of thousand years are enough for even very important changes in the system of human society.[19]

For all its dogmatic and formulaic tone, Stalin's passage captures an assumption perhaps common to historians of the mid-twentieth century: man's environment did change but changed so slowly as to make the history of man's relation to his environment almost timeless and thus not a subject of historiography at all. Even when Fernand Braudel rebelled against the state of the discipline of history as he found it in the late 1930s and proclaimed his rebellion later in 1949 through his great book *The Mediterranean*, it was clear that he rebelled mainly against historians who treated the environment simply as a silent and passive backdrop to their historical narratives, something dealt with in the introductory chapter but forgotten thereafter, as if, as Braudel put it, "the flowers did not come back every spring, the flocks of sheep migrate every year, or the ships sail on a real sea that changes with the seasons." In composing *The Mediterranean*, Braudel wanted to write a history in which the seasons—"a history of constant repetition, ever-recurring cycles"—and other recurrences in nature played an active role in molding human actions.[20] The environment, in that sense, had an agentive presence in Braudel's pages, but the idea that nature was mainly repetitive had a long and ancient history in European thought, as Gadamer showed in his discussion of Johann Gustav Droysen.[21] Braudel's position was no doubt a great advance over the kind of nature-as-a-backdrop argument that Stalin developed. But it shared a fundamental assumption, too, with the stance adopted by Stalin: the history of "man's relationship to the environment" was so slow as to be "almost timeless."[22] In today's climatologists' terms, we could say that Stalin and Braudel and others who thought thus did not have available to them the idea, now widespread in the literature on global warming, that the climate, and hence the overall environment, can sometimes reach a tipping point at which this slow and apparently timeless backdrop for human actions transforms itself with a speed that can only spell disaster for human beings.

If Braudel, to some degree, made a breach in the binary of natural/human history, one could say that the rise of environmental history in the late twentieth century made the breach wider. It could even be argued that environmental historians have sometimes indeed progressed towards producing what could be called natural histories of man. But there is a very important difference between the understanding of the human being that these histories have been based on and the agency of the human now being proposed by scientists writing on climate change. Simply put, environmental history, where it was not straightforwardly cultural, social, or economic history, looked upon human beings as biological agents. Alfred Crosby, Jr., whose book *The Columbian Exchange* did much to pioneer the "new" environmental histories in the early 1970s, put the point thus in his original preface: "Man is a biological entity before he is a Roman Catholic or a capitalist or anything else."[23] The recent book by Daniel Lord Smail, *On Deep History and the Brain*, is adventurous in attempting to connect knowledge gained from evolutionary science and neurosciences with human histories. Smail's book pursues possible connections between biology and culture—between the history of the human brain and cultural history, in particular—while being sensitive to the limits of biological reasoning. But it is the history of human biology and not any recent theses about the newly acquired geological agency of humans that concerns Smail.[24]

Scholars writing on the current climate-change crisis are indeed saying something significantly different from what environmental historians have said so far. In unwittingly destroying the artificial but time-honored distinction between natural and human histories, climate scientists posit that the human being has become something much larger than the simple biological agent that he or she always has been. Humans now wield a geological force. As Oreskes puts it: "To deny that global warming is real is precisely to deny that humans have become geological agents, changing the most basic physical processes of the earth."

> For centuries, [she continues,] scientists thought that earth processes were so large and powerful that nothing we could do could change them. This was a basic tenet of geological science: that human chronologies were insignificant compared with the vastness of geological time; that human activities were insignificant compared with the force of geological processes. And once they were. But no more. There are now so many of us cutting down so many trees and burning so many billions of tons of fossil fuels that we have indeed become geological agents. We have changed the chemistry of our atmosphere, causing sea level to rise, ice to melt, and climate to change. There is no reason to think otherwise.[25]

Biological agents, geological agents—two different names with very different consequences. Environmental history, to go by Crosby's masterful survey of the origins and the state of the field in 1995, has much to do with biology and geography but hardly ever imagined human impact on the planet on a geological scale. It was still a vision of man "as a prisoner of climate," as Crosby put it quoting Braudel, and not of man as the maker of it.[26] To call human beings geological agents is to scale up our imagination of the human. Humans are biological agents, both collectively and as individuals. They have always been so. There was no point in human history when humans were not biological agents. But we can become geological agents only historically and collectively, that is, when we have reached numbers and invented technologies that are on a scale large enough to have an impact on the planet itself. To call ourselves geological agents is to attribute to us a force on the same scale as that released at other times when there has been a mass extinction of species. We seem to be currently going through that kind of a period. The current "rate in the loss of species diversity," specialists argue, "is similar in intensity to the event around 65 million years ago which wiped out the dinosaurs."[27] Our footprint was not always that large. Humans began to acquire this agency only since the Industrial Revolution, but the process really picked up in the second half of the twentieth century. Humans have become geological agents very recently in human history. In that sense, we can say that it is only very recently that the distinction between human and natural histories—much of which had been preserved even in environmental histories that saw the two entities in interaction—has begun to collapse. For it is no longer a question simply of man having an interactive relation with nature. This humans have always had, or at least that is how man has been imagined in a large part of what is generally called the Western tradition.[28] Now it is being claimed that humans are a force of nature in the geological sense. A fundamental assumption of Western (and now universal) political thought has come undone in this crisis.[29]

Thesis 2: The Idea of the Anthropocene, the New Geological Epoch When Humans Exist as a Geological Force, Severely Qualifies Humanist Histories of Modernity/Globalization

How to combine human cultural and historical diversity with human freedom has formed one of the key underlying questions of human histories written of the period from 1750 to the years of present-day globalization. Diversity, as Gadamer pointed out with reference to Leopold von Ranke, was itself a figure of freedom in the historian's imagination of the historical process.[30] Freedom has, of course, meant different things at different times, ranging from ideas of human and citizens' rights to those of decolonization and self-rule. Freedom, one could say, is a blanket category for diverse imaginations of human autonomy and sovereignty. Looking at the works of Kant, Hegel, or Marx; nineteenth-century ideas of progress and class struggle; the struggle against slavery; the Russian and Chinese revolutions; the resistance to Nazism and Fascism; the decolonization movements of the 1950s and 1960s and the revolutions in Cuba and Vietnam; the evolution and explosion of the rights discourse; the fight for civil rights for African Americans, indigenous peoples, Indian *Dalits*, and other minorities; down to the kind of arguments that, say, Amartya Sen put forward in his book *Development as Freedom*, one could say that freedom has been the most important motif of written accounts of human history of these 250 years. Of course, as I have already noted, freedom has not always carried the same meaning for everyone. Francis Fukuyama's understanding of freedom would be significantly different from that of Sen. But this semantic capaciousness of the word only speaks to its rhetorical power.

In no discussion of freedom in the period since the Enlightenment was there ever any awareness of the geological agency that human beings were acquiring at the same time as and through processes closely linked to their acquisition of freedom. Philosophers of freedom were mainly, and understandably, concerned with how humans would escape the injustice, oppression, inequality, or even uniformity foisted on them by other humans or human-made systems. Geological time and the chronology of human histories remained unrelated. This distance between the two calendars, as we have seen, is what climate scientists now claim has collapsed. The period I have mentioned, from 1750 to now, is also the time when human beings switched from wood and other renewable fuels to large-scale use of fossil fuel—first coal and then oil and gas. The mansion of modern freedoms stands on an ever-expanding base of fossil-fuel use. Most of our freedoms so far have been energy-intensive. The period of human history usually associated with what we today think of as the institutions of civilization—the beginnings of agriculture, the founding of cities, the rise of the religions we know, the invention of writing—began about 10,000 years ago, as the planet moved from one geological period, the last ice age or the Pleistocene, to the more recent and warmer Holocene. The Holocene is the period we are supposed to be in; but the possibility of anthropogenic climate change has raised the question of its termination. Now that humans—thanks to our numbers, the burning of fossil fuel, and other related activities—have become a geological agent on the planet, some scientists have proposed that we recognize the beginning of a new geological era, one in which humans act as a main determinant of the environment of the planet. The name they have coined for this new geological age is Anthropocene. The proposal was first made by the Nobel-winning chemist Paul J. Crutzen and his collaborator, a marine science specialist, Eugene F. Stoermer. In a short statement published in 2000, they said, "Considering . . . [the] major and still growing impacts of human activities on earth and atmosphere, and at all, including global, scales, it seems to us more than appropriate to emphasize the central role of mankind in geology and ecology by proposing to use the term 'anthropocene' for the current geological epoch."[31] Crutzen elaborated on the proposal in a short piece published in *Nature* in 2002:

> For the past three centuries, the effects of humans on the global environment have escalated. Because of these anthropogenic emissions of carbon dioxide, global climate may depart significantly from natural behaviour for many millennia to come. It seems appropriate to assign the term "Anthropocene" to the present, . . . human-dominated, geological epoch, supplementing the Holocene—the warm period of the past 10–12

millennia. The Anthropocene could be said to have started in the latter part of the eighteenth century, when analyses of air trapped in polar ice showed the beginning of growing global concentrations of carbon dioxide and methane. This date also happens to coincide with James Watt's design of the steam engine in 1784.[32]

It is, of course, true that Crutzen's saying so does not make the Anthropocene an officially accepted geologic period. As Mike Davis comments, "in geology, as in biology or history, periodization is a complex, controversial art," involving, always, vigorous debates and contestation.[33] The name Holocene for "the post-glacial geological epoch of the past ten to twelve thousand years" ("A," p. 17), for example, gained no immediate acceptance when proposed—apparently by Sir Charles Lyell—in 1833. The International Geological Congress officially adopted the name at their meeting in Bologna after about 50 years in 1885 (see "A," p. 17). The same goes for Anthropocene. Scientists have engaged Crutzen and his colleagues on the question of when exactly the Anthropocene may have begun. But the February 2008 newsletter of the Geological Society of America, *GSA Today*, opens with a statement signed by the members of the Stratigraphy Commission of the Geological Society of London accepting Crutzen's definition and dating of the Anthropocene.[34] Adopting a "conservative" approach, they conclude: "Sufficient evidence has emerged of stratigraphically significant change (both elapsed and imminent) for recognition of the Anthropocene—currently a vivid yet informal metaphor of global environmental change—as a new geological epoch to be considered for formalization by international discussion."[35] There is increasing evidence that the term is gradually winning acceptance among social scientists as well.[36]

So, has the period from 1750 to now been one of freedom or that of the Anthropocene? Is the Anthropocene a critique of the narratives of freedom? Is the geological agency of humans the price we pay for the pursuit of freedom? In some ways, yes. As Edward O. Wilson said in his *The Future of Life*: "Humanity has so far played the role of planetary killer, concerned only with its own short-term survival. We have cut much of the heart out of biodiversity. . . . If Emi, the Sumatran rhino could speak, she might tell us that the twenty-first century is thus far no exception."[37] But the relation between Enlightenment themes of freedom and the collapsing of human and geological chronologies seems more complicated and contradictory than a simple binary would allow. It is true that human beings have tumbled into being a geological agent through our own decisions. The Anthropocene, one might say, has been an unintended consequence of human choices. But it is also clear that for humans any thought of the way out of our current predicament cannot but refer to the idea of deploying reason in global, collective life. As Wilson put it: "We know more about the problem now. . . . We know what to do" (FL, p. 102). Or, to quote Crutzen and Stoermer again:

Mankind will remain a major geological force for many millennia, maybe millions of years, to come. To develop a world-wide accepted strategy leading to sustainability of ecosystems against human-induced stresses will be one of the great future tasks of mankind, requiring intensive research efforts and wise application of knowledge thus acquired. . . . An exciting, but also difficult and daunting task lies ahead of the global research and engineering community to guide mankind towards global, sustainable, environmental management. ("A," p. 18)

Logically, then, in the era of the Anthropocene, we need the Enlightenment (that is, reason) even more than in the past. There is one consideration though that qualifies this optimism about the role of reason and that has to do with the most common shape that freedom takes in human societies: politics. Politics has never been based on reason alone. And politics in the age of the masses and in a world already complicated by sharp inequalities between and inside nations is something no one can control. "Sheer demographic momentum," writes Davis, "will increase the world's urban population by 3 billion people over the next 40 years (90% of them in poor cities), and no one—absolutely no one [including, one might say, scholars on the Left]—has a clue how a planet of slums, with growing food and energy crises, will accommodate their biological survival, much less their inevitable aspirations to basic happiness and dignity" ("LIS").

It is not surprising then that the crisis of climate change should produce anxieties precisely around futures that we cannot visualize. Scientists' hope that reason will guide us out of the present predicament is reminiscent of the social opposition between the myth of Science and the actual politics of the sciences that Bruno Latour discusses in his *Politics of Nature*.[38] Bereft of any sense of politics, Wilson can only articulate his sense of practicality as a philosopher's hope mixed with anxiety: "Perhaps we will act in time" (FL, p. 102). Yet the very science of global warming produces of necessity political imperatives. Tim Flannery's book, for instance, raises the dark prospects of an "Orwellian nightmare" in a chapter entitled "2084: The Carbon Dictatorship?"[39] Mark Maslin concludes his book with some gloomy thoughts: "It is unlikely that global politics will solve global warming. Technofixes are dangerous or cause problems as bad as the ones they are aimed at fixing. . . . [Global warming] requires nations and regions to plan for the next 50 years, something that most societies are unable to do because of the very short-term nature of politics." His recommendation, "we must prepare for the worst and adapt," coupled with Davis's observations about the coming "planet of slums" places the question of human freedom under the cloud of the Anthropocene.[40]

Thesis 3: The Geological Hypothesis regarding the Anthropocene Requires Us to Put Global Histories of Capital in Conversation with the Species History of Humans

Analytic frameworks engaging questions of freedom by way of critiques of capitalist globalization have not, in anyway, become obsolete in the age of climate change. If anything, as Davis shows, climate change may well end up accentuating all the inequities of the capitalist world order if the interests of the poor and vulnerable are neglected (see "LIS"). Capitalist globalization exists; so should its critiques. But these critiques do not give us an adequate hold on human history once we accept that the crisis of climate change is here with us and may exist as part of this planet for much longer than capitalism or long after capitalism has undergone many more historic mutations. The problematic of globalization allows us to read climate change only as a crisis of capitalist management. While there is no denying that climate change has profoundly to do with the history of capital, a critique that is only a critique of capital is not sufficient for addressing questions relating to human history once the crisis of climate change has been acknowledged and the Anthropocene has begun to loom on the horizon of our present. The geologic now of the Anthropocene has become entangled with the now of human history.

Scholars who study human beings in relation to the crisis of climate change and other ecological problems emerging on a world scale make a distinction between the recorded history of human beings and their deep history. Recorded history refers, very broadly, to the 10,000 years that have passed since the invention of agriculture but more usually to the last 4,000 years or so for which written records exist. Historians of modernity and "early modernity" usually move in the archives of the last 400 years. The history of humans that goes beyond these years of written records constitutes what other students of human pasts—not professional historians—call deep history. As Wilson, one of the main proponents of this distinction, writes: "Human behavior is seen as the product not just of recorded history, 10,000 years recent, but of deep history, the combined genetic and cultural changes that created humanity over hundreds of [thousands of] years."[41] It, of course, goes to the credit of Smail that he has attempted to explain to professional historians the intellectual appeal of deep history.[42]

Without such knowledge of the deep history of humanity it would be difficult to arrive at a secular understanding of why climate change constitutes a crisis for humans. Geologists and climate scientists may explain why the current phase of global warming—as distinct from the warming of the planet that has happened before—is anthropogenic in nature, but the ensuing crisis for humans is not understandable unless one works out the consequences of that warming. The consequences make sense only if we think of humans as a form of life and look on human history as part of the history of life on this planet. For, ultimately, what the warming of the planet threatens is not the geological planet itself but the very conditions, both biological and geological, on which the survival of human life as developed in the Holocene period depends.

The word that scholars such as Wilson or Crutzen use to designate life in the human form—and in other living forms—is *species*. They speak of the human being as a species and find that category useful in thinking about the nature of the current crisis. It is a word that will never occur in any standard history or political-economic analysis of globalization by scholars on the Left, for the analysis of globalization refers, for good reasons, only to the recent and recorded history of humans. Species thinking, on the other hand, is connected to the enterprise of deep history. Further, Wilson and Crutzen actually find such thinking essential to visualizing human well-being. As Wilson writes: "We need this longer view . . . not only to understand our species but more firmly to secure its future" (*SN*, p. x). The task of placing, historically, the crisis of climate change thus requires us to bring together intellectual formations that are somewhat in tension with each other: the planetary and the global; deep and recorded histories; species thinking and critiques of capital.

In saying this, I work somewhat against the grain of historians' thinking on globalization and world history. In a landmark essay published in 1995 and entitled "World History in a Global Age," Michael Geyer and Charles Bright wrote, "At the end of the twentieth century, we encounter, not a universalizing and single modernity but an integrated world of multiple and multiplying modernities." "As far as world history is concerned," they said, "there is no universalizing spirit. . . . There are, instead, many very specific, very material and pragmatic practices that await critical reflection and historical study." Yet, thanks to global connections forged by trade, empires, and capitalism, "we confront a startling new condition: humanity, which has been the subject of world history for many centuries and civilizations, has now come into the purview of all human beings. This humanity is extremely polarized into rich and poor."[43] This humanity, Geyer and Bright imply in the spirit of the philosophies of difference, is not one. It does not, they write, "form a single homogenous civilization." "Neither is this humanity any longer a mere species or a natural condition. For the first time," they say, with some existentialist flourish, "we as human beings collectively constitute ourselves and, hence, are responsible for ourselves" ("WH," p. 1059). Clearly, the scientists who advocate the idea of the Anthropocene are saying something quite the contrary. They argue that because humans constitute a particular kind of species they can, in the process of dominating other species, acquire the status of a geologic force. Humans, in other words, have become a natural condition, at least today. How do we create a conversation between these two positions?

It is understandable that the biological-sounding talk of species should worry historians. They feel concerned about their finely honed sense of contingency and freedom in human affairs having to cede ground to a more deterministic view of the world. Besides, there are always, as Smail recognizes, dangerous historical examples of the political use of biology.[44] The idea of species, it is feared, in addition, may introduce a powerful degree of essentialism in our understanding of humans. I will return to the question of contingency later in this section, but, on the issue of essentialism, Smail helpfully points out why species cannot be thought of in essentialist terms:

> Species, according to Darwin, are not fixed entities with natural essences imbued in them by the Creator. . . . Natural selection does not homogenize the individuals of a species. . . . Given this state of affairs, the search for a normal . . . nature and body type [of any particular species] is futile. And so it goes for the equally futile quest to identify "human nature." Here, as in so many areas, biology and cultural studies are fundamentally congruent.[45]

It is clear that different academic disciplines position their practitioners differently with regard to the question of how to view the human being. All disciplines have to create their objects of study. If medicine or biology reduces the human to a certain specific understanding of him or her, humanist historians often do not realize that the protagonists of their stories—persons—are reductions, too. Absent personhood, there is no human subject of history. That is why Derrida earned the wrath of Foucault by pointing out that any desire to enable or allow madness itself to speak in a history of madness would be "the *maddest* aspect" of the project.[46] An object of critical importance to humanists of all traditions, personhood is nevertheless no less of a reduction of or an abstraction from the embodied and whole human being than, say, the human skeleton discussed in an anatomy class.

The crisis of climate change calls on academics to rise above their disciplinary prejudices, for it is a crisis of many dimensions. In that context, it is interesting to observe the role that the category of species has begun to play among scholars, including economists, who have already gone further than historians in investigating and explaining the nature of this crisis. The economist Jeffrey Sachs's book, *Common Wealth*, meant for the educated but lay public, uses the idea of species as central to its argument and devotes a whole chapter to the Anthropocene.[47] In fact, the scholar from whom Sachs solicited a foreword for his book was none other than Edward Wilson. The concept of species plays a quasi-Hegelian role in Wilson's foreword in the same way as the multitude or the masses in Marxist writings. If Marxists of various hues have at different times thought that the good of humanity lay in the prospect of the oppressed or the multitude realizing their own global unity through a process of coming into self-consciousness, Wilson pins his hope on the unity possible through our collective self-recognition as a species: "Humanity has consumed or transformed enough of Earth's irreplaceable resources to be in better shape than ever before. We are smart enough and now, one hopes, well informed enough to achieve self-understanding as a unified species. . . . We will be wise to look on ourselves as a species."[48]

Yet doubts linger about the use of the idea of species in the context of climate change, and it would be good to deal with one that can easily arise among critics on the Left. One could object, for instance, that all the anthropogenic factors contributing to global warming—the burning of fossil fuel, industrialization of animal stock, the clearing of tropical and other forests, and so on—are after all part of a larger story: the unfolding of capitalism in the West and the imperial or quasi-imperial domination by the West of the rest of the world. It is from that recent history of the West that the elite of China, Japan, India, Russia, and Brazil have drawn inspiration in attempting to develop their own trajectories toward superpower politics and global domination through capitalist economic, techno-logical, and military might. If this is broadly true, then does not the talk of species or mankind simply serve to hide the reality of capitalist production and the logic of imperial—formal, informal, or machinic in a Deleuzian sense—domination that it fosters? Why should one include the poor of the world—whose carbon footprint is small anyway—by use of such all-inclusive terms as *species* or *mankind* when the blame for the current crisis should be squarely laid at the door of the rich nations in the first place and of the richer classes in the poorer ones?

We need to stay with this question a little longer; otherwise the difference be-tween the present historiography of globalization and the historiography demanded by anthropogenic theories of climate change will not be clear to us. Though some scientists would want to date the Anthropocene from the time agriculture was invented, my readings mostly suggest that our falling into the Anthropocene was neither an ancient nor an inevitable happening. Human civilization surely did not begin on condition that, one day in his history, man would have to shift from wood to coal and from coal to petroleum and gas. That there was much historical con-tingency in the transition from wood to coal as the main source of energy has been demonstrated powerfully by Kenneth Pomeranz in his pathbreaking book *The Great Divergence*.[49] Coincidences and historical accidents similarly litter the sto-ries of the "discovery" of oil, of the oil tycoons, and of the automobile industry as they do any other histories.[50] Capitalist societies themselves have not remained the same since the beginning of capitalism.[51]

Human population, too, has dramatically increased since the Second World War. India alone is now more than three times more populous than at indepen-dence in 1947. Clearly, nobody is in a position to claim that there is something inherent to the human species that has pushed us finally into the Anthropocene. We have stumbled into it. The way to it was no doubt through industrial civiliza-tion. (I do not make a distinction here between the capitalist and socialist socie-ties we have had so far, for there was never any principled difference in their use of fossil fuel.)

If the industrial way of life was what got us into this crisis, then the question is, Why think in terms of species, surely a category that belongs to a much longer history? Why could not the narrative of capitalism—and hence its critique—be sufficient as a framework for interrogating the history of climate change and understanding its consequences? It seems true that the crisis of climate change has been necessitated by the high-energy consuming models of society that capitalist industrialization has created and promoted, but the current crisis has brought into view certain other conditions for the existence of life in the human form that have no intrinsic connection to the logics of capitalist, nationalist, or socialist identities. They are connected rather to the history of life on this planet, the way different life-forms connect to one another, and the way the mass extinction of one species could spell danger for another. Without such a history of life, the crisis of climate change has no human "meaning." For, as I have said before, it is not a crisis for the inorganic planet in any meaningful sense.

In other words, the industrial way of life has acted much like the rabbit hole in Alice's story; we have slid into a state of things that forces on us a recognition of some of the parametric (that is, boundary) conditions for the existence of institutions central to our idea of modernity and the meanings we derive from them. Let me explain. Take the case of the agricultural revolution, so called, of 10,000 years ago. It was not just an expression of human inventiveness. It was made possible by certain changes in the amount of carbon dioxide in the atmosphere, a certain stability of the climate, and a degree of warming of the planet that followed the end of the Ice Age (the Pleistocene era)—things over which human beings had no control. "There can be little doubt," writes one of the editors of *Humans at the End of the Ice Age*, "that the basic phenomenon—the waning of the Ice Age—was the result of the Milankovich phenomena: the orbital and tilt relationships between the Earth and the Sun."[52] The temperature of the planet stabilized within a zone that allowed grass to grow. Barley and wheat are among the oldest of such grasses. Without this lucky "long summer" or what one climate scientist has called an "extraordinary" "fluke" of nature in the history of the planet, our industrial-agricultural way of life would not have been possible.[53] In other words, whatever our socioeconomic and technological choices, whatever the rights we wish to celebrate as our freedom, we cannot afford to destabilize conditions (such as the temperature zone in which the planet exists) that work like boundary parameters of human existence. These parameters are independent of capitalism or socialism. They have been stable for much longer than the histories of these institutions and have allowed human beings to become the dominant species on earth. Unfortunately, we have now ourselves become a geological agent disturbing these parametric conditions needed for our own existence.

This is not to deny the historical role that the richer and mainly Western nations of the world have played in emitting greenhouse gases. To speak of species thinking is not to resist the politics of "common but differentiated responsibility" that China, India, and other developing countries seem keen to pursue when it comes to reducing greenhouse gas emissions.[54] Whether we blame climate change on those who are retrospectively guilty—that is, blame the West for their past performance—or those who are prospectively guilty (China has just surpassed the United States as the largest emitter of carbon dioxide, though not on a per capita basis) is a question that is tied no doubt to the histories of capitalism and modernization.[55] But scientists' discovery of the fact that human beings have in the process become a geological agent points to a shared catastrophe that we have all fallen into. Here is how Crutzen and Stoermer describe that catastrophe:

> The expansion of mankind . . . has been astounding. . . . During the past 3 centuries human population increased tenfold to 6000 million, accompanied e.g. by a growth in cattle population to 1400 million (about one cow per average size family). . . . In a few generations mankind is exhausting the fossil fuels that were generated over several hundred million years. The release of SO_2 . . . to the atmosphere by coal and oil burning, is at least two times larger than the sum of all natural emissions . . . ; more than half of all accessible fresh water is used by mankind; human activity has increased the species extinction rate by thousand to ten thousand fold in the tropical rain forests. . . . Furthermore, mankind releases many toxic substances in the environment. . . . The effects documented include modification of the geochemical cycle in large freshwater systems and occur in systems remote from primary sources. ("A," p. 17)

Explaining this catastrophe calls for a conversation between disciplines and between recorded and deep histories of human beings in the same way that the agricultural revolution of 10,000 years ago could not be explained except through a convergence of three disciplines: geology, archaeology, and history.[56]

Scientists such as Wilson or Crutzen may be politically naive in not recognizing that reason may not be all that guides us in our effective collective choices—in other words, we may collectively end up making some unreasonable choices—but I find it interesting and symptomatic that they speak the language of the Enlightenment. They are not necessarily anticapitalist scholars, and yet clearly they are not for business-as-usual capitalism either. They see knowledge and reason providing humans not only a way out of this present crisis but a way of keeping us out of harm's way in the future. Wilson, for example, speaks of devising a "wiser use of resources" in a manner that sounds distinctly Kantian (SN, p. 199). But the knowledge in question is the knowledge of humans as a species, a species dependent on other species for its own existence, a part of the general history of life. Changing the climate, increasingly not only the average temperature of the planet but also the acidity and the level of the oceans, and destroying the food chain are actions that cannot be in the interest of our lives. These parametric conditions hold irrespective of our political choices. It is therefore impossible to understand global warming as a crisis without engaging the propositions put forward by these scientists. At the same time, the story of capital, the contingent history of our falling into the Anthropocene, cannot be denied by recourse to the idea of species, for the Anthropocene would not have been possible, even as a theory, without the history of industrialization. How do we hold the two together as we think the history of the world since the Enlightenment? How do we relate to a universal history of life—to universal thought, that is—while retaining what is of obvious value in our postcolonial suspicion of the universal? The crisis of climate change calls for thinking simultaneously on both registers, to mix together the immiscible chronologies of capital and species history. This combination, however, stretches, in quite fundamental ways, the very idea of historical understanding.

Thesis 4: The Cross-Hatching of Species History and the History of Capital Is a Process of Probing the Limits of Historical Understanding

Historical understanding, one could say following the Diltheyan tradition, entails critical thinking that makes an appeal to some generic ideas about human experience. As Gadamer pointed out, Dilthey saw "the individual's private world of experience as the starting point for an expansion that, in a living transposition, fills out the narrowness and fortuitousness of his private experience with the infinity of what is available by re-experiencing the historical world." "*Historical consciousness*," in this tradition, is thus "*a mode of self-knowledge*" garnered through critical reflections on one's own and others' (historical actors') experiences.[57] Humanist histories of capitalism will always admit of something called the experience of capitalism. E. P. Thompson's brilliant attempt to reconstruct working-class experience of capitalist labor, for instance, does not make sense without that assumption.[58] Humanist histories are histories that produce meaning through an appeal to our capacity not only to reconstruct but, as Collingwood would have said, to reenact in our own minds the experience of the past.

When Wilson then recommends in the interest of our collective future that we achieve self-understanding as a species, the statement does not correspond to any historical way of understanding and connecting pasts with futures through the assumption of there being an element of continuity to human experience. (See Gadamer's point mentioned above.) Who is the we? We humans never experience ourselves as a species. We can only intellectually comprehend or infer the existence of the human species but never experience it as such. There could be no phenomenology of us as a species. Even if we were to emotionally identify with a word like *mankind*, we would not know what being a species is, for, in species history, humans are only an instance of the concept species as indeed would be any other life form. But one never experiences being a concept.

The discussion about the crisis of climate change can thus produce affect and knowledge about collective human pasts and futures that work at the limits of historical understanding. We experience specific effects of the crisis but not the whole phenomenon. Do we then say, with Geyer and Bright, that "humanity no longer comes into being through 'thought'" ("WH," p. 1060) or say with Foucault that "the human being no longer has any history"?[59] Geyer and Bright go on to write in a Foucaultian spirit: "Its [world history's] task is to make transparent the lineaments of power, underpinned by information, that compress humanity into a single humankind" ("WH," p. 1060).

This critique that sees humanity as an effect of power is, of course, valuable for all the hermeneutics of suspicion that it has taught postcolonial scholarship. It is an effective critical tool in dealing with national and global formations of domination. But I do not find it adequate in dealing with the crisis of global warming. First, inchoate figures of us all and other imaginings of humanity invariably haunt our sense of the current crisis. How else would one understand the title of Weisman's book, *The World without Us*, or the appeal of his brilliant though impossible attempt to depict the experience of New York after we are gone![60] Second, the wall between human and natural history has been breached. We may not experience ourselves as a geological agent, but we appear to have become one at the level of the species. And without that knowledge that defies historical understanding there is no making sense of the current crisis that affects us all. Climate change, refracted through global capital, will no doubt accentuate the logic of inequality that runs through the rule of capital; some people will no doubt gain temporarily at the expense of others. But the whole crisis cannot be reduced to a story of capitalism. Unlike in the crises of capitalism, there are no lifeboats here for the rich and the privileged (witness the drought in Australia or recent fires in the wealthy neighborhoods of California). The anxiety global warming gives rise to is reminiscent of the days when many feared a global nuclear war. But there is a very important difference. A nuclear war would have been a conscious decision on the part of the powers that be. Climate change is an unintended consequence of human actions and shows, only through scientific analysis, the effects of our actions as a species. Species may indeed be the name of a placeholder for an emergent, new universal history of humans that flashes up in the moment of the danger that is climate change. But we can never *understand* this universal. It is not a Hegelian universal arising dialectically out of the movement of history, or a universal of capital brought forth by the present crisis. Geyer and Bright are right to reject those two varieties of the universal. Yet climate change poses for us a question of a human collectivity, an us, pointing to a figure of the universal that escapes our capacity to experience the world. It is more like a universal that arises from a shared sense of a catastrophe. It calls for a global approach to politics without the myth of a global identity, for, unlike a Hegelian universal, it cannot subsume particularities. We may provisionally call it a "negative universal history."[61]

Notes

This essay is dedicated to the memory of Greg Dening.

1. Alan Weisman, *The World without Us* (New York: Harper Perennial, 2007), 3–5.

2. See C. A. Bayly, *The Birth of the Modern World, 1780–1914: Global Connections and Comparisons* (Maiden, MA: Blackwell, 2004).

3. The prehistory of the science of global warming going back to nineteenth-century European scientists like Joseph Fourier, Louis Agassiz, and Arrhenius is recounted in many popular publications. See, for example, the book by Bert Bolin, the chairman of the UN's Intergovernmental Panel on Climate Change (1988–1997), *A History of the Science and Politics of Climate Change: The Role of the Intergovernmental Panel on Climate Change* (Cambridge: Cambridge University Press, 2007), pt. 1.

4. Quoted in Mark Bowen, *Censoring Science: Inside the Political Attack on Dr. James Hansen and the Truth of Global Warming* (New York: Dutton, 2008), 1.

5. Quoted in ibid., 228. See also "Too Hot to Handle: Recent Efforts to Censor Jim Hansen," *Boston Globe*, February 5, 2006, E1.

6. See, for example, Walter K. Dodds, *Humanity's Footprint: Momentum, Impact, and Our Global Environment* (New York: Columbia University Press, 2008), 11–62.

7. Giovanni Arrighi, *The Long Twentieth Century: Money, Power, and the Origins of Our Times* (1994; repr., London: Verso, 2006), 356; see Arrighi, *Adam Smith in Beijing: Lineages of the Twenty-First Century* (London: Verso, 2007), 227–389.

8. An indication of the growing popularity of the topic is the number of books published in the last 4 years with the aim of educating the general reading public about the nature of the crisis. Here is a random list of some of the most recent titles that inform this essay: Mark Maslin, *Global Warming: A Very Short Introduction* (Oxford: Oxford University Press, 2004); Tim Flannery, *The Weather Makers: The History and Future Impact of Climate Change* (Melbourne: Text Publishing, 2005); David Archer, *Global Warming: Understanding the Forecast* (Maiden, MA: Blackwell, 2007); *Global Warming*, ed. Kelly Knauer (New York: Time Books, 2007); Mark Lynas, *Six Degrees: Our Future on a Hotter Planet* (Washington, DC: National Geographic, 2008); William H. Calvin, *Global Fever: How to Treat Climate Change* (Chicago: University of Chicago Press, 2008); James Hansen, "Climate Catastrophe," *New Scientist*, July 28–August 3, 2007, 30–34; Hansen et al., "Dangerous Human-Made Interference with Climate: A CISS Model E Study," *Atmospheric Chemistry and Physics* 7, no. 9 (2007): 2287–2312; and Hansen et al., "Climate Change and Trace Gases," *Philosophical Transactions of the Royal Society*, July 15, 2007, 1925–54. See also Nicholas Stern, *The Economics of Climate Change: The "Stem Review"* (Cambridge: Cambridge University Press, 2007).

9. Naomi Oreskes, "The Scientific Consensus on Climate Change: How Do We Know We're Not Wrong?," in *Climate Change: What It Means for Us, Our Children, and Our Grandchildren*, ed. Joseph F. C. Dimento and Pamela Doughman (Cambridge, MA: MIT Press, 2007), 73, 74.

10. A long history of this distinction is traced in Paolo Rossi, *The Dark Abyss of Time: The History of the Earth and the History of Nations from Hooke to Vico*, trans. Lydia G. Cochrane (1979; repr., Chicago: University of Chicago Press, 1984).

11. Benedetto Croce, *The Philosophy of Giambattista Vico*, trans. R. G. Collingwood (1913; repr., New Brunswick, NJ: H. Latimer, 2002), 5. Carlo Ginzburg has alerted me to problems with Collingwood's translation.

12. See the discussion in Perez Zagorin, "Vico's Theory of Knowledge: A Critique," *Philosophical Quarterly* 34 (January 1984): 15–30.

13. Karl Marx, "The Eighteenth Brumaire of Louis Bonaparte," in Marx and Frederick Engels, *Selected Works*, trans. pub., 3 vols. (Moscow: International, 1969), 1:398. See V. Gordon Childe, *Man Makes Himself* (London: Watts, 1941). Indeed, Althusser's revolt in the 1960s against humanism in Marx was in part a jihad against the remnants of Vico in the savant's texts; see Etienne Balibar, personal communication to author, December 1, 2007. I am grateful to Ian Bedford for drawing my attention to complexities in Marx's connections to Vico.

14. David Roberts describes Collingwood as "the lonely Oxford historicist . . . , in important respects a follower of Croce's" (David D. Roberts, *Benedetto Croce and the Uses of Historicism* [Berkeley: University of California Press, 1987], 325).

15. On Croce's misreading of Vico, see the discussion in general in Cecilia Miller, *Giambattista Vico: Imagination and Historical Knowledge* (Basingstoke: St. Martin's, 1993), and James G. Morrison, "Vico's Principle of *Verum* is *Factum* and the Problem of Historicism," *Journal of the History of Ideas* 39 (October–December 1978): 579–95.

16. Collingwood, *The Idea of History* (1946; repr., New York: Oxford University Press, 1976), 214, 212, 213, 216.

17. Ibid., 193.

18. Roberts, *Benedetto Croce and the Uses of Historicism*, 59, 60, 62.

19. Joseph Stalin, *Dialectical and Historical Materialism* (Moscow: International, 1938), www.marxists.org.

20. Fernand Braudel, "Preface to the First Edition," *The Mediterranean and the Mediterranean World in the Age of Philip II*, trans. Siân Reynolds, 2 vols. (1949; repr., London: Collins, 1972), 1:20. See also Peter Burke, *The French Historical Revolution: The "Annales" School, 1929–89* (Stanford: Stanford University Press, 1990), 32–64.

21. See Hans-Georg Gadamer, *Truth and Method*, 2nd ed., trans. Joel Weinsheimer and Donald G. Marshall (1975, 1979; repr., London: Sheed & Ward, 1988), 214–18. See also Bonnie G. Smith, "Gender and the Practices of Scientific History: The Seminar and Archival Research in the Nineteenth Century," *American Historical Review* 100 (October 1995): 1150–76.

22. Braudel, "Preface to the First Edition," 20.

23. Alfred W. Crosby Jr., *The Columbian Exchange: Biological and Cultural Consequences of 1492* (1972; repr., London: Praeger, 2003), xxv.

24. See Daniel Lord Smail, *On Deep History and the Brain* (Berkeley: University of California Press, 2008), 74–189.

25. Oreskes, "Scientific Consensus," 93.

26. Crosby Jr., "The Past and Present of Environmental History," *American Historical Review* 100 (October 1995): 1185.

27. Will Steffen, director of the Centre for Resource and Environmental Studies at the Australian National University, quoted in "Humans Creating New 'Geological Age,'" *The*

Australian, March 31, 2008, www.theaustralian.news.com.au. Steffen's reference was the Millennium Ecosystem Assessment Report of 2005. See also Neil Shubin, "The Disappearance of Species," *Bulletin of the American Academy of Arts and Sciences* 61 (Spring 2008): 17–19.

28. Bill McKibben's argument about the "end of nature" implied the end of nature as "a separate realm that had always served to make us feel smaller" (Bill McKibben, *The End of Nature* [1989; repr., New York: Random House, 2006], xxii).

29. Bruno Latour's *Politics of Nature: How to Bring the Sciences into Democracy*, trans. Catherine Porter (1999; repr., Cambridge, MA: Harvard University Press, 2004), written before the intensification of the debate on global warming, calls into question the entire tradition of organizing the idea of politics around the assumption of a separate realm of nature and points to the problems that this assumption poses for contemporary questions of democracy.

30. Gadamer, *Truth and Method*, 206: the historian "knows that everything could have been different, and every acting individual could have acted differently."

31. Paul J. Crutzen and Eugene F. Stoermer, "The Anthropocene," *IGBP [International Geosphere-Biosphere Programme] Newsletter* 41 (2000): 17; hereafter abbreviated "A."

32. Crutzen, "Geology of Mankind," *Nature*, January 3, 2002, 23.

33. Mike Davis, "Living on the Ice Shelf: Humanity's Meltdown," June 26, 2008, tomdispatch.com/post/174949; hereafter abbreviated "LIS." I am grateful to Lauren Berlant for bringing this essay to my attention.

34. See William F. Ruddiman, "The Anthropogenic Greenhouse Era Began Thousands of Years Ago," *Climatic Change* 61, no. 3 (2003): 261–93; Crutzen and Steffen, "How Long Have We Been in the Anthropocene Era?" *Climatic Change* 61, no. 3 (2003): 251–57; and Jan Zalasiewicz et al., "Are We Now Living in the Anthropocene?" *GSA Today* 18 (February 2008): 4–8. I am grateful to Neptune Srimal for this reference.

35. Zalasiewicz et al., "Are We Now Living in the Anthropocene?," 7. Davis described the London Society as "the world's oldest association of Earth scientists, founded in 1807" ("LIS").

36. See, for instance, Libby Robin and Will Steffen, "History for the Anthropocene," *History Compass* 5, no. 5 (2007): 1694–1719, and Jeffrey D. Sachs, "The Anthropocene," *Common Wealth: Economics for a Crowded Planet* (New York: Penguin, 2008), 57–82. Thanks to Debjani Ganguly for drawing my attention to the essay by Robin and Steffen, and to Robin for sharing it with me.

37. Edward O. Wilson, *The Future of Life* (New York: Knopf, 2002), 102; hereafter abbreviated *FL*.

38. See Latour, *Politics of Nature*.

39. Flannery, *Weather Makers*, xiv.

40. Maslin, *Global Warming*, 147. For a discussion of how fossil fuels created both the possibilities for and the limits of democracy in the twentieth century, see Timothy Mitchell, "Carbon Democracy," forthcoming in *Economy and Society*. I am grateful to Mitchell for letting me cite this unpublished paper.

41. Wilson, *In Search of Nature* (Washington, DC: Island Press, 1996), ix–x; hereafter abbreviated *SN*.

42. See Smail, *On Deep History and the Brain*.

43. Michael Geyer and Charles Bright, "World History in a Global Age," *American Historical Review* 100 (October 1995): 1058–59; hereafter abbreviated "WH."

44. See Smail, *On Deep History and the Brain*, 124.

45. Ibid., 124–25.

46. Jacques Derrida, "Cogito and the History of Madness," *Writing and Difference*, trans. Alan Bass (Chicago: University of Chicago Press, 1978), 34.

47. See Sachs, *Common Wealth*, 57–82.

48. Wilson, foreword to Sachs, *Common Wealth*, xii. Students of Marx may be reminded here of the use of the category "species being" by the young Marx.

49. See Kenneth Pomeranz, *The Great Divergence: Europe, China, and the Making of the Modern World Economy* (Princeton, NJ: Princeton University Press, 2000).

50. See Mitchell, "Carbon Democracy." See also Edwin Black, *Internal Combustion: How Corporations and Governments Addicted the World to Oil and Derailed the Alternatives* (New York: St. Martin's, 2006).

51. Arrighi's *The Long Twentieth Century* is a good guide to these fluctuations in the fortunes of capitalism.

52. Lawrence Guy Straus, "The World at the End of the Last Ice Age," in *Humans at the End of the Ice Age: The Archaeology of the Pleistocene–Holocene Transition*, ed. Lawrence Guy Straus et al. (New York: Plenum, 1996), 5.

53. Flannery, *Weather Makers*, 63, 64.

54. Ashish Kothari, "The Reality of Climate Injustice," *Hindu*, November 18, 2007.

55. I have borrowed the idea of "retrospective" and "prospective" guilt from a discussion led at the Franke Institute for the Humanities by Peter Singer during the Chicago Humanities Festival, November 2007.

56. See Colin Tudge, *Neanderthals, Bandits, and Farmers: How Agriculture Really Began* (New Haven, CT: Yale University Press, 1999), 35–36.

57. Gadamer, *Truth and Method*, 232, 234. See also Michael Ermarth, *Wilhelm Dilthey: The Critique of Historical Reason* (Chicago: University of Chicago Press, 1978), 310–22.

58. See E. P. Thompson, *The Making of the English Working Class* (Harmondsworth: Penguin, 1963).

59. Michel Foucault, *The Order of Things: An Archaeology of Human Knowledge*, trans. pub. (1966; repr., New York: Vintage, 1973), 368.

60. See Weisman, *World without Us*, 25–28.

61. I am grateful to Antonio Y. Vasquez-Arroyo for sharing with me his unpublished paper "Universal History Disavowed: On Critical Theory and Postcolonialism," where he has tried to develop this concept of negative universal history on the basis of his reading of Theodor Adorno and Walter Benjamin.

❱❱❱ System Failure: Oil, Futurity, and the Anticipation of Disaster

Imre Szeman

> Siri sipped at her coffee. "I would have thought that your Hegemony was far beyond a petroleum economy." I laughed. . . . "Nobody gets beyond a petroleum economy. Not while there's petroleum there."
>
> —Dan Simmons, *Hyperion*

> And lurking behind any possible reconfiguration of world politics would be questions of access to energy and to water, in a world beset by ecological dilemmas and potentially producing vastly more than existing capacities of capitalist accumulation. Here could be the most explosive issues of all, for which no geopolitical manoeuvring or reshuffling offers any solution.
>
> —Immanuel Wallerstein, "The Curve of American Power"

The way one establishes epochs or defines historical periods inevitably shapes how one imagines the direction the future will take. And so it is with the dominant periodization of the history of capital, which has been organized primarily around moments of hegemonic economic imperium: Dutch mercantilism, British imperialism, US transnationalism. All the effort in reading the tea leaves of contemporary capitalism is thus directed at determining when the current hegemonic formation will collapse and which new one (or ones) will come in its stead. According to Giovanni Arrighi, David Harvey, Immanuel Wallerstein, and others, the US moment is at an end; the new hegemonic formation will emerge only after a turbulent and violent interregnum that is already upon us, even if we do not yet recognize it. Through it all, it seems, capitalism emerges largely unscathed: different in content, perhaps, and no doubt occupying a different space on the globe, but essentially the same in form—a system organized around limitless accumulation, at whatever social cost.

What if we were to think about the history of capital not exclusively in geopolitical terms, but in terms of the forms of energy available to it at any given historical moment? So steam capitalism in 1765 creates the conditions for the first great subsumption of agricultural labor into urban factories (a process of proletarianization that is only now coming to a completion), followed by the advent of oil capitalism in 1859 (with its discovery in Titusville, Pennsylvania), which enabled powerful and forceful new modalities of capitalist reproduction and expansion.

South Atlantic Quarterly 106, no. 4 (Fall 2007): 805–23. Copyright 2007, Duke University Press. Republished by permission.

From oil flows capitalism as we still know it: the birth of the first giant multi-nationals—Standard Oil (whose component elements still persist in Exxon Mobil, Texaco, and British Petroleum), DuPont, and the Big Three automobile makers; the defining social system of private transportation—cars, air travel, freeways, and with these, suburbs, "white flight," malls, inner-city ghettoization, and so on; and the environmental and labor costs that come with access to a huge range of rela-tively inexpensive consumer goods, most of which contain some product of the petrochemical industry (plastics, artificial fibers, paints, etc.) and depend on the possibility of mass container shipping. No petroleum, no modern war machine, no global shipping industry, no communication revolution. Imagined in geopolitical terms, the future is one in which US hegemony gives way to (say) a Sino-Russian bloc or perhaps to some hydra-headed creature made up of economies that have passed from socialist caterpillar to capitalist butterfly (Brazil–India–China). But if we think of capital in terms of energy, what do the tea leaves tell us about what comes next? Wind capital? Solar capital? Biomass capital? A seemingly impossible conjunction of terms. Nuclear capital? Hydrogen capital? These are somehow more imaginable—even if the technological problems of the latter have yet to be worked out and the nuclear option would require a staggering and unprecedented investment in building new reactors, an expenditure that is not on the horizon anywhere.[1]

Oil capital seems to represent a stage that neither capital nor its opponents can think beyond. Oil and capital are linked inextricably, so much so that the looming demise of the petrochemical economy has come to constitute perhaps the biggest disaster that "we" collectively face. The success of capital is dependent on contin-uous expansion, which enables not only profit taking but investment in the repro-duction of capital that is a necessary condition for its continuation on into the fu-ture. During the period of oil capital, this expansion and reproduction was fueled by cheap and readily available sources of oil, not least (until the early 1970s) in the United States itself. In "Critique of the Gotha Program," Marx reminds us that "labor is *not the source* of all wealth. *Nature* is just as much a source of use values (and it is surely of such that material wealth consists!) as labor which is itself only the manifestation of a force of nature, human labor power."[2] The discourse of di-saster around the end of oil recognizes that, unimaginably, at least one part of the use values originating in nature (the one that only seems to come for free) is on the verge of being exhausted. What happens to capital now that oil is (at best) likely to remain expensive or is (at worst) actually running out? What future can be imagined not only for oil capital, but for capital as such?

There seems to be general agreement that even if we have yet to reach Hub-bert's peak—the point of maximum global oil supply prior to its downward decline to zero—the point at which we will is coming soon.[3] More optimistically (at least from one perspective), taking account of all possible sources of fossil fuels—tar sands, hard-to-access or currently off-limits sources of oil and gas (in the deep sea, in natural reserves and national parks), and especially coal—some economists and resource experts have estimated that the global economy could continue to be hydrocarbon based for 200 to 500 years, depending on the levels of growth in energy usage.[4] Whether oil is disappearing or in relative abundance, the recent triangulation of military adventurism, the demands of rapidly expanding developing economies as a result of globalization, and the hard cold facts of global warming and ecological catastrophe has led to a feverish explosion of discourses about the probable future of oil—and, to a lesser degree, of oil capitalism. It is the orientation of these discourses toward the "disaster" of the end of oil and the potential futures with which I will concern myself in this essay. Though the scientific veracity of the claims made on behalf of this or that narrative of the likely fate and future direction of oil capital is not incidental, it is also the case that the power of these narratives and the likelihood that one or another is adopted (to whatever degree and however incompletely) as a way of precluding the collapse of oil capital depend less on the judiciousness of the notoriously shaky predictions of petroleum geologists or traders in commodity futures than on the way in which they mobilize and intersect with existing social narratives of expertise, technology, progress, consumption, nature, and politics. In the case of such narratives, precise statistics and measurements hardly begin to capture the social anxieties, fears, and hopes embodied in discourses that try to imagine the shape of future social formations.

Is the end of oil a disaster? This depends, of course, on the perspective one has on the system in danger of collapse: capitalism. The disaster discourses of the end of oil are necessarily anticipatory, future-oriented ones—narratives put into play in the present in order to enable the imagined disaster at the end of oil to be averted through geopolitical strategy, rational planning, careful management of resources, the mobilization of technological and scientific energies, and so on. What is all too frequently absent from these quintessentially modern discourses is the shape and configuration of the political. Eco-dystopians and techno-utopians alike take the current configuration of the political and economic as given. Because of this, it seems impossible from these perspectives to envision a systemic revolution. This deficit within existing narratives of the end of oil should alert us to the largely unarticulated political possibilities that lurk within them. The task here, then, will be to critically assess existing "end of oil" narratives in order to consider their lessons for a Left that has the difficult task of generating and articulating alternatives to oil capital. While the equation "blood for oil" effectively draws attention to one dimension of the geopolitics of oil, it leaves unaddressed how one conceptualizes energy demands for a human polity that is expected to grow to 9 billion by midcentury. Indeed, in celebrating the possibilities of the potentialities of South America's "Bolivarian Revolution" or the continued attractions of even latter-day Scandinavian social democracy (which, especially in the case of Norway, is fueled by oil), the Left has seemed to resist thinking too deeply about the larger consequences of petroeconomies, of their sustainability as social and political models, and of what, if anything, comes after.

What might a Left position on oil capital—and its aftermath—look like? There are three dominant narratives circulating today concerning what is to be done about the disaster of oil: strategic realism, techno-utopianism, and eco-apocalypse. In what follows, I take each one up in turn in order to see what lessons they have to offer for the Left, before concentrating on one of the most contentious recent confrontations with oil capitalism: the "Blood for Oil?" chapter of Retort's *Afflicted Powers*.[5]

Strategic Realism

It has become a given that contemporary geopolitical maneuvering is driven by access to goods and resources, chief among these being access to oil. In describing the actions and motivations of imperialist jockeying between the major powers at the turn of the last century, Lenin evokes the name Standard Oil, but only as one of the many capitalist monopolies that had established themselves by the beginning of the twentieth century: there is nothing to distinguish it from any of the others he lists, such as the Rhine-Westphalian Coal Syndicate, United States Steel, the Tobacco Trust, and so on.[6] Today, steel and cigarettes have receded, and oil has come to the fore as a prime factor guiding the political decision making and military actions of both advanced capitalist countries and developing ones. As Daniel Yergin notes, oil arrives on the geopolitical stage at the outset of World War I when Winston Churchill, First Lord of the Admiralty, decides to power Britain's navy by oil from Persia as opposed to coal from Wales—a shift designed to improve the speed of the navy, but at the expense of national energy security.[7] This founding equation between oil and military power has been consistently in force ever since. The political character of the Middle East in particular has been shaped throughout the past century by the military and political struggle of Britain, France, the United States, and other powers to secure access to a commodity essential to the smooth operation of their economies.[8]

The Advanced Energy Initiative (AEI), announced early in 2006 by President George W. Bush's administration, is intended to reduce dependency on foreign oil by promoting the clean use of oil, nuclear power, natural gas, and a variety of renewable resources.[9] "Let me put it bluntly. . . . We are too dependent on oil," Bush stated at a 2006 conference organized by the US Departments of Energy and Agriculture to promote the use of biofuels (such as ethanol) in support of the AEI.[10] This bluntness and the announced aim by the administration to support new forms of energy have little impact on the necessity, at the moment and for the foreseeable future, to do whatever it takes to keep oil flowing into the US economy. As the AEI notes, it is not just the US economy that requires oil, but countries such as China and India, which are consuming more oil and at an accelerating rate.[11] Even if the United States and other major consumers of oil (China, Japan, Russia, Germany, etc.) should manage to reduce consumption and develop alternative sources of energy, there is no question that it remains an essential commodity, growing in demand even as its supply decreases.

What I have termed *strategic realism* is a relatively common discourse around oil that derives from a strict realpolitik approach to energy. Those who employ it—and it is a discourse employed widely by government and the media alike—suspend or minimize concerns about the cumulative environmental disaster of oil or the fact that oil is disappearing altogether, and focus instead on the potential political and economic tensions that will inevitably arise as countries pursue their individual energy security in an era of scarcity.[12] What is of prime interest in strategic realism is engaging in the geopolitical maneuverings required to keep economies floating in oil. At the heart of strategic realism stands the blunt need for nations to protect themselves from energy disruptions by securing and maintaining steady and predictable access to oil.

These maneuverings around energy can and do take multiple forms, from military intervention intended to shore up existing "power interdependencies"[13] (due to the US invasion of Iraq, military intervention has come mistakenly to stand as the prime mode through which access to oil is secured) to economic agreements between states, and from the creation of new trade and security arrangements of mutual benefit to the big users of oil (looking down the road, the United States, China, and India)[14] to even the (largely) quixotic attempt to create energy independence by promoting the use of alternative fuels. What ties these various approaches together is an element so obvious that it appears hardly to need mentioning: the centrality of the *nation-state* itself in the calculations of oil accessibility and security. When it comes to the potential disaster of oil, in the discourse of strategic realism the figures, concepts, and protagonists that we have all come to love in the discourses of globalization—the withered nation, Colossus-like transnational corporations, the mixed sovereignty of empire—seem not only to fade to the background but to disappear altogether. Strategic realism is a discourse that makes the nation-state the central actor in the drama of the looming disaster of oil, an actor that engages in often brutal geopolitical calculations in order to secure the stability of national economies and communities. While oil is hardly divorced from the operations of global finances, its political value as a commodity is such that it is apparently not permitted to slosh autonomously through markets that we have been repeatedly told take little note of borders today: the state must be present in order to ensure that every day the right amount of oil flows in the right direction.

Discussions of the strategic calculations at work when it comes to oil are hardly limited to the Right. While right-wing discourses, especially those that adopt a "might is right" approach to the defense of the homeland, are both more prominent and less troubled by the ambiguous and unpalatable outcome of petrorealism—support for antidemocratic oligarchs being the least of these—there are both liberal and Left responses to and employment of the discourse of strategic realism. In *Blood and Oil*, for instance, Michael Klare explores the consequences of the US dependency on foreign oil, drawing attention to the huge sums of money that are spent annually to keep access to oil open. He argues that "ultimately, the cost of oil will be measured in blood: the blood of American soldiers who die in combat, and the blood of many other casualties of oil-related violence."[15] For Klare, the proposed solution is for Americans to "adopt a new attitude toward petroleum—a conscious decision to place basic values and the good of the country ahead of immediate personal convenience."[16] The reality of continued growth in energy use in circumstances in which oil is disappearing isn't at issue. Rather, what is proposed is a potentially less violent and more stable way of managing the geopolitical realities created by struggles over access to energy, including vast reductions by Americans in their individual energy usage. The nation remains the central actor, and the misfit between supply and demand for oil is one that needs to be seriously considered so that existing differentials of national power are maintained into the indefinite future. As for the larger consequences of oil usage for the environment or for humanity as a whole? Strategic realism recognizes only that oil is essential to capital and capital is essential for the status quo to remain in place in the future. The disaster in this discourse is figured as the mismanagement or misrecognition of geopolitical strategy, such that a commodity essential to state power is no longer available in the abundance necessary for economic growth. On the Left, meanwhile, there continues to be an abiding fascination with the dynamics of capitalist geopolitics, not, it seems, to plot weaknesses and to imagine something beyond it, but because of the inherent interest in ceaseless rearrangement of deck chairs on a capitalist ship that seems in little real danger of sinking.

Techno-utopianism

Strategic realism sees the disaster of oil as a problem primarily for the way in which nations preserve or enhance their geopolitical status. A founding assumption is that the political future will look more or less like the present: strategy can't be developed around the promise of new sources of energy but emerges out of plans to capture and control (economically, diplomatically, or militarily) existing ones. A second narrative related to the looming disaster of the end of oil looks to science and technology to develop energy alternatives that will mitigate the end of oil. This form of techno-utopianism can be used as an element of strategic realism, but in practice these narratives are kept discursively distinct. For instance, the text of the AEI barely mentions oil, focusing instead on nuclear energy, clean coal, natural gas, and renewable energies; the strategic military intervention in the Persian Gulf lies outside of this narrative of future alternatives. Whereas in strategic realism the future is imagined as a continuation of the present, the AEI announces a belief in the new future, albeit one secured by existing sources of energy: "It will take time for America to move from a hydrocarbon economy to a hydrogen economy. In the meantime, there are billions of barrels of oil and enormous amounts of natural gas off the Alaskan coast and in the Gulf of Mexico."[17]

What I am calling techno-utopianism is a discourse employed by government officials, environmentalists, and scientists from across the political spectrum. With respect to the end of oil, it proposes two solutions: either scientific advances will enable access to oil resources hitherto too expensive to develop (the Alberta tar sands, deep-sea reserves, etc.) while simultaneously devising solutions for carbon emissions (exhaust scrubbers, carbon sequestering, etc.), *or* technological innovations will create entirely new forms of energy, such as hydrogen fuel cells for space-age automobiles. As with strategic realism, its ubiquity today makes techno-utopianism a familiar discourse. It can be employed as mere political rhetoric to defer difficult decisions with negative economic impacts to some distant future, as in Canadian Prime Minister Stephen Harper's recent announcement concerning "intensity-based" emissions standards: "With technological change, massive reduction in emissions are possible. . . . We have reason to believe that by harnessing technology we can make large-scale reductions in other types of emissions. But this will take time. It will have to be done as part of technological turnover."[18] Somewhat more convincingly, techno-utopianism also underwrites the activities of those working actively in science and technology, who hope through their work to offset the civilization blunder of hitching a complex global economy to a nonrenewable dirty fuel source fast evaporating from the earth. The utopia I have in mind here is the "bad utopia" of future dreamscapes and fanciful political confections—"utopia" not quite just as an insulting slur against one's enemies, but rather as a projection of an alternative future that is, in fact, anything but a "conception of systematic otherness."[19] In "The Politics of Utopia," Fredric Jameson speaks of "one of the most durable oppositions in utopian projection"—that between city and country. He asks: "Did your fantasies revolve around a return to the countryside and the rural commune, or were they on the other hand incorrigibly urban, unwilling and unable to do without the excitement of the great metropolis, with its crowds and its multiple offerings, from sexuality and consumer goods to culture?"[20] Techno-utopian discourses of future alternatives to oil magically resolve this opposition: since the future is undeniably urban, great metropolises are envisioned as leafy green oases, filled with mid-twenty-first-century flaneurs and cyclists who move between buildings crowned with solar sails.[21] All of our worst fears about the chaos that will ensue when oil runs out are resolved through scientific innovations that are in perfect synchrony with the operations of the capitalist economy: problem solved, without the need for radical ruptures or alterations in political and social life.

An excellent example of such techno-utopianism can be found in a 2006 special issue of *Scientific American*, "Energy's Future: Beyond Carbon." The issue's subtitle announces its politics directly: "How to Power the Economy and Still Fight Global Warming." The issue presents technological strategies for carbon reduction, new transportation fuels, efficient building design, clean options for coal, possibilities for nuclear power, and so on.[22] The long-term impact of existing energy use—primarily oil—on the environment is the focus here; each article provides a potential solution based on current scientific research and technological innovation. The articles all begin in much the same way, noting first the deleterious environmental effects of existing social and cultural practices, especially those in the developed world, followed by the failures at the level of politics to mobilize and enforce necessary changes to environmental laws and standards. In his introduction to the special issue, Gary Stix writes: "The slim hope for keeping atmospheric carbon below 500 ppm hinges on aggressive programs of energy efficiency instituted by national governments."[23] But since such programs don't seem to be on the horizon, scientific innovation rushes into the gap vacated by public policy. In the coming disaster of oil, technology absorbs and mediates all the risks that might normally unfold at the level of the political. A profusion of developments from the astonishing to the relatively banal—new refrigerators use one-quarter of the energy of their 1974 counterparts, LCD computer screens 60% less than CRT monitors—will bring about not only a cleaner environment but a soft landing for oil capital. If the various timescale charts and projections for reductions in oil usage are less than comforting, we are reminded of the following: "Deeply ingrained in the patterns of technological evolution is the substitution of cleverness for energy."[24] The natural temporal flow of scientific discovery will resolve the energy and environmental problems we have produced for ourselves.

The notion of technological evolution lies at the heart not only of techno-utopian solutions to the disaster of oil but of modern imaginings of science more generally. Technology is figured as just around the corner, as always just on the verge of arriving. Innovation can be hurried along (through increased grants, for instance), but only slightly: technological solutions arrive just in time and never fail to come. In a perversion of Marx's comments in the preface to *A Contribution to a Critique of Political Economy*, it would appear that mankind produces only such disasters as technology can solve; the disaster arises only when the conditions in which to repair it are already in the process of formation.[25] This is, as we see above in Harper's comment, certainly part of the political dream of techno-utopianism. It is equally part of the scientists' self-imaginings as well: "The vast potential of this new industry underscores the importance of researching, developing, and demonstrating hydrogen technologies now, so they will be ready when we need them."[26]

At the core of the notion that technological developments are on the horizon to address even such massive, global problems as the end of oil lies a further temporal imagining. If technological developments are thought to be poised to imminently bring about a change from oil capital to (in this case) hydrogen capital, it is because technological developments in the past have always appeared in the nick of time to help push modernity along. But where? And how? History offers no models whatsoever: the fantasy of past coincidence between technological discovery and historic necessity simply reinforces the bad utopianism of hope in technological solutions to the looming end of oil.

Apocalyptic Environmentalism

In his editorial in *Scientific American*, Gary Stix writes:

> Sustained marshalling of cross-border engineering and political resources over the course of a century or more to check the rise of carbon emissions makes a moon mission or a Manhattan Project appear comparatively straightforward. . . . Maybe a miraculous new energy technology will simultaneously solve our energy and climate problems during that time, but another scenario is at least as likely: a perceived failure of Kyoto or international bickering over climate questions could foster the burning of abundant coal for electricity and synthetic fuels for transportation, both without meaningful checks on carbon emissions.[27]

Narratives of the end of oil that focus on this other scenario are best described as eco-apocalypse discourses. If strategic realism is largely a discourse of the Right, its Left complement is located largely in eco-apocalypse discourse. These take the disaster of oil capitalism head on: the deep political and economic investments in oil are assessed, the dire social-political environmental consequences of inaction on oil are laid out, and because it becomes obvious that avoiding these results would require changing everything, apocalyptic narratives and statistics are trotted out. Strategic realism and techno-utopianism remain committed to capitalism and treat the future as one in which change has to occur (new geopolitical realignments, innovations in energy use) if change at other levels is to be deferred (fundamental social and political changes). Eco-apocalypse sees the future more grimly: unlike the other two discourses, it understands that social and political change is fundamental to genuinely addressing the disaster of the end of oil—a disaster that it relates to the environment before economics. However, since such change is not on the horizon or is difficult to imagine, it sees the future as Bosch-like—a hell on earth, obscured by a choking carbon dioxide smog.

The volume *The Final Energy Crisis*, edited by Andrew McKillop and Sheila Newman, is but one of many books and articles in this genre.[28] With great care, clarity, and attention to the scientific evidence about fossil fuel depletion and environmental impacts, the volume lays out the case for getting serious about the looming disaster. The statistics pile up to paint an alarming picture of the disaster. Fertilizers are impossible to produce without fossil fuels; in their absence, the earth's carrying capacity for human life will necessarily fall by 50% to 60%; the growth in car ownership in India and China to Western levels, even with conservative estimates as to distance traveled, would require 10 billion barrels of oil each year, *"three times the total oil imports of all EU countries in 2002,* nearly three times the maximum possible production capacity of Saudi Arabia"; the postoil population carrying capacity of France is estimated as 20 to 25 million and in Australia less than 1.5 million; and so on.[29] Everything in the volume points to the coming disaster that is the only possible outcome of oil capitalism.

At issue is not the veracity of such claims, which are here always presented relatively conservatively, but what such information is intended to accomplish. All three of the discourses delineated in this essay make claims on the social, inviting it to participate in the framing of a response to the end of the energy source around which we structure social reality—and social hope, and social fantasy. Unlike the other two, the discourse of eco-apocalypse understands itself as a *pedagogic* one, a genre of disaster designed to modify behavior and transform the social. The McKillop and Newman book is exemplary in this regard, combining serious scientific articles (replete with charts and even equations) with Spinozian *scholie*-like passages by McKillop that narrate the coming end.[30] Even while recognizing the potential traumas for human communities and for capital, strategic realism and techno-utopianism operate within existing understandings of the way the world operates. Eco-apocalyptic discourse makes it clear that disaster cannot be avoided without fundamental changes to human social life. With hope for a new way of doing things, the conditions for avoiding disaster are put forward: "A simpler, non-affluent way of life"; "more communal, cooperative and participatory practices"; "new values" ("a much more collective, less individualistic social philosophy and outlook"); and, of course, "an almost totally new economic system. There is no chance whatsoever of making these changes while we retain the present consumer-capitalist economic system."[31]

The difficult question of *how* such a complete transformation of social life is to be brought about remains open. At best, the reality of a coming future disaster is imagined as being enough on its own to produce the shift in everything from values to economic systems that would be necessary to counter it. There is a form of "bad" utopianism at work here too. Although a new social system is outlined in utopian fashion (down to what kinds of houses should be on a single street and the kinds of animals that we might find in our suburbs[32]), the subject roaming through this landscape is none other than a liberal one, motivated by pleasure, convenience, and comfort. Despite the demands and claims for changing individual behavior and social reality, at the heart of eco-apocalyptic discourses is a recognition that even if its coming can be established, nothing can be done to stop the disaster from coming. Indeed, there is a sense in which disaster is all but welcome: the end of oil might well be a case of capitalism digging its own grave, since without oil, current configurations of capital are impossible.

The Left and Oil Capitalism: Retort and Disaster

National futures, technological futures, and apocalyptic ones. We can, as a form of critical activity, point to the limits of such discourses—to the revival, for instance, of nations and nationalism in strategic realism, or to the shaky temporality of techno-utopianism, or to the political limits of eco-apocalyptic discourses. However valuable such criticisms might be, the issue of what kind of response would frame this disaster in a manner that would create alternatives to oil capitalism still needs to be addressed more forcefully.

The very possibility of a disaster on the scale of the end of oil seems, on its own, not to be able to generate the kind of social transformation one might expect would be needed in order to head off a crisis that would be felt at every level—including that of capital accumulation and reproduction. Jacob Lund Fisker notes:

> The increase in human wealth and well-being during the past few centuries is often attributed to such things as state initiatives, governmental systems and economic policies, but the real and underlying cause has been a massive increase in energy consumption. . . . Discovering and extracting fossil fuels requires little effort when resources are abundant, before their depletion. It is this cheap "surplus energy" that has enabled classical industrial, urban and economic development.[33]

With the end of "surplus energy" thus comes the collapse of surplus profit—or so one would think. It may be that the disaster of oil is already prefigured in the temporal shift of the capitalist economy that goes by the name of neoliberalism. The ferocious return of primitive accumulation, now directed not only toward the last remaining vestiges of the public sector (such as universities and hospitals) but also inward into subjectivity, announces, too, a temporal recalibration of capital away from the future to the present. There is no longer any wait for surplus or any attention to the reproduction of capital for the future; instead, as if the future of capital is in doubt, profit taking has to occur as close to immediately as possible, whatever the long-term consequences.

Something like this view of contemporary capital informs the collective Retort's arguments in *Afflicted Powers: Capital and Spectacle in a New Age of War*. The book as a whole is intended to be a rallying cry for a new Left vanguardism that emerges out of the book's framing of the post-9/11 political landscape. This landscape is one structured by a "military neoliberalism" (72) that is described as "no more than primitive accumulation in disguise" (75); this neoliberalism in turn operates largely unopposed due to the dynamics of the "spectacle," which, as in so many appropriations of Guy Debord's concept, appears as a social situation defined by advertising and consumer images—that is, ideology through image form as well as image content.[34] Oil figures as a prominent part of Retort's account of the contemporary political situation. While the authors are struck by the bluntness of the slogan "No blood for oil," as it appears to directly name the reasons for the use of US military force in Iraq, they take pains to argue that placing oil at the causal center of the war is misleading. "Oil's powers," they write, "are drawn from a quite specific force field having a capitalist core that must periodically reconstitute the conditions of its own profitability" (54). The idea that the US invasion was prompted by a kind of petro-Malthusianism—of the kind, it must be said, that informs discourses of strategic realism—is premised on a false assumption about the market for oil. "The history of twentieth-century oil is *not* the history of shortfall and inflation, but of the constant menace—for the industry and the oil states—of excess capacity and falling prices, of surplus and glut" (59).

For Retort, the argument against oil as the cause of the war in Iraq allows the authors to draw out the broader motive driving the use of the military today, which is to support " 'extra-economic' restructuring of the conditions necessary for expanded profitability—paving the way, in short, for new rounds of American-led dispossession and capital accumulation" (72). This in turn permits them to consider our contemporary political options against capitalism. Rhetorically, the book makes use of anxiety about the war in Iraq to draw its readers into broader consideration of the dynamics of neoliberal globalization and possible responses to it. The final chapter, "Modernity and Terror," is both where Retort comes clean about its political aims and where it runs up against the limits of envisioning an end to oil capital. The argument the authors make in this chapter is a powerful one. For the Left, the opponent is nothing less than the "disenchantment of the world"— modernity itself. There are two central processes that they associate with capitalist modernity. The first is a consumerism that functions by seeming to offer a solution to modernity's disenchantment: "It promises to fill the life-world with meanings again, with magical answers to deep wishes, with models of having and being and understanding (undergoing) Time itself " (178). In other words, commodity fetishism, figured here as the lack of social resources that would allow us to recall the "mere instrumentality" of objects "in a world of meanings vastly exceeding those that any *things* can conjure up" (179). The second process is the "process of endless *enclosure*" (193), a continuation of the long process by which natural and human resources were taken from the common for the exclusive use of capital. The goal they set for themselves is to set out a "non-nostalgic, non-anathematizing, non-regressive, non-fundamental, non-apocalyptic critique of the modern" (185). They admit: "The Left has a long way to go even to lay the groundwork of such a project . . . but it is still only from the Left that a real opposition to modernity could come" (185).

Despite the grandeur of such a goal, who could disagree with such a project?

Or perhaps just as important: how is this modernity any different from the capitalism that the Left has been opposed to all along, even if consumerism and the processes of enclosure are both more intensive and more extensive than in previous eras? One thing that is glaringly absent is any consideration of future disaster. Though Retort pushes oil to the sidelines in its attempt to bring those chanting "No blood for oil!" into its larger critique of neoliberalism, when it considers the function of oil in relation to capital it only looks backward at the history of the twentieth century and not toward the horizon of the disaster that oil's absence will create. Recall Fisker: "It is this cheap 'surplus energy' that has enabled classical industrial, urban and economic development."[35] Oil is hardly incidental to capital or to modernity—which is not the same as saying that it is the prime mover of all decision making by nation-states or other actors in the global economy. At the same time, the growing sense of this coming horizon and the necessity of having to respond to it—whether through the machinations of resource strategy or by leaving it to technology to figure things out—cannot be simply left aside in shaping responses to the dark modernity sketched out by Retort. It is telling, for instance, that there is not even an appeal to the discourse of eco-apocalypse—barely anything at all about environmental limits, population carrying capacity, the need to think up to and beyond oil capitalism. Retort proposes a Left response—typically and understandably sketchy and open-ended (how could it not be?)—to the violence of military neoliberalism. But as for a Left discourse on oil capitalism that would go beyond the pedagogic gestures of eco-apocalyptic discourse, we have yet to find it.

Can such a discourse even exist? Retort suggests that opposition to what it terms "consumer metaphysics" is rooted in a crisis of time. "What is the current all-invasive, portable, minute-by-minute apparatus of mediation," the authors ask, ". . . if not an attempt to expel the banality of the present moment?" (183). The hope, drifting throughout the social, is for "*another* present—a present with genuine continuities with a retrieved past, and therefore one opening onto some *non-empty, non-fantastical* vision of the future" (183). Such futures—futures that are in a very real sense "post"-modernity—are in the process of being created planetwide and in those very spaces where enclosure is violently taking place and consumer metaphysics is at its weakest. As Mary Louise Pratt points out, "Where identities cannot be organized around salaried work, consumption, or personal projects like upward mobility, life has to be lived, organized and understood by other means. People generate ways of life, values, knowledges and wisdoms, pleasures, meanings, hopes, forms of transcendence relatively independent of the ideologies of the market."[36] These narratives of meaning can take many forms, from classic Left narratives to wild new religions like the one Pratt discusses, Alfa y Omega, whose two primary symbols are the lamb of god and a flying saucer.[37] Whether such futures are "non-empty" or "non-fantastical" is open to question, even if one was careful to resist measuring them by the standard of whether they figure disaster, much less imagine a way of addressing it.

Whither capital? Will the end of oil capital bring an end to capital as such (and thus, potentially, in its wake, bring new political possibilities)? The expectation that haunts the future is not the end of capital, but that, despite everything, oil capital will not end until every last drop of oil (or atom of fossil fuels) is burned and released into the atmosphere. Fredric Jameson's often-repeated suggestion that "it seems to be easier for us today to imagine the thoroughgoing deterioration of the earth and of nature than the breakdown of late capitalism" points to a limit in how, to date, we have framed the coming future and its disasters.[38] It is not that we can't name or describe, anticipate or chart the end of oil and the consequences for nature and humanity. It is rather that because these discourses are unable to mobilize or produce any response to a disaster we know is a direct result of the law of capitalism—limitless accumulation—it is easy to see that nature will end before capital. As Jan Oosthoek and Barry Gills write, "What is most urgently needed . . . is not short-term technological fixes but a different paradigm of political economy. This new political economy must take our impact on the planet's environment fully and realistically into account."[39] Easy enough to say, but much, much harder to produce when what is called for is full-scale retraction against the flow of a social whose every element moves toward accumulation and expansion.

Notes

1. "A new nuclear power plant would have to open every few days to replace the world's fossil fuel use in a century, and the problems of renewable, low-density, hard-to-store, distant renewable energy sources will take a lot of time and money to overcome on the scale needed." Julie Jowett, "Fossilised Myths: Fresh Thinking on 'Dirty' Coal," *Guardian Weekly*, March 17–23, 2006, 5.

2. Karl Marx, "Critique of the Gotha Program," in *The Marx-Engels Reader*, 2nd ed., ed. Robert C. Tucker (New York: W. W. Norton, 1978), 525–41, at 525.

3. The literature addressing the end of oil is now extensive. For some representative studies, see Kenneth Deffeyes, *Beyond Oil: The View from Hubbert's Peak* (New York: Hill & Wang, 2005); David Goodstein, *Out of Gas: The End of the Age of Oil* (New York: W. W. Norton, 2005); Richard Heinberg, *The Party's Over: Oil, War, and the Fate of Industrial Societies* (New York: New Society, 2005); and Paul Roberts, *The End of Oil: On the Edge of a Perilous New World* (New York: Mariner Books, 2005). For a contrary view, see the study by John H. Wood, Gary R. Long, and David F. Morehouse, "Long-Term World Oil Supply Scenarios: The Future Is Neither as Bleak or Rosy as Some Assert," posted on the US government's Energy Information Administration website, www.eia.doe.gov/pub/oil_gas/petroleum/feature_arti cles/2004/worldoilsupply/oilsupply04.html (accessed September 19, 2006).

4. These figures come from Mark Jaccard, *Sustainable Fossil Fuels: The Unusual Suspect in the Quest for Clean and Enduring Energy* (Cambridge: Cambridge University Press, 2006). For other studies that take the view that oil remains relatively abundant, see Wood, Long,

and Morehouse, "Long-Term World Oil Supply Scenarios," and the especially influential "2000 U.S. Geological Survey World Petroleum Assessment," http://pubs.usgs.gov/dds/dds -060/ (accessed September 19, 2006). Wood, Long, and Morehouse place peak oil production as late as 2047, while the USGS estimates that there remain as much as 2.3 trillion barrels of usable oil on earth (including reserves, reserve growth, and undiscovered reserves).

5. Retort, "Blood for Oil?," in *Afflicted Powers: Capital and Spectacle in a New Age of War* (New York: Verso, 2005), 38–77. Hereafter cited parenthetically by page number.

6. V. I. Lenin, *Imperialism: The Highest Stage of Capitalism* (London: Pluto Press, 1996), 11–26.

7. Daniel Yergin, "Ensuring Energy Security," *Foreign Affairs* 85, no. 2 (2006): 67–82.

8. David Harvey, *The New Imperialism* (New York: Oxford University Press, 2003), 1–25 and 183–212; and Neil Smith, *The Endgame of Globalization* (New York: Routledge, 2004), 177–209.

9. See "Energy Security for the 21st Century," www.whitehouse.gov/infocus/energy/ (accessed March 26, 2007).

10. Alexei Barrionuevo, "Bush Says Lower Oil Prices Won't Blunt New-Fuel Push," *New York Times*, October 13, 2006, http://select.nytimes.com/gst/abstract.html?res=F10910FE3 A540C708DDDA90994DE404482.

11. "China's share of the world oil market is about 8 percent, but its share of total growth in demand since 2000 has been 30 percent. World oil demand has grown by 7 million barrels per day since 2000; of this growth, 2 million barrels each day have gone to China. India's oil consumption is currently less than 40 percent of China's, but because India has now embarked on what the economist Vijay Kelkar calls the 'growth turnpike,' its demand for oil will accelerate." Yergin, "Ensuring Energy Security," 72.

12. As just one example, consider the recent report of a "blue ribbon" task force on US energy policy prepared for the Council on Foreign Relations, which focuses on solutions to US dependency on foreign oil. "The task force suggests that energy security has not been a central focus of U.S. foreign policy, though it noted the widespread perception that the invasion of Iraq and other interventions in the Middle East have been driven by the desire to control the region's oil supplies." Shawn McCarthy, "Report Slams U.S. Domestic Energy Policy," *Globe and Mail*, October 13, 2006.

13. Smith, *Endgame of Globalization*, 188.

14. Yergin, "Ensuring Energy Security."

15. Michael Klare, *Blood and Oil: How America's Thirst for Petrol Is Killing Us* (New York: Penguin, 2004), 183.

16. Ibid., 182.

17. See the summary of the Advanced Energy Initiative on the White House's Energy Security web page, www.whitehouse.gov/infocus/energy/ (accessed March 26, 2007).

18. Prime Minister Stephen Harper, quoted in Bill Curry and Mark Hume, "PM Plans 'Intensity' Alternative to Kyoto," *Globe and Mail*, October 11, 2006.

19. Fredric Jameson, "The Politics of Utopia," *New Left Review*, no. 25 (2004): 35–54, at 36.

20. Ibid., 48.

21. See Kenn Brown's illustration in *Scientific American* 295, no. 3 (2006): 51.

22. "Energy's Future: Beyond Carbon," *Scientific American* 295, no. 3 (2006): 46–114.

23. Gary Stix, "A Climate Repair Manual," *Scientific American* 295, no. 3 (2006): 46–49, at 49.

24. Robert H. Socolow and Stephen W. Pacala, "A Plan to Keep Carbon in Check," *Scientific American* 295, no. 3 (2006): 50–57, at 52.

25. The original reads, "Mankind only sets itself such tasks as it can solve; since, looking at the matter more closely, it will always be found that the task itself arises only when the material conditions for its solution already exist or are at least in the process of formation." *Marx-Engels Reader*, 5.

26. Joan Ogden, "High Hopes for Hydrogen," *Scientific American* 295, no. 3 (2006): 94–101, at 101.

27. Stix, "Climate Repair Manual," 49.

28. Andrew McKillop with Sheila Newman, ed., *The Final Energy Crisis* (London: Pluto Press, 2005). See also Julian Darley, *High Noon for Natural Gas* (White River Junction, VT: Chelsea Green, 2004); Kenneth Deffeyes, *Hubbert's Peak: The Impending World Oil Shortage* (Princeton, NJ: Princeton University Press, 2001); David Goodstein, *Out of Gas*; Richard Heinberg, *The Party's Over*; James Howard Kunstler, *The Long Emergency: Surviving the Converging Catastrophes of the Twenty-First Century* (New York: Atlantic Monthly, 2005); and Roberts, *End of Oil*, among others.

29. McKillop and Newman, *Final Energy Crisis*, figures from 7, 232, and 265–73, respectively.

30. See, for instance, the following chapters from McKillop and Newman, *Final Energy Crisis*: "Apocalypse 2035," 186–90; "The Chinese Car Bomb," 228–32; "The Last Oil Wars," 259–64; and "Musing Along," 289–94.

31. Ted Trainer, "The Simpler Way," in McKillop and Newman, *Final Energy Crisis*, 279–88; quotes from 280, 283, 286–87, and 284, respectively.

32. Ibid., 281.

33. Jacob Lund Fisker, "The Laws of Energy," in McKillop and Newman, *Final Energy Crisis*, 74–86, at 74.

34. For a more philosophically thorough treatment of the concept of the "spectacle," see Anselm Jappe, *Guy Debord* (Berkeley: University of California Press, 1999), 5–30.

35. Fisker, "Laws of Energy," 74.

36. Mary Louise Pratt, "Planetary Longings: Sitting in the Light of the Great Solar TV," in *World Writing: Poetics, Ethics, Globalization*, ed. Mary Gallagher (Toronto: University of Toronto Press, 2008), 207–23.

37. Ibid.

38. Fredric Jameson, *The Seeds of Time* (New York: Columbia University Press, 1996), xii.

39. Jan Oosthoek and Barry K. Gills, "Humanity at the Crossroads: The Globalization of Environmental Crisis," *Globalizations* 2, no. 3 (2005): 283–91, at 285.

))) The Great White Way

David Nye

In Times Square spectacular lighting had moved far beyond functional necessity. Like the displays at world fairs, it served as an instrument of cultural expression providing symbolic validation of the urban industrial order. Electric advertising signs, intensive street lighting, spotlights, searchlights, and beacons could turn any landscape, from Main Street to Niagara Falls, into a text without words. Dramatic lighting transformed the gloomy night city into a scintillating, sublime landscape, and presented a startling contrast to the world seen by daylight. The development of White Ways and illuminated buildings, the triumph of electrical advertising, and the visual simplification of the city into a glamorous pattern of light, together signaled more than the triumph of a new technology. Spectacular lighting had become a sophisticated cultural apparatus. The corporations and public officials who used it could commemorate history, encourage civic pride, simulate natural effects, sell products, highlight public monuments, and edit both natural and urban landscapes.

Yet this new visual discourse was not unified; there were tensions developing between genteel architects and mass advertisers, and also between the older horizontal city of buildings that seldom rose above six stories and the vertical thrust of corporate structures. Reformers striving to recreate Chicago's White City in the actual city deplored both the spread of huge advertising signs and the scale of the skyscrapers with their spectacular lighting effects; they wanted a much different kind of civic art. After 1900 the emerging electric landscape increasingly challenged writers, photographers, and painters to grapple with its heterogeneity, and many cultural critics were drawn to "The Night Glow of the City." New York in particular became a site for meditations on the electrical sublime. Writing for *Harper's* in 1910, Edward Hungerford rhapsodized, "At the first soft breath of twilight, windows in the office buildings become radiant, store-fronts blaze forth in brilliancy. . . . The surface cars . . . are as gayly brilliant as a king's palace one hundred years ago—we do progress in a century. Any well-born citizen of 1810 might rub his eyes in wonder—the cold white iridescence of a myriad puttering arcs makes the city street at night as brilliant as midday." He called electricity "the nerve force of the modern city" and lionized the engineers and their machines that made "the unseen force that vibrates out to light the town."

The city looked its best from a distance in the twilight, when many found it

Electrifying America: Social Meaning of a New Technology (Cambridge, MA: MIT Press, 1990), 73–84

mysterious, beautiful, and powerful. Like many others, as a young man Lewis Mumford had a personal epiphany when crossing Brooklyn Bridge from the Brooklyn side. "Three quarters of the way across the Bridge I saw the skyscrapers in the deepening darkness become slowly honeycombed with lights until, before I reached the Manhattan end, these buildings piled up in a dazzling mass against the indigo sky. . . . Here was my city, immense, overpowering, flooded with energy and light. . . . The world at that moment opened before me, challenging me, beckoning me. . . . In that sudden revelation of power and beauty all the confusions of adolescence dropped from me, and I trod the narrow, resilient boards of the footway with a new confidence."[1] Ezra Pound, back from Europe for a visit in 1910, had a similar positive response to the night skyline of New York, finding it "the most beautiful city in the world" in the evening. "It is then that the great buildings lose reality and take on their magical powers. They are immaterial; that is to say one sees but the lighted windows. Squares after squares of flame, set and cut into the aether. Here is our poetry, for we have pulled down the stars to our will."[2]

Other writers were less appreciative. The painter Louis Eilshemius wrote a poem about New York City, which he called a "Godless monster . . . merciless to man and beast."[3] Willa Cather's narrator in "Behind the Singer Tower" was almost as harsh when from a ship he observed New York's illuminated skyline. "As I looked at the great encandescent [sic] signs along the Jersey shore, blazing across the night the names of beer and perfumes and corsets, it occurred to me that, after all, that kind of thing could be overdone; a single name, a single question, could be blazed too far. Our whole scheme of life and progress was perpendicular."[4] Nor was the English author Rupert Brooke enchanted by the Great White Way during a visit in 1913; rather he found it to be a debased version of Greek mythology.

> Out of the gulfs of night, spring two vast fiery toothbrushes, erect, leaning toward each other, and hanging on to the bristles of them a little devil, little but gigantic, who kicks and wriggles and glares. After a few moments the devil, baffled by the firmness of the bristles, stops, hangs still, and rolls his eyes, moon-large, and in a fury of disappointment goes out. . . . [Nearby] come forth a youth and a man-boy, flaming in celestial underwear, box a short round, vanish, reappear for another round, and again disappear. Night after night they wage this combat. . . . Beyond, a Spanish goddess, some minor deity in the Dionysian theogony, dances continually, rapt and mysterious, to the music of the spheres. . . . And near her, Orion, archer no longer, releases himself from his strained posture to drive a sidereal golf ball out of sight through the meadows of paradise.[5]

In "A Rhyme about an Electrical Advertising Sign" Vachel Lindsay was more ambivalent. Lindsay called the electric light "specious," "blatant, mechanical, crawling and white, / wickedly red or malignantly green." He recognized that the electric advertisement was

Starting new fads for the shame-weary girl,
By maggoty motions in sickening line.
Proclaiming a hat or a soup or a wine,
While there far above the steep cliffs of the street
The stars sing a message elusive and sweet.

Yet after drawing this stark contrast between the natural light of the stars and the specious light of man, in a characteristically hopeful conclusion Lindsay suggests that "some day this old Broadway shall climb to the skies," join the planets and the stars, and create a "new Zodiac." Rather than accepting Brooke's bleak view that the White Way had effaced the constellations, Lindsay blurred heaven and earth, man and nature. For him, the invention of the electric light "leads on to the marvelous change beyond change."[6] Lindsay's hopeful confusion was a more common attitude than Brooke's parodic contrast, as can be seen in the public preference for postcards of the urban skyline at night, widely sold by the 1920s, that became icons of the new visual order.

Indeed, photographers had long used the apparent realism of their medium to impose symmetry on the city to the satisfaction of their customers. Peter Hales has shown how in earlier decades commercial cameramen produced urban panoramas that depicted "the city not as a place of chaos, darkness, and danger, but of order, light, and intelligibility," reassuring their upper-middle-class purchasers that American civilization was progressing.[7] Toward the end of the century, however, Jacob Riis, Lewis Hine, and other reform-minded photographers presented far less pleasant views of the city, which emphasized housing problems, poverty, and the plight of new immigrants. In a counter to this unsettling vision, pictorialist photography provided a new aestheticization of the city, making its outlines gentle, its skyscrapers ethereal, its lights luminous. Alfred Stieglitz produced *Night, New York* in 1897, an image that reduces the city to a row of electric lights softly reflected on wet pavement and darkly outlining the bare branches of a tree. Working in the same vein, the art photographers who clustered around Stieglitz created widely admired images of the city at twilight. One of the most famous of these was Edward J. Steichen's *The Flatiron Building* from 1909, taken with a soft-focus lens that turned one of the tallest buildings in the city into a shrouded enigma. Many similar images were made in the first two decades of the century, aestheticizing the city and dissolving it into a blurred impressionism.[8]

The work of painters between 1900 and 1930 shows even more clearly how electric lighting began to suffuse the American sense of place. The night cityscape was a new visual reality signifying a break in the continuity of lived experience that the painter confronted along with the rest of society. It challenged artists with a new subject, and most of the important painters of the early twentieth century attempted to portray the city at night, despite the considerable difficulty of doing so.[9] The subject was hard to observe or to sketch; colors under artificial light were not the same as in sunlight; and a landscape with many black areas and a multitude of light sources presented new problems of perspective. The city painted at the end of the late nineteenth century was at first nothing like the electrified utopias of the great expositions, but was rather that of the Ashcan School—untidy, unnatural, ebullient, complex, mechanical, and often ugly.[10] But as the technological sublime of the lighting engineers was deployed, it appealed to many artists, who saw the new night landscape as a world at once softer and more glamorous than the everyday world, and poignantly removed from it. Significantly, whatever style painters used to depict the electrified cityscape, they tended to adopt an Olympian perspective; the painting's point of view was no longer at street level with the average citizen but high in the air, looking at the city as a vast design. In abandoning the common ground for elevated vantage points, artists could easily abandon all reference to people and become engrossed in the patterns of skyscrapers, lights, and machines. Aside from this general preference for an elevated perspective, painters adopted different idioms as the electrification of the city proceeded from Stieringer's subdued pointillism toward Ryan's flood lighting that would dominate by the 1920s.

The earlier canvases (until about 1910) emphasize how light transformed an ordinary scene in the city into something picturesque. They often deal with dusk, when electric lighting mingles with the departing sunlight. The buildings are still seen in natural light, but at twilight shadows become indistinct, and electric lights provide new patterns, accents, and contours. Impressionism was well adapted to such subjects, as Childe Hassam showed in *Broadway and 42nd Street*, a canvas from 1902 which depicts the crowds, coaches, and street cars in Times Square on a snowy evening. In the *New York Sun* he noted that the city was best "viewed in the early evening when just a few flickering lights are seen here and there and the city is a magical evocation of blended strength and mystery."[11] J. A. Weir also painted the city at dusk in *The Plaza: Nocturne* (1911) and *The Bridge, Nocturne*, (1910), taking a high vantage point on a rooftop and deploying a pattern of greys, browns, and shadows, punctuated by the lights in windows. His approach reflected Whistler's influence, and transformed the city into a veiled and melancholy mystery spangled with patches of light. In later years the presentation of the city at twilight became a formula. *Painting Cityscapes*, a how-to book, explained: "Cityscapes are often quite beautiful in the early evening, when lights begin to appear in windows, and street lights go on. One can still see the main details of buildings, and the soft, gray-blue tones of dusk make the lamp lights look soft. Contrasts are very delicate."[12] As these remarks suggest, such paintings of the electrified city tended toward the sentimental and the picturesque, adopting an aesthetic appropriate to Stieringer's pointillistic lighting style.

But as lighting engineers added a new array of powerful electric lights to the urban landscape, creating a brighter world of intense colors, other artists began to paint the city after nightfall, when full darkness made contrasts more absolute. Such canvases tend to exploit the ways that the electric light edits the urban landscape. New problems arise here because "everything is dark, but never black, while the lights acquire greater splendor. Such paintings can hardly be done on the spot, because . . . you couldn't see your own colors clearly enough . . . you'd have to do the layout by daylight, adding the colors from memory."[13] Because of these difficulties, many night landscapes were made of scenes intensely familiar to the artist, as was the case with Georgia O'Keeffe's *New York, Night, 1929* and *The American Radiator Building*. Both were painted after she had moved into rooms on the thirtieth floor of the Shelton Building. She recalled, "I had never lived so high before and was so excited that I began talking about trying to paint New York. Of course I was told that it was an impossible idea—even the men hadn't done too well with it."[14] Her paintings are a paradoxical combination of the urban and the pastoral. The jumbled and chaotic elements of urban life disappear in a pattern of lights that give the buildings a warm quality, emphasized by the use of yellows, soft reds, deep purples, and other appealing colors. She adopted clean, symmetrical, but not excessively sharp lines that together with her warm palette give an inviting quality to the skyscraper windows. *New York, Night, 1929* contains more than one hundred small spots of light, representing windows in yellow and red, streetlights in white, and automobile headlights in a yellow-white. The few lines are created by the contrasts between the dark edges of buildings in the middle distance and the artificial lights behind them. The scale, the Olympian perspective, and the darkness combine to eliminate human beings from the composition, leaving moot the relationship between them and the city. In feeling, these two paintings are related to photographic pictorialism, particularly that done using special papers and washes to make a tinted print, such as a bluish-green image Stiechen produced of the Flatiron Building.[15]

In the same years that O'Keeffe and Stieglitz depicted a city of softened heroic dimensions, Charles Burchfield and Edward Hopper focused on smaller corners of urban space that had been transformed by the electric light. Just before World War I Burchfield repeatedly sketched New York in the evening, each time making a large halo around the streetlights, like the halation in night photographs. Lighting becomes intrusive as it obliterates the objects it is meant to illuminate, as in *Plaza Hotel* or *Dark Tower with Lamppost*. He restated the theme of electrification as obstruction more forcefully in a watercolor from the same period, *Chimney and Powerlines*, in which a view over the rooftops consists of fragments of dark buildings and trees, sliced by a tangle of power lines, each represented not as a black line but as a thick, bright yellow swatch that distracts the eye from the rest of the landscape.[16] If Burchfield depicted electric lines as an obstruction, Hopper depicted individuals. In the etching *Night in the Park* a man alone on a park bench at night reads a newspaper under a street light. *Night Shadows* shows a man walking toward a street corner, seen from above, his small shadow dwarfed by that of the street light. In both, streetlights provide the only light source, in a literal sense making these scenes possible. They do not beautify their surroundings, however, but create harsh contrasts and cast unusual shadows. In *Night Windows* and *Room in New York* Hopper painted the interiors of homes and apartments glimpsed from the street on hot summer nights when the windows were open. Each shows an unposed private moment not meant for general scrutiny, a moment bathed in intense light from a hidden ceiling fixture that seems insensitively harsh. More emphatic suggestion of Hopper's distaste for commercial lighting emerges in *Chop Suey* (1929), where an otherwise tranquil scene of intimacy between two women at lunch is marred by a garish red electric sign mounted outside the window.[17]

By the 1920s, however, most artists were not concerned with these fleeting impressions of private life but with the dramatic patterns of the larger night landscape. The Great White Way had more electric signs, bridges were outlined in lights, and most skyscrapers, monuments, and public buildings had floodlighting. Painters working in the tradition of European modernism, such as Stefan Hirsch and George Ault, responded to this brighter night city, exploring how the lighting engineers had given it an almost cubist skyline. This in itself did not necessarily make the city glamorous; some were struck by how light had become a way to sell products, commercializing urban space, as in one of Stefan Hirsch's canvases where a giant billboard advertising toothpaste contains the slogan, "It's best because it's better."[18] An implied criticism also lies in several of Ault's paintings where the stark, powerful lines virtually efface the human subject in a city built on the largest scale.

In contrast, Joseph Stella developed an emphatically positive image of the electrified city. Of Italian origin, he had studied in Italy and France at the time futurism emerged. The futurists "came to think of the picture zone as a flickering network of multiple stresses, charged with magnets and electric currents, invisible forces now controlled by man."[19] Though Stella's personal contact with the futurists was limited, he knew their work and shared their enthusiasm for the city. The paintings he did after returning to the United States just before World War I reflect his rediscovery of the American urban landscape, now seen as an expression of the new machine age. He was enchanted by New York, and his paintings stressed its monumental perpendicularity and kaleidoscopic sense of movement, light, and strong color. Stella proclaimed that "steel and electricity are the creative factors of modern Art," and to embody this dictum created five canvases with the collective title, *The Voice of the City of New York Interpreted*. Peter Conrad has argued that "in Stella's New York paintings, the genius of electricity expresses itself as color, that of steel as taut and supple line. Stella sought to electrify his palette, wanting his colors to glare as blindingly as the lights to which they paid homage."[20] Electricity pervades all of these canvases, most obviously in two entitled *The Great White Way*. Working with the vocabulary provided by Broadway's advertising signs, footlights, and mirrors, Stella recreated an illusory perspective, playing shafts of light over the surface, leading the viewer's eye into one logical cul-de-sac after another, baffling attempts to establish the classical perspective of front and back. These White Way paintings frame *The Skyscrapers*, towers of steel enhanced by light, and are flanked by studies of Brooklyn Bridge and the New York Harbor. At the bottom of all five canvases, Stella placed abstract shapes derived from the electrified subway system, suggesting how electricity and steel had penetrated the earth as well as the sky. As Oliver Larkin noted, "Stella composed the movements of a visual symphony, in which night, the glare of electricity, the sheen of steel, and the roar of the subway train played their parts." The five panels together have "the intensity of stained glass and the monumentality of murals."[21] Stella put on canvas the visionary technology that Ryan had created for the world's fairs—the searchlights, scintillator effects, and flood lighting that had transfigured the city into a futurist vision.

If Stella embraced the machine age and the special effects of the lighting engineers, not only making electrical technologies a central theme of his five canvases but adopting its colors and lines as part of his visual vocabulary, the realist John Sloan took an entirely different approach to the night cityscape. The difference between them was more than one of technique; they disagreed about the moral implications of the new landscape, and their differences can be summed up by comparing Stella's *The Voice of the City of New York Interpreted* with Sloan's city from *Greenwich Village*, 1922. As early as 1905 Sloan had seen how electric light was altering New York, in *Sunset, West 23rd Street*. Like Hassam and Weir at this time he explored the visual richness of the city at twilight. Later, in 1910 he trained his eye on smaller details in an etching, *Night Windows*, showing glimpses of domestic life on a hot summer night that is similar to Hopper's treatment of the same subject. In *Bleecker Street, Saturday Night* (1918) he dealt with a series of streetlights and lighted windows. All of this previous work prepared him for the much more complex *City from Greenwich Village*, which contains all of these forms of electrification as they shaped the New York skyline and daily life. Indeed, the canvas depicts New York almost entirely in terms of electricity, creating a work that not only represents the night cityscape but summarizes the responses of other artists to the same subject.

As is characteristic of most night cityscapes, this one presents the city from a high elevation, in this case the view from the roof of Sloan's studio on a wet night, which he could paint largely from memory with the help of a few sketches. In the distance glows the Great White Way, occupying the upper left corner with the largest concentration of bright colors. The rest of the canvas is done in warm, dark tones, interrupted by the contrasting lights from storefronts and apartment windows. The sense of space is created only partially through the use of conventional perspective; equally important is a careful recession of the colors, so that although the Great White Way in the distance seems to contain the brightest colors in the painting, it does not. To achieve these effects, Sloan had studied the problem of color in night landscapes; in *Gist of Art* he explained how to create the "spacial recession of colors. If you are painting a street at night, make the lights in the foreground yellow-orange, running in a sequence back to red-orange in the background, with similar changes in the colors of the buildings. If you were to put a strong yellow-orange on a plane in the background that is made with neutral purples, it might jump out of place." Sloan had found, "In painting night-lighted subjects, it is very difficult to carry the color of the light through the picture; to carry not only the tonal quality but also the color character. It may be done by mixing all the lights with a colored white."[22] Sloan applied these principles in *The City from Greenwich Village*, using a yellow-orange in the foreground for the store windows and for reflections on the wet pavement, and carefully splitting these intense patches of light with the elevated line that literally points back toward the White Way, as does the line of shop windows running from the lower right corner toward the yellow-orange skyscrapers in the upper left-hand corner. To increase the sense of depth, the contrasts in the foreground, particularly in the area below the White Way, are much stronger than those along the skyline. Throughout the composition electric lighting gives structure to the painting, creating a visual tension between the neighborhood built on a human scale in the foreground and the glitter and power of the White Way in the distance. Despite the ubiquity of electrical advertising by 1922, the canvas includes only one legible word—"moonshine"—at once a comment on the hypocrisy of Prohibition, which was scarcely enforced by 1922, and a reminder of the natural light missing from the night cityscape.

Not only does electricity shape and organize this night landscape, but the painting summarizes contemporary artistic responses to electric lighting, as much as that is possible within one coherent style.[23] The incandescent White Way in the distance appears as a gauzy version of the futurist city, while Greenwich Village, the home of artists and intellectuals, is a retreat from its enormous towers. Here, in the right foreground, Sloan can depict the still horizontal city of the pictorialists, including a wedge-shaped building much like the Flatiron. To the left, on the other side of the elevated line, is a view across the rooftops reminiscent of Hopper's approach to similar subjects. The lower half of the painting also stresses the lighted advertising signs and shop windows, which are more subdued than those of the White Way. As the former art editor of *The Masses*, Sloan was acutely aware of how New York was being transformed and commercialized. Thus it is appropriate that he used an ensemble of techniques to capture both the local picturesque qualities of Greenwich Village and the electrical sublime of the distant skyscrapers, including the floodlighted Singer Tower and Woolworth Building, which stand out as the two tallest structures. Sloan commented on the canvas, "The picture makes a record of the beauty of the older city which is giving way to the chopped-out towers of modern New York."[24] He thus located it in a visual history of the city: the foreground of Greenwich Village has not yet been entirely transformed by lighting specialists and the verticality of downtown Manhattan. It is a place still built to human scale, and so encourages the strategies of realism or impressionism appropriate to the modest lighting of an earlier era. But these buildings, this scale, and the visual vocabulary itself are all marginalized by the advancing towers and brilliance of the White Way, whose chopped-out towers lend themselves to the modernist aesthetics of futurism and cubism. The painting does not merely contrast the two regions, but shows how the Great White Way's commercial energies stream out along the arteries of the bright streets and the track of the electric railway, proclaiming the triumph of the lighting engineers and the electrical sublime they invented.

Notes

1. Edward Hungerford, "The Night Glow of the City," *Harper's* 54 (1910). Also see George Fife, "A Fantasy of City Light," *Harper's* 57 (1913): 60–63; Lewis Mumford, *Sketches from Life: The Autobiography of Lewis Mumford: The Early Years* (New York: Dial Press, 1982), 129–30.

2. Cited in Hugh Kenner, *A Homemade World: The American Modernist Writers* (New York: William Morrow, 1975), 5.

3. Louis Eilshemius, *Fragments and Flashes of Thought* (New York: E. Lewis, 1907). And see his canvas *New York at Night*, c. 1910.

4. Willa Cather, "Behind the Singer Tower," in *Collected Short Fiction, 1892–1912* (Lincoln: University of Nebraska Press, 1965).

5. Rupert Brooke, "The Lights of Broadway," *Literary Digest* 47 (November 15, 1913): 977.

6. Vachel Lindsay, *Collected Poems* (New York: Macmillan, 1952), 339–40.

7. Peter Hales, *Silver Cities* (Philadelphia: Temple University Press, 1985), 120.

8. *America and Alfred Stieglitz: A Collective Portrait* (New York: Aperture, 1975), reprint of original 1934 edition, plate 6. A fine reproduction of Steichen's Flatiron Building photograph is in William Crawford, *The Keepers of Light* (Dobbs Ferry, NY: Morgan & Morgan, 1979), plate 38.

9. Electric light made it possible for artists to work at night in the studio, though colors and shadows did not appear the same under artificial illumination as they did in sunlight. Museums and galleries also found that some canvases changed appearance under artificial light.

10. In the words of J. B. Jackson, it was an "existential landscape, without absolutes, without prototypes, devoted to change and mobility and the free confrontation of men. . . . It has vitality but is neither physically beautiful nor socially just." In *Landscapes: Selected Writings of J. B. Jackson*, ed. Ervin H. Zube (Amherst: University of Massachusetts, 1970), 9.

11. *New York Sun*, February 13, 1913, sec. 4, 16. E. Redfield painted a similar image, *Between Day and Night*.

12. Ralph Fabri, *Painting Cityscapes* (New York: Watson Guptill, n.d., c. 1975), 34.

13. Ibid., 34.

14. Georgia O'Keeffe, *Georgia O'Keeffe* (New York: Viking, 1976), text facing plate 17. Her first image of the night city was *New York with Moon* in 1925. See also her austere *City Night* (1926), which looks up at dark windowless skyscrapers against a darkening blue sky at evening. The city here is almost reduced to an abstraction.

15. Georgia O'Keeffe, *New York-Night, 1928–1929*; and *The American Radiator Building*. Also compare Stieglitz's similar photograph of a skyscraper at dusk, *New York, Night, 1931*, in *America and Alfred Stieglitz*, plate 83.

16. Charles E. Burchfield's work is documented in the catalog "Charles E. Burchfield at the Kennedy Galleries" (New York, n.d.), which contains the following: *Chimney and Powerlines*, c. 1917; *Hotel Plaza, Night Scene*, 1916, pencil; *Dark Tower with Lamppost*, 1916; *New York Street Study*, 1916 sketchbook; *New York Scene at Night*, 1916, pencil, sketchbook. Two important works not discussed here are: *Winter Twilight*, 1930, Whitney Museum; and *Rainy Night*, 1929–1930. San Diego Museum.

17. Paintings and etchings reproduced in Lloyd Goodrich, *Edward Hopper* (New York: Harry Abrams, 1983). The generalizations made here could also be supported by discussing *Drug Store* (1927), *Summer Evening* (1947), and *Rooms for Tourists* (1945).

18. Stefan Hirsch, *Night Terminal*; Charles Sheeler, *Church Street El*; George Ault, *Construction, Night, 1922*. Also see Paul Outerbridge, *42nd Street Elevated*, and Therese Bernstein, *Searchlights on the Hudson, 1915*.

19. See Max Kozloff, who calls the Futurists "the most committed urbanophiles yet seen in modern art." *Cubism/Futurism* (New York: Charterhouse, 1973), 119, 142.

20. Peter Conrad, *The Art of the City* (New York: Oxford University Press, 1984), 133. Stella quotation ibid. Stella's *The Voice of the City of New York Interpreted* hangs in the Newark Museum.

21. Oliver Larkin, *Art and Life in America* (New York: Holt, Rinehart, & Winston, 1964), rev. ed., 384.

22. John Sloan, *Gist of Art* (New York: American Artists Group, 1939), 128.

23. John Sloan's *City from Greenwich Village, 1922* and three preliminary sketches, National Gallery, Washington, DC.

24. *John Sloan, 1871–1951* (Washington, DC: National Gallery of Art, 1972), 169–70.

⟫ Standard Oil Co.

Pablo Neruda

When the drill bored down toward the stony fissures
and plunged its implacable intestine
into the subterranean estates,

Canto General (Berkeley: University of California Press, 2011), trans. Jack Schmitt, 176–77. Originally written in 1940.

and dead years,
eyes of the ages,
imprisoned plants' roots
and scaly systems
became strata of water,
fire shot up through the tubes
transformed into cold liquid,
in the customs house of the heights,
issuing from its world of sinister depth,
it encountered a pale engineer
and a title deed.

However entangled the petroleum's arteries may be,
however the layers may change their silent site
and move their sovereignty amid the earth's bowels,
when the fountain gushes its paraffin foliage,
Standard Oil arrived beforehand with its checks and its guns,
with its governments and its prisoners

Their obese emperors from New York
are suave smiling assassins
who buy silk, nylon, cigars
petty tyrants and dictators.

They buy countries, people, seas, police, county councils,
distant regions where the poor hoard their corn
like misers their gold:
Standard Oil awakens them,
clothes them in uniforms, designates
which brother is the enemy.
The Paraguayan fights its war,
and the Bolivian wastes away
in the jungle with its machine gun.

A President assassinated for a drop of petroleum,
a million-acre mortgage,
a swift execution on a morning mortal with light,
petrified,
a new prison camp for subversives,
in Patagonia, a betrayal, scattered shots
beneath a petroliferous moon,

a subtle change of ministers
in the capital, a whisper
like an oil tide, and zap, you'll see
how Standard Oil's letters shine above the clouds,
above the seas, in your home,
illuminating their dominions.

))) The Petrol Pump

Italo Calvino

I should have thought of it before, it's too late now. It's after twelve thirty and I didn't remember to fill up; the service stations will be closed until three. Every year two million tons of crude are brought up from the earth's crust where they have been stored for millions of centuries in the folds of rocks buried between layers of sand and clay. If I set off now there's a danger I'll run out on the way; the gauge has been warning me for quite a while that the tank is in reserve. They have been warning us for quite a while that underground global reserves can't last more than twenty years or so. I had plenty of time to think about it, as usual I've been irresponsible; when the red light begins to wink on the dashboard I don't pay attention, or I put things off, I tell myself there's still the whole reserve to use up, and then I forget about it. No, maybe that's what happened in the past, being careless like that and forgetting about it: in the days when petrol still seemed as plentiful as the air itself. Now when the light comes on it transmits a sense of alarm, of menace, at once vague and impending; that is the message I pick up and record along with the many angst-ridden signs sedimenting down among the folds of my consciousness, dissolving in a state of mind that I can't shake off, but that doesn't prompt me to any precise practical action as a consequence, such as, for example, stopping at the first pump I find and filling up. Or is it an instinct for making savings that has gripped me, a reflex miserliness: as I become aware that my tank is about to run out, so I sense refinery stocks dwindling, and likewise oil pipeline flows, and the loads in tankers ploughing the seas; drill-bits probe the depths of the earth and bring up nothing but dirty water; my foot on the accelerator grows conscious of the fact that its slightest pressure can burn up the last squirts of energy our planet has stored; my attention focuses on sucking up the last dribbles of fuel; I press the pedal as if the tank were a lemon that must be squeezed

Numbers in the Dark and Other Stories, trans. Tim Parks (New York: Vintage, 1996), 170–75

without wasting a drop; I slow down; no: I accelerate, my instinctive reaction being that the faster I go, the further I'll get with this squeeze, which could be the last.

I don't want to risk leaving town without having filled up. Surely I'll find one station open. I start patrolling the avenues, searching the pavements and flower-beds where the coloured signs of different petroleum companies bristle, though less aggressively than they used to, in the days when tigers and other mythical animals blew flames into our engines. Again and again I'm fooled by the "Open" signs which only mean that the station is open today during regular hours, and hence closed during the lunch break. Sometimes there's a pump attendant sitting on a folding chair eating a sandwich or half asleep: he spreads his arms in apology, the rules are the same for everybody and my questioning gestures are pointless as I knew they would be. The time when everything seemed easy is over, the time when you could believe that human energy like natural energy was uncondition-ally and endlessly at your service: when filling stations blossomed enticingly in your path all in a line with the attendant in green or blue or striped dungarees, dripping sponge at the ready to cleanse a windscreen contaminated by the massa-cre of swarms of gnats.

Or rather: between the end of the times when people with certain jobs worked round the clock and the end of the times when you imagined that certain com-modities would never be used up, lies a whole era of history whose length varies from country to country, person to person. So let me say that right now I am ex-periencing simultaneously the rise, apex, and decline of the so-called opulent so-cieties, the same way a rotating drill pushes in an instant from one millennium to the next as it cuts through the sedimentary rocks of the Pliocene, the Cretaceous, the Triassic.

I take stock of my situation in space and time, confirming the data supplied by the kilometre clock, just returned to zero, the fuel gauge now steady on zero, and the time clock, whose short hand is still high in the meridian quadrant. In the meridian hours, when the Water Truce brings thirsty tiger and stag to the same muddy pool, my car searches in vain for refreshment as the Oil Truce sends it scurrying from pump to pump. In the meridian hours of the Cretaceous living creatures surged on the surface of the sea, swarms of minute algae and thin shells of plankton, soft sponges and sharp corals, simmering in the heat of a sun that will go on living through them in the long circumnavigation life begins after death, when reduced to a light rain of animal and vegetal detritus they sediment down in shallow waters, sink in the mud, and with the passing cataclysms are chewed up in the jaws of calcareous rocks, digested in the folds of syncline and anticline, liquefied in dense oils that push upward through dark subterranean porosities until they spurt out in the midst of the desert and burst into flames that once again warm the earth's surface in a blaze of primordial noon.

And here in the middle of the noonday urban desert I've spotted an open service station: a swarm of cars surges around it. There are no attendants; it's one of those self-service pumps that take notes in a machine. The drivers are busy pulling chrome pump nozzles out of their sheaths, they stop in mid-gesture to read instructions, uncertain hands push buttons, snakes of rubber arch their retractable coils. My hands fiddle with a pump, hands that grew up in a period of transition, that are used to waiting for other hands to perform those actions indispensable for my survival. That this state of affairs wasn't permanent I was always aware, in theory; in theory my hands would like nothing better than to regain their role of performing all the manual operations of the race, just as in the past when inclement nature beset a man armed with no more than his two bare hands, so today we are beset by a mechanical world that is doubtless more easily manipulated than brute nature, the world in which our hands will henceforth have to go back to managing on their own, no longer able to pass on to other hands the mechanical labour our daily life depends on.

As it turns out my hands are a little disappointed: the pump is so easy to work you wonder why on earth self-service didn't become commonplace ages ago. But the satisfaction of doing it yourself isn't much greater than that of using an automatic chocolate bar dispenser, or any other money munching device. The only operations that require some attention are those involved in paying. You have to place a thousand-lire note in the right position in a little drawer, so that a photoelectric eye can recognize the effigy of Giuseppe Verdi, or perhaps just the thin metal strip that crosses every banknote. It seems the value of the thousand lire is entirely concentrated in that strip; when the note is swallowed up a light goes on, and I have to hurry to push the nozzle of the pump into the mouth of the tank and send the jet gushing in, compact and trembling in its iridescent transparency, have to hurry to enjoy this gift that is incapable of gratifying my senses but nevertheless avidly craved by that part of me which is my means of locomotion. I have just enough time to think all this when with a sharp click the flow stops, the lights go off. The complicated mechanism set in motion a few seconds ago is already stilled and inert, the stirring of those telluric powers my rituals called to life lasted no more than an instant. In return for a thousand lire reduced to a meagre metal strip the pump will concede only a meagre quantity of petrol. Crude costs eleven dollars a barrel.

I have to start all over again, feed in another note, then others again, a thousand lire a go. Money and the subterranean world are family and they go back a long way; their relationship unfolds in one cataclysm after another, sometimes desperately slow, sometimes quite sudden; as I fill my tank at the self-service station a bubble of gas swells up in a black lake buried beneath the Persian Gulf, an emir silently raises hands hidden in wide white sleeves and folds them on his chest, in a skyscraper an Exxon computer is crunching numbers, far out to sea a cargo fleet gets the order to change course, I rummage in my pockets, the puny power of paper money evaporates.

I look around: I'm the only one left by the deserted pumps. The to-ing and fro-ing of cars round the only filling station open at this hour has unexpectedly stopped, as if at this very moment the convergence of creeping cataclysms had suddenly produced the ultimate cataclysm, the simultaneous drying up perhaps of oilwells pipelines tanks pumps carburettors oil sumps. Progress does have its risks, what matters is being able to say you foresaw them. For a while now I've been getting used to imagining the future without flinching, I can already see rows of abandoned cobweb-draped cars, the city reduced to a plastic scrap heap, people running with sacks on their backs chased by rats.

All of a sudden I'm seized by a craving to get out of here; but to go where? I don't know, it doesn't matter; perhaps I just want to burn up what little energy is left us and finish off the cycle. I've dug out a last thousand lire to siphon off one more shot of fuel.

A sports car stops at the filling station. The driver, a girl wrapped in the spirals of her flowing hair, scarf and woollen turtle-neck, lifts a small nose from this tangled mass and says: "Fill her up."

I'm standing there with the nozzle in the air; I may as well dedicate the last octanes to her, so they at least leave a memory of pleasant colours when they burn, in a world where everything is so unattractive: the operations I perform, the materials I use, the salvation I can hope for. I unscrew the fuel cap on the sports car, slip in the pump's slanted beak, press the button, and as I feel the jet penetrate, I at last experience something like the memory of a distant pleasure, the sort of vital strength that establishes a relationship, a liquid flow is passing between myself and the stranger at the wheel.

She has turned to look at me, she lifts the big frames of her glasses, she has green eyes of iridescent transparency. "But you're not a pump attendant. . . . What are you doing. . . . Why . . ." I want her to understand that this is an extreme act of love on my part, I want to involve her in the last blast of heat the human race can make its own, an act of love that is an act of violence too, a rape, a mortal embrace of subterranean powers.

I make a sign for her to shush and point down with my hand in the air as though to warn her that the spell could break any second, then I make a circular gesture as if to say it's all the same, and what I mean is that through me a black Pluto is reaching up from the underworld to carry off, through her, a blazing Persephone, because that's how that ruthless devourer of living substances, the Earth, starts her cycle over again.

She laughs, revealing two pointed young incisors. She's uncertain. The search for oil deposits in California has brought to light skeletons of animal species extinct these fifty thousand years, including a sabre-toothed tiger, doubtless attracted by a stretch of water lying on the surface of a black lake of pitch which sucked the animal in and swallowed it up.

But the short time granted me is over: the flow stops, the pump is still, the embrace is broken off. There's a deep silence, as if all engines everywhere had ceased their firing and the wheeling life of the human race had stopped. The day the earth's crust reabsorbs the cities, this plankton sediment that was humankind will be covered by geological layers of asphalt and cement until in millions of years' time it thickens into oily deposits, on whose behalf we do not know.

I look into her eyes: she doesn't understand, perhaps she's only just beginning to get scared. Well, I'll count to a hundred: if the silence goes on, I'll take her hand and we'll start to run.

))) Reading Wordsworth in the Tar Sands

Stephen Collis

Am I fanciful if I extend the obligation of gratitude to insensate things?
—Wordsworth 1799 letter to Coleridge

An evening walk
An afternoon tripping
A landscape with

No there left there
And who knows how
To negate a negation

Turn our cups upside down
And pour sand into this
Sea of sand

Up north where woods
Are wet and moosey
Except not here not

A single green thing
In sight the site like
An abandoned beehive

The Goose 13, no. 2 (2015): 37

Broken open its grey
Papery layers scattered
Around on the ground

The small spaces where
The bees dark bodies
Should have been

Occupied by the things
We have already forgotten
About the pastoral tradition

We were walkers
In a dangerous time
Of storm and thaw

Took damage in our
Stride—the vacant
Air the wildered mind
Ensnares—beat down
And scraped clean
Of the burden
Of overwhelming being

Wordsworth—I feel you too!
Though there is no mechanism
To nuance this conversation
Across the years—so I brought
Your ruined cottages your
Evening walks and Grasmere
Homing here to the Tar Sands
To stroll across northern desarts
Not knowing how well you fit—
The method of our walking
From seeing to contemplating
To remembering—is yours
Though no solitary haunts
Are here—no birds that scud
The flood—here we tread
Together the shadowy ground

Bright in the sun round
The sharp place absence occupies

The place from which I looked
The plane descending on Fort
Mac or the road we walked

Around the bounds of one dry lake
And if I thought I thought of dying
Of stone and tombs and pits

No profit but one thought
The lot of others could be mined
Yet—aerial—we might business

Halt—tempting notions—wind
Over dead water—I thought of
Clouds where none lay

Grey billows of moneyed dust
Nickel and naught caught up
In tracks of trucks—shadows

Brittle butterflies and the liquid
Crystal depths of dry grass—
Benzene and naphthenic acid sands

Without restraints or bounds
Blowing out and over this
Huge sand ensnared world

Upper limit elegy
Lower limit pastoral
Fader glide between

Walking—we were seeing
Silvered shunts of sand lakes
Like salt flats wondering what
Winkles out in yonder mercury
Sheen? No ponds pretend to

Lighten belief—air canon and
Scarecrow miners surround
These tailings are desolation's
Dream of crumbling decor
Whoever it was saw boreal
Swept it clean in cold accounts
Before land wastes were
Fenced former forests of sand
Thick dark thoughts leaching
Heavy metal music machines
Or death metal bands screaming
Unfathomable ruination inside
A sealed steel cube in space

Dear imagination—lighten up!
Your part is human protest
But there are no visionary scenes
Of lofty beauties uplifting to see
No picturesque prospects
Even if Burtynsky might
Shoot them from the air chromatic as
Abstract patterns of chemical dirt
No matter!—When in service
Of monetary gain and increasing
Industries of land liquidation
This world is anvil entertainment
Bashed first peoples flat land home
Still springing up thrust midst the
Fossilized dead on whose ancestral
Heat we strange grammar feed
As strange accumulations folk
Pummel pores and veins of
Saturated soils coiled up in the
Barrage we make making roads
And the slow bombardment
Of never ending development

Tell me I'm preaching to the choir
I'll tell you I cannot live
Without the choir's solidarity

Tell me I'm flogging a dead horse
I'll tell you I feel every
Lash landing on the galloping land

Perhaps I digress—the occasion
Was a public walk on a
Public road—but the aesthetics
Of the place is pure negation—
Open maw is no landscape
Ripped wound no terrain
There is no viewpoint despite
The signs and picnic tables of
Doom's treeless playground
No play of light at sunset on
Tumescent swath of an earth
Heaving its golden breast towards
A slate sky where gawkers careen
In tin cans winged while in utter
Foundries of digital light
Pounding out templates of data
We break to browse disaster porn
Look death in its vertiginous eye
One house sized truck after another
Blanket ourselves in perspectival
Air of vanished relations—no
This is just the vast insides
Of machine whose impetus
Money tells—no point from which
To see it whole or unveil its grasp
On brow of yonder nonexistent hill—
Just a moving power that moves itself
And us tempest tossed within it
Sloughing boreal off its bitumen back
The calculus which compels
Its animate limbs for alien power
Assembled from our loathing
And slouching now towards Fort
McMurray and Fort McKay to
Deliver a world of dead birds
And unquenchable thirsts

Walking—we were old
Technology
Biotic and slow moving

Dropped into circuit
Pilgrims circling on a
Healing walk walking

All day beating the bounds
Of a single vast and dry
Tailings pond

Edge of the largest mine in the
World past Syncrude and Suncor
Refineries and the vast desart

Tar Sands pastoral
Between upper limit lament
Lower limit complaint

Like we needed a new thing
Could sing ourselves to
Like disappear in where

Our appearance was a trapeze
Over leisure a pratfall for
Liquid asset junk pile and

Property gave out maps
Territory an escape hatch
For animals and others outside

This is where we walked
This is where we swam
Voice again humming
Drum and song Indigenous
To keep us timed timeless
Moving beneath bullets of
Economic praise spraying
Billboards and the birdless

Lakes on our left not
Lakes but pools of poison
Doing *what* beneath their beds
We can only guess leaching
Towards the Athabasca River
Flowing wide nearby on
To Fort Chip and the toxins
Captured in animal flesh there
Last human tenant imagined
Ruined shack planet home
Barren of all future good
Water scarred skin and wooden
Buffalo of Wood Buffalo
Cigar shop life and mines
And ponds where ancestors lie
Don't let the new houses fool you

It is surrounded by fencing
And air canons and clearly
Owns the police

Its money is heaped
In deep black banks
It has broken every treaty with life

Its ceremony is poison
It seems to have eaten the ducks
Or at least their feathered inner lives

Where ghost flight soared
Radar pond to pond
Its magnetic sojourn is lacking

Its clime is coming fast
And is difficult to resist
Though merchants ship it so

This is where we walked
This is where we swam
Wrapped animal bone in
Sweet dry grass offering

And now stand in dry grass
Beside the road offering
Prayer on this first stop
First of four directions west
Second stop north drumming
Singing between two tailings
Ponds not ponds edged by sands
Remembrance that came and went
Like a bird to its grave in the water
Not water—third stop east
Past the refinery smoke and tanks
Fourth stop south and fourth
Direction—still drumming and
Still singing the elders praying
Should earth be wrenched
Throughout or fire wither all
Her pleasant habitations and
Dry up ocean left singed
And bare or the waters
Of the deep gather upon us
Fleet waters of the drowning
World—know that kindlings
Like the morning still
Foretell—though slow—
A returning day lodged
In the frail shrine of us aglow
Old technology of people together
Holding the line against changing weather

Wordsworth there are things
That are fucked up
That we live among

That we are
New life we wanted
All the clear particulars perceived

In active water and airshafts
Struck by slanting light
Energy of forests

Leaped out of animal form
And run into the quiet
Of our empty developments

Caught there electric on cctv
Buffering then lighting up the net
And darkness under the earth

We didn't put there though
We dug its inherent
Capacity to burn and

There are banks and there
Are signs posted up high
That say "we are banks"

Dear common—lowest
Denominator—highest right
Lift light of future foliage
Here where bright burnt
Sands hinge chemical ponds
Over loosest leaves of boreal
Burnt brooks and forests for
Fatter fuel in bitumen beds
Beneath everything we see
Remove everything we can see
To reveal it—paucity of
Ideas for making homes
Making lives led as ghosts
Already haunting doomed
Earth we split and devour

It's elders brings us back
Living idea elders drumming
Singing and walking Indigenous
To all the overburden which
Is no burden but carries
Itself echolocaic through
Leaves of this living and
Wakes while walking still
Breathing in dreamt shade

I could almost gather
Intuitive hopes for spring
Heap method of gleaning

Against Google Chrome of
Most expensive trucks
Or cheap flights to Vegas

Or the women who—treated bare
Commodities—are brought here
Or the single battered yellow bus

Bringing migrants to clean
The factories of empty futures
Grinding horizon I'm drawn to

Having stood wondering at
Far end of pipeline then
Salmon like travelled

To its source in boreal we
Cannot erase the
Colonial continent's bitumen heart

But we can know what arteries
Liquefying histories we
Walk along coasts to mountain view

Stopped here near the
Blasted vale or just after
Lift off on gas wing south
Over seeming endless forest
I find I still need a little
Language of the Tar Sands
The knowing by walking
That tells how boreal grew
And gathered animal cohort
And plant polity over bitumen
Deposit and didn't once think
Noxious profit gas even when
Bubbling surface bogs leached

And aspen trembled—even when
Drinking its life from waters
Just thin surfaces veiling the
Pitch coppered tight beneath

What strange adaptors we are!
That things will grow again
Is no consolation—the difference
Between *this* situation and
The situation of the old growth
On top of bitumen base is the
Difference between a happen
And the ecological capacity
To bear this happening and
A making and the ecological
Capacity to bear this human
Act and choice—what strange
Adaptors we are—moving
Swifter than old accumulations
To chemical our hues where we
Are still that vitality that springs
A weed beside the poison road
Banks of the poison pond
Beneath arch of poison sky

Will we—delimit—ourselves
Or—ova storm of digital increase
Uncap our climate and trade
Mere earth to reach residual heights
Of the value form and receive—
A new dispensation of finitude
Forced from the very ground we have removed
And the sky we have spilt our angers on?

Let me walk a little longer at
Bodily scale—beings have always
Been here—contemplating this
Landscape and letting the flood
Of memories of the future in
Recollecting that time to come
When none of us will be disposable waste

That time somewhere near
Where the road turns at the guarded
Edge of the refinery that we were
Circumambulating a common to come
Curling towards stillness at all scales
Having walked one amongst many
Through a dangerous time and place
The withering land turning towards
Each animal's unrecountable face

))) The Visible Hand of the Sun: Blueprint for a Solar World

Hermann Scheer

Social developments may seem unpredictable, but that does not mean that the course they take is wholly determined by chance. Provided that no great war or natural disaster plunges everything into chaos, they follow a clearly discernible pattern of events. The fossil energy system has shaped the global economy from its inception, leading the world to the lip of the abyss. A global shift towards renewable resources will overturn the structure of the fossil global economy and draw mankind back from the abyss, towards a sustainable future. Sooner or later there will be general recognition of the need for fundamental change.

When the way to a lasting supply of clean energy becomes clear, people will not let the opportunity slip through their fingers. It could be a long time before the optimum path is known, and the route chosen will determine the pace of change. But the capacity of existing norms and structures to hold up the transition that the planet and its people so badly need may mean that it comes too late.

Over the course of history, many civilizations have fallen victim to their failure to wake up to mortal dangers. The fossil fuel crisis places the entire world in such a life-or-death predicament. Political and business leaders have so far shown an inability to rise to the challenge. They shrug their shoulders, content to blame anonymous market forces, to the approval of all those who hold that political institutions are in any case terminally sclerotic. There is some truth in this, as today's politicians seem to have no stomach for decisions that run counter to established business interests. Never before has political failure been so comfortable. This, too, we owe to the market—for now.

Political initiatives can and must drive and accelerate the replacement of fossil

The Solar Economy: Renewable Energy for a Sustainable Global Future (London: Earthscan, 2002), 311–25.

resources. Waiting for reserves to be exhausted is not an option. Their primary task must be to end privileged consumption of fossil fuels—which means abolishing direct and indirect subsidies and absurd tax exemptions—and to blaze the trail for renewable resources. If renewables can achieve rapid market penetration and revolutionize energy consumption, then there is hope for the world even in the declining years of the Industrial Revolution.

History tells us that the Industrial Revolution did not take place everywhere at the same time or even to the same extent, and that it was anything but harmonious. It was not the result of any political plan; it took off because at the time, the new technologies offered by far the best use of resources and thus the greatest potential for economic development. For this reason, it became the preeminent model for economic development. But the number of losers is growing exponentially as the disastrous environmental and social consequences become ever more acute, and greater numbers of the former winners now figure among the losers.

Whether the world can make the transition from fossil and nuclear energy to renewable energy in time will ultimately determine the historical status of the Industrial Revolution: a new era of opportunity, or the first step on the road to doom. Idealism alone will not bring about a solar technological revolution in energy supplies. We must recognize and exploit the economic potential of solar resources, and have no truck with the unscrupulous motives of nuclear and fossil energy companies. The road to the solar global economy is like a stream that takes a straight or winding course depending on the topographical hurdles to be overcome, flowing faster or slower depending on the size and speed of its tributaries, and finally swelling to a river with the power to reshape the landscape that surrounds it.

The first priority of every society must be to secure essential supplies of water, energy, raw materials, and food. This truth is only overlooked in societies (and in the scientific community) where easy access to necessary resources has for a long time been a matter of course. This era of plenty is coming to an end. If a source of vital supplies dries up or becomes polluted, then people are forced to move to a new source of supplies or make a concerted effort to find a replacement. For this reason alone, merely moving from a recently exhausted source of supplies to another on the edge of exhaustion can only provide a temporary respite. The seemingly boundless capacity of the global market still leads people to rely on switching at will from one fossil fuel to another, although the limits to global supplies of these resources are clear to see. Companies are now beginning to see a major business opportunity in the area of water supplies, precisely because many regions are already feeling the pinch or can at least see the limits to their supplies. In view of the disastrous practices of the expanding agribusiness industry, anyone who relies on a boundless global market for food supplies is clearly living in cloud-cuckoo-land. The widespread belief that global transport capacities mean that dependence on anonymous suppliers for the basic needs of society is no longer a problem just goes to show how thoroughly the world has been led up the garden path. Whether they choose to see it or not, societies everywhere must focus their efforts on the use of environmentally sustainable and inexhaustible resources—and in particular on those that require the least effort to extract while offering the greatest economic benefit. The laws of economics compel us to seek lower real costs and hence to return to supplying humanity's essential needs from primarily local sources.

It is in the area of energy supplies that humanity has moved furthest from the natural world that sustains us, and it is thus in this area that past errors will be the most difficult to rectify. Hence our top priority must be a realignment towards renewable energy. In agriculture, it is only in recent decades, albeit at an accelerating rate, that ever-greater swathes of humanity have become estranged from local supplies. This has become the second crucial issue that we face, as sustainable long-term food supplies come under increasing threat. Once fossil and nuclear have been replaced with renewable energy, and regional agriculture has undergone its crucially necessary revitalization, a broad shift towards renewable resources will also follow.

In a solar global economy, water supplies will be secure in the long term; farm- and woodland will be managed sustainably; the essential needs of humanity for energy and raw materials will be met from inexhaustible sources; and energy will be supplied almost exclusively—and food and raw materials to a far greater extent than hitherto—from regional sources. The solar global economy is the economically and environmentally superior model, and it is the world's greatest social and cultural opportunity.

It is the task of the modern, environmentally conscious age to force this transition to a solar global economy, thereby overcoming the doomed fossil industrial age that has not only closed its eyes to the life-and-death choices that confront it, but utterly denies that such choices exist. The US philosopher Arran Gare writes in his book *Postmodernism and the Environmental Crisis* that disorientation "has been made a virtue, and the absence of fixed reference points is celebrated."[1] He portrays a generation mistrustful of the wider picture and of large-scale solutions. Such crises of identity and the loss of confidence in the future of society have always made an appearance when the existing social model has lost its credibility. But that is no excuse to abandon all convincing models.

How can we believe that there can be no more grand designs, no convictions, and no more strength at a point when the global environmental crisis threatens to bring megacatastrophes of untold scale? Such ideas speak of intellectual and moral poverty. Why is it that even social democratic parties, which owe their origins to a faith in human progress, and even green parties are showing an unmistakable preference for other topics than environmental planning for the future? Why does the modern age so lack courage and conviction? Why is this capitulation celebrated as "facing up to reality," and how is it possible for cloud-cuckoo delusions to pass themselves off as neorealism without being laughed out of court? Disorientation and the lack of a moral compass are the symptoms of an age in which the multitudes who are embedded within the current system have utterly lost the power to imagine an alternative. That goes particularly for the many people who have recently had their illusions shattered by the failure of truly unrealistic utopian ideals, ideals to which they clung by censoring out unwanted truths. We are left with people who speak vaguely of renewal, but whose real aim is to preserve the comforts that the present still affords, at least for themselves, all in the knowledge and acceptance that that is something that ever-decreasing numbers of people can aspire to.

Forwards—Towards the Primary Economy

Bringing about the vitally necessary displacement of fossil by renewable energy and resources will open up a future that gives a new impetus to the primary productive economy. No longer the economic leftovers, agriculture and forestry will become the new and lasting motors of the economy as a whole; not picture-postcard nostalgia, but modern, forward-looking enterprise; not a declining industry, but a major source of new employment.

The industrial and, to an even greater extent, the postindustrial age saw the classical primary sector of the national economy—most notably agriculture—being marginalized as the secondary manufacturing and tertiary service sectors expanded. Employment statistics showing the ever-declining proportion of the workforce employed in the agriculture and forestry sector can, however, lead to erroneous analyses and correspondingly erroneous conclusions. The current primary sector actually employs considerably more workers than the statistics would suggest. This includes all those employed in industries further up or down the agricultural supply chain, who would formerly have been directly employed by agricultural enterprises: in the production of fertilizers and pesticides, seeds and animal feeds, in energy supply and in marketing. Then there is employment in shipping agricultural inputs and products, food processing, and the manufacture of agricultural machinery. None of these occupations find their way into statistics on the agricultural sector, although without it none of them would exist.

The received wisdom of the industrial and post-industrial modern age is that this development is irreversible. This view, common as it is, demonstrates both prejudice and a lack of imagination. If a country has to import technological products or services on a large scale, nobody would dream of concluding that the industries concerned have no future. The response is rather to attempt to re-establish them on domestic soil. Yet when agricultural production moves abroad, it is thought to be gone for good. Even dire news of global environmental trends cannot shake these negative attitudes to agriculture, although the obvious and swift consequence may be that the global trade in agricultural produce is no guarantee of stable food supplies.

If all those who would naively place all economic activity in the hands of the global market only understood the immutable laws of nature that override all ideology and dogma, then they could not fail to see that, while technology can be globalized, in the long term, resource supplies cannot. Manufacturing plant and services that do not require direct personal contact can be expanded and relocated almost at will. Croplands and other natural resources cannot be so easily moved. Agricultural productivity is not simply a function of education, efficient organization of labour, and optimal use of machinery. It also depends heavily on the invariables of the local geographical and climatic conditions. This is the crucial difference between the primary and all other sectors, which the undifferentiated ideology of the global market simply ignores. If the current state of affairs continues, innumerable countries risk losing their agricultural sector as global production concentrates on the most geographically and climatically favourable arable land, the fertility of which will then be lost all the more quickly to overproduction.

The assumption that agriculture must develop business structures equipped to meet global market conditions, following existing developmental trends, is also criminally negligent in its acceptance of unhealthy consequences for society, especially in the countries of the developing world. If the greater part of the world's 600 million farmers and their families, 3 billion people in total, were to make their way to the cities, leaving the land to be worked by agrifactories producing ever fewer products, the economic and cultural consequences would be incalculable. This model of economic development is jeopardizing our future. Much is made of the knowledge economy as a strategy for dealing with the individual and social challenges to come. What most people fail to realize, however, is that we will be in no position to face these challenges if the existing practical knowledge of land management and plant husbandry of hundreds of millions of farmers were to vanish along with their livelihoods.

The future of society can no longer be secured if the economy is not structured around primary production. Reintegrating primary production back into the national and regional economy is of paramount importance. It will become indispensable as renewables begin to replace fossil resources. Economic realignment towards renewable resources and the biotechnology that drives their use will give further impetus to this process. As agriculture transforms itself into an integrated food, energy, and resources business, it will start to grow rather than continue to shrink.

The real business opportunity for the agricultural sector lies in this cross-sectoral synergy. This, together with the concomitant ability to produce fertilizers and pest deterrents on site, is what will liberate farming from its suppliers, the chemicals, and energy industries. It will also bring jobs back out into the countryside, partly with new roles, but also with wholly new opportunities. The person specification for a farmer capable of dealing with the whole spectrum of plant life is as demanding as any: the adaptability to learn the land, climate, and nutrient requirements of a variety of plants; and a solid grounding in biology and biochemistry, and in the latest harvesting technology. In a solar global economy there will be a need for more independent farmers and more farm enterprises; agriculture will once again offer secure jobs to many more people. Agriculture will also deliver a huge number of less demanding jobs, of the kind that the service economy is crying out for. But as long as driving a JCB in an open-cast lignite mine, assembly-line work, or the new domestic service jobs are seen as more valuable than the comparatively demanding and varied work of sowing seed, operating harvesting machinery, managing woodland, or operating desiccation or biogas plants, the modern world with its cultural blinkers will fail to see this future.

This model of a rediscovered primary productive economy has several important implications, which show the agricultural trends of recent decades, along with agricultural policy and the current direction of the biotech industry, to be thoroughly short-sighted:

- There must be an immediate stop to the continued haemorrhage in the farming industry. Otherwise we will have to painfully rebuild in the near future what is now carelessly being sacrificed to the market. That is not an argument for maintaining the current subsidies indefinitely. Farming could be more effectively supported with help to organize the independent or communal production of energy and nutrients and to establish structures for regional direct marketing, such that through reduced costs and greater profitability farmers can once again support themselves. Subsidies should come with a "sunset clause," and be redirected to help finance the installation of bio-energy processing plant and the conversion of agricultural machinery to run on bio-fuels. There must be support for regional marketing cooperatives for food, energy, and solar resources.

- Seed supplies must be secure and freely accessible. No country will be able to manage without national seed banks providing immediate and universal access to the whole variety of plants likely to be demanded in future. Plant seed is a cultural treasure-chest that belongs to the nation, and seed stores from which seed can be purchased must be an integral part of the public infrastructure of the agriculture of the future, thus providing agricultural businesses with as level a playing field as possible.

If these opportunities are not to be squandered before they can be properly seized, there must be political action to call a halt to the recent growing wave of patents on genes and gene sequences. The natural fruits of evolution must remain in the public domain, freely available to all farmers. At most, only processing techniques should be patentable. It is in the vital interest of every country to put a stop to the patenting of genes, not just for reasons of human and bio-ethics, but also for the future development of the national economy. There must be an end to this most heinous case of dispossession in history, the dispossession of society by private companies. Only a ban on gene patents can prevent the rich potential of biological resources coming under the control of a few global firms before that potential has been realized on a significant scale. Ceding control of biological resources to bio-tech companies accords them a level of global power that all the global and colonial empires in history cannot rival. It is essential that no effort be spared to break this power if there is to be a renaissance of the primary productive economy.

Work and the Solar Economy

In his book *Die energetischen Grundlagen der Kulturwissenschaften* (The Energy Basis of the Humanities), published at the beginning of the twentieth century, the winner of the Nobel Prize for chemistry and expert in the sociology of energy systems, Wilhelm Ostwald, characterized energy as "everything that comes from or can be converted back into work."[2] This understanding of energy covers three different types of work: human labour, native energy, and the work done by mechanical devices and machinery. The question of the "end of work"[3] must therefore also be seen in the context of energy and resource structures. The transition to solar resources will have effects on the future of work as a social institution over and above the new jobs it creates.

In the industrialized countries, machines have replaced human labour. For a long time workers saw no benefits from this, as the technology was primarily used to increase productivity, not to make the remaining jobs easier for people to do. The burden on workers was only ever lightened as the result of action by politicians or trade unions. The future will be no different: technology may make improvements in living standards possible, but this is not guaranteed to happen when others have other aims in mind. New computer and information technologies make it possible as never before for people to let motors and machinery do the work for them; mental and technical skills are replacing physical skills faster and more comprehensively than ever before.

This brings a new edge to the dispute over how the product of labour should be distributed across society so that everybody can make a living for themselves. If everybody has a sufficiently well-paid job, then the product of labour is distributed through employment incomes—with all the squabbling over fair and adequate pay that inevitably ensues. If there is not enough work to go around or growing numbers of people are excluded from the direct redistribution of income, then income must be redistributed across society through shorter working weeks and/or state minimum income guarantees. This is the new big issue for social policy. At the same time, there is the increasingly interesting question of how people spend their time outside work. Mathias Greffrath talks of a "three-shift society," in which people spend a third of their time in paid work, a third in unpaid voluntary activities, and a third on their own needs.[4] Johano Strasser emphasizes the indisputable need for a redistribution of work in his book *Geht der Arbeitsgesellschaft die Arbeit aus?* (When the Working Society Runs Out of Work).[5]

As long as this redistribution can be only imperfectly realized, the question of how people with no income from paid work can make a living will be the key issue facing any community. In theory, the output from energy and technology is so great that ever-decreasing numbers of workers are required. But the question that becomes more urgent by the day is how political institutions can siphon off the necessary income from highly productive firms, which are increasingly organized on a transnational basis and which effectively operate beyond the borders of all national governments. It is sheer fantasy to suppose that international political institutions could impose taxes on excess profits to redistribute on an individual basis. This makes it all the more necessary to take a close look at the energy component of work.

Industrial society has forgotten that the sun is the greatest and most versatile source of energy available to life on Earth, and that the power of the sun can be used to save human labour. This is equally true of agriculture and forestry. Here, too, human labour has been replaced by machinery and fossil energy. If the work done by fossil energy were to be done by the sun, then agriculture would reach the quintessence of its actual economic potential. The same can be said of work in nonagricultural sectors. The Industrial Revolution was reduced to replacing human with mechanical labour while also increasing the work done by fossil energy. Replacing fossil fuel work with solar technology will radically reshape the work society. With the sun doing the work formerly performed by fossil fuels, the total cost of work to society is reduced to the cost of human labour and technology.

Leveraging the sun will, as has been described, result in a more equitable distribution of jobs across the regions. The concomitant regionalization of the economy will make it easier for governments to use taxes to finance public services. Individual living costs will be permanently lowered, thereby making it easier to find a solution to the big question of how to provide everybody with the opportunity to live a dignified life free of destitution, as the cost of state income guarantees falls. Renewable energy helps us to escape the environmental catastrophes of the fossil industrial age, and brings with it lower energy costs. At the same time, it also reduces the risk of damage to human health—the costs of which are borne not by the perpetrators but by society as a whole—and thereby reduces the cost of maintaining a health service. Without the shift to renewable energy, disaster prevention and disaster relief will consume ever-greater proportions of public and private funds, until increasingly frequent catastrophes overload our capacity to cope, and social and civil order can no longer be maintained by even the best-equipped security forces. Society also foots the bill for the current bout of spending on military equipment to ensure the security of the remaining fossil fuel reserves, and the public is led to believe that this is in their interest.

Even if the proportion of work done by people continues to fall in comparison to the proportion done by machines and by the sun, society will face this issue from a wholly different starting point than today. There will be less potential for aggression, in part because the environmental dangers that now threaten us will have been averted. The only remaining question is how people spend their free time—and the answer will come not from legislation, not from market rules, and not from energy systems or technology, but rather from traditions, cultural norms, the human capacity for education and social interaction, and from the cultural achievements of society.

From the Bounty of the Sun to Global Economic Prosperity

The Earth is rich, and it owes its wealth to the sun. That this wealth is today more often burnt than used and preserved for the future is the greatest economic non-sense imaginable. And then to call this destruction of resources "economic growth" makes a mockery of the phrase. This is not economic growth, but economic destruction, and it leads not to Adam Smith's "wealth of nations," but rather to Elmar Altvater's "poverty of nations."[6]

The fundamental problem with today's global economy is not globalization per se, but that this globalization is not based on the sun—the only global force that is equally available to all and whose bounty is so great that it need never be fully tapped. Only with solar in place of fossil energy can the world reach the pinnacle of its potential. As long as economic progress depends on resources found only in a few regions, there will inevitably be increasingly bitter conflicts in which national interests will come before the interests of the planet, national economies before the economy as a whole, short-term before long-term interests, and individuals and companies before society. The global hierarchies that have grown and continue to grow out of fossil energy supplies stand in the way of a new era in which people can make as close to an independent living as can be achieved, and in which people can make their contribution to global output according to the measure of their ability and need. The existing hierarchies, however, are ironing out economic and cultural differences, depriving the world of its vibrant diversity. Cultural destitution is following hard on the heels of its economic twin.

It is because the global flow of fossil resources has for a long time been widening the scope of possibility and opportunity for increasing numbers of people in the industrialized nations that people now fail to see that the same resource flow now has the opposite effect, narrowing the range of opportunities for increasing numbers of people, and ultimately for everybody. Global resource conflict, environmental catastrophe, fossil energy prices that are unaffordable for most of the world's population, the economic crises to come as supplies dwindle—all these put the world in grave danger of turning back the clock. Hard-won achievements of civilization may be lost: the UN and international law, international treaties, the global economy itself. The most likely consequence of the struggle to control dwindling fossil reserves is a deep decline in the global economy, leading ultimately to the fall of global civilization itself.

The solar global economy makes possible a new global division of labour. Each national economy exploits the resources directly afforded them by the sun, resources that no one can take away; all other needs are satisfied by the free interaction of supply and demand. Only in this way can the rich diversity of global culture be maintained and revitalized, or further developed through mutual enrichment.

The globalization process is a roller-coaster ride driven by fossil fuels. The faster it goes, the more frightening and bruising a ride it is for human passengers and the natural world alike. Dwindling numbers of people are able to climb aboard, while growing proportions are tossed out of the carriage. By contrast, the new division of labour in the solar global economy to come encompasses a whole variety of swings and roundabouts, some small, some large, all offering a much calmer ride, much less violent to—and more under the control of—their passengers. There will always be room for more attractions, with plenty of space for all comers. The solar global economy affords much greater freedom and scope for the productive use of technology because of the countless individual practical applications that, in combination with the immediate availability of the sun's power, it makes possible. Technology will no longer be the preserve of the few, who use it to impose technocratic constraints on everybody. The universal accessibility of technology will open the floodgates for many more new ideas and innovative applications. Growing numbers of independent producers and more diverse resource use will give rise to a whole range of new products. The solar global economy is an economy that does not wantonly destroy its resources, and which is thus free of constraints on its development.

By taking hold of the visible hand of the sun and producing from sustainable resources, the world remains close to the land, and its inhabitants meet in a freer and more just environment. From riches for the few, be they individuals, companies, or societies, will increasingly come wealth for all, more justly and more equally distributed. Renewable resources will bring a new era of wealth-creating economic development, initiated not by bureaucratic fiat, but by the free choices of individuals.

Notes

1. Arran E Gare, *Postmodernism and the Environmental Crisis* (London: Routledge, 1995), 34.

2. Wilhelm Ostwald, *Die energetischen Grundlagen der Kulturwissenschaften* (The Energy Basis of the Humanities) (Leipzig, 1909), 2 et seq.

3. Jeremy Rifkin, *The End of Work: The Decline of the Global Workforce and the Dawn of the Post-market Era* (London: Penguin, 2000).

4. Mathias Greffrath, "Freizeit, die sie meinen" (Freedom They Mean), *Süddeutsche Zeitung* 24 (June 1998).

5. Johano Strasser, *Wenn der Arbeitsgesellschaft die Arbeit ausgeht* (When the Working Society Runs Out of Work) (Zürich: Pendo, 1999).

6. Elmar Altvater, *The Poverty of Nations: A Guide to the Debt Crisis from Argentina to Zaire* (London: Zed Books, 1991).

))) The Frenzy of Fossil Fuels

Naomi Oreskes and Erik M. Conway

In the early Penumbral Period, physical scientists who spoke out about the potentially catastrophic effects of climate change were accused of being "alarmist" and of acting out of self-interest—to increase financial support for their enterprise, gain attention, or improve their social standing. At first, these accusations took the form of public denunciations; later they included threats, thefts, and the subpoena of private correspondence.[1] A crucial but under-studied incident was the legal seizing of notes from scientists who had documented the damage caused by a famous oil spill of the period, the 2011 British Petroleum Deepwater Horizon. Though leaders of the scientific community protested, scientists yielded to the demands, thus helping set the stage for further pressure on scientists from both governments and the industrial enterprises that governments subsidized and protected.[2] Then legislation was passed (particularly in the United States) that placed limits on what scientists could study and how they could study it, beginning with the notorious House Bill 819, better known as the "Sea Level Rise Denial Bill," passed in 2012 by the government of what was then the US state of North Carolina (now part of the Atlantic Continental Shelf).[3] Meanwhile the Government Spending Accountability Act of 2012 restricted the ability of government scientists to attend conferences to share and analyze the results of their research.[4]

Though ridiculed when first introduced, the Sea Level Rise Denial Bill would become the model for the US National Stability Protection Act of 2025, which led to the conviction and imprisonment of more than three hundred scientists for "endangering the safety and well-being of the general public with unduly alarming threats." By exaggerating the threat, it was argued, scientists were preventing the economic development essential for coping with climate change. When the scientists appealed, their convictions were upheld by the US Supreme Court under the

The Collapse of Western Civilization (New York: Columbia University Press, 2014), 11–33

Clear and Present Danger doctrine, which permitted the government to limit speech deemed to represent an imminent threat.

Had scientists exaggerated the threat, inadvertently undermining the evidence that would later vindicate them? Certainly, narcissistic fulfillment played a role in the public positions that some scientists took, and in the early part of the Penumbral Period, funds flowed into climate research at the expense of other branches of science, not to mention other forms of intellectual and creative activity. Indeed, it is remarkable how little these extraordinarily wealthy nations spent to support artistic production; one explanation may be that artists were among the first to truly grasp the significance of the changes that were occurring. The most enduring literary work of this time is the celebrated science "fiction" trilogy by American writer Kim Stanley Robinson—*Forty Signs of Rain, Fifty Degrees Below*, and *Sixty Days and Counting*.[5] Sculptor Dario Robleto also "spoke" to the issue, particularly species loss; his material productions have been lost, but responses to his work are recorded in contemporary accounts.[6] Some environmentalists also anticipated what was to come, notably the Australians Clive Hamilton and Paul Gilding. (Perhaps because Australia's population was highly educated and living on a continent at the edge of habitability, it was particularly sensitive to the changes under way.)[7] These "alarmists"—scientists and artists alike—were correct in their forecasts of an imminent shift in climate; in fact, by 2010 or so, it was clear that scientists had been underestimating the threat, as new developments outpaced early predictions of warming, sea level rise, and Arctic ice loss, among other parameters.[8]

It is difficult to understand why humans did not respond appropriately in the early Penumbral Period, when preventive measures were still possible. Many have sought an answer in the general phenomenon of *human adaptive optimism*, which later proved crucial for survivors. Even more elusive to scholars is why scientists, whose job it was to understand the threat and warn their societies—and who thought that they *did* understand the threat and that they *were* warning their societies— failed to appreciate the full magnitude of climate change.

To shed light on this question, some scholars have pointed to the epistemic structure of Western science, particularly in the late nineteenth and twentieth centuries, which was organized both intellectually and institutionally around "disciplines" in which specialists developed a high level of expertise in a small area of inquiry. This "reductionist" approach, sometimes credited to the seventeenth-century French philosophe René Descartes but not fully developed until the late nineteenth century, was believed to give intellectual power and vigor to investigations by focusing on singular elements of complex problems. "Tractability" was a guiding ideal of the time: problems that were too large or complex to be solved in their totality were divided into smaller, more manageable elements. While reductionism proved powerful in many domains, particularly quantum physics and medical diagnostics, it impeded investigations of complex systems. Reductionism also made it difficult for scientists to articulate the threat posed by climatic change, since many experts did not actually know very much about aspects of the problem beyond their expertise. (Other environmental problems faced similar challenges. For example, for years, scientists did not understand the role of polar stratospheric clouds in severe ozone depletion in the still glaciated Antarctic region because "chemists" working on the chemical reactions did not even know that there were clouds in the polar stratosphere!) Even scientists who had a broad view of climate change often felt it would be inappropriate for them to articulate it, because that would require them to speak beyond their expertise, and seem to be taking credit for other people's work.

Responding to this, scientists and political leaders created the IPCC to bring together the diverse specialists needed to speak to the whole problem. Yet, perhaps because of the diversity of specialist views represented, perhaps because of pressures from governmental sponsors, or perhaps because of the constraints of scientific culture already mentioned, the IPCC had trouble speaking in a clear voice. Other scientists promoted the ideas of systems science, complexity science, and, most pertinent to our purposes here, earth systems science, but these so-called holistic approaches still focused almost entirely on natural systems, omitting from consideration the social components. Yet in many cases, the social components were the dominant system drivers. It was often said, for example, that climate change was caused by increased atmospheric concentrations of greenhouse gases. Scientists understood that those greenhouse gases were accumulating because of the activities of human beings—deforestation and fossil fuel combustion—yet they rarely said that the cause was people, and their patterns of conspicuous consumption.

Other scholars have looked to the roots of Western natural science in religious institutions. Just as religious orders of prior centuries had demonstrated moral rigor through extreme practices of asceticism in dress, lodging, behavior, and food—in essence, practices of physical self-denial—so, too, did physical scientists of the twentieth and twenty-first centuries attempt to demonstrate their intellectual rigor through practices of intellectual self-denial.[9] These practices led scientists to demand an excessively stringent standard for accepting claims of any kind, even those involving imminent threats. In an almost childlike attempt to demarcate their practices from those of older explanatory traditions, scientists felt it necessary to prove to themselves and the world how strict they were in their intellectual standards. Thus, they placed the burden of proof on novel claims—even empirical claims about phenomena that their theories predicted. This included claims about changes in the climate.

Some scientists in the early twenty-first century, for example, had recognized that hurricanes were intensifying. This was consistent with the expectation—based on physical theory—that warmer sea surface temperatures in regions of cyclogenesis could, and likely would, drive either more hurricanes or more intense ones.

However, they backed away from this conclusion under pressure from their scientific colleagues. Much of the argument surrounded the concept of *statistical significance*. Given what we now know about the dominance of nonlinear systems and the distribution of stochastic processes, the then-dominant notion of a 95% confidence limit is hard to fathom. Yet overwhelming evidence suggests that twentieth-century scientists believed that a claim could be accepted only if, by the standards of Fisherian statistics, the possibility that an observed event could have happened by chance was less than 1 in 20. Many phenomena whose causal mechanisms were physically, chemically, or biologically linked to warmer temperatures were dismissed as "unproven" because they did not adhere to this standard of demonstration. Historians have long argued about why this standard was accepted, given that it had neither epistemological nor substantive mathematical basis. We have come to understand the 95% confidence limit as a social convention rooted in scientists' desire to demonstrate their disciplinary severity.

Western scientists built an intellectual culture based on the premise that it was worse to fool oneself into believing in something that did not exist than not to believe in something that did. Scientists referred to these positions, respectively, as "type I" and "type II" errors, and established protocols designed to avoid type I errors at almost all costs. One scientist wrote, "A type I error is often considered to be more serious, and therefore more important to avoid, than a type II error." Another claimed that type II errors were not errors at all, just "missed opportunities."[10] So while the pattern of weather events was clearly changing, many scientists insisted that these events could not yet be attributed with certainty to anthropogenic climate change. Even as lay citizens began to accept this link, the scientists who studied it did not.[11] More important, political leaders came to believe that they had more time to act than they really did. The irony of these beliefs need not be dwelt on; scientists missed the most important opportunity in human history, and the costs that ensued were indeed nearly "all costs."

By 2012, more than 365 billion tons of carbon had been emitted to the atmosphere from fossil fuel combustion and cement production. Another 180 were added from deforestation and other land use changes. Remarkably, more than half of these emissions occurred *after* the mid-1970s—that is, *after* scientists had built computer models demonstrating that greenhouse gases would cause warming. Emissions continued to accelerate even after the UNFCCC was established: between 1992 and 2012, total CO_2 emissions increased by 38%.[12] Some of this increase was understandable, as energy use grew in poor nations seeking to raise their standard of living. Less explicable is why, at the very moment when disruptive climate change was becoming apparent, wealthy nations dramatically increased their production of fossil fuels. The countries most involved in this enigma were two of the world's richest: the United States and Canada.

A turning point was 2005, when the US Energy Policy Act exempted shale gas drilling from regulatory oversight under the Safe Drinking Water Act. This statute opened the floodgates (or, more precisely, the wellheads) to massive increases in shale gas production.[13] US shale gas production at that time was less than 5 trillion cubic feet (Tef, with "feet" an archaic imperial unit roughly equal to a third of a meter) per annum. By 2035, it had increased to 13.6 Tef. As the United States expanded shale gas production and exported the relevant technology, other nations followed. By 2035, total gas production had exceeded 250 Tef per annum.[14]

This bullish approach to shale gas production penetrated Canada as well, as investor-owned companies raced to develop additional fossil fuel resources; "frenzy" is not too strong a word to describe the surge of activity that occurred. In the late twentieth century, Canada was considered an advanced nation with a high level of environmental sensitivity, but this changed around the year 2000 when Canada's government began to push for development of huge tar sand deposits in the province of Alberta, as well as shale gas in various parts of the country. The tar sand deposits (which the government preferred to call oil sands, because liquid oil had a better popular image than sticky tar) had been mined intermittently since the 1960s, but the rising cost of conventional oil now made sustained exploitation economically feasible. The fact that 70% of the world's known reserves were in Canada explains the government's reversed position on climate change: in 2011, Canada withdrew from the Kyoto Protocol to the UNFCCC.[15] Under the protocol, Canada had committed to cut its emissions by 6%, but its actual emissions instead increased more than 30%.[16]

Meanwhile, following the lead of the United States, the government began aggressively to promote the extraction of shale gas, deposits of which occurred throughout Canada. Besides driving up direct emissions of both CO_2 and CH_4 to the atmosphere (since many shale gas fields also contained CO_2 and virtually all wells leaked), the resulting massive increase in supply of natural gas led to a collapse in the market price, driving out nascent renewable energy industries everywhere except China, where government subsidies and protection for fledgling industries enabled the renewable sector to flourish.

Cheap natural gas also further undermined the already ailing nuclear power industry, particularly in the United States. To make matters worse, the United States implemented laws forbidding the use of biodiesel fuels first by the military and then by the general public, undercutting that emerging market as well.[17] Bills were passed on both the state and federal level to restrict the development and use of other forms of renewable energy—particularly in the highly regulated electricity generation industry—and to inhibit the sale of electric cars, maintaining the lock that fossil fuel companies had on energy production and use.[18]

Meanwhile, Arctic sea ice melted, and seaways opened that permitted further exploitation of oil and gas reserves in the north polar region. Again, scientists noted what was happening. By the mid-2010s, the Arctic summer sea had lost about 30% of its areal extent compared to 1979, when high-precision satellite measurements were first made; the average loss was rather precisely measured at 13.7% per decade from 1979 to 2013.[19] When the areal extent of summer sea ice was compared to earlier periods using additional data from ships, buoys, and airplanes, the total summer loss was nearly 50%. The year 2007 was particularly worrisome, as the famous Northwest Passage—long sought by Arctic explorers—opened, and the polar seas became fully navigable for the first time in recorded history. Scientists understood that it was only a matter of time before the Arctic summer would be ice-free, and that this was a matter of grave concern. But in business and economic circles it was viewed as creating opportunities for further oil and gas exploitation.[20] One might have thought that governments would have stepped in to prevent this ominous development—which could only exacerbate climate change—but governments proved complicit. One example: in 2012 the Russian government signed an agreement with American oil giant ExxonMobil, allowing the latter to explore for oil in the Russian Arctic in exchange for Russian access to American shale oil drilling technology.[21]

How did these wealthy nations—rich in the resources that would have enabled an orderly transition to a zero-net-carbon infrastructure—justify the deadly expansion of fossil fuel production? Certainly, they fostered the growing denial that obscured the link between climate change and fossil fuel production and consumption. They also entertained a second delusion: that natural gas from shale could offer a "bridge to renewables." Believing that conventional oil and gas resources were running out (which they were, but at a rate insufficient to avoid disruptive climate change), and stressing that natural gas produced only half as much CO_2 as coal, political and economic leaders—and even many climate scientists and "environmentalists"—persuaded themselves and their constituents that promoting shale gas was an environmentally and ethically sound approach.

This line of reasoning, however, neglected several factors. First, *fugitive emissions* —CO_2 and CH_4 that escaped from wellheads into the atmosphere—greatly accelerated warming. (As with so many climate-related phenomena, scientists had foreseen this, but their predictions were buried in specialized journals.) Second, most analyses of the greenhouse gas benefits of gas were based on the assumption that it would replace coal in electricity generation where the benefits, if variable, were nevertheless fairly clear. However, as gas became cheap, it came to be used increasingly in transportation and home heating, where losses in the distribution system negated many of the gains achieved in electricity generation. Third, the calculated benefits were based on the assumption that gas would replace coal, which it did in some regions (particularly in the United States and some parts of Europe), but elsewhere (for example, Canada) it mostly replaced nuclear and hydropower. In many regions cheap gas simply became an additional energy source, satisfying expanding demand without replacing other forms of fossil fuel energy production. As new gas-generating power plants were built, infrastructures based on fossil fuels were further locked in, and total global emissions continued to rise. The argument for the climatic benefits of natural gas presupposed that net CO_2 emissions would fall, which would have required strict restrictions on coal and petroleum use in the short run and the phase-out of gas as well in the long run.[22] Fourth, the analyses mostly omitted the cooling effects of aerosols from coal, which although bad for human health had played a significant role in keeping warming below the level it would otherwise have already reached. Fifth, and perhaps most important, the sustained low prices of fossil fuels, supported by continued subsidies and a lack of external cost accounting, undercut efficiency efforts and weakened emerging markets for solar, wind, and biofuels (including crucial liquid biofuels for aviation).[23] Thus, the bridge to a zero-carbon future collapsed before the world had crossed it.

The net result? Fossil fuel production escalated, greenhouse gas emissions increased, and climate disruption accelerated. In 2001, the IPCC had predicted that atmospheric CO_2 would double by 2050.[24] In fact, that benchmark was met by 2042. Scientists had expected a mean global warming of 2 to 3 degrees Celsius; the actual figure was 3.9 degrees. Though originally merely a benchmark for discussion with no particular physical meaning, the doubling of CO_2 emissions turned out to be quite significant: once the corresponding temperature rise reached 4 degrees, rapid changes began to ensue. By 2040, heat waves and droughts were the norm. Control measures—such as water and food rationing and Malthusian "one-child" policies—were widely implemented. In wealthy countries, the most hurricane- and tornado-prone regions were gradually but steadily depopulated, putting increased social pressure on areas less subject to those hazards. In poor nations, conditions were predictably worse: rural portions of Africa and Asia began experiencing significant depopulation from out-migration, malnutrition-induced disease and infertility, and starvation. Still, sea level had risen only 9 to 15 centimeters around the globe, and coastal populations were mainly intact.

Then, in the Northern Hemisphere summer of 2041, unprecedented heat waves scorched the planet, destroying food crops around the globe. Panic ensued, with food riots in virtually every major city. Mass migration of undernourished and dehydrated individuals, coupled with explosive increases in insect populations, led to widespread outbreaks of typhus, cholera, dengue fever, yellow fever, and viral and retroviral agents never before seen. Surging insect populations also destroyed huge swaths of forests in Canada, Indonesia, and Brazil. As social order began to break down in the 2050s, governments were overthrown, particularly in Africa, but also in many parts of Asia and Europe, further decreasing social capacity to deal with increasingly desperate populations. As the Great North American Desert surged north and east, consuming the High Plains and destroying some of the world's most productive farmland, the US government declared martial law to prevent food riots and looting. A few years later, the United States announced plans with Canada for the two nations to begin negotiations toward the creation of the United States of North America, to develop an orderly plan for resource-sharing and northward population relocation. The European Union announced similar plans for voluntary northward relocation of eligible citizens from its south-ernmost regions to Scandinavia and the United Kingdom.

While governments were straining to maintain order and provide for their people, leaders in Switzerland and India—two countries that were rapidly losing substantial portions of their glacially sourced water resources—convened the First International Emergency Summit on Climate Change, organized under the rubric of Unified Nations for Climate Protection (the former United Nations having been discredited and disbanded over the failure of the UNFCCC). Political, business, and religious leaders met in Geneva and Chandigarh to discuss emergency action. Many said that the time had come to make the switch to zero-carbon energy sources. Others argued that the world could not wait the 10 to 50 years required to alter the global energy infrastructures, much less the 100 years it would take for atmospheric CO_2 to diminish. In response, participants hastily wrote and signed the Unified Nations Convention on Climate Engineering and Protection (UNCCEP), and began preparing blueprints for the International Climate Cooling Engineering Projects (ICCEP).

As a first step, ICCEP launched the International Aerosol Injection Climate Engineering Project (IAICEP, pronounced ay-yi-yi-sep) in 2052.[25] Sometimes called the Crutzen Project after the scientist who first suggested the idea in 2006, projects like this engendered heated public opposition when first proposed in the early twenty-first century but had widespread support by mid-century—from wealthy nations anxious to preserve some semblance of order, from poor nations desperate to see the world do something to address their plight, and from frantic low-lying Pacific Island nations at risk of being submerged by rising sea levels.[26]

IAICEP began to inject submicrometer-size sulfate particles into the stratosphere at a rate of approximately 2.0 teragrams per year, expecting to reduce mean global temperature by 0.1 degrees Celsius annually from 2059 to 2079. (In the meantime, a substantial infrastructural conversion to renewable energy could have been achieved.) Initial results were encouraging: during the first 3 years of implementation, temperature decreased as expected and the phase-out of fossil fuel production commenced. However, in the project's fourth year, an anticipated but discounted side effect occurred: the shutdown of the Indian monsoon. (By decreasing incoming solar radiation, IAICEP also decreased evaporation over the Indian Ocean, and hence the negative impact on the monsoon.) As crop failures and famine swept across India, one of IAICEP's most aggressive promoters now called for its immediate cessation.

IAICEP was halted in 2063, but a fatal chain of events had already been set in motion. It began with *termination shock*—that is, the abrupt increase in global temperatures following the sudden cessation of the project. Once again, this phenomenon had been predicted, but IAICEP advocates had successfully argued that, given the emergency conditions, the world had no choice but to take the risk.[27] In the following 18 months, temperature rapidly rebounded, regaining not just the 0.4 degrees Celsius that had been reduced during the project but an additional 0.6 degrees. This rebound effect pushed the mean global temperature increase to nearly 5 degrees Celsius.

Whether it was caused by this sudden additional heating or was already imminent is not known, but the greenhouse effect then reached a global tipping point. By 2060, Arctic summer ice was completely gone. Scores of species perished, including the iconic polar bear—the Dodo bird of the twenty-first century. While the world focused on these highly visible losses, warming had meanwhile accelerated the less visible but widespread thawing of Arctic permafrost. Scientists monitoring the phenomenon observed a sudden increase in permafrost thaw and CH_4 release. Exact figures are not available, but the estimated total carbon release of Arctic CH_4 during the next decade may have reached over 1,000 gigatonnes, effectively doubling the total atmospheric carbon load.[28] This massive addition of carbon led to what is known as the Sagan effect (sometimes more dramatically called the Venusian death): a strong positive feedback loop between warming and CH_4 release. Planetary temperature increased by an additional 6 degrees Celsius over the s-degree rise that had already occurred.

The ultimate blow for Western civilization came in a development that, like so many others, had long been discussed but rarely fully assimilated as a realistic threat: the collapse of the West Antarctica Ice Sheet. Technically, what happened in West Antarctica was not a collapse; the ice sheet did not fall in on itself, and it did not happen all at once. It was more of a rapid disintegration. Post hoc failure analysis shows that extreme heat in the Northern Hemisphere disrupted normal patterns of ocean circulation, sending exceptionally warm surface waters into the southern ocean that destabilized the ice sheet from below. As large pieces of ice shelf began to separate from the main ice sheet, removing the bulwark that had kept the sheet on the Antarctic Peninsula, sea level began to rise rapidly.

Social disruption hampered scientific data-gathering, but some dedicated individuals—realizing the damage could not be stopped—sought, at least, to chronicle it. Over the course of the next two decades (from 2073 to 2093), approximately 90% of the ice sheet broke apart, disintegrated, and melted, driving up sea level approximately five meters across most of the globe. Meanwhile, the Greenland Ice Sheet, long thought to be less stable than the Antarctic Ice Sheet, began its own disintegration. As summer melting reached the center of the Greenland Ice Sheet, the east side began to separate from the west. Massive ice breakup ensued, adding another two meters to mean global sea level rise.[29] These cryogenic events were soon referred to as the Great Collapse, although some scholars now use the term more broadly to include the interconnected social, economic, political, and demographic collapse that ensued.

Analysts had predicted that an eight-meter sea level rise would dislocate 10% of the global population. Alas, their estimates proved low: the reality was closer to 20%. Although records for this period are incomplete, it is likely that during the Mass Migration 1.5 billion people were displaced around the globe, either directly from the impacts of sea level rise or indirectly from other impacts of climate change, including the secondary dislocation of inland peoples whose towns and villages were overrun by eustatic refugees. Dislocation contributed to the Second Black Death, as a new strain of the bacterium *Yersinia pestis* emerged in Europe and spread to Asia and North America. In the Middle Ages, the Black Death killed as much as half the population of some parts of Europe; this second Black Death had similar effects.[30] Disease also spread among stressed nonhuman populations. Although accurate statistics are scant because twentieth-century scientists did not have an inventory of total global species, it is not unrealistic to estimate that 60% to 70% of species were driven to extinction. (Five previous mass extinctions were known to scientists of the Penumbral Period, each of which was correlated to rapid greenhouse gas level changes, and each of which destroyed more than 60% of identifiable species—the worst reached 95%. Thus, 60%–70% is a conservative estimate insofar as most of these earlier mass extinctions happened more slowly than the anthropogenic mass extinction of the late Penumbral Period.)[31]

There is no need to rehearse the details of the human tragedy that occurred; every schoolchild knows of the terrible suffering. Suffice it to say that total losses—social, cultural, economic, and demographic—were greater than any in recorded human history. Survivors' accounts make clear that many thought the end of the human race was near. Indeed, had the Sagan effect continued, warming would not have stopped at 11 degrees, and a runaway greenhouse effect would have followed.

However, around 2090 (the date cannot be determined from extant records), something occurred whose exact character remains in dispute. Japanese genetic engineer Akari Ishikawa developed a form of lichenized fungus in which the photosynthetic partner consumed atmospheric CO_2 much more efficiently than existing forms, and was able to grow in a wide diversity of environmental conditions. This pitch-black lichen, dubbed *Pannaria ishikawa*, was deliberately released from Ishikawa's laboratory, spreading rapidly throughout Japan and then across most of the globe. Within two decades, it had visibly altered the visual landscape and measurably altered atmospheric CO_2, starting the globe on the road to atmospheric recovery and the world on the road to social, political, and economic recovery.

In public pronouncements, the Japanese government has maintained that

Ishikawa acted alone, and cast her as a criminal renegade. Yet many Japanese citizens have seen her as a hero, who did what their government could not, or would not, do. Most Chinese scholars reject both positions, contending that the Japanese government, having struggled and failed to reduce Japan's own carbon emissions, provided Ishikawa with the necessary resources and then turned a blind eye toward its dangerous and uncertain character. Others blame (or credit) the United States, Russia, India, or Brazil, as well as an international consortium of financiers based in Zurich. Whatever the truth of this matter, Ishikawa's actions slowed the increase of atmospheric CO_2 dramatically.

Humanity was also fortunate in that a so-called "Grand Solar Minimum" reduced incoming solar radiation during the twenty-second century by 0.5%, offsetting some of the excess CO_2 that had accumulated, and slowing the rise of surface and oceanic temperatures for nearly a century, during which time survivors in northern inland regions of Europe, Asia, and North America, as well as inland and high-altitude regions of South America, were able to begin to regroup and rebuild. The human populations of Australia and Africa, of course, were wiped out.

Notes

1. Michael Mann, *The Hockey Stick and the Climate Wars: Dispatches from the Front Lines* (New York: Columbia University Press, 2012).

2. See http://www.wired.com/2012/06/bp-scientist-emails/. .

3. Seth Cline, "Sea Level Bill Would Allow North Carolina to Stick Its Head in the Sand," *U.S. News and World Report*, June 1, 2012, http://www.usnews.com/news/articles/2012/06/01/sea-level-bill-would-allow-north-carolina-to-stick-its-head-in-the-sand. Stephen Colbert made a satire of the law (see Stephen Colbert, "The Word—Sink or Swim," *The Colbert Report*, June 4, 2012, http://www.colbertnewshub.com/2012/06/05/june-4-2012-rep-john-lewis/).

4. Government Spending Accountability Act of 2012, 112th Cong., 2012, H.R. 4631, http://oversight.house.gov/wp-content/uploads/2012/06/WALSIL_032_xml.pdf.

5. Kim Stanley Robinson, *Forty Signs of Rain, Fifty Degrees Below, and Sixty Days and Counting* (New York: Spectra, 2005–2007).

6. Naomi Oreskes, "Seeing Climate Change," in *Dario Robleto: Survival Does Not Lie in the Heavens*, ed. Gilbert Vicario (Des Moines, IA: Des Moines Art Center, 2011).

7. See Clive Hamilton, *Requiem for a Species: Why We Resist the Truth about Climate Change* (Sydney: Allen & Unwin, 2010), http://clivehamilton.com/books/requiem-for-a-species/; and Paul Gilding, *The Great Disruption: Why the Climate Crisis Will Bring On the End of Shopping and the Birth of a New World* (New York: Bloomsbury, 2010).

8. For an electronic archive of predictions and data as of 2012, see http://www.columbia.edu/~mhs119/Temperature/. An interesting paper from the University of California, San Diego, addresses the issue of under-prediction; see Keynyn Brysse et al., "Climate Change Prediction: Erring on the Side of Least Drama?," *Global Environmental Change* 23 (2013): 327–37.

9. David F. Noble, *A World without Women: The Christian Clerical Culture of Western Science* (New York: Knopf, 1992); and Lorraine Daston and Peter L. Galison, *Objectivity* (Cambridge, MA: Zone Books, 2007).

10. Naomi Oreskes and Erik M. Conway, *Merchants of Doubt: How a Handful of Scientists Obscured the Truth on Issues from Tobacco to Climate Change* (New York: Bloomsbury, 2010), chap. 5, esp. 157nn91–92. See also Aaron M. McCright and Riley E. Dunlap, "Challenging Global Warming as a Social Problem: An Analysis of the Conservative Movement's Counterclaims," *Social Problems* 47 (2000): 499–522; Aaron M. McCright and Riley E. Dunlap, "Cool Dudes: The Denial of Climate Change among Conservative White Males in the United States," *Global Environmental Change* 21 (2011): 1163–72.

11. Justin Gillis, "In Poll, Many Link Weather Extremes to Climate Change," *New York Times*, April 17, 2012, http://www.nytimes.com/2012/04/18/science/earth/americans-link -global-warming-to-extreme-weather-poll-says.html.

12. Tom A. Boden, Gregg Marland, and Robert J. Andres, "Global, Regional, and National Fossil-Fuel CO_2 Emissions," Carbon Dioxide Information Analysis Center (Oak Ridge, TN: Oak Ridge National Laboratory, 2011), http://cdiac.ornl.gov/trends/emis/overview_2008 .html.

13. Sarah Collins and Tom Kenworthy, "Energy Industry Fights Chemical Disclosure," Center for American Progress, April 6, 2010, https://www.americanprogress.org/issues/ green/news/2010/04/06/7613/energy-industry-fights-chemical-disclosure/; Jad Mouawad, "Estimate Places Natural Gas Reserves 35% Higher," *New York Times*, June 17, 2009, http:// www.nytimes.com/2009/06/18/business/energy-environment/18gas.html?_r=0.

14. See http://www.eia.gov/naturalgas.

15. Emil D. Attanasi and Richard F. Meyer, "Natural Bitumen and Extra-Heavy Oil," in *Survey of Energy Resources*, 22nd ed. (London: World Energy Council, 2010), 123–40.

16. David W. Schindler and John P. Smol, "After Rio, Canada Lost Its Way," *Ottawa Citizen*, June 20, 2012, http://www.ottawacitizen.com/opinion/op-ed/Opinion/6814332/story .html.

17. See http://security.blogs.cnn.com/2012/06/08/militarys-plan-for-a-green-future-has -congress-seeing-red/.

18. "Georgia Power Opposes Senate Solar Power Bill," *Augusta Chronicle*, February 18, 2012, http://chronicle.augusta.com/news/metro/2012-02-18/georgia-power-opposes-senate -solar-power-bill.

19. Arctic Sea Ice Extent, IARC-JAXA Information System (IJIS), accessed May 15, 2016, https://ads.nipr.ac.jp/vishop/vishop-extent.html; Arctic Sea Ice News and Analysis, National Snow & Ice Data Center, accessed May 15, 2016, https://nsidc.org/arcticseaice news/; Christine Dell'Amore, "Ten Thousand Walruses Gather on Island as Sea Ice Shrinks," *National Geographic*, October 2, 2013; William M. Connolley, "Sea Ice Extent in Million Square Kilometers," accessed May 15, 2016, https://commons.wikimedia.org/wiki/File :Seaice-1870-part-2009.png.

20. Gerald A. Meehl and Thomas F. Stocker, "Global Climate Projections," in *Fourth Assessment Report of the Intergovernmental Panel on Climate Change*, "Climate Change 2007— The Physical Science Basis," February 2, 2007.

21. Clifford Krauss, "Exxon and Russia's Oil Company in Deal for Joint Projects," *New York Times*, April 16, 2012.

22. For statistics on continued coal and oil use in the mid-twentieth century, see US Energy Information Administration, *International Energy Outlook 2011* (Washington, DC: US Department of Energy, 2011), 139, Figures 110–11, http://www.eia.gov/pressroom/pre sentations/howard_09192011.pdf.

23. On twentieth- and twenty-first-century subsidies to fossil fuel production, see https://www.oecd.org/site/tadffss/; and John Vidal, "World Bank: Ditch Fossil Fuel Sub- sidies to Address Climate Change," *Guardian*, September 21, 2011, http://www.theguardian .com/environment/2011/sep/21/world-bank-fossil-fuel-subsidies.

24. "Canada, Out of Kyoto, Must Still Cut Emissions: U.N.," *Reuters*, December 13, 2011, http://www.reuters.com/article/ 2011/12/13/us-climate-canada-idUSTRE7BC2BW20111213; Adam Vaughan, "What Does Canada's Withdrawal from Kyoto Protocol Mean?," *Guardian*, December 13, 2011, http://www.theguardian.com/environment/2011/dec/13/canada-with drawal-kyoto-protocol; James Astill and Paul Brown, "Carbon Dioxide Levels Will Double by 2050, Experts Forecast," *Guardian*, April 5, 2001, http://www.guardian.co.uk/environment/ 2001/apr/06/usnews.globalwarming.

25. Acknowledgments to https://www.epsrc.ac.uk/newsevents/news/spiceprojectupdate/.

26. Paul Crutzen, "Albedo Enhancement by Stratospheric Sulfur Injections: A Contri- bution to Resolve a Policy Dilemma?," *Climatic Change* 77 (2006): 211–19. See also Daniel Bodansky, "May We Engineer the Climate?," *Climatic Change* 33 (1996): 309–21. Also see http://www.etcgroup.org/sites/www.etcgroup.org/files/publication/pdf_file/SPICE-Opposition %20Letter.pdf.

27. Andrew Ross and H. Damon Matthews, "Climate Engineering and the Risk of Rapid Climate Change," *Environmental Research Letters* 4, no. 4 (2009), http://iopscience .iop.org/article/10.1088/1748-9326/4/4/045103/pdf.

28. Ian Allison, et al., *The Copenhagen Diagnosis: Updating the World on the Latest Cli- mate Science* (Sydney: University of New South Wales Climate Change Research Centre, 2009), esp. 21; Jonathan Adams, "Estimates of Total Carbon Storage in Various Important Reservoirs," Oak Ridge National Laboratory, http://www.esd.ornl.gov/projects/qen/carbon2 .html.

29. See http://www.sciencedaily.com/releases/2012/031 120312003232.htm; http://www .pnas.org/content/105/38/14245.

30. Philip Ziegler, *The Black Death* (London: Folio Society, 1997).

31. A. Hallam and P. B. Wignall, *Mass Extinctions and Their Aftermath* (New York: Ox- ford University Press, 1997), gives the Big 5 as occurring at the end of the Devonian, Or- dovician, Permian, Triassic, and Cretaceous periods, on the classical Western geological timescale.

>>> Excerpt from *The Windup Girl*

Paolo Bacigalupi

"No! I don't want the mangosteen." Anderson Lake leans forward, pointing. "I want that one, there. *Kaw pollamai nee khap.* The one with the red skin and the green hairs."

The peasant woman smiles, showing teeth blackened from chewing betel nut, and points to a pyramid of fruits stacked beside her. "*Un nee chai mai kha?*"

"Right. Those. *Khap.*" Anderson nods and makes himself smile. "What are they called?"

"*Ngaw.*" She pronounces the word carefully for his foreign ear, and hands across a sample.

Anderson takes the fruit, frowning. "It's new?"

"*Kha.*" She nods an affirmative.

Anderson turns the fruit in his hand, studying it. It's more like a gaudy sea anemone or a furry puffer fish than a fruit. Coarse green tendrils protrude from all sides, tickling his palm. The skin has the rust-red tinge of blister rust, but when he sniffs he doesn't get any stink of decay. It seems perfectly healthy, despite its appearance.

"*Ngaw,*" the peasant woman says again, and then, as if reading his mind. "New. No blister rust."

Anderson nods absently. Around him, the market *soi* bustles with Bangkok's morning shoppers. Mounds of durians fill the alley in reeking piles and water tubs splash with snakehead fish and red-fin *plaa.* Overhead, palm-oil polymer tarps sag under the blast furnace heat of the tropic sun, shading the market with hand-painted images of clipper ship trading companies and the face of the revered Child Queen. A man jostles past, holding vermilion-combed chickens high as they flap and squawk outrage on their way to slaughter, and women in brightly colored *pha sin* bargain and smile with the vendors, driving down the price of pirated U-Tex rice and new-variant tomatoes.

None of it touches Anderson.

"*Ngaw,*" the woman says again, seeking connection.

The fruit's long hairs tickle his palm, challenging him to recognize its origin. Another Thai genehacking success, just like the tomatoes and eggplants and chiles that abound in the neighboring stalls. It's as if the Grahamite Bible's prophecies are coming to pass. As if Saint Francis himself stirs in his grave, restless, preparing to stride forth onto the land, bearing with him the bounty of history's lost calories.

"*And he shall come with trumpets, and Eden shall return . . .*"

The Windup Girl (San Francisco: Night Shade Books, 2009), 2–21

Anderson turns the strange hairy fruit in his hand. It carries no stink of cibiscosis. No scab of blister rust. No graffiti of genehack weevil engraves its skin. The world's flowers and vegetables and trees and fruits make up the geography of Anderson Lake's mind, and yet nowhere does he find a helpful signpost that leads him to identification.

Ngaw. A mystery.

He mimes that he would like to taste and the peasant woman takes back the fruit. Her brown thumb easily tears away the hairy rind, revealing a pale core. Translucent and veinous, it resembles nothing so much as the pickled onions served in martinis at research clubs in Des Moines.

She hands back the fruit. Anderson sniffs tentatively. Inhales floral syrup. *Ngaw.* It shouldn't exist. Yesterday, it didn't. Yesterday, not a single stall in Bangkok sold these fruits, and yet now they sit in pyramids, piled all around this grimy woman where she squats on the ground under the partial shading of her tarp. From around her neck, a gold glinting amulet of the martyr Phra Seub winks at him, a talisman of protection against the agricultural plagues of the calorie companies.

Anderson wishes he could observe the fruit in its natural habitat, hanging from a tree or lurking under the leaves of some bush. With more information, he might guess genus and family, might divine some whisper of the genetic past that the Thai Kingdom is trying to excavate, but there are no more clues. He slips the *ngaw's* slick translucent ball into his mouth.

A fist of flavor, ripe with sugar and fecundity. The sticky flower bomb coats his tongue. It's as though he's back in the HiGro fields of Iowa, offered his first tiny block of hard candy by a Midwest Compact agronomist when he was nothing but a farmer's boy, barefoot amid the cornstalks. The shell-shocked moment of flavor—real flavor—after a lifetime devoid of it.

Sun pours down. Shoppers jostle and bargain, but nothing touches him. He rolls the *ngaw* around in his mouth, eyes closed, tasting the past, savoring the time when this fruit must once have flourished, before cibiscosis and Nippon genehack weevil and blister rust and scabis mold razed the landscape.

Under the hammer heat of tropic sun, surrounded by the groan of water buffalo and the cry of dying chickens, he is one with paradise. If he were a Grahamite, he would fall to his knees and give ecstatic thanks for the flavor of Eden's return.

Anderson spits the black pit into his hand, smiling. He has read travelogues of history's botanists and explorers, the men and women who pierced the deepest jungle wildernesses of the earth in search of new species—and yet their discoveries cannot compare to this single fruit.

Those people all sought discoveries. He has found a resurrection.

The peasant woman beams, sure of a sale. "*Ao gee kilo kha?*" How much?

"Are they safe?" he asks.

She points at the Environment Ministry certificates laid on the cobbles beside her, underlining the dates of inspection with a finger. "Latest variation," she says. "Top grade."

Anderson studies the glinting seals. Most likely, she bribed the white shirts for stamps rather than going through the full inspection process that would have guaranteed immunity to eighth-generation blister rust along with resistance to cibiscosis 111.mt7 and mt8. The cynical part of him supposes that it hardly matters. The intricate stamps that glitter in the sun are more talismanic than functional, something to make people feel secure in a dangerous world. In truth, if cibiscosis breaks out again, these certificates will do nothing. It will be a new variation, and all the old tests will be useless, and then people will pray to their Phra Seub amulets and King Rama XII images and make offerings at the City Pillar Shrine, and they will all cough up the meat of their lungs no matter how many Environment Ministry stamps adorn their produce.

Anderson pockets the *ngaw*'s pit. "I'll take a kilo. No. Two. *Song.*"

He hands over a hemp sack without bothering to bargain. Whatever she asks, it will be too little. Miracles are worth the world. A unique gene that resists a calorie plague or utilizes nitrogen more efficiently sends profits sky-rocketing. If he looks around the market right now, that truth is everywhere displayed. The alley bustles with Thais purchasing everything from generipped versions of U-Tex rice to vermilion-variant poultry. But all of those things are old advances, based on previous genehack work done by AgriGen and PurCal and Total Nutrient Holdings. The fruits of old science, manufactured in the bowels of the Midwest Compact's research labs.

The *ngaw* is different. The *ngaw* doesn't come from the Midwest. The Thai Kingdom is clever where others are not. It thrives while countries like India and Burma and Vietnam all fall like dominoes, starving and begging for the scientific advances of the calorie monopolies.

A few people stop to examine Anderson's purchase, but even if Anderson thinks the price is low, they apparently find it too expensive and pass on.

The woman hands across the *ngaw*, and Anderson almost laughs with pleasure. Not a single one of these furry fruits should exist; he might as well be hefting a sack of trilobites. If his guess about the *ngaw*'s origin is correct, it represents a return from extinction as shocking as if a Tyrannosaurus were stalking down Thanon Sukhumvit. But then, the same is true of the potatoes and tomatoes and chiles that fill the market, all piled in such splendid abundance, an array of fecund nightshades that no one has seen in generations. In this drowning city, all things seem possible. Fruits and vegetables return from the grave, extinct flowers blossom on the avenues, and behind it all, the Environment Ministry works magic with the genetic material of generations lost.

Carrying his sacked fruit, Anderson squeezes back down the *soi* to the avenue beyond. A seethe of traffic greets him, morning commuters clogging Thanon Rama

IX like the Mekong in flood. Bicycles and cycle rickshaws, blue-black water buffaloes and great shambling megodonts.

At Anderson's arrival, Lao Gu emerges from the shade of a crumbling office tower, carefully pinching off the burning tip of a cigarette. Nightshades again. They're everywhere. Nowhere else in the world, but here they riot in abundance. Lao Gu tucks the remainder of the tobacco into a ragged shirt pocket as he trots ahead of Anderson to their cycle rickshaw.

The old Chinese man is nothing but a scarecrow, dressed in rags, but still, he is lucky. Alive, when most of his people are dead. Employed, while his fellow Malayan refugees are packed like slaughter chickens into sweltering Expansion towers. Lao Gu has stringy muscle on his bones and enough money to indulge in Singha cigarettes. To the rest of the yellow card refugees he is as lucky as a king.

Lao Gu straddles the cycle's saddle and waits patiently as Anderson clambers into the passenger seat behind. "Office," Anderson says. "*Bai khap.*" Then switches to Chinese. "*Zou ba.*"

The old man stands on his pedals and they merge into traffic. Around them, bicycle bells ring like cibiscosis chimes, irritated at their obstruction. Lao Gu ignores them and weaves deeper into the traffic flow.

Anderson reaches for another *ngaw*, then restrains himself. He should save them. They're too valuable to gobble like a greedy child. The Thais have found some new way to disinter the past, and all he wants to do is feast on the evidence. He drums his fingers on the bagged fruit, fighting for self-control.

To distract himself, he fishes for his pack of cigarettes and lights one. He draws on the tobacco, savoring the burn, remembering his surprise when he first discovered how successful the Thai Kingdom had become, how widely spread the nightshades. And as he smokes, he thinks of Yates. Remembers the man's disappointment as they sat across from one another with resurrected history smoldering between them.

* * *

"Nightshades."

Yates' match flared in the dimness of Spring Life's offices, illuminating florid features as he touched flame to a cigarette and drew hard. Rice paper crackled. The tip glowed and Yates exhaled, sending a stream of smoke ceilingward to where crank fans panted against the sauna swelter.

"Eggplants. Tomatoes. Chiles. Potatoes. Jasmine. Nicotiana." He held up his cigarette and quirked an eyebrow. "Tobacco."

He drew again, squinting in the cigarette's flare. All around, the shadowed desks and treadle computers of the company sat silent. In the evening, with the factory closed, it was just possible to mistake the empty desks for something other than the topography of failure. The workers might have only gone home, resting in anticipation of another hard day at their labors. Dust-mantled chairs and treadle computers put the lie to it—but in the dimness, with shadows draped across furniture and moonlight easing through mahogany shutters, it was possible to imagine what might have been.

Overhead, the crank fans continued their slow turns, Laotian rubber motorbands creaking rhythmically as they chained across the ceiling, drawing a steady trickle of kinetic power from the factory's central kink-springs.

"The Thais have been lucky in their laboratories," Yates said, "and now here you are. If I were superstitions, I'd think they conjured you along with their tomatoes. Every organism needs a predator, I understand."

"You should have reported how much progress they were making," Anderson said. "This factory wasn't your only responsibility."

Yates grimaced. His face was a study in tropic collapse. Broken blood vessels mapped rosy tributaries over his cheeks and punctuated the bulb of his nose. Watery blue eyes blinked back at Anderson, as hazy as the city's dung-choked air. "I should have known you'd cut my niche."

"It's not personal."

"Just my life's work." He laughed, a dry rattling reminiscent of early onset cibiscosis. The sound would have had Anderson backing out of the room if he didn't know that Yates, like all of AgriGen's personnel, had been inoculated against the new strains.

"I've spent years building this," Yates said, "and you tell me it's not personal." He waved toward the office's observation windows where they overlooked the manufacturing floor. "I've got kink-springs the size of my fist that hold a gigajoule of power. Quadruple the capacity-weight ratio of any other spring on the market. I'm sitting on a revolution in energy storage, and you're throwing it away." He leaned forward. "We haven't had power this portable since gasoline."

"Only if you can produce it."

"We're close," Yates insisted. "Just the algae baths. They're the only sticking point."

Anderson said nothing. Yates seemed to take this as encouragement. "The fundamental concept is sound. Once the baths are producing in sufficient quantities—"

"You should have informed us when you first saw the nightshades in the markets. The Thais have been successfully growing potatoes for at least five seasons. They're obviously sitting on top of a seedbank, and yet we heard nothing from you."

"Not my department. I do energy storage. Not production."

Anderson snorted. "Where are you going to get the calories to wind your fancy kink-springs if a crop fails? Blister rust is mutating every three seasons now. Recreational generippers are hacking into our designs for TotalNutrient Wheat and SoyPRO. Our last strain of HiGro Corn only beat weevil predation by sixty percent, and now we suddenly hear you're sitting on top of a genetic gold mine. People are starving—"

Yates laughed. "Don't talk to me about saving lives. I saw what happened with the seedbank in Finland."

"We weren't the ones who blew the vaults. No one knew the Finns were such fanatics."

"Any fool on the street could have anticipated. Calorie companies do have a certain reputation. "

"It wasn't my operation."

Yates laughed again. "That's always our excuse, isn't it? The company goes in somewhere and we all stand back and wash our hands. Pretend like we weren't the ones responsible. The company pulls SoyPRO from the Burmese market, and we all stand aside, saying intellectual property disputes aren't our department. But people starve just the same." He sucked on his cigarette, blew smoke. "I honestly don't know how someone like you sleeps at night."

"It's easy. I say a little prayer to Noah and Saint Francis, and thank God we're still one step ahead of blister rust."

"That's it then? You'll shut the factory down?"

"No. Of course not. The kink-spring manufacturing will continue."

"Oh?" Yates leaned forward, hopeful.

Anderson shrugged. "It's a useful cover."

* * *

The cigarette's burning tip reaches Anderson's fingers. He lets it fall into traffic. Rubs his singed thumb and index finger as Lao Gu pedals on through the clogged streets. Bangkok, City of Divine Beings, slides past.

Saffron-robed monks stroll along the sidewalks under the shade of black umbrellas. Children run in clusters, shoving and swarming, laughing and calling out to one another on their way to monastery schools. Street vendors extend arms draped with garlands of marigolds for temple offerings and hold up glinting amulets of revered monks to protect against everything from infertility to scabis mold. Food carts smoke and hiss with the scents of frying oil and fermented fish while around the ankles of their customers, the flicker-shimmer shapes of cheshires twine, yowling and hoping for scraps.

Overhead, the towers of Bangkok's old Expansion loom, robed in vines and mold, windows long ago blown out, great bones picked clean. Without air conditioning or elevators to make them habitable, they stand and blister in the sun. The black smoke of illegal dung fires wafts from their pores, marking where Malayan refugees hurriedly scald *chapatis* and boil *kopi* before the white shirts can storm the sweltering heights and beat them for their infringements.

In the center of the traffic lanes, northern refugees from the coal war prostrate themselves with hands upstretched, exquisitely polite in postures of need. Cycles and rickshaws and megodont wagons flow past them, parting like a river around boulders. The cauliflower growths of *fa' gan* fringe scar the beggars' noses and mouths. Betel nut stains blacken their teeth. Anderson reaches into his pocket and tosses cash at their feet, nodding slightly at their *wais* of thanks as he glides past.

A short while later, the whitewashed walls and alleys of the *farang* manufacturing district come into view. Warehouses and factories all packed together along with the scent of salt and rotting fish. Vendors scab along the alley lengths with bits of tarping and blankets spread above to protect them from the hammer blast of the sun. Just beyond, the dike and lock system of King Rama XII's seawall looms, holding back the weight of the blue ocean.

It's difficult not to always be aware of those high walls and the pressure of the water beyond. Difficult to think of the City of Divine Beings as anything other than a disaster waiting to happen. But the Thais are stubborn and have fought to keep their revered city of Krung Thep from drowning. With coal-burning pumps and leveed labor and a deep faith in the visionary leadership of their Chakri Dynasty, they have so far kept at bay that thing which has swallowed New York and Rangoon, Mumbai and New Orleans.

Lao Gu forges down an alley, ringing his bell impatiently at the coolie laborers who clot the artery. WeatherAll crates rock on brown backs. Logos for Chaozhou Chinese kink-springs, Matsushitaanti-bacterial handlegrips, and Bo Lok ceramic water filters sway back and forth, hypnotic with shambling rhythm. Images of the Buddha's teachings and the revered Child Queen splash along the factory walls, jostling with hand-painted pictures of *muay thai* matches past.

The SpringLife factory rises over the traffic press, a high-walled fortress punctuated by huge fans turning slowly in its upper story vents. Across the *soi* a Chaozhou bicycle factory mirrors it, and between them, the barnacle accretion of jumbled street carts that always clog around the entrances of factories, selling snacks and lunches to the workers inside.

Lao Gu brakes inside the SpringLife courtyard and deposits Anderson before the factory's main doors. Anderson climbs down from the rickshaw, grabs his sack of *ngaw*, and stands for a moment, staring up at the eight-meter wide doors that facilitate megodont access. The factory ought to be renamed Yates' Folly. The man was a terrible optimist. Anderson can still hear him arguing the wonders of gene-hacked algae, digging through desk drawers for graphs and scrawled notes as he protested.

"You can't pre-judge my work just because the Ocean Bounty project was a failure. Properly cured, the algae provides exponential improvements in torque absorption. Forget its calorie potential. Focus on the industrial applications. I can deliver the entire energy storage market to you, if you'll just give me a little more time. Try one of my demo springs at least, before you make a decision . . ."

The roar of manufacturing envelopes Anderson as he enters the factory, drowning out the last despairing howl of Yates' optimism.

Megodonts groan against spindle cranks, their enormous heads hanging low, prehensile trunks scraping the ground as they tread slow circles around power spindles. The genehacked animals comprise the living heart of the factory's drive system, providing energy for conveyor lines and venting fans and manufacturing machinery. Their harnesses clank rhythmically as they strain forward. Union handlers in red and gold walk beside their charges, calling out to the beasts, switching them occasionally, encouraging the elephant-derived animals to greater labor.

On the opposite side of the factory, the production line excretes newly packaged kink-springs, sending them past Quality Assurance and on to Packaging where the springs are palletized in preparation for some theoretical time when they will be ready for export. At Anderson's arrival on the floor, workers pause in their labors and *wai*, pressing their palms together and raising them to their foreheads in a wave of respect that cascades down the line.

Banyat, his head of QA, hurries over smiling. He *wais*.

Anderson gives a perfunctory *wai* in return. "How's quality?"

Banyat smiles. "*Dee khap.* Good. Better. Come, look." He signals up the line and Num, the day foreman, rings the warning bell that announces full line stop. Banyat motions Anderson to follow. "Something interesting. You will be pleased."

Anderson smiles tightly, doubting that anything Banyat says will be truly pleasing. He pulls a *ngaw* out of the bag and offers it to the QA man. "Progress? Really?"

Banyat nods as he takes the fruit. He gives it a cursory glance and peels it. Pops the semi-translucent heart into his mouth. He shows no surprise. No special reaction. Just eats the damn thing without a second thought. Anderson grimaces. *Farang* are always the last to know about changes in the country, a fact that Hock Seng likes to point out when his paranoid mind begins to suspect that Anderson intends to fire him. Hock Seng probably already knows about this fruit as well, or will pretend when he asks.

Banyat tosses the fruit's pit into a bin of feed for the megodonts and leads Anderson down the line. "We fixed a problem with the cutting press," he says.

Num rings his warning bell again and workers step back from their stations. On the third sounding of the bell, the union *mahout* tap their charges with bamboo switches and the megodonts shamble to a halt. The production line slows. At the far end of the factory, industrial kink-spring drums tick and squeal as the factory's flywheels shed power into them, the juice that will restart the line when Anderson is done inspecting.

Banyat leads Anderson down the now silent line, past more *wai*-ing workers in their green and white livery, and pushes aside the palm oil polymer curtains that mark the entrance to the fining room. Here, Yates' industrial discovery is sprayed with glorious abandon, coating the kink-springs with the residue of genetic serendipity. Women and children wearing triple-filter masks look up and tear away their breathing protection to *wai* deeply to the man who feeds them. Their faces are streaked with sweat and pale powder. Only the skin around their mouths and noses remains dark where the filters have protected them.

He and Banyat pass through the far side and into the swelter of the cutting rooms. Temper lamps blaze with energy and the tide pool reek of breeding algae clogs the air. Overhead, tiered racks of drying screens reach the ceiling, smeared with streamers of generipped algae, dripping and withering and blackening into paste in the heat. The sweating line techs are stripped to nearly nothing—just shorts and tanks and protective head gear. It is a furnace, despite the rush of crank fans and generous venting systems. Sweat rolls down Anderson's neck. His shirt is instantly soaked.

Banyat points. "Here. See." He runs his finger along a disassembled cutting bar that lies beside the main line. Anderson kneels to inspect the surface. "Rust," Banyat murmurs.

"I thought we inspected for that."

"Saltwater." Banyat smiles uncomfortably. "The ocean is close."

Anderson grimaces at the dripping algae racks overhead. "The algae tanks and drying racks don't help. Whoever thought we could just use waste heat to cure the stuff was a fool. Energy efficient my ass."

Banyat gives another embarrassed smile, but says nothing.

"So you've replaced the cutting tools?"

"Twenty-five percent reliability now."

"That much better?" Anderson nods perfunctorily. He signals to the tool leader and the man shouts out through the fining room to Num. The warning bell rings again and the heat presses and temper lamps begin to glow as electricity pours into the system. Anderson shies from the sudden increase in heat. The burning lamps and presses represent a carbon tax of fifteen thousand baht every time they begin to glow, a portion of the Kingdom's own global carbon budget that SpringLife pays handsomely to siphon off. Yates' manipulation of the system was ingenious, allowing the factory to use the country's carbon allocation, but the expense of the necessary bribes is still extraordinary.

The main flywheels spin up and the factory shivers as gears beneath the floor engage. The floorboards vibrate. Kinetic power sparks through the system like adrenaline, a tingling anticipation of the energy about to pour into the manufacturing line. A megodont screams protest and is lashed into silence. The whine of the flywheels rises to a howl, and then cuts off as joules gush into the drive system.

The line boss' bell rings again. Workers step forward to align the cutting tools. They're producing two-gigajoule kink-springs, and the smaller size requires extra care with the machinery. Further down the line, the spooling process begins and the cutting press with its newly repaired precision blades rises into the air on hydraulic jacks, hissing.

"*Khun*, please." Banyat motions Anderson behind a protection cage.

Num's bell rings a final time. The line grinds into gear. Anderson feels a brief thrill as the system engages. Workers crouch behind their shields. Kink-spring filament hisses out from alignment flanges and threads through a series of heated rollers. A spray of stinking reactant showers the rust-colored filament, greasing it in the slick film that will accept Yates' algae powder in an even coat.

The press slams down. Anderson's teeth ache with the crush of weight. The kink-spring wire snaps cleanly and then the severed filament is streaming through the curtains and into the fining room. Thirty seconds later it reemerges, pale gray and dusty with the algae-derived powder. It threads into a new series of heated rollers before being tortured into its final structure, winding in on itself, torquing into a tighter and tighter curl, working against everything in its molecular structure as the spring is tightened down. A deafening shriek of tortured metal rises. Lubricants and algae residue shower from the sheathing as the spring is squeezed down, spattering worker sand equipment, and then the compressed kink-spring is being whisked away to be installed in its case and sent on to QA.

A yellow LED flashes all-clear. Workers dash out from their cages to reset the press as a new stream of rust-colored metal hisses out of the bowels of the tempering rooms. Rollers chatter, running empty. Stoppered lubricant nozzles cast a fine mist into the air as they self-cleanse before the next application. The workers finish aligning the presses then duck again behind their barriers. If the system breaks, the kink-spring filament will become a high energy blade, whipping uncontrollably through the production room. Anderson has seen heads carved open like soft mangoes, the shorn parts of people and the Pollack-spatter of blood that comes from industrial system failures—

The press slams down, clipping another kink-spring among the forty per hour that now, apparently, will have only a seventy-five percent chance of ending up in a supervised disposal fill at the Environment Ministry. They're spending millions to produce trash that will cost millions more to destroy—a double-edged sword that just keeps cutting. Yates screwed something up, whether by accident or by spiteful sabotage, and it's taken more than a year to realize the depths of the problem, to examine the algae baths that breed the kink-springs' revolutionary coatings, to rework the corn resins that enclose the springs' gear interfaces, to change the QA practices, to understand what a humidity level that hovers near 100% year-round does to a manufacturing process conceptualized in drier climes.

A burst of pale filtering dust kicks into the room as a worker stumbles through the curtains from the fining chamber. His dark face is a sweat-streaked combination of grit and palm-oil spray. The swinging curtains reveal a glimpse of his colleagues encased in pale dust clouds, shadows in a snowstorm as the kink-spring filament is encased in the powder that keeps the springs from locking under intense compression. All that sweat, all those calories, all that carbon allotment—all to present a believable cover for Anderson as he unravels the mystery of night-shades and *ngaw*.

A rational company would shut down the factory. Even Anderson, with his limited understanding of the processes involved in this next-generation kink-spring manufacture would do so. But if his workers and the unions and the white shirts and the many listening ears of the Kingdom are to believe that he is an aspiring entrepreneur, the factory must run, and run hard.

Anderson shakes Banyat's hand and congratulates him on his good work.

It's a pity, really. The potential for success is there. When Anderson sees one of Yates' springs actually work, his breath catches. Yates was a madman, but he wasn't stupid. Anderson has watched joules pour out of tiny kink-spring cases, tick-ing along contentedly for hours when other springs wouldn't have held a quarter of the energy at twice the weight, or would simply have constricted into a single molecularly bound mass under the enormous pressure of the joules being dumped into them. Sometimes, Anderson is almost seduced by the man's dream.

Anderson takes a deep breath and ducks back through the fining room. He comes out on the other side in a cloud of algae powder and smoke. He sucks air redolent with trampled megodont dung and heads up the stairs to his offices. Be-hind him one of the megodonts shrieks again, the sound of a mistreated animal. Anderson turns, gazing down on the factory floor, and makes a note of the *mahout*. Number Four spindle. Another problem in the long list that SpringLife presents. He opens the door to the administrative offices.

Inside, the rooms are much as they were when he first encountered them. Still dim, still cavernously empty with desks and treadle computers sitting silent in shadows. Thin blades of sunlight ease between teak window shutters, illuminating smoky offerings to whatever gods failed to save Tan Hock Seng's Chinese clan in Malaya. Sandalwood incense chokes the room, and more silken streamers rise from a shrine in the corner where smiling golden figures squat over dishes of U-Tex rice and sticky fly-covered mangoes.

Hock Seng is already sitting at his computer. His bony leg ratchets steadily at the treadle, powering the microprocessors and the glow of the 12cm screen. In its gray light, Anderson catches the flicker of Hock Seng's eyes, the twitch of a man fearing bloody slaughter every time a door opens. The old man's flinch is as hallu-cinogenic as a cheshire's fade—one moment there, the next gone and doubted—but Anderson is familiar enough with yellow card refugees to recognize the sup-pressed terror. He shuts the door, muting the manufacturing roar, and the old man settles.

Anderson coughs and waves at the swirling incense smoke. "I thought I told you to quit burning this stuff."

Hock Seng shrugs, but doesn't stop treadling or typing. "Shall I open the windows?" His whisper is like bamboo scraping over sand.

"Christ, no." Anderson grimaces at the tropic blaze beyond the shutters. "Just burn it at home. I don't want it here. Not any more."

"Yes. Of course."

"I mean it."

Hock Seng's eyes flick up for a moment before returning to his screen. The jut of his cheek bones and the hollows of his eyes show in sharp relief under the glow of the monitor. His spider fingers continue tapping at the keys. "It's for luck," he murmurs. A low wheezing chuckle follows. "Even foreign devils need luck. With all the factory troubles, I think maybe you would appreciate the help of Budai."

"Not here." Anderson dumps his newly acquired *ngaw* on his desk and sprawls in his chair. Wipes his brow. "Burn it at home."

Hock Seng inclines his head slightly in acknowledgment. Overhead, the rows of crank fans rotate lazily, bamboo blades panting against the office's swelter. The two of them sit marooned surrounded by the map of Yates' grand design. Ranks of empty desks and workstations sit silent, the floor plan that should have held sales staff, shipping logistics clerks, HR people, and secretaries.

Anderson sorts through the *ngaw*. Holds up one of his green-haired discoveries for Hock Seng. "Have you ever seen one of these before?"

Hock Seng glances up. "The Thai call them *ngaw*." He returns to his work, treadling through spreadsheets that will never add and red ink that will never be reported.

"I know what the Thai call them." Anderson gets up and crosses to the old man's desk. Hock Seng flinches as Anderson sets the *ngaw* beside his computer, eyeing the fruit as if it is a scorpion. Anderson says, "The farmers in the market could tell me the Thai name. Did you have them down in Malaya, too?"

"I—" Hock Seng starts to speak, then stops. He visibly fights for self-control, his face working through a flicker-flash of emotions. "I—" Again, he breaks off.

Anderson watches fear mold and re-mold Hock Seng's features. Less than one percent of the Malayan Chinese escaped the Incident. By any measure, Hock Seng is a lucky man, but Anderson pities him. A simple question, a piece of fruit, and the old man looks as if he's about to flee the factory.

Hock Seng stares at the *ngaw*, breath rasping. Finally he murmurs, "None in Malaya. Only Thais are clever with such things." And then he is working again, eyes fixed on his little computer screen, memories locked away.

Anderson waits to see if Hock Seng will reveal anything more but the old man doesn't raise his eyes again. The puzzle of the *ngaw* will have to wait.

Anderson returns to his own desk and starts sifting through the mail. Receipts and tax papers that Hock Seng has prepared sit at one corner of his desk, demanding attention. He begins working through the stack, adding his signature to Megodont Union paychecks and the SpringLife chop to waste disposal approvals. He tugs at his shirt, fanning himself against the increasing heat and humidity.

Eventually Hock Seng looks up. "Banyat was looking for you."

Anderson nods, distracted by the forms. "They found rust on the cutting press. The replacement improved reliability by five percent."

"Twenty-five percent, then?"

Anderson shrugs, flips more pages, adds his chop to an Environment Ministry carbon assessment. "That's what he says." He folds the document back into its envelope.

"Still not a profitable statistic. Your springs are all wind and no release. They keep joules the way the Somdet Chaopraya keeps the Child Queen."

Anderson makes a face of irritation but doesn't bother defending the erratic quality.

"Did Banyat also tell you about the nutrient tanks?" Hock Seng asks. "For the algae?"

"No. Just the rust. Why?"

"They have been contaminated. Some of the algae is not producing the . . ." Hock Seng hesitates. "The skim. It is not productive."

"He didn't mention it to me."

Another slight hesitation. Then, "I'm sure he tried."

"Did he say how bad it was?"

Hock Seng shrugs. "Just that the skim does not meet specifications."

Anderson scowls. "I'm firing him. I don't need a QA man who can't actually tell me the bad news."

"Perhaps you were not paying close attention."

Anderson has a number of words for people who try to raise a subject and then somehow fail, but he's interrupted by a scream from the megodont downstairs. The noise is loud enough to make the windows shake. Anderson pauses, listening for a follow-up cry.

"That's the Number Four power spindle," he says. "The *mahout* is incompetent."

Hock Seng doesn't look up from his typing. "They are Thai. They are all incompetent."

Anderson stifles a laugh at the yellow card's assessment. "Well, that one is worse." He goes back to his mail. "I want him replaced. Number Four spindle. Remember that."

Hock Seng's treadle loses its rhythm. "This is a difficult thing, I think. Even the Dung Lord must bow before the Megodont Union. Without the labor of the megodonts, one must resort to the joules of men. Not a powerful bargaining position."

"I don't care. I want that one out. We can't afford a stampede. Find some polite way to get rid of him." Anderson pulls over another stack of paychecks waiting for his signature.

Hock Seng tries again. "*Khun*, negotiating with the union is a complicated thing."

"That's why I have you. It's called delegating." Anderson continues flipping the papers.

"Yes, of course." Hock Seng regards him drily. "Thank you for your management instruction."

"You keep telling me I don't understand the culture here," Anderson says. "So take care of it. Get rid of that one. I don't care if you're polite or if everyone loses face, but find a way to axe him. It's dangerous to have someone like that in the power train."

Hock Seng's lips purse, but he doesn't protest any more. Anderson decides to assume that he will be obeyed. He flips through the pages of another permit letter from the Environment Ministry, grimacing. Only Thais would spend so much time making a bribe look like a service agreement. They're polite, even when they're shaking you down. Or when there's a problem with the algae tanks. Banyat . . .

Anderson shuffles through the forms on his desk. "Hock Seng?"

The old man doesn't look up. "I will take care of your *mahout*," he says as he keeps typing. "It will be done, even if it costs you when they come to bargain again for bonuses."

"Nice to know, but that's not my question." Anderson taps his desk. "You said Banyat was complaining about the algae skim. Is he having problems with the new tanks? Or the old ones?"

"I . . . He was unclear."

"Didn't you tell me we had replacement equipment coming off the anchor pads last week? New tanks, new nutrient cultures?" Hock Seng's typing falters for a moment. Anderson pretends puzzlement as he shuffles through his papers again, already knowing that the receipts and quarantine forms aren't present. "I should have a list here somewhere. I'm sure you told me it was arriving." He looks up. "The more I think about it, the more I think I shouldn't be hearing about any contamination problems. Not if our new equipment actually cleared Customs and got installed."

Hock Seng doesn't answer. Presses on with his typing as though he hasn't heard.

"Hock Seng? Is there something you forgot to tell me?"

Hock Seng's eyes remain fixed on the gray glow of his monitor. Anderson waits. The rhythmic creak of the crank fans and the ratchet of Hock Seng's treadle fills the silence.

"There is no manifest," the old man says, finally. "The shipment is still in Customs."

"It was supposed to clear last week."

"There are delays."

"You told me there wouldn't be any problem," Anderson says. "You were certain. You told me you were expediting the Customs personally. I gave you extra cash to be sure of it."

"The Thai keep time in their own method. Perhaps it will be this afternoon. Perhaps tomorrow." Hock Seng makes a face that resembles a grin. "They are not like we Chinese. They are lazy."

"Did you actually pay the bribes? The Trade Ministry was supposed to get a cut, to pass on to their pet white shirt inspector."

"I paid them."

"Enough?"

Hock Seng looks up, eyes narrowed. "I paid."

"You didn't pay half and keep half for yourself?"

Hock Seng laughs nervously. "Of course I paid everything."

Anderson studies the yellow card a moment longer, trying to determine his honesty, then gives up and tosses down the papers. He isn't even sure why he cares, but it galls him that the old man thinks he can be fooled so easily. He glances again at the sack of *ngaw*. Perhaps Hock Seng senses just how secondary the factory is . . . He forces the thought away and presses the old man again. "Tomorrow then?"

Hock Seng inclines his head. "I think this is most likely."

"I'll look forward to it."

Hock Seng doesn't respond to the sarcasm. Anderson wonders if it even translates. The man speaks English with an extraordinary facility, but every so often they reach an impasse of language that seems more rooted in culture than vocabulary.

Anderson returns to the paperwork. Tax forms here. Paychecks there. The workers cost twice as much as they should. Another problem of dealing with the Kingdom. Thai workers for Thai jobs. Yellow card refugees from Malaya are starving in the street, and he can't hire them. By rights, Hock Seng should be out in the job lines starving with all the other survivors of the Incident. Without his specialized skills in language and accountancy and Yates' indulgence, he would be starving.

Anderson pauses on a new envelope. It's posted to him, personally, but true to form the seal is broken. Hock Seng has a hard time respecting the sanctity of other people's mail. They've discussed the problem repeatedly, but still the old man makes "mistakes."

Inside the envelope, Anderson finds a small invitation card. Raleigh, proposing a meeting.

Anderson taps the invitation card against his desk, thoughtful. Raleigh. Flotsam of the old Expansion. An ancient piece of driftwood left at high tide, from the time when petroleum was cheap and men and women crossed the globe in hours instead of weeks.

When the last of the jumbo jets rumbled off the flooded runways of Suvarnabhumi, Raleigh stood knee-deep in rising seawater and watched them flee. He squatted with his girlfriends and then outlived them and then claimed new ones, forging a life of lemongrass and baht and fine opium. If his stories are to be believed, he has survived coups and counter-coups, calorie plagues and starvation. These days, the old man squats like a liver-spotted toad in his Ploenchit "club," smiling in self-satisfaction as he instructs newly arrived foreigners in the lost arts of pre-Contraction debauch.

Anderson tosses the card on the desk. Whatever the old man's intentions, the invitation is innocuous enough. Raleigh hasn't lived this long in the Kingdom without developing a certain paranoia of his own. Anderson smiles slightly, glancing up at Hock Seng. The two would make a fine pair: two uprooted souls, two men far from their homelands, each of them surviving by their wits and paranoia . . .

"If you are doing nothing other than watching me work," Hock Seng says, "the Megodont Union is requesting a renegotiation of their rates."

Anderson regards the expenses piled on his desk. "I doubt they're so polite."

Hock Seng's pen pauses. "The Thai are always polite. Even when they threaten."

The megodont on the floor below screams again.

Anderson gives Hock Seng a significant look. "I guess that gives you a bargaining chip when it comes to getting rid of the Number Four *mahout*. Hell, maybe I just won't pay them anything at all until they get rid of that bastard."

"The union is powerful."

Another scream shakes the factory, making Anderson flinch. "And stupid!" He glances toward the observation windows. "What the hell are they doing to that animal?" He motions at Hock Seng. "Go check on them."

Hock Seng looks as if he will argue, but Anderson fixes him with a glare. The old man gets to his feet.

A resounding trumpet of protest interrupts whatever complaint the old man is about to voice. The observation windows rattle violently.

"What the—"

Another trumpeted wail shakes the building, followed by a mechanical shriek: the power train, seizing. Anderson lurches out of his chair and runs for the window but Hock Seng reaches it ahead of him. The old man stares through the glass, mouth agape.

Yellow eyes the size of dinner plates rise level with the observation window. The megodont is up on its hind legs, swaying. The beast's four tusks have been sawn off for safety, but it is still a monster, fifteen feet at the shoulder, ten tons of muscle and rage, balanced on its hind legs. It pulls against the chains that bind it to the winding spindle. Its trunk lifts, exposing a cavernous maw. Anderson jams his hands over his ears.

The megodont's scream hammers through the glass. Anderson collapses to his knees, stunned. "Christ!" His ears are ringing. "Where's that *mahout*?"

Hock Seng shakes his head. Anderson isn't even sure the man has heard. Sounds in his own ears are muffled and distant. He staggers to the door and yanks it open just as the megodont crashes down on Spindle Four. The power spindle shatters. Teak shards spray in all directions. Anderson flinches as splinters fly past and his skin burns with needle slashes.

Down below, the *mahouts* are frantically unchaining their beasts and dragging them away from the maddened animal, shouting encouragement, forcing their will on the elephantine creatures. The megodonts shake their heads and groan protest, tugging against their training, overwhelmed by the instinctual urge to aid their cousin. The rest of the Thai workers are fleeing for the safety of the street.

The maddened megodont launches another attack on its winding spindle. Spokes shatter. The *mahout* who should have controlled the beast is a mash of blood and bone on the floor.

Anderson ducks back into his office. He dodges around empty desks and jumps another, sliding over its surface to land before the company's safes.

His fingers slip as he spins combination dials. Sweat drips in his eyes. 23-right. 106-left . . . His hand moves to the next dial as he prays that he won't screw up the pattern and have to start again. More wood shatters out on the factory floor, accompanied by the screams of someone who got too close.

Hock Seng appears at his elbow, crowding.

Anderson waves the old man away. "Tell the people to get out of here! Clear everyone out! I want everyone out!"

Hock Seng nods but lingers as Anderson continues to struggle with the combinations.

Anderson glares at him. "Go!"

Hock Seng ducks acquiescence and runs for the door, calling out, his voice lost in the screams of fleeing workers and shattering hardwoods. Anderson spins the last of the dials and yanks the safe open: papers, stacks of colorful money, eyes-only records, a compression rifle . . . a spring pistol.

Yates.

He grimaces. The old bastard seems to be everywhere today, as if his *phii* is riding on Anderson's shoulder. Anderson pumps the handgun's spring and stuffs it in his belt. He pulls out the compression rifle. Checks its load as another scream echoes behind. At least Yates prepared for this. The bastard was naïve, but he wasn't stupid. Anderson pumps the rifle and strides for the door.

Down on the manufacturing floor, blood splashes the drive systems and QA lines. It's difficult to see who has died. More than just the one *mahout*. The sweet stink of human offal permeates the air. Gut streamers decorate the megodont's circuit around its spindle. The animal rises again, a mountain of genetically engineered muscle, fighting against the last of its bonds.

Anderson levels his rifle. At the edge of his vision, another megodont rises onto its hind legs, trumpeting sympathy. The *mahouts* are losing control. He forces himself to ignore the expanding mayhem and puts his eye to the scope.

His rifle's crosshairs sweep across a rusty wall of wrinkled flesh. Magnified with the scope, the beast is so vast he can't miss. He switches the rifle to full automatic, exhales, and lets the gas chamber unleash.

A haze of darts leaps from the rifle. Blaze orange dots pepper the megodont's skin, marking hits. Toxins concentrated from AgriGen research on wasp venom pump through the animal's body, gunning for its central nervous system.

Anderson lowers the rifle. Without the scope's magnification, he can barely make out the scattered darts on the beast's skin. In another few moments it will be dead.

The megodont wheels and fixes its attention on Anderson, eyes flickering with Pleistocene rage. Despite himself, Anderson is impressed by the animal's intelligence. It's almost as if the animal knows what he has done.

The megodont gathers itself and heaves against its chains. Iron links crack and whistle through the air, smashing into conveyor lines. A fleeing worker collapses. Anderson drops his useless rifle and yanks out the spring gun. It's a toy against ten tons of enraged animal, but it's all he has left. The megodont charges and Anderson fires, pulling the trigger as quickly as his finger can convulse. Useless bladed disks spatter against the avalanche.

The megodont slaps him off his feet with its trunk. The prehensile appendage coils around his legs like a python. Anderson scrabbles for a grip on the door jam, trying to kick free. The trunk squeezes. Blood rushes into his head. He wonders if the monster simply plans to pop him like some blood-bloated mosquito, but then the beast is dragging him off the balcony. Anderson scrabbles for a last handhold as the railing slides past and then he's airborne. Flying free.

The megodont's exultant trumpeting echoes as Anderson sails through the air. The factory floor rushes up. He slams into concrete. Blackness swallows him. *Lie down and die.* Anderson fights unconsciousness. *Just die.* He tries to get up, to roll away, to do anything at all, but he can't move. Colorful shapes fill his vision, trying to coalesce. The megodont is close. He can smell its breath.

Color blotches converge. The megodont looms, rusty skin and ancient rage. It raises a foot to pulp him. Anderson rolls onto his side but can't get his legs to work. He can't even crawl. His hands scrabble against the concrete like spiders on ice. He can't move quickly enough. Oh Christ, I don't want to die like this. *Not here. Not like this.* . . . He's like a lizard with its tail caught. He can't get up, he can't get away, he's going to die, jelly under the foot of an oversized elephant.

The megodont groans. Anderson looks over his shoulder. The beast has lowered its foot. It sways, drunken. It snuffles about with its trunk and then abruptly its hindquarters give out. The monster settles back on its haunches, looking ridiculously like a dog. Its expression is almost puzzled, a drugged surprise that its body no longer obeys.

Slowly its forelegs sprawl before it and it sinks, groaning, into straw and dung. The megodont's eyes sink to Anderson's level. They stare into his own, nearly human, blinking confusion. Its trunk stretches out for him again, slapping clumsily, a python of muscle and instinct, all uncoordinated now. Its maw hangs open, panting. Sweet furnace heat gusts over him. The trunk prods at him. Rocks him. Can't get a grip.

Anderson slowly drags himself out of reach. He gets to his knees, then forces himself upright. He sways, dizzy, then manages to plant his feet and stand tall. One of the megodont's yellow eyes tracks his movement. The rage is gone. Long-lashed eyelids blink. Anderson wonders what the animal is thinking. If the neural havoc tearing through its system is something it can feel. If it knows its end is imminent. Or if it just feels tired.

Standing over it, Anderson can almost feel pity. The four ragged ovals where its tusks once stood are grimy foot-diameter ivory patches, savagely sawed away. Sores glisten on its knees and scabis growths speckle its mouth. Close up and dying, with its muscles paralyzed and its ribs heaving in and out, it is just an ill-used creature. The monster was never destined for fighting.

The megodont lets out a final gust of breath. Its body sags.

People are swarming all around Anderson, shouting, tugging at him, trying to help their wounded and find their dead. People are everywhere. Red and gold union colors, green SpringLife livery, the *mahouts* clambering over the giant corpse.

For a second, Anderson imagines Yates standing beside him, smoking a nightshade and gloating at all the trouble. *"And you said you'd be gone in a month."* And then Hock Seng is beside him, whisper voice and black almond eyes and a bony hand that reaches up to touch his neck and comes away drenched red.

"You're bleeding," he murmurs.

⟩⟩⟩ It's Not Climate Change—It's Everything Change

Margaret Atwood

> *Oil! Our secret god, our secret sharer, our magic wand, fulfiller of our every desire, our co-conspirator, the sine qua non in all we do! Can't live with it, can't—right at this moment—live without it. But it's on everyone's mind.*
>
> *Back in 2009, as fracking and the mining of the oil/tar sands in Alberta ramped up—when people were talking about Peak Oil and the dangers of the supply giving out—I wrote a piece for the German newspaper Die Zeit. In English it was called "The Future without Oil." It went like this:*

The future without oil! For optimists, a pleasant picture: let's call it Picture One. Shall we imagine it?

There we are, driving around in our cars fueled by hydrogen, or methane, or solar, or something else we have yet to dream up. Goods from afar come to us by solar-and-sail-driven ship—the sails computerized to catch every whiff of air—or else by new versions of the airship, which can lift and carry a huge amount of freight with minimal pollution and no ear-slitting noise. Trains have made a comeback. So have bicycles, when it isn't snowing; but maybe there won't be any more winter.

We've gone back to small-scale hydropower, using fish-friendly dams. We're eating locally, and even growing organic vegetables on our erstwhile front lawns, watering them with greywater and rainwater, and with the water saved from using low-flush toilets, showers instead of baths, water-saving washing machines, and other appliances already on the market. We're using low-draw lightbulbs— incandescents have been banned—and energy-efficient heating systems, including pellet stoves, radiant panels, and long underwear. Heat yourself, not the room is no longer a slogan for nutty eccentrics: it's the way we all live now.

Due to improved insulation and indoor-climate-enhancing practices, including heatproof blinds and awnings, air-conditioning systems are obsolete, so they no longer suck up huge amounts of power every summer. As for power, in addition to hydro, solar, geothermal, wave, and wind generation, and emissions-free coal plants, we're using almost foolproof nuclear power. Even when there are accidents it isn't all bad news, because instant wildlife refuges are created as Nature invades those high-radiation zones where Man now fears to tread. There's said to be some remarkable wildlife and botany in the area surrounding Chernobyl.

Matter, June 27, 2015, https://medium.com/matter/it-s-not-climate-change-it-s-everything -change-8fd9aa671804#.7vpcsosrz; an earlier version of this article appeared in the Norwegian magazine *Samtiden*

What will we wear? A lot of hemp clothing, I expect: hemp is a hardy fiber source with few pesticide requirements, and cotton will have proven too costly and destructive to grow. We might also be wearing a lot of recycled tinfoil—keeps the heat in—and garments made from the recycled plastic we've harvested from the island of it twice the size of Texas currently floating around in the Pacific Ocean. What will we eat, besides our front-lawn vegetables? That may be a problem—we're coming to the end of cheap fish, and there are other shortages looming. Abundant animal protein in large hunks may have had its day. However, we're an inventive species, and when push comes to shove we don't have a lot of fastidiousness: being omnivores, we'll eat anything as long as there's ketchup. Looking on the bright side: obesity due to overeating will no longer be a crisis, and diet plans will not only be free, but mandatory.

That's Picture One. I like it. It's comforting. Under certain conditions, it might even come true. Sort of. More or less.

Then there's Picture Two. Suppose the future without oil arrives very quickly. Suppose a bad fairy waves his wand, and poof! Suddenly there's no oil, anywhere, at all.

Everything would immediately come to a halt. No cars, no planes; a few trains still running on hydroelectric, and some bicycles, but that wouldn't take very many people very far. Food would cease to flow into the cities, water would cease to flow out of the taps. Within hours, panic would set in.

The first result would be the disappearance of the word "we": except in areas with exceptional organization and leadership, the word "I" would replace it, as the war of all against all sets in. There would be a run on the supermarkets, followed immediately by food riots and looting. There would also be a run on the banks— people would want their money out for black market purchasing, although all currencies would quickly lose value, replaced by bartering. In any case the banks would close: their electronic systems would shut down, and they'd run out of cash.

Having looted and hoarded some food and filled their bathtubs with water, people would hunker down in their houses, creeping out into the backyards if they dared because their toilets would no longer flush. The lights would go out. Communication systems would break down. What next? Open a can of dog food, eat it, then eat the dog, then wait for the authorities to restore order. But the authorities— lacking transport—would be unable to do this.

Other authorities would take over. These would at first be known as thugs and street gangs, then as warlords. They'd attack the barricaded houses, raping, pillaging, and murdering. But soon even they would run out of stolen food. It wouldn't take long—given starvation, festering garbage, multiplying rats, and putrefying corpses—for pandemic disease to break out. It will quickly become apparent that the present world population of six and a half billion people is not only dependent on oil, but was created by it: humanity has expanded to fill the space made possible to it by oil, and without that oil it would shrink with astounding rapidity. As for the costs to "the economy," there won't be any "economy." Money will vanish: the only items of exchange will be food, water, and most likely—before everyone topples over—sex.

Picture Two is extreme, and also unlikely, but it exposes the truth: we're hooked on oil, and without it we can't do much of anything. And since it's bound to run out eventually, and since cheap oil is already a thing of the past, we ought to be investing a lot of time, effort, and money in ways to replace it.

Unfortunately, like every other species on the planet, we're conservative: we don't change our ways unless necessity forces us. The early lungfish didn't develop lungs because it wanted to be a land animal, but because it wanted to remain a fish even as the dry season drew down the water around it. We're also self-interested: unless there are laws mandating conservation of energy, most won't do it, because why make sacrifices if others don't? The absence of fair and enforceable energy-use rules penalizes the conscientious while enriching the amoral. In business, the laws of competition mean that most corporations will extract maximum riches from available resources with not much thought to the consequences. Why expect any human being or institution to behave otherwise unless they can see clear benefits?

In addition to Pictures One and Two, there's Picture Three. In Picture Three, some countries plan for the future of diminished oil, some don't. Those planning now include—not strangely—those that don't have any, or don't need any. Iceland generates over half its power from abundant geothermal sources: it will not suffer much from an oil dearth. Germany is rapidly converting, as are a number of other oil-poor European countries. They are preparing to weather the coming storm.

Then there are the oil-rich countries. Of these, those who were poor in the past, who got rich quick, and who have no resources other than oil are investing the oil wealth they know to be temporary in technologies they hope will work for them when the oil runs out. But in countries that have oil, but that have other resources too, such foresight is lacking. It does exist in one form: as a Pentagon report of 2003 called "An Abrupt Climate Change Scenario and Its Implications for United States National Security" put it, "Nations with the resources to do so may build virtual fortresses around their countries, preserving resources for themselves." That's already happening: the walls grow higher and stronger every day.

But the long-term government planning needed to deal with diminishing oil within rich, mixed-resource countries is mostly lacking. Biofuel is largely delusional: the amount of oil required to make it is larger than the payout. Some oil companies are exploring the development of other energy sources, but by and large they're simply lobbying against anything and anyone that might cause a decrease in consumption and thus impact on their profits. It's gold-rush time, and oil is the gold, and short-term gain outweighs long-term pain, and madness is afoot, and anyone who wants to stop the rush is deemed an enemy.

My own country, Canada, is an oil-rich country. A lot of the oil is in the Athabasca oil sands, where licenses to mine oil are sold to anyone with the cash, and where CO_2 is being poured into the atmosphere, not only from the oil used as an end product, but also in the course of its manufacture. Also used in its manufacture is an enormous amount of water. The water mostly comes from the Athabasca River, which is fed by a glacier. But due to global warming, glaciers are melting fast. When they're gone, no more water, and thus no more oil from oil sands. Maybe we'll be saved—partially—by our own ineptness. But we'll leave much destruction in our wake.

The Athabasca oil-sand project has now replaced the pyramids as the must-see man-made colossal sight, although it's not exactly a monument to hopes of immortality. There has even been a tour to it: the venerable Canadian company Butterfield & Robinson ran one in 2008 as part of its series "Places on the Verge."

Destinations at risk: first stop, the oil sands. Next stop, the planet. If we don't start aiming for Picture One, we'll end up with some version of Picture Two. So hoard some dog food, because you may be needing it.

It's interesting to look back on what I wrote about oil in 2009, and to reflect on how the conversation has changed in a mere 6 years. Much of what most people took for granted back then is no longer universally accepted, including the idea that we could just go on and on the way we were living then, with no consequences. There was already some alarm back then, but those voicing it were seen as extreme. Now their concerns have moved to the center of the conversation. Here are some of the main worries.

Planet Earth—the Goldilocks planet we've taken for granted, neither too hot nor too cold, neither too wet nor too dry, with fertile soils that accumulated for millennia before we started to farm them—that planet is altering. The shift towards the warmer end of the thermometer that was once predicted to happen much later, when the generations now alive had had lots of fun and made lots of money and gobbled up lots of resources and burned lots of fossil fuels and then died, are happening much sooner than anticipated back then. In fact, they're happening now.

Here are three top warning signs. First, the transformation of the oceans. Not only are these being harmed by the warming of their waters, in itself a huge affector of climate. There is also the increased acidification due to CO_2 absorption, the ever-increasing amount of oil-based plastic trash and toxic pollutants that human beings are pouring into the seas, and the overfishing and destruction of marine ecosystems and spawning grounds by bottom-dragging trawlers. Most lethal to us—and affected by warming, acidification, toxins, and dying marine ecosystems—would be the destruction of the bluegreen marine algae that created our present oxygen-rich atmosphere 2.45 billion years ago, and that continue to make the majority of the oxygen we breathe. If the algae die, that would put an end to us, as we would gasp to death like fish out of water.

A second top warning sign is the drought in California, said to be the worst for 1,200 years. This drought is now in its fourth year; it is mirrored by droughts in other western US states, such as Utah and Idaho. The snowpack in the mountains that usually feeds the water supplies in these states was only 3% of the norm this winter. It's going to be a long, hot, dry summer. The knock-on effect of such widespread drought on such things as the price of fruit and vegetables has yet to be calculated, but it will be extensive. As drought conditions spread elsewhere, we may expect water wars as the world's supply of fresh water is exhausted.

A third warning sign is the rise in ocean levels. There have already been some noteworthy flooding events, the most expensive in North America being Hurricane Katrina and the inundation of lower Manhattan at the time of Hurricane Sandy in 2012. Should the predicted sea-level rise of a foot to two feet take place, the state of Florida stands to lose most of its beaches, and the city of Miami will be wading. Many other low-lying cities around the world will be affected.

This result, however, is not accepted by some of the politicians who are supposed to be alert to dangers threatening the welfare of their constituents. The present governor of Florida, Rick Scott, is said to have issued a memo to all government of Florida employees forbidding them to use the terms "climate change" and "global warming," because he doesn't believe in them (though Scott has denied this to the press). I myself would like to disbelieve in gravitational forces, because then I could fly, and also in viruses, because then I would never get colds. Makes sense: you can't see viruses or gravity, and seeing is believing, and when you've got your head stuck in the sand you can't see a thing, right?

The Florida government employees also aren't allowed to talk about sea-level rise: when things get very wet inside people's houses, it's to be called "nuisance flooding." (If the city of Miami gets soaked, as it will should the level rise the two feet predicted in the foreseeable future, it will indeed be a nuisance, especially in the real-estate sector; so the governor isn't all wrong.) What a practical idea for solving pesky problems: let's not talk about it, and maybe it will go away.

The Canadian federal government, not to be outdone in the area of misleading messages, has just issued a new map that shows more Arctic sea ice than the previous map did. Good news! The sea ice is actually increasing! So global warming and climate change doesn't exist? How reassuring for the population, and how convenient for those invested in carbon fuels!

But there's some fine print. It seems that this new map shows an *average* amount of sea ice, and the averaging goes back 30 years. As the *Globe and Mail* article on this new map puts it:

> In reality, climate change has been gnawing away at the planet's permanent polar ice cap and it is projected to continue doing so.
>
> "It's a subtle way, on a map, to change the perspective on the way something is viewed," said Christopher Storie, an assistant professor of geography at the University of Winnipeg and president of the Canadian Cartographic Association.
>
> Whereas the older version of the map showed only that part of the sea ice that permanently covered Arctic waters year round at that time, the new edition uses a 30-year median of September sea-ice extent from 1981 through 2010. September sea ice hit a record low in 2012 and is projected to decline further. The change means there is far more ice shown on the 2015 version of the map than on its predecessor.
>
> "Both are correct," Dr. Storie said. "They've provided the right notation for the representation, but not many people will read that or understand what it means."

Cute trick, wouldn't you say? Not as cute as Florida's trick, but cute. And both tricks emphasize the need for scientific literacy. Increasingly, the public needs to know how to evaluate the worth of whatever facts they're being told. Who's saying it? What's their source? Do they have a bias? Unfortunately, very few people have the expertise necessary to decode the numbers and statistics that are constantly being flung at us.

Both the Florida cute trick and the Canadian map one originate in worries about the Future, and the bad things that may happen in that future; also the desire to deny these things or sweep them under the carpet so business can go on as usual, leaving the young folks and future generations to deal with the mess and chaos that will result from a changed climate, and then pay the bill. Because there will be a bill: the cost will be high, not only in money but in human lives. The laws of chemistry and physics are unrelenting, and they don't give second chances. In fact, that bill is already coming due.

There are many other effects, from species extinction to the spread of diseases to a decline in overall food production, but the main point is that these effects are not happening in some dim, distant future. They are happening now.

In response to our growing awareness of these effects, there have been some changes in public and political attitudes, though these changes have not been universal. Some acknowledge the situation, but shrug and go about their daily lives taking a "What can I do?" position. Some merely despair. But only those with their heads stuck so firmly into the sand that they're talking through their nether ends are still denying that reality has changed.

Even if the deniers can be brought reluctantly to acknowledge the facts on the ground, they display two fallback positions: (1) The changes are natural. They have nothing to do with humankind's burning of fossil fuels. Therefore we can keep on having our picnic, such as it is, perhaps making a few gestures in the direction of "adaptation"—a seawall here, the building of a desalination plant there—without worrying about our own responsibility. (2) The changes are divine. They are punishments being inflicted on humankind for its sins by supernatural agency. In extreme form, they are part of a divine plan to destroy the world, send most of its inhabitants to a hideous death, and make a new world for those who will be saved. People who believe this kind of thing usually number themselves among the lucky few. It would, however, be a mistake to vote for them, as in a crisis they would doubtless simply head for higher ground or their own specially equipped oxygen shelters, and then cheer while billions die, rather than lifting a finger to save their fellow citizens.

Back in 2009, discussion of the future of energy and thus of civilization as we know it tended to be theoretical. Now, however, action is being taken and statements are being made, some of them coming from the usual suspects— "left-wingers" and "artists" and "radicals," and other such dubious folks—but others now coming from directions that would once have been unthinkable. Some are even coming—mirabile dictu!—from politicians. Here are some examples of all three kinds:

In September 2014, the international petition site Avaaz (over 41 million members) pulled together a Manhattan climate march of 400,000 people, said to be the largest climate march in history. On April 11, 2015, approximately 25,000 people congregated in Quebec City to serve notice on Canadian politicians that they want them to start taking climate change seriously. Five years ago, that number would probably have been 2,500. Just before that date, Canada's most populous province, Ontario, announced that it was bringing in a cap-and-trade plan. The chances of that happening 5 years ago were nil.

In case anyone thinks that it's only people on the so-called political left that are concerned, there are numerous straws in the wind that's blowing from what might once have been considered the resistant right. Henry Paulson, Secretary of the Treasury under George W. Bush, has just said that there are two threats to our society that are even greater than the 2008 financial meltdown he himself helped the world navigate: environmental damage due to climate change, and the possible failure of China. (Chinese success probably means China can tackle its own carbon emissions and bring them under control; Chinese failure means it probably can't.)

In Canada, an organization called the Ecofiscal Commission has been formed; it includes representatives from the erstwhile Reform Party (right), the Liberal Party (centrist), and the NDP (left), as well as members from the business community. Its belief is that environmental problems can be solved by business sense and common sense, working together; that a gain for the environment does not have to be a financial loss, but can be a gain. In America, the Tesla story would certainly bear this out: this electric plug-in is doing a booming business among the rich. Meanwhile, there are other changes afoot. Faith-based environmental movements such as A Rocha are gaining ground; others, such as Make Way For Monarchs, engage groups of many vocations and political stripes. The coalition of the well-intentioned and action-oriented from finance, faith, and science could prove to be a very powerful one indeed.

But will all of this, in the aggregate, be enough?

Two writers have recently contributed some theorizing about overall social and energy systems and the way they function that may be helpful to us in our slowly unfolding crisis. One is from art historian and energetic social thinker Barry Lord; it's called *Art and Energy* (AAM Press). Briefly, Lord's thesis is that the kind of art a society makes and values is joined at the hip with the kind of energy that society depends on to keep itself going.[1] He traces the various forms of energy we have known as a species throughout our pre-history—our millennia spent in the Pleistocene—and in our recorded history—sexual energy, without which societies can't continue; the energy of the body while hunting and foraging; wood for fire; slaves; wind and water; coal; oil; and "renewables"—and makes some cogent observations about their relationship to art and culture. In his Prologue, he says:

> Everyone knows that all life requires energy. But we rarely consider how dependent art and culture are on the energy that is needed to produce, practice and sustain them. What we fail to see are the usually invisible sources of energy that make our art and culture(s) possible and bring with them fundamental values that we are all constrained to live with (whether we approve of them or not). Coal brought one set of values to all industrialized countries; oil brought a very different set. . . . I may not approve of the culture of consumption that comes with oil . . . but I must use [it] if I want to do anything at all.

Those living within an energy system, says Lord, may disapprove of certain features, but they can't question the system itself. Within the culture of slavery, which lasted at least 5,000 years, nobody wanted to be a slave, but nobody said slavery should be abolished, because what else could keep things going?

Coal, says Lord, produced a culture of production: think about those giant steel mills. Oil and gas, once they were up and running, fostered a culture of consumption. Lord cites "the widespread belief of the 1950s and early '60s in the possibility of continuing indefinitely with unlimited abundance and economic growth, contrasted with the widespread agreement today that both that assumption and the world it predicts are unsustainable." We're in a transition phase, he says: the next culture will be a culture of "stewardship," the energy driving it will be renewables, and the art it produces will be quite different from the art favored by production and consumption cultures.

What are the implications for the way we view both ourselves and the way we live? In brief: in the coal energy culture—a culture of workers and production—you are your job. "I am what I make." In an oil and gas energy culture—a culture of consumption—you are your possessions. "I am what I buy." But in a renewable energy culture, you are what you conserve. "I am what I save and protect." We aren't used to thinking like this, because we can't see where the money will come from. But in a culture of renewables, money will not be the only measure of wealth. Well-being will factor as an economic positive, too.

The second book I'll mention is by anthropologist, classical scholar, and social thinker Ian Morris, whose book *Foragers, Farmers, and Fossil Fuels: How Human Values Evolve* has just appeared from Princeton University Press.[2] Like Barry Lord, Morris is interested in the link between energy-capture systems and the cultural values associated with them, though in his case it's the moral values, not only the aesthetic ones—supposing these can be separated—that concern him. Roughly, his argument runs that each form of energy capture favors values that maximize the chance of survival for those using both that energy system and that package of moral values. Hunter-gatherers show more social egalitarianism, wealth-sharing, and more gender equality than do farmer societies, which subordinate women—men are favored, as they must do the upper-body-strength heavy lifting—tend to practice some form of slavery, and support social hierarchies, with peasants at the low end and kings, religious leaders, and army commanders at the high end. Fossil fuel societies start leveling out gender inequalities—you don't need upper body strength to operate keyboards or push machine buttons—and also social distinctions, though they retain differences in wealth.

The second part of his argument is more pertinent to our subject, for he postulates that each form of energy capture must hit a "hard ceiling," past which expansion is impossible; people must either die out or convert to a new system and a new set of values, often after a "great collapse" that has involved the same five factors: uncontrolled migration, state failure, food shortages, epidemic disease, and "always in the mix, though contributing in unpredictable ways—climate change." Thus, for hunting societies, their way of life is over once there are no longer enough large animals to sustain their numbers. For farmers, arable land is a limiting factor. The five factors of doom combine and augment one another, and people in those periods have a thoroughly miserable time of it, until new societies arise that utilize some not yet exhausted form of energy capture.

And for those who use fossil fuels as their main energy source—that would be us, now—is there also a hard ceiling? Morris says there is. We can't keep pouring carbon into the air—nearly 40 billion tons of CO_2 in 2013 alone—without the consequences being somewhere between "terrible and catastrophic." Past collapses have been grim, he says, but the possibilities for the next big collapse are much grimmer.

We are all joined together globally in ways we have never been joined before, so if we fail, we all fail together: we have "just one chance to get it right." This is not the way we will inevitably go, says he, though it is the way we will inevitably go unless we choose to invent and follow some less hazardous road.

But even if we sidestep the big collapse and keep on expanding at our present rate, we will become so numerous and ubiquitous and densely packed that we will transform both ourselves and our planet in ways we can't begin to imagine. "The 21st century," he says, "shows signs of producing shifts in energy capture and social organization that dwarf anything seen since the evolution of modern humans."

Science fiction? you may say. Or you may say "speculative fiction." For a final straw in the wind, let's turn to what the actual writers of these kinds of stories (and films, and television series, and video games, and graphic novels) have been busying themselves with lately.

A British author called Piers Torday has just come out with a Young Adult book called *The Wild Beyond*. In April, he wrote a piece in the *Guardian* that summarizes the field, and explains the very recent term, "cli-fi":

> "Cli-fi" is a term coined by blogger Dan Bloom to describe fiction dealing with the current and projected effects of climate change. . . . Cli-fi as a new genre has taken off in a big way and is now being studied by universities all over the world. But don't make the mistake of confusing it with sci-fi. If you think stories showing the effects of climate change are still only futuristic fantasies, think again. For example, I would argue that the only truly fantastical element in my books is that the animals talk. To one boy. Other cli-fi elements of my story that are often described as fantastical or dystopian, include the death of nearly all the animals in the world. That's just me painting an extreme picture, right, to make a good story? I wish.
>
> The recent 2014 WWF Living Planet Report revealed that the entire animal population of the planet had in fact halved over the last 40 years. 52% of our wildlife, gone, just like that. Whether through the effects of climate change to the growth in human population to the depredation of natural habitats, the children reading my books now might well find themselves experiencing middle-age in a world without the biodiversity we once took for granted. A world of humans and just a few pigeons, rats and cockroaches scratching around. . . . So, how about the futuristic vision of a planet where previously inhabited areas become too hot and dry to sustain human life? That's standard dystopian world-building fare, surely?
>
> Yes, except that right now, as you read this, super developed and technological California—the eighth largest economy in the world, bigger than Russia—is suffering a record breaking drought. The lowest rainfall since 1885 and enforced water restrictions of up to 25%. They can track every mouse click ever made from Palo Alto apparently, but they can't figure out how to keep the taps running. That's just California—never mind Africa or Australia.
>
> Every effect of climate change in the books—from the rising sea levels of *The Dark*

Wild to the acidic and jelly-fish filled oceans in *The Wild Beyond*, is happening right
now, albeit on a lesser level.[3]

Could cli-fi be a way of educating young people about the dangers that face them,
and helping them to think through the problems and divine solutions?[4] Or will it
become just another part of the "entertainment business"? Time will tell. But if
Barry Lord is right, the outbreak of such fictions is in part a response to the transition now taking place—from the consumer values of oil to the stewardship values of renewables. The material world should no longer be treated as a bottomless
cornucopia of use-and-toss endlessly replaceable mounds of "stuff": supplies are
limited, and must be conserved and treasured.

Can we change our energy system? Can we change it fast enough to avoid
being destroyed by it? Are we clever enough to come up with some viable plans?
Do we have the political will to carry out such plans? Are we capable of thinking
about longer-term issues, or, like the lobster in a pot full of water that's being
brought slowly to the boil, will we fail to realize the danger we're in until it's too
late?

Not that the lobster can do anything about it, once in the pot. But we might.
We're supposed to be smarter than lobsters. We've committed some very stupid
acts over the course of our history, but our stupidity isn't inevitable. Here are three
smart things we've managed to do:

First, despite all those fallout shelters built in suburban backyards during the
Cold War, we haven't yet blown ourselves up with nuclear bombs. Second, thanks
to Rachel Carson's groundbreaking book on pesticides, *Silent Spring*, not all the
birds were killed by DDT in the '50s and '60s. And, third, we managed to stop the
lethal hole in the protective ozone layer that was being caused by the chlorofluo-
rocarbons in refrigerants and spray cans, thus keeping ourselves from being radiated to death. As we head towards the third decade of the 21st century, it's hopeful
to bear in mind that we don't always act in our own worst interests.

"For everything to stay the same, everything has to change," says a character
in Giuseppe di Lampedusa's 1963 novel, *The Leopard*. What do we need to change
to keep our world stable? How do we solve for $X+Y+Z$—X being our civilization's
need for energy, without which it will fall swiftly into anarchy; Y being the finite
nature of the earth's atmosphere, incapable of absorbing infinite amounts of CO_2
without destroying us; and Z being our understandable wish to live full and
happy lives on a healthy planet, followed by future human generations doing the
same. One way of solving this equation is to devise more efficient ways of turning
sunlight into electrical energy. Another way is to make oil itself—and the CO_2 it
emits—part of a cyclical process rather than a linear one. Oil, it seems, does not
have to come out of the ground, and it doesn't have to have pollution as its end
product.

There are many smart people applying themselves to these problems, and many new technologies emerging. On my desk right now is a list of 15 of them. Some take carbon directly out of the air and turn it into other materials, such as cement. Others capture carbon by regenerating degraded tropical rainforests—a fast and cheap method—or sequestering carbon in the soil by means of biochar, which has the added benefit of increasing soil fertility. Some use algae, which can also be used to make biofuel. One makes a carbon-sequestering asphalt. Carbon has been recycled ever since plant life emerged on earth; these technologies and enterprises are enhancing that process.

Meanwhile, courage: homo sapiens sapiens sometimes deserves his double plus for intelligence. Let's hope we are about to start living in one of those times.

Notes

1. See Barry Lord, "The Culture of a World without Oil," https://medium.com/@blord/the-culture-of-a-world-without-oil-130df6e7d63a#.3716wsoc8.

2. Ian Morris, *Foragers, Farmers, and Fossil Fuels: How Human Values Evolve* (Princeton, NJ: Princeton University Press, 2015).

3. Piers Torday, "Why Writing Stories about Climate Change Isn't Fantasy or Sci-fi," *Guardian*, April 21, 2015, http://www.theguardian.com/childrens-books-site/2015/apr/21/climate-change-isnt-fantasy-sci-fi-piers-torday.

4. See Dan Bloom, "Can 'Cli-Fi' Help Keep Our Planet Livable?," https://medium.com/@clificentral/can-cli-fi-help-keep-our-planet-livable-8b053bd4aa35#.alu099x6d.

PART TWO

Energy, Power, and Politics

Building upon our investigation of energy and modernity, the second section of the volume offers a deeper sounding of the relationship between energy and political power. In particular, we examine here how different forms and infrastructures of energy enable different arrangements of political power. Electricity and fuel are revealed to be something more than inert instruments and resources for the work of political regimes. Rather, energy is very often the precondition for political power, especially at a large scale; power grids and pipelines set important conditions of possibility on the various operations and inventions of statecraft and governance. This conditionality points toward a second striking revelation of the work of energy humanities—that energy also influences the content of political ideas and philosophy, although often in an aptly subterranean fashion. In other words, patterns of energy use exert a kind of ideological effect, shaping the terms of public discourse and seeping into the *doxa* of various cultures of expertise. This helps to explain why it is often very difficult for both citizens and experts invested in the status quo to imagine post-carbon, post-nuclear futures given the extent to which popular and expert knowledge incorporates key assumptions of energic materiality, process, and magnitude.

Timothy Mitchell's landmark "Carbon Democracy" project, excerpted here, captures these points in a vivid way. On the one hand, Mitchell shows us how the shifting materiality and circulatory systems of fossil fuels across the nineteenth and twentieth centuries contributed first to the rise of social democracy and then to its decline. "Political possibilities were opened up or narrowed down by different ways of organizing the flow and concentration of energy, and these possibilities were enhanced or limited by arrangements of people, finance, expertise, and violence that were assembled in relationship to the distribution and control of energy." On the other hand, Mitchell argues that the Keynesian theory of "the economy" as an autonomous domain of transactional activity that could constantly and infinitely grow was a form of "petroknowledge" in that it took for granted the existence of oil as an inexhaustible, cheap, sustainable resource. That petroknowledge was, in turn, predicated on a historically specific neo-imperial regime of Anglo-American control over the massive fossil fuel reserves in the Middle East. When that regime began to unravel in the 1970s with the formation of the Organization of the Petroleum Exporting Countries (OPEC), the oil shocks, and the Iranian Revolution, the oily infrastructure of Keynesian economic theory was compromised, thus opening the door to the neoliberal revolution.

Dominic Boyer's discussion of "energopower" broadens Mitchell's historical analysis to address the entanglement of energy systems and political power more generally. Energopower suggests an alternative political genealogy to that proposed by Foucault's influential concept of biopower. Instead of solely focusing on power over life as the core propellant of modern governmentality, Boyer argues that "enablement" through fuel and electricity deserves parallel consideration. He proposes that energopower sets conditions of possibility for biopolitical regimes, even as biopolitical desires and imaginaries reciprocally shape and reinforce energopolitical regimes. In the Anthropocene, as the world faces increasing tension between biopolitical (and ecopolitical) aspirations and energopolitical inertia, he highlights the potential for new energopolitical visions and realities.

Using a comparative historical method, Jean-François Mouhot provocatively juxtaposes modern western civilization's foundational dependence on slavery to contemporary society's reliance on fossil fuels. Mouhot discusses how slavery provided "human engines" to do the work of a modernizing society that had not yet fully developed coal-powered "automated engines" to replace human labor. Even as fossil fuels helped negate the need for slavery, they conjured a new form of dependence on carbon energy magnitudes, promising the widespread availability of a lifestyle that in a pre-carbon era would have necessitated the labor of dozens if not hundreds of bonded humans. "There are many similarities between our dependence on fossil fuels and slave societies' dependence on bonded labor," Mouhot writes. "The human exploitation and suffering resulting (directly) from slavery and (indirectly) from the excessive burning of fossil fuels are now morally comparable, even though they operate in a different way." Both regimes cause harm to others and present severe moral challenges, Mouhot argues, but the impersonality and universality of the damage caused by fossil fuel make it paradoxically more difficult to mobilize against politically than slavery, in which the suffering of human beings is personal and visible in the scars of violence. Thus, he argues, "it is difficult to recognize the potential connection between a coal-fired power plant in Europe and a refugee camp in Africa today and even more complicated to grasp the effects of climate change for future generations."

What proves easier to recognize, Michael Watts argues in his essay "Imperial Oil," is how the development of a multibillion-dollar oil industry in Nigeria "proved to be little more than a nightmare." With 85% of oil revenues accruing to just 1% of the population and with as much as 70% of industry profits simply disappearing, the Nigerian oil industry has become, contrary to its promises of economic prosperity, a juggernaut of corruption, violence, and dispossession held together by an autocratic security state and its international partners. Watts details the elements of this "oil complex" but also the anatomy of a popular insurgency in the Niger delta that has produced its own oil-fed militias, leading Nigeria toward an "enormously unstable and volatile mix of political, economic, and social forces." Watts offers a chilling indictment of the extractive logics of development in Africa today, where "the vast bulk of private transnational investment—the hallmark of success for the neoliberal project—[is] monopolized by a quartet of mining-energy economies."

John McGrath's play (excerpted here) *The Cheviot, the Stag and the Black, Black Oil*—which toured community centers across Scotland in the 1970s to great acclaim—offers a more darkly humorous but no less cautionary approach to understanding the dispossessive impact of the "oil complex" on local communities in resource-rich landscapes. McGrath's sketches position Scotland's oil boom in a historical continuum dating back to the mid-eighteenth-century Clearances that systematically evicted Highland crofters from their lands in order to create the basis for more intensive agricultural production. Now, in the twentieth century, figures like Texas Jim—himself of Scottish ancestors likely forced to become diasporic by changing land tenure—have returned to repeat the process, transforming Scottish lives and labors, making them dance to the tune of his bluegrass guitar:

> So leave your fishing, and leave your soil,
> Come work for me, I want your oil.
>
> Screw your landscape, screw your bays
> I'll screw you in a hundred ways—
>
> Take your partner by the hand
> Tiptoe through the oily sand
>
> Honour your partner, bow real low
> You'll be honouring me in a year or so.

The relationship between energy and power exceeds carbon autocracy and carbon democracy, however. It is all too often forgotten that contemporary statecraft in some parts of the world is as strongly influenced by nuclear power as it is by the politics of carbon fuel. Based on her research on nuclear energy and society in France and uranium-producing African countries, Gabrielle Hecht analyzes different understandings of "nuclearity," arguing that what counts as being nuclear "can never be defined in simple, clear-cut, scientific terms. Rather, nuclearity is a technopolitical spectrum that shifts in time and space. It is a historical and geographical condition, *as well as* a scientific and technological one." Hecht makes of nuclearity an ontological question—one that might reciprocally challenge the easy use of terms such as "carbon" and "petro" too—and she shows how "ontologies of nuclearity are ever-shifting" given that massive new investments in nuclear energy continue to be put forward in some scientific and political quarters as the only reasonable path toward rapid decarbonization of the global economy. The Fukushima Daiichi tragedy took some of the luster off of the so-called nuclear renaissance movement, but it would be mistaken to think that the ontology of nuclearity has been resolved once and for all. Hecht insists that we must think nuclear ontology in the plural, and her argument has been proven once more over recent political arguments over whether Iranian nuclearity means peaceful development or weaponized threat.

In her essay "A Dark Art," Gökçe Günel looks at the politics surrounding another proposed salvational technology, carbon capture and storage (CCS), whose proponents allege that it can maintain a global carbon energy system while drastically reducing greenhouse gas emissions. Günel focuses on the controversial effort to make CCS projects eligible for funding through the UN's Clean Development Mechanism (CDM) from COP11 through to COP17. She highlights the "constructive ambiguity" CCS lobbyists sought to write into CDM policy-making sessions to manage liability concerns and to allow CCS projects to be connected to enhanced oil recovery (EOR) projects (pumping captured carbon dioxide into oil reservoirs to increase output). Günel also notes how the CCS debates activated North/South tensions concerning responsibility for climate change remediation. Countries such as China, India, and more obviously Saudi Arabia were especially vocal in their support of CDM funding for projects that seemed to promise a solution to global warming with no need to wean themselves off fossil fuel production and consumption.

Sheena Wilson similarly investigates recent efforts to reimagine and remoralize fossil fuels through her analysis of the Canadian Ethical Oil billboard and television campaign in 2011. The campaign offered images contrasting "conflict oil" from the Middle East (depicting women being stoned to death) with "ethical oil" produced by the Canadian tar sands. Wilson draws attention to the tropes of saving Muslim women that have been deployed to justify military interventions in the Middle East as well. Those interventions—whose own logic was more often than not to secure petroleum resources—intersect aesthetically and narratologically with the new geopolitics imagined by North American unconventional oil producers and their allies, in which greater energy security, justice, and equity can be won through the development of resources like tar sands and fracking by more ethical liberal-democratic societies. Wilson concludes, "the Ethical Oil campaign is written into an accepted history of petropolitics that justified the suppression of Other peoples (domestically and internationally) and the invasion of foreign nations—for the purpose of gaining control of petroleum resources—by invoking the national myth of Canada as a defender of human rights. This billboard implies, through its false logic, that Canadians can choose continued security and freedom not only for Canadian women but also for foreign women by supporting Canadian oil."

Cymene Howe's article "Anthropocenic Ecoauthority" offers a powerful reminder that the politics of veridiction regarding energy futures is by no means limited to contemporary debates over fossil fuels. Working amid intensifying social and political conflicts in the densest area of onshore wind park development anywhere in the world, Howe navigates through the opposing ethical imaginations and "ecoauthorities" offered by indigenous campesinos, fisherfolk, renewable energy developers, and state and federal political officials in Mexico's Isthmus of Tehuantepec. Howe helps us to understand the different ethical and epistemic positions these stakeholders in renewable energy development take and why their frequently incommensurable ideas regarding environmental and social benefits generate friction that slows the process of energy transition. "The competing scales of ecological remedy appear incommensurate," she argues, "because each of them sounds out a different audience and each of them focuses on slowing different sorts of environmental distress and harm." In Howe's account, what is needed is greater cultural awareness, equity, and reconciliation between the parties to Mexican wind development such that local and global imaginaries of climatological and environmental good can be coordinated in the "future of renewability."

We include here a short excerpt from Pope Francis's influential encyclical *Laudato Si'* as a reminder that there are powerful humanist voices outside of the academy and the arts who are starting to make energy and environmental futures into objects of moral reflection and political intervention. The encyclical is in many respects a conservative document, in the true meaning of *conservare* (to preserve or to keep watch). But its message nonetheless resonates strongly with contemporary liberal and radical critics of petrocapitalism who argue that our consumer society is a fuel society to its core, having no mission or agenda other than to accelerate the growth of consumption until a systemic collapse occurs. Pope Francis makes the acceptance of finitude an ethical problem: "environmental deterioration and human and ethical degradation are closely linked. Many people will deny doing anything wrong because distractions constantly dull our consciousness of just how limited and finite our world really is." And he also does not shy away from criticizing the world's secular leaders for failing to act to preserve the welfare of their nations and their shared terrestrial home: "What would induce anyone, at this stage, to hold on to power only to be remembered for their inability to take action when it was urgent and necessary to do so?" It is a striking statement on the urgency of our contemporary environmental condition that one of the oldest and most conservative cultural and political institutions in the world has become an activist for radical change.

The last word should belong to Ken Saro-Wiwa, the Ogoni writer and environmental activist who paid with his life for his political resistance against Royal Dutch Shell and the Nigerian government's devastation of Ogoniland. His short story "Night Ride" draws our attention to the war machines and collateral damage of the oil complex, in addition to the frailty of human lives and the weakness of human character in the face of oil-fueled violence. But Saro-Wiwa's night riders also remind us of the constant potential of remaking our humanity in the struggle toward alternative futures. As Saro-Wiwa wrote shortly before his execution, "I predict that the scene here will be played and replayed by generations yet unborn. Some have already cast themselves in the role of villains, some are tragic victims, some still have a chance to redeem themselves. The choice is for each individual."

❱❱❱ Carbon Democracy

Timothy Mitchell

Fossil fuels helped create both the possibility of twentieth-century democracy and its limits. To understand the limits, I propose to explore what made the emergence of a certain kind of democratic politics possible, the kind I will call carbon democracy. Before turning to the past, however, let me mention some of the contemporary limits I have in mind.

In the wake of the US invasion of Iraq in 2003, one of those limits was widely discussed. A distinctive feature of the Middle East, many said, is the region's lack of democracy. In several of the scholarly accounts, the lack has something to do with oil. Countries that depend upon petroleum resources for a large part of their earnings from exports tend to be less democratic.[1] However, most of those who write about the question of the "rentier state" or the "oil curse," as the problem is known, have little to say about the nature of oil and how it is produced, distributed, and used.[2] They merely discuss the oil rents, the income that accrues after the petroleum is converted into government revenue.[3] So the reasons proposed for the anti-democratic properties of oil—that it gives government the resources to relieve social pressures, buy political support, or repress dissent—have little to do with the ways oil is extracted, processed, shipped, and consumed, the forms of agency and control these processes involve, or the powers of oil as a concentrated source of energy.

Ignoring the properties of oil itself reflects an underlying conception of democracy. This is the conception shared by the American democracy expert who addressed a local council in southern Iraq: "Welcome to your new democracy," he said. "I have met you before. I have met you in Cambodia. I have met you in Russia. I have met you in Nigeria." At which point, we are told, two members of the council walked out (Stewart, 2006, p. 280). It is to see democracy as fundamentally the same everywhere, defined by universal principles that are to be reproduced in every successful instance of democratization, as though democracy occurs only as a carbon copy of itself. If it fails, as it seems to in oil states, the reason must be that some universal element is missing or malfunctioning.

Failing to follow the oil itself, accounts of the oil curse diagnose it as a malady located within only one set of nodes of the networks through which oil flows and is converted into energy, profits, and political power—in the decision-making organs of individual producer states. Its aetiology involves isolating the symptoms found in producer states that are not found in non-oil states. But what if democracies have not been carbon copies, but carbon-based? Are they tied in specific

Economy and Society 38, no. 3 (August 2009): 399–432 [excerpt, 399–409, 415–32]

ways to the history of carbon fuels? Can we follow the carbon itself, the oil, so as to connect the problem afflicting oil-producing states to other limits of carbon democracy?

The leading industrialized countries are also oil states. Without the energy they derive from oil their current forms of political and economic life would not exist. Their citizens have developed ways of eating, travelling, housing themselves, and consuming other goods and services that require very large amounts of energy from oil and other fossil fuels. These ways of life are not sustainable, and they now face the twin crises that will end them: although calculating reserves of fossil fuels is a political process involving rival calculative techniques, there is substantial evidence that those reserves are running out;[4] and in the process of using them up we have taken carbon that was previously stored underground and placed it in the atmosphere, where it is causing increases in global temperatures that may lead to catastrophic climate change (IPCC, 2007).[5] A larger limit that oil represents for democracy is that the political machinery that emerged to govern the age of fossil fuels may be incapable of addressing the events that will end it.

To follow the carbon does not mean substituting a materialist account for the idealist schemes of the democracy experts, or tracing political outcomes back to the forms of energy that determine them—as though the powers of carbon were transmitted unchanged from the oil well or coalface to the hands of those who control the state. The carbon itself must be transformed, beginning with the work done by those who bring it out of the ground. The transformations involve establishing connections and building alliances—connections and alliances that do not respect any divide between material and ideal, economic and political, natural and social, human and non-human, or violence and representation. The connections make it possible to translate one set of resources and powers into another. Understanding the relations between fossil fuels and democracy requires tracing how these connections are built, the vulnerabilities and opportunities they create, and the narrow points of passage where control is particularly effective.[6] Political possibilities were opened up or narrowed down by different ways of organizing the flow and concentration of energy, and these possibilities were enhanced or limited by arrangements of people, finance, expertise, and violence that were assembled in relationship to the distribution and control of energy.

Buried Sunshine

Like mass democracy, fossil fuels are a relatively recent phenomenon. The histories of the two kinds of forces have been connected in several ways. This article traces four sets of connections, the first two concerned with coal and the rise of mass politics in the late nineteenth and early twentieth centuries, the second two with oil and organizing limits to democratic politics in the mid-twentieth century.

The first connection is that fossil fuel allowed the reorganization of energy systems that made possible, in conjunction with other changes, the novel forms of collective life out of which late nineteenth-century mass politics developed. Until 200 years ago, the energy needed to sustain human existence came almost entirely from renewable sources, which obtain their force from the sun. Solar energy was converted into grain and other crops to provide fuel for humans, into grasslands to raise animals for labour and further human fuel, into woodlands to provide firewood, and into wind and water power to drive transportation and machinery (Sieferle, 2001).

For most of the world, the capture of solar radiation in replenishable forms continued to be the main source of energy until perhaps the mid-twentieth century.[7] From around 1800, however, these renewable sources were steadily replaced with highly concentrated stores of buried solar energy, the deposits of carbon laid down 150 to 350 million years ago when the decay of peat-bog forests and of marine organisms in particular oxygen-deficient environments converted biomass into the relatively rare but extraordinarily potent deposits of coal and oil.[8]

The earth's stock of this "capital bequeathed to mankind by other living beings," as Sartre (1977, p. 154) once described it, will be exhausted in a remarkably short period—most of it, by some calculations, in the 100 years between 1950 and 2050 (Aleklett and Campbell, 2003; Deffeyes, 2005).[9] To give an idea of the concentration of energy we will be exhausting, compared to the plant-based and other forms of captured solar energy that preceded the hydrocarbon age: a single litre of petrol used today needed about twenty-five metric tons of ancient marine life as precursor material, and organic matter the equivalent of the earth's entire production of plant and animal life for 400 years was required to produce the fossil fuels we burn in a single year (1997 figures from Dukes, 2003; Haberl, 2006).

Compared to these concentrated hydrocarbon stores, solar radiation is a weak form of energy. However, it is very widely distributed. Historically its use encouraged relatively dispersed forms of human settlement—along rivers, close to pastureland, and within reach of large reserves of land set aside as woods to provide fuel. The switch to coal over the last two centuries enabled the concentration of populations in cities, in part because it freed urban populations from the need for adjacent pastures and woods. In Great Britain, the substitution of wood by coal created a quantity of energy that would have required forests many times the size of existing wooded areas if energy had still depended on solar radiation. By the 1820s, coal "freed," as it were, an area of land equivalent to the total surface area of the country. By the 1840s, coal was providing energy that to obtain from timber would have required forests covering twice the country's area, double that amount by the 1860s, and double again by the 1890s (Sieferle, 2001; Pomeranz, 2000). Thanks to coal, Great Britain, the United States, Germany, and other coal-producing regions could be catapulted into a new "energetic metabolism," based on cities and large-scale manufacturing.[10]

We associate industrialization with the growth of cities, but it was equally an agrarian phenomenon—and a colonial one. Production on a mass scale required access to large new territories for growing crops, both to supply the food on which the growth of cities and manufacturing depended and to produce industrial raw materials, especially cotton. By freeing land previously reserved as woodland for the supply of fuel, fossil energy contributed to this agrarian transformation. As Pomeranz (2000) argues, the switch to coal in north-west Europe interacted with another land-releasing factor, the acquisition of colonial territories. Colonies in the New World provided the land to grow industrial crops. They also generated a direct and indirect demand for European manufacturing, by creating populations of enslaved Africans who were prevented from producing for their own needs. Europe now controlled surplus land that could be used to produce agricultural goods in quantities that, together with arrangements of the slave plantation, allowed the development of coal-based mass production, centred in cities. This relationship between coal, colonization, and industrialization points to the first set of connections between fossil fuels and democracy. Limited forms of representative government had developed in parts of Europe and its settler colonies in the eighteenth and nineteenth centuries. From the 1870s, however, the emergence of mass political movements and organized political parties shaped the period that Eric Hobsbawm calls both "the age of democratization" and "the age of empire."[11] The mobilization of new political forces depended upon the concentration of population in cities and in manufacturing, enabled in part by the control of colonized territories and enslaved labour forces, but equally associated with the forms of mass collective life made possible by organizing the flow of unprecedented concentrations of non-renewable stores of carbon.

Controlling Carbon Channels

Fossil fuels are connected with the mass democracy of the late nineteenth and early twentieth centuries in a second way. Large stores of high-quality coal were discovered and developed in relatively few sites: central and northern England, South Wales, the Ruhr Valley, Upper Silesia, and Appalachia.[12] Most of the world's industrial regions grew above or adjacent to supplies of coal (Pollard, 1981; Rodgers, 1998, p. 45). However, coal was so concentrated in carbon content that it became cost-effective to transport energy overland or on waterways in much greater quantities than timber or other renewable fuel supplies. In Britain, the first Canal Acts were passed to dig waterways for the movement of coal (Jevons, 1865, pp. 87–88). The development of steam transport, whose original function was to serve coal-mining and which in turn was fuelled by coal, facilitated this movement. Large urban and industrial populations could now accumulate at sites that were no longer adjacent to sources of energy. By the end of the nineteenth century, industrialized regions had built networks that moved concentrated carbon stores from the underground coal face to the surface, to railways, to ports, to cities, and to sites of manufacturing and electrical power generation.

Great quantities of energy now flowed along very narrow channels. Large numbers of workers had to be concentrated at the main junctions of these channels. Their position and concentration gave them, at certain moments, a new kind of political power. The power derived not just from the organizations they formed, the ideas they began to share, or the political alliances they built, but from the extraordinary concentrations of carbon energy whose flow they could now slow, disrupt, or cut off. Coal-miners played a leading role in contesting labour regimes and the powers of employers in the labour activism and political mobilization of the 1880s onwards. Between 1881 and 1905, coal-miners in the United States went on strike at a rate about three times the average for workers in all major industries, and double the rate of the next highest industry, tobacco manufacturing. Coal-mining strikes also lasted much longer than strikes in other industries.[13] The same pattern existed in Europe. Podobnik (2006) has documented the wave of industrial action that swept across the world's coal-mining regions in the later nineteenth century and early twentieth century, and again after the First World War.[14]

The militancy of the miners can be attributed in part to the fact that moving carbon stores from the coal seam to the surface created unusually autonomous places and methods of work. The old argument that mining communities enjoyed a special isolation compared to other industrial workers, making their militancy "a kind of colonial revolt against far-removed authority," misrepresents this autonomy (Kerr and Siegel, 1934, p. 192). More recent accounts stress the diversity of mining communities and the complexity of their political engagements with other groups, with mine-owners, and with state authorities (Church, Outram, and Smith, 1991; Fagge, 1996; Harrison, 1978). As Goodrich had argued, "the miner's freedom" was a product not of the geographical isolation of coal mining regions from political authority but of "the very geography of the working places inside a mine" (1925, p. 19). In the traditional room-and-pillar method of mining, a pair of miners worked a section of the coal seam, leaving pillars or walls of coal in place between their own chamber and adjacent chambers to support the roof. They usually made their own decisions about where to cut and how much rock to leave in place to prevent cave-ins (Podobnik, 2006, pp. 82–85). Before the widespread mechanization of mining, Goodrich wrote, "the miner's freedom from supervision is at the opposite extreme from the carefully ordered and regimented work of the modern machine-feeder" (1925, p. 14).[15]

The militancy that formed in these workplaces was typically an effort to defend this autonomy against the threats of mechanization or against the pressure to accept more dangerous work practices, longer working hours, or lower rates of pay. Strikes were effective, not because of mining's colonial isolation, but on the contrary because of the flows of carbon that connected chambers beneath the ground to every factory, office, home, or means of transportation that depended on steam or electric power.

The power of the miner-led strikes appeared unprecedented. In Germany, a wave of coal-mining strikes in early 1889 and again in December of that year shocked the new Kaiser, Wilhelm II, into abandoning Bismarck's hard-line social policy and supporting a programme of labour reforms (Canning, 1996, pp. 130–33). The Kaiser convened an international conference in March 1890 that called for international standards to govern labour in coal-mining, together with limits on the employment of women and children. By a "curious and significant coincidence," as the New York Times reported, on the same day the conference opened in Berlin, "by far the biggest strike in the history of organized labour" was launched by the coal-miners of England and Wales. The number of men, women, and children on strike reached "the bewildering figure of 260,000." With the great manufacturing enterprises of the north about to run out of coal, the press reported, "the possibilities of a gigantic and ruinous labor conflict open before us."[16]

Large coal strikes could trigger wider mobilizations, as with the violent strike that followed the Courrières colliery disaster of 1906 in northern France, which helped provoke a general strike that paralysed Paris.[17] The commonest pattern, however, was for strikes to spread through the interconnected industries of coal-mining, railways, dock workers, and shipping.[18] By the turn of the twentieth century, the vulnerability of these connections made the general strike a new kind of weapon.

A generation earlier, in 1873, Engels had rejected the idea of a general strike as a political instrument, likening it to ineffectual plans for the "holy month," a nationwide suspension of work, that the Chartist movement had preached in the 1840s (Engels, 1939 [1873]). Workers lacked the resources and organization to carry out a general strike, he argued. Were they to acquire such resources and powers of organization, he said, they would already be powerful enough to overthrow the state, so the general strike would be an unnecessary detour.

Thirty years later, Rosa Luxemburg developed an alternative view. After witnessing the wave of strikes that paralysed Russia in the 1905 Revolution, she argued in The Mass Strike (1906) that workers could now organize a revolution without a unified political movement, because isolated economic struggles were connected into a single force. This force, she wrote, "flows now like a broad billow over the whole kingdom, and now divides into a gigantic network of narrow streams" (1925 [1906], ch. 4).[19] Luxemburg's language tried to capture the dispersed yet interconnected power that workers had now acquired. However, her use of a fluvial metaphor missed the fact that it was railways and canals, more than streams and tides, and the coal and coal-based products they carried that assembled workers together into a new kind of political force.

During the First World War, US and British coalfields and railways were placed under the direction of government administrators, and coal and rail workers were in some cases excused conscription and integrated into the war effort. Strikes were reduced, but the critical role of these energy networks became more visible (Corbin, 1981; Reifer, 2004). After the war, from the West Virginia coal strikes of 1919 to the British General Strike of 1926, one can trace the development of the "triple alliance," of mine workers, dockers, and railwaymen, with the power to shut down energy nodes. The dispersed energy systems of solar radiation had never allowed groups of workers this kind of power.

The strikes were not always successful, but the new vulnerability experienced by the owners of mines, railways, and docks, together with the steel mills and other large manufacturing enterprises dependent on coal, had its effects. In 1914, the massacre of striking coal-miners in Ludlow, Colorado, caused a political crisis that threatened the power of the Rockefeller family, which owned the mines (Chernow, 1998, pp. 571–90). Rockefeller hired Mackenzie King, a Harvard-trained political economist who had helped resolve more than forty coal, railway, shipping, and other strikes as Canadian minister of labour, to devise a new method of managing workers (King, 1918, p. 13). King's report on the crisis, *Industry and Humanity: A Study in the Principles Underlying Industrial Reconstruction* (1918), explained the new vulnerability:

> If the recent past has revealed the frightful consequences of industrial strife, do not present developments all over the world afford indications of possibilities infinitely worse? Syndicalism aims at the destruction by force of existing organization, and the transfer of industrial capital from present possessors to syndicates or revolutionary trades unions. This it seeks to accomplish by the "general strike." What might not happen, in America or in England, if upon a few days' or a few weeks' notice, the coal mines were suddenly to shut down, and the railways to stop running! . . . Here is power which, once exercised, would paralyze the . . . nation more effectively than any blockade in time of war. (King, 1918, pp. 494–95)

King's report provided a blueprint for the corporate management of labour. After working as an industrial relations consultant to Rockefeller and other firms, he returned to politics in Canada, where he served as prime minister for 22 years and became the architect of the country's welfare state.[20] The difficult fight against the resources of a labour movement that, for the period of a few decades, could threaten a country's carbon energy networks helped impel the owners of large industrial firms and their political allies into accepting the forms of welfare democracy and universal suffrage that would weaken working-class mobilization.

From Coal to Oil

After the Second World War, the coal-miners of Europe again appeared as the core of a militant threat to corporatist democratic politics. As US planners worked to engineer the post-war political order in Europe, they came up with a new mechanism to defeat the coal-miners: to convert Europe's energy system from one based on coal to one based predominantly on oil. Western Europe had no oilfields, so the additional oil would come from the Middle East.[21] Scarce supplies of steel and construction equipment were shipped from the United States to the Persian Gulf, to build a pipeline from eastern Saudi Arabia to the Mediterranean, to enable a rapid increase in oil supplies to Europe. The diversion of steel and of Marshall Plan funds for this purpose was justified in part by the need to undermine the political power of Europe's coalminers (Forrestal, 1941–49, p. 2005, 1951, vols. 9–10, January 6, 1948; Painter, 1984, p. 361).[22]

Like coal, oil gave workers new kinds of power. Decisive industrial action was organized at Baku in 1905 in the Russian-controlled Caucasus, in the Maracaibo strikes of 1922 and 1936 in Venezuela, in the 1937 Mexican oil strike, and in the 1945–46 strike in Iran. These conflicts were a training ground for later confrontations. A young labour activist in Baku, Joseph Stalin, later said that the advanced organizing skills of the Azeri oil workers and the intensity of their conflict with the oil industrialists gave him an experience that qualified him as "a journeyman for the revolution."[23]

However, the material qualities and physical locations of oil made things different from with coal. Since it comes to the surface driven by underground pressure, either from the water trapped beneath it or the gas above it, oil required a smaller workforce than coal in relation to the quantity of energy produced.[24] Workers remained above ground, under the continuous supervision of managers. Since the carbon occurs in liquid form, pumping stations and pipelines could replace railways as a means of transporting energy from the site of production to the places where it was used or shipped abroad. Pipelines were vulnerable, as we will see, but not as easy to incapacitate through strike actions as were the railways that carried coal. In addition, diesel oil and petrol are lighter than coal and vaporize more easily, and their combustion leaves little residue compared to the burning of coal. For these reasons, as Lewis Mumford noted:

> they could be stowed away easily, in odds and ends of space where coal could not be placed or reached: being fed by gravity or pressure the engine had no need for a stoker. The effect of introducing liquid fuel and of mechanical stokers for coal, in electric steam plants, and on steamships, was to emancipate a race of galley slaves, the stokers. (Mumford, 1934, p. 235)

The relative lightness and fluidity of oil made it feasible to ship it in large quantities across oceans. Historically, very little coal crossed oceans.[25] In 1912, Britain exported one-third of its coal and was responsible for two-thirds of the world's seaborne exported coal. But almost 90% of its exports went to the adjacent regions of Europe and the Mediterranean (Jevons, 1915, pp. 676–84).[26] Over the course of the twentieth century, the proportion of coal exported internationally stabilized at about 15%.[27] By contrast, from the 1920s onwards about 60% to 80% of oil was exported (Podobnik, 2006, p. 79). So much oil was moved across oceans that, by 1970, 60% of world seaborne cargo consisted of oil (Parker, 2001; UNCTAD, 2007, table 5, p. 8).[28]

Compared to carrying coal by rail, moving oil by sea eliminated the labour of coal-heavers and stokers, and thus the power of organized workers to withdraw their labour from a critical point in the energy system. Transoceanic shipping operated beyond the territorial spaces governed by the labour regulations and democratic rights won in the era of widespread coal and railway strikes. In fact, shipping companies could escape the regulation of labour laws all together (as well as the payment of taxes) by resorting to international registry, or so-called "flags of convenience," removing whatever limited powers of labour organizing might have been left.

Unlike railways, ocean shipping was not constrained by the need to run on a network of purpose-built tracks of a certain capacity, layout, and gauge. Oil tankers frequently left port without knowing their final destination. They would steam to a waypoint, and then receive a destination determined by the level of demand in different regions. This flexibility carried risks (in March 1967 it was one of the causes of the world's first giant oil spill, the Torrey Canyon disaster, which helped trigger the emergence of the environmental movement, a later threat to the carbon fuel industry), but it further weakened the powers of local forces that tried to control sites of energy production.[29] If a labour strike, for example, or the nationalization of an industry affected one production site, oil tankers could be quickly re-routed to supply oil from alternative sites. In other words, whereas the movement of coal tended to follow dendritic networks, with branches at each end but a single main channel, creating potential choke points at several junctures, oil flowed along networks that often had the properties of a grid, like an electrical grid, where there is more than one possible path and the flow of energy can switch to avoid blockages or overcome breakdowns.

These changes in the way forms of fossil energy were extracted, transported, and used made energy networks less vulnerable to the political claims of those whose labour kept them running. At the same time, the fluidity and flexibility of oil presented new problems for those who owned or managed the production sites and distribution networks. It was no longer sufficient to control production and distribution in one particular region. Since oil could move easily from one region to another, petroleum companies were always vulnerable to the arrival of cheaper oil from elsewhere. This vulnerability, seldom recognized in accounts of the oil industry, set further limits to the democratizing potential of petroleum.

Market competition destroyed profits and ruined companies and had if possible to be prevented. The difficulty of transporting coal across oceans meant that coal producers faced competition only within their own region. They prevented it either by forming cartels, as in Germany and the United States, or creating new organizations to regulate production, such as the postwar European Coal and Steel Community. In Britain, producers were ruined by competition and taken over by the state.

Oil companies faced similar threats, but on a transoceanic scale. The two world wars helped restrict the supply and movement of oil, but between the wars both domestic firms in the United States, where most world oil was still produced, and the handful of oil companies seeking to control international trade needed a new set of mechanisms to limit the production and distribution of energy. The devices they developed included government quotas and price controls in the United States, consortium agreements to restrict the development of new oil discoveries in the Middle East, and cartel arrangements to govern the worldwide distribution and marketing of oil. These controls shaped the development of the transnational oil corporation, which emerged as a long-distance machinery for maintaining limits to the supply of oil (Mitchell, 2002a). One could think of this development as the formation of what Barry (2006) calls a "technical zone," a set of coordinated but widely dispersed regulations, calculative arrangements, infrastructures, and technical procedures that render certain objects or flows governable.[30]

After the Second World War, new devices were added to this machinery for the production of scarcity. There were two important techniques for transforming post-war carbon energy abundance into a system of limited supplies. The first was the new apparatus of peacetime "national security."[31] The war had given US oil companies the opportunity to reduce or shut down most of their production in the Middle East. In 1943, when Ibn Saud demanded funds to compensate for the loss of oil revenues, the oil companies persuaded Washington to extend Lend Lease loans to the Saudi monarch. These payments for not producing oil were presented as something necessary for America's national security. They marked the start of a long relationship in which Saudi collaboration in restricting the flow of oil was organized as though it were a system for "protecting" the oil against others.

The second method of preventing energy abundance involved the rapid construction of lifestyles in the United States organized around the consumption of extraordinary quantities of energy. In January 1948, James Forrestal, recently appointed as the country's first Secretary of Defense under the new National Security Act, discussed with Brewster Jennings, President of Socony-Vacuum (later renamed Mobil Oil), how "unless we had access to Middle East oil, American motorcar companies would have to design a four-cylinder motorcar sometime within the next five years" (1951, January 6, 1948, p. 2005).[32] In the following years the US car companies helped out by replacing standard six-cylinder engines with the new V8s as the dream of every middle-class family.[33] As Forrestal spoke, the Morris Motor Company in Britain was preparing to challenge the successful four-cylinder

Volkswagen Beetle with the four-cylinder Morris Minor, Citroen to do the same with the two-cylinder 2CV, and the German engine maker BMW with its first post-war passenger car, the one-cylinder Isetta 250. The European vehicles outsold and outlasted most of the badly engineered American models. But the latter helped engineer something larger—carbon-heavy forms of middle-class American life that, combined with new political arrangements in the Middle East, would help the oil companies keep oil scarce enough to allow their profits to thrive.[34]

If the ability of organized workers to disrupt the networks and nodal points of a coal-based energy system shaped the kinds of mass politics that emerged, or threatened to emerge, in the first half of the twentieth century, this post-war re-organization of fossil fuel networks altered the energetics of democracy.

* * *

The Carbon Economy

In a memorable passage in *The General Theory*, John Maynard Keynes explains his novel theory of the economy in terms of bank notes buried in disused coal mines:

> If the Treasury were to fill old bottles with bank notes, bury them at suitable depths in disused coal mines which are then filled up to the surface with town rubbish, and leave it to private enterprise on well-tried principles of laissez-faire to dig the notes up again . . . there need be no more unemployment and, with the help of the repercussions, the real income of the community, and its capital wealth also, would probably become a great deal greater than it actually is. (Keynes, 1936a, p. 129)

British coal production had passed its peak in the 1920s. By the time Keynes wrote *The General Theory*, coal mines were being exhausted at an unprecedented rate.[35] William Stanley Jevons, the author of an earlier revolution in British economic thinking, marginalist theory of the 1870s, had published a book warning of the coming exhaustion of coal reserves (Jevons, 1865).[36] Keynes was reading that book as he published *The General Theory*, and gave a lecture on Jevons in 1936 to the Royal Statistical Society (1936b).[37] It is indicative of the transformation in economic thinking in which Keynes played a role that the exhaustion of coal reserves no longer appeared as a crisis. The management of coal reserves could now be replaced in the mind, and in the textbooks of economics, with reserves of currency. In the era that Keynes's thinking helped to shape, the supply of carbon energy was no longer a practical limit to economic possibility. What mattered was the proper circulation of bank notes.

A fourth set of connections between oil and mid-twentieth-century democratic politics concerns the role of economic expertise and the economy. Like twentieth-century democracy, twentieth-century economic expertise developed in a specific relationship to the hydrocarbon age.

The shaping of Western democratic politics from the 1930s onwards was carried out in part through the application of new kinds of economic expertise: the development and deployment of Keynesian economic knowledge, its expansion into different areas of policy and debate, its increasingly technical nature, and the efforts to claim an increasing variety of topics as subject to determination not by democratic debate but by economic planning and expertise.

The Keynesian and New Deal elaboration of economic knowledge was a response to the threat of populist politics, especially in the wake of the 1929 financial crisis and the labour militancy that accompanied the crisis and re-emerged a decade later. It provided a method of setting limits to democratic practice and maintaining them.

The deployment of expertise requires, and encourages, the making of worlds that it can master. In this case, what had to be made was "the economy." This was an object that no economist or planner prior to the 1930s spoke of or knew to exist. Of course the term "economy" existed prior to the 1930s, but it referred to a process, not a thing. It meant "government" or the proper management of people and resources, as in the phrase "political economy" (Mitchell, 1998, 2005, pp. 126–41, 2008, pp. 447–66). The economy became the central object of democratic politics in the West (paralleled by the emergence of "development" outside the West): an object whose management was the central task of government, and which required the deployment of specialist knowledge.

The peculiar nature of the project of the "national economy" deployed by Keynesian planners and colonial development officers and its relationship to forms of democracy can be seen by comparing it with a rival project, formulated at the same time and destined to overtake it: neo-liberalism. Launched at a colloquium in Paris organized in August of 1938 to discuss the work of Walter Lippmann criticizing the New Deal, as a movement against this new object of planning, the economy, and against planning itself as a method of concentrating and deploying expert knowledge, neo-liberalism proposed an alternative ordering of knowledge, expertise, and political technology that it named "the market" (Denord, 2001). This was not the market of David Ricardo or William Jevons, but a term that began to take on new meanings in the hands of the nascent neo-liberal movement. Drawing on Lippmann's warnings in *The Phantom Public* (1925) and *The Good Society* (1938) about the dangers of public opinion and the need to expand the areas of concern that are reserved to the decisions of experts, neo-liberalism was launched by Hayek and his collaborators as an alternative project to defeat the threat of populist democracy.

The development of neo-liberalism was delayed by the war and the programmes of post-war reconstruction. Its political challenge to the Keynesian consensus got under way a decade later, with the founding of a think tank called the Institute of Economic Affairs in London in 1955. The launch was triggered by the first post-war crisis in the oil-currency system: Britain's attempt to preserve the sterling area as a mechanism of currency regulation, despite losing its control of the hub of that mechanism, the Anglo-Iranian oilfields in Iran. The desperate measures with which London tried to retain the pound's value despite the loss of the oil wells through which this value had been manufactured provided the point of vulnerability where the neo-liberal movement first aimed its weapons.[38]

Larger connections can be drawn between the assembling of "the economy" and the transition from a coal-based to a predominantly oil-based energy system. The conception of the economy depended upon abundant and low-cost energy supplies, making post-war Keynesian economics a form of "petroknowledge." The economy was conceived in a particular way. It was not the total of the nation's wealth, something that had proven impossible to calculate. (There seemed no way to avoid continually counting everything twice, for example when wholesale goods were resold as retail.) It was imagined and measured, rather simply, as the phenomenon of bank notes changing hands. Even if it was the same money, every time it changed hands it was measured as part of the economy. The economy was the sum total of those monetary transactions (Mitchell, 2008).

This re-conceptualization defined the main feature of the new object: it could expand without getting physically bigger. Older ways of thinking about wealth were based upon physical processes that suggested limits to growth: the expansion of cities and factories, the colonial enlargement of territory, the accumulation of gold reserves, the growth of population and absorption of migrants, the exploitation of new mineral reserves, increasing volumes of trade in commodities. All these were spatial and material processes that had physical limits. By the 1930s, many of those limits seemed to be approaching: population growth in the West was levelling off, the colonial expansion of the United States and the European imperial powers had ended and was threatened with reversal, coal mines were being exhausted, and agriculture and industry were facing gluts of overproduction. The economy, on the other hand, measured by the new calculative device of national income accounting, had no obvious limit. National income, later renamed the gross national product, was a measure not of the accumulation of wealth but of the speed and frequency with which paper money changed hands. It could grow without any problem of physical or territorial limits.

Oil contributed to the new conception of the economy as an object that could grow without limit in two ways. First, oil declined continuously in price. Adjusting for inflation, the price of a barrel of oil in 1970 was one-third of what it sold for in 1920.[39] So, although increasing quantities of energy were consumed, the cost of energy did not appear to represent a limit to growth. Second, thanks to its relative abundance and the ease of shipping it across oceans, oil could be treated as something inexhaustible. Its cost included no calculation for the exhaustion of reserves. The growth of the economy, measured in terms of GNP, had no need to account for the depletion of energy resources. The leading contributions to the academic formulation of the economy—Keynes's *General Theory* (1936a), Hicks's *Value and Capital* (1939), Samuelson's *Foundations* (1947), and the Arrow and Debreu (1954) model—paid no attention to the depletion of energy. The economics of growth of the 1950s and 1960s could conceive of long-run growth as something unrestrained by the availability of energy (Heal and Dasgupta, 1979, p. 1). Moreover, the costs of air pollution, environmental disaster, climate change, and other negative consequences of using fossil fuels were not deducted from the measurement of GNP. Since the measurement of the economy made no distinction between beneficial and harmful costs, the increased expenditure required to deal with the damage caused by fossil fuels appeared as an addition rather than an impediment to growth (Daly, 1991). In all these ways, the availability and supply of oil contributed to the shaping of the economy and its growth as the new object of mid-twentieth-century politics.

The oil wells and pipelines of the Middle East and the political arrangements that were built with them helped make possible the idea of the Keynesian economy and the forms of democracy in which it played a central part.

The 1967–74 Reorganization

With all this in mind, we can turn briefly to the 1967–74 dollar–oil crisis, a pivotal episode in the story of post-war carbon democracy. The linked crises of the US dollar and the nationalization of oil in the Middle East brought into play and reconfigured the intersecting elements of carbon democracy.

Again, by following the oil one can trace how relations between oil production, the gold standard, the circulation of dollars, and Keynesian economic expertise were all transformed in the crisis, along with the possibilities for democratic politics in the Middle East. Following the balance of payments crisis of the late 1950s, and the pressure created by the accumulation of the unregulated offshore dollar reserves known as Euro-dollars, Washington had introduced oil import quotas to protect the value of the dollar and later tried to support its pegged gold price by interventions in the London gold market. When this scheme collapsed in November 1968, the US tried to transform Bretton Woods into a mechanism that allowed the gold peg to float. In an effort to lower domestic oil prices, Washington removed the controls on oil imports in 1970, but this caused more dollars to flow abroad. By the following year, the US had used up most of its non-gold reserves and only 22% of its currency reserves were backed by gold. When European banks requested payment for their dollars in gold, the US defaulted. Described as "the abandoning of the gold standard," it amounted to a declaration of bankruptcy by the US government.[40]

These developments coincided with the emergence of a politics of "the limits to growth" as an alternative project to that of "the economy" in which the oil companies helped trigger the production of the environment as a rival object of politics (Meadows, Meadows, Randers, and Behrens, 1972; Schumacher, 1973). They did this in part inadvertently, by adopting ways of drilling and transporting oil that led to giant oil spills, around which environmentalists were able to organize. But they also helped produce the environment as a matter of political concern, by changing the way they calculated the world's reserves of oil.

In 1971 the oil companies abruptly abandoned their cornucopian calculations of oil as an almost limitless resource (calculations that had underpinned post-war theories of the economy as an object capable of limitless growth), and began to forecast the end of oil (Bowden, 1985). The recalculations were needed to deal with the threat posed by the new Ba'athist government of Iraq, which was developing the first major oil production in the region independent of any Western oil company. When the oil majors tried to punish Iraq by cutting their own production in the country, Baghdad responded by nationalizing their assets.[41] To dissuade other Gulf states from following Iraq's lead, the oil companies now sought to accommodate or even encourage their demand for an unprecedented increase in the price of oil, a goal already supported by agencies of the US government (Blair, 1976; Bromley, 1991; Oppenheim, 1976–77, pp. 24–57). A doubling or tripling of the price of oil would enable the major oil companies to survive the transition to a much lower share of Middle Eastern oil revenue, and would make it feasible to develop the less accessible, high-cost oilfields of the North Sea and northern Alaska. No model of the economy or its future growth could rationalize such an unprecedented transformation in costs of energy or flows of finance. But if the world was reconfigured as a system of finite resources, rapidly running out, then entirely new calculations became possible.

The need to conserve environmental resources and protect them for the long term also helped with another calculation. For the oil companies, the large increase in oil prices carried a risk. It threatened to make affordable a rival source of energy, nuclear power. However, if the oil companies could force producers of nuclear power to introduce into the price of the energy they sold a payment to cover its long-term environmental effects—the cost of decontaminating reactors when they went out of service and of storing spent fuel for millennia—it would remain more expensive than oil. To promote such calculations, the oil companies joined the effort to frame the environment as a new object of politics, and to define it and calculate it in particular ways. Like the economy, the environment was not simply an external reality principle—against which the oil industry had to contend. It was a set of forces and calculations that rival groups attempted to mobilize.

The role of oil companies in framing the politics of the environment suggests another dimension of the relationship between oil and democracy: compared to the production of coal, oil production has a different way of deploying and distributing expertise. Earlier, I suggested that the democratic militancy of coal-miners could be traced in part to the autonomy that miners exercised at the coal face, especially prior to the large-scale mechanization of production. The autonomy of those who mined the ore placed a significant amount of expertise in their hands. Oil, in contrast, leaves its workers on the surface and distributes more of the expertise of production into the offices of managers and engineers.

This difference extends further. Once mined, coal is ready to use. It may require cleaning and sorting, but it needs no chemical transformation. Oil comes out of the ground in an unusable form, known as crude oil. The crude must be heated in a furnace, separated into its different hydrocarbons by fractional distillation, and further processed into useable and uniform products. Initially its main use was in the form of heavy oils (kerosene) for domestic lighting and for lubrication. Gasoline and other lighter by-products of the refining process were treated as waste. To increase their profit margin, oil companies developed large research and development divisions to find uses for these unused by-products, distribution and marketing divisions to promote their use, and political and public relations departments to help build the kinds of societies that would demand them.[42] The major oil companies also collaborated to deny expertise to others, including the coal industry. The 1928 oil cartel, as Nowell (1994) has shown, was actually a broader hydrocarbon cartel, because it was an agreement not just to restrict the production of oil, but to prevent the use of patents that would allow coal companies to move into the production of synthetic oils.

Compared to coal companies, oil companies developed much larger and more extended networks for the production of expertise, which became increasingly involved in making of the wider world a place where its products could thrive. For this reason, the international oil industry was well equipped to meet the challenge of the 1967–74 crisis. Facing both the demand from producer states for a much larger share of oil revenues and the rise of environmentalist challenges to carbon democracy, the major oil companies could draw upon a wide array of resources in public relations, marketing, planning, energy research, international finance, and government relations, all of which could be used to help define the nature of the crisis and promote a particular set of solutions.

One other element in the 1967–74 reorganization deserves mention, given its importance to the question of democracy. As Nitzan and Bichler (2002) have shown, the crisis gave rise to a new set of relations between oil-producing countries and the United States, based on the selling of arms. The export of weapons by American and other manufacturers, previously a relatively small trade financed mostly through US overseas development aid, was transformed into a highly profitable commercial industry.[43] The real value of US arms exports more than doubled between 1967 and 1975, with most of the new market in the Middle East (SIPRI, n.d.). The commercialization of weapons exports was made possible by establishing a series of linkages between the Western import of oil from the Middle East, the flow of dollars to the producer countries, the production of political vulnerabilities and military threats to the further flow of oil, and the use of the petrodollars to purchase arms from the West as protection against those threats. The 1967–74 crisis represented the work of connecting together these elements. The flow of weapons, and related opportunities in construction, consulting, military assistance, and banking, now depended on new levels of militarism, and indeed on a US policy of prolonging and exacerbating local conflicts in the Middle East and on an increasingly disjunctive relationship with the Salafist forms of Islam that had helped defend the mid-twentieth-century oil order against nationalist and popular pressures in the region. The tensions between militarism, Salafism, and armed conflict would render the prospects for a more democratic politics of oil production even weaker in the post-1974 period.[44]

Conclusion

This article has not attempted to draw up a general theory of democracy. General theories of democracy, of which there are many, have no place for oil, except as an exception. Rather, the goal has been to follow closely a particular set of connections that were engineered between carbon fuels and certain kinds of democratic and undemocratic politics.

The forms of democracy that emerged in leading industrialized countries by the middle decades of the twentieth century were enabled and shaped by the extraordinary concentrations of energy obtained from the world's limited stores of hydrocarbons and the socio-technical arrangements required for extracting and distributing that energy. When the production of energy shifted to oil from the Middle East, however, the transformation provided opportunities to weaken rather than extend, in both the West and the Middle East, the forms of carbon-based political mobilization on which the emergence of industrial democracy had depended. Exploring the properties of oil, the networks along which it flowed, and the connections established between flows of energy, finance, and other objects provides a way of understanding how the relations among these different elements and forces were constructed. The relations we have followed connected energy and politics, materials and ideas, humans and non-humans, calculations and the objects of calculation, representations and forms of violence, and the present and the future.

Democratic politics developed, thanks to oil, with a peculiar orientation towards the future: the future was a limitless horizon of growth. This horizon was not some natural reflection of a time of plenty. It was the result of a particular way of organizing expert knowledge and its objects, in terms of a novel world called "the economy." Innovations in methods of calculation, the use of money, the measurement of transactions, and the compiling of national statistics made it possible to imagine the central object of politics as an object that could expand without any form of ultimate material constraint. In the 1967–74 crisis, the relations among these disparate elements were all transformed. Those relations are being transformed again in the present.

In their book *Afflicted Powers* (Retort, 2005), Ian Boal and his colleagues have suggested that understanding the contemporary politics of oil involves the difficult task of bringing together the violence that has been repeatedly deployed to secure arrangements for the production of oil and the forms of spectacle and representation that seem somehow an equally effective aspect of the undemocratic politics of oil—not least the representation of the latest rounds of US militarism as a project to bring democracy to the Middle East.

We can better understand the relationship between spectacle and violence, and between other apparently disparate or discordant features of the politics of oil, by following closely the oil itself; not because the material properties or strategic necessity of oil determine everything else (on the contrary, as I suggested, a lot of hard work went into producing America's "strategic dependence" on the control of Middle Eastern oil, starting with those V8 engines), but because, in tracing the connections that were made between pipelines and pumping stations, refineries and shipping routes, road systems and automobile cultures, dollar flows and economic knowledge, weapons experts and militarism, one discovers how a peculiar set of relations was engineered among oil, violence, finance, expertise, and democracy. These relations are quite different from those of the coal age. If the emergence of the mass politics of the early twentieth century, out of which certain sites and episodes of welfare democracy were achieved, should be understood in relation to coal, the limits of contemporary democratic politics can be traced in relation to oil. The possibility of more democratic futures, in turn, depends on the political tools with which we address the passing of the era of fossil fuel.

Acknowledgments

I am grateful for the help received from Robert Vitalis, Munir Fakher Eldin, Katayoun Shafiee, Andrew Barry, three anonymous referees for *Economy and Society*, and seminar participants at Rutgers, Cornell, Princeton, Binghamton, NYU, SOAS, the University of Illinois at Urbana-Champaign, and the University of California, Berkeley, where parts of this work were presented.

Notes

1. Ross (2001) demonstrates a negative correlation between oil exports as a percentage of GDP and degree of democracy, as estimated in the Polity dataset compiled by Keith Jaggers and Ted Robert Gurr (1995). The data are derived from an evaluation of the institutional procedures by which the candidate for chief executive is selected, elected, and held accountable. The narrowness of this conception of democracy, the unreliability of its measurement, and the assumption that diverse institutional arrangements can be compared and

ranked as differing degrees of a universal principle of democracy are among the many prob-lems presented by the data. Ross is unable to establish reasons for the statistical relation-ship between oil exports and Polity data ranking.

2. The problem of the rentier state was first formulated in Hussein Mahdavy (1970); subsequent contributions on the Middle East include Isam al-Khafaji (2004), Hazem Beb-lawi and Giacomo Luciani (1987), and Ghassan Salamé (1994). For other regions, see Karl (1997), Rosser (2007), Wantchekon (2002), and Yates (1996). Among economists, the problem of natural resources is posed in terms of obstacles to economic growth rather than democracy (Sachs and Warner, 1995). Goldberg, Wibbels, and Mvukiyehe (2008) examine the impact of oil and other mineral wealth on American politics, showing that in states like Texas and Louisiana oil appears to cause lower rates of growth and makes it less likely that opposition groups win elections.

3. An important exception to this tendency to ignore the materiality of oil in discus-sions of the rentier state is Fernando Coronil (1997), where the problem is connected to a wider erasure of nature in understanding the formation of wealth. See also Michael Watt's discussion (2004) of "the oil complex" and the "governable spaces" it requires and Robert Vitalis's examination (2006) of the labour regime and imagemaking that organized the production of oil in Saudi Arabia.

4. See Aleklett and Campbell (2003), Deffeyes (2005), Hubbert (1956), and Robelius (2007); on the history of peak oil estimates, see Bowden (1985) and Dennis (1985).

5. Research by James Hansen and his colleagues (2007) on palaeoclimate data suggests that feedback loops in the melting of ice can cause a rapid acceleration in the loss of ice cover, forcing much more extreme climate change with potentially cataclysmic conse-quences. These findings make even the dire warnings from the IPCC look absurdly optimis-tic. See also Hansen et al. (2008).

6. On the sociology of translation and "obligatory passage points," see Callon (1986). See also Mitchell (2002b, ch. 1).

7. Podobnik (2006, p. 5) calculates that coal replaced wood and other biomass materi-als as the main source of the world's commercial energy as early as the 1880s. But until well into the twentieth century the bulk of this fossil energy was consumed by just a handful of countries.

8. The use of coal (as well as peat, another fossil fuel) was already known in antiquity. But its use was generally restricted to the localities where it was found, and to particular trades that required large quantities of process heat, such as limestone burning and metal-smithing. Shortages of wood, especially in Britain, led to a gradual rise in the use of coal as a general substitute for wood from the sixteenth century (Sieferle, 2001, pp. 78–89).

9. Until recently it was assumed that coal reserves would long outlast oil, with plentiful supplies for hundreds of years. Recent studies suggest that estimates of coal reserves are even less reliable than those for oil, that production in the US, the country with the largest reserves, has already peaked and begun to decline, and that global production may peak as early as 2025 (Zittel and Schindler, 2007). See also Iain Boal's warnings about risks of Mal-thusianism in discussions of oil depletion (Boal and Martinez, 2006).

10. Pollard (1981) documents the link between coal-producing regions (rather than states) and industrial development in Europe. See also Haberl (2006).

11. As Hobsbawm (1989, p. 88) points out, democratization came slowly. In most countries with systems of representative rule, property qualifications and registration procedures restricted the electorate to between 30% and 40% of adult males. Voting rights for the majority of men, and for women, were won only in the twentieth century. For the restrictions in the British case, see Blewett (1965, pp. 27–56).

12. Britain also developed coal resources in the colonies—Natal and the Transvaal, parts of Queensland and New South Wales, and West Bengal. Coal production was also developed on a large scale in the Donets Basin in Russia, in the Illinois and Rocky Mountain basins in the United States, and in China.

13. The strike rate per 1,000 employees for coal-mining and for all industries was, respectively, 134 and 72 (1881–86), 241 and 73.3 (1887–99), 215 and 66.4 (1894–1900), and 208 and 86.9 (1901–5) (Edwards, 1981, p. 106).

14. On the central role of the left in creating democracy in Europe, see Eley (2002). Coal was also associated with labour militancy beyond the main centres of the industrialized world. Quataert (2006) notes the repeated strikes among the workers of the Zonguldak coalfield on the Black Sea coast of Ottoman Anatolia. In Egypt, a strike by the coal-heavers at Port Said, the world's largest coaling station, in April 1882 is recorded as the first collective action by indigenous workers in the country. See Beinin and Lockman (1987, pp. 23, 27–31). However, without the linkages that connected coal to centres of industrial production within the country, these actions could not paralyse local energy systems and gain the political force they enjoyed in northern Europe and the United States.

15. Other discussions of relative autonomy of coal-miners and its loss under mechanization include Dix (1988) and Tilly and Tilly (1998, pp. 43–51).

16. "Labor's cause in Europe: The Kaiser's conference and the English strike," *New York Times*, March 16, 1890, p. 1.

17. In one of the world's worst pit disasters, a gas explosion destroyed the Courrières mine on March 10, 1906, leaving 1,100 dead (Neville, 1978).

18. Silver (2003, fig. 3.3, p. 98) shows that strikes were concentrated in these industries rather than in manufacturing.

19. Georges Sorel (1914 [1908]) offered another contemporary reflection on the new power of the general strike.

20. See Chernow (1998, pp. 581–90) and "William Lyon Mackenzie King," *Dictionary of Canadian Biography Online*, retrieved June 2, 2008, from http://www.biographi.ca.

21. Prior to the development of North Sea oil in the 1970s, the only significant oilfields in Europe were in the Carpathian basin extending from southern Poland to Rumania. See Frank (2007).

22. See also Forrestal (1951, vols. 7–8, May 2, 1947), Citino (2000, 2006), and Block (1977).

23. Stalin's words, from a 1926 speech to railway workers, are cited in Suny (1972, p. 373).

24. As oil is extracted, the pressure in the reservoir drops. Pumps may then be used to bring more oil to the surface or to increase the reservoir pressure by driving water or gas into secondary wells.

25. The main exception was high-quality steam coal from South Wales, essential for the

navy and fast liners, which was shipped to British coaling stations around the world (Jevons, 1915, p. 684). Historically, long-distance coal shipments from Britain could be used as ballast or make-weight and benefited from low rates for back-carriage (Jevons, 1865, p. 227).

26. Charles P. Kindleberger, an economist with the Office of Strategic Services in 1942–44, recalled that, at the outbreak of the Second World War, "coal was regarded as something that didn't move across big bodies of water. It was shipped to British coaling stations but you wouldn't expect international transoceanic trade as a regular thing. And yet when the war came along, and we needed to get coal to Europe we started to move coal out. . . . They were loading it in clam shell buckets on to barges in Puget Sound to go to Europe, a landing in Texas, Portland, Maine, everywhere" (McKinzie, 1973, pp. 108–9). After the Second World War, Japan built a steel industry based on coal and ore shipped from Australia.

27. In 2005, 86% of world coal production was consumed within the country of production (IEA, 2005).

28. The figure refers to ton-miles of crude oil and oil products. In 1970 coal accounted for less than 5% of seaborne trade.

29. The Torrey Canyon, an oil tanker owned by a Bermuda-based subsidiary of the Union Oil Company of California, registered in Liberia, chartered to BP, built in 1959 and rebuilt in 1966 in a Japanese shipyard to increase her size from 66,000 to 119,000 deadweight tons, ran aground off the coast of Cornwall in March 1967. The tanker had set sail without knowing its final destination, and lacked detailed navigation charts for the coast of south-west England. The damage to the coastline and to wildlife was exacerbated by the lack of methods to handle large oil spills. The British government tried to set fire to the oil by having air defence forces bomb it with napalm, creating further damage and inadvertently revealing their possession of the controversial weapon and the inaccuracy of the bombers (more than a quarter of the bombs missed their target) (Sheail, 2007; Cabinet Office, 1967).

30. Other raw materials presented similar problems of regulating global production to prevent competition. None of them, however, were as cheap to produce and transport as oil, or usable in such vast quantities, so they did not generate a need on the same scale for techniques for the production of scarcity.

31. Critical accounts of US international oil policy tend to accept "national security" as the concept with which to frame the history of oil, exposing its true meaning either in terms of the logic of capitalist expansion that confronts an inevitable scarcity of resources, as in Klare (2001, 2008), or in terms of the need for an imperial power to secure the conditions for capitalist expansion, as in Bromley (1991, 2005). Explaining oil in terms of the logics of capitalist expansion leads such accounts to overlook the socio-technical work that must be done to turn the multiple struggles over oil into the singular narrative of the unfolding and stabilizing of the logic of capital. On the ability of the US oil majors to frame their programme in terms of "national security," and the reproduction of this perspective in scholarship, see Vitalis (2006).

32. Forrestal made the same argument at a Cabinet meeting on January 16, 1948 (1951, p. 2026).

33. On the history of American attitudes towards energy, see Nye (1999).

34. Much more could be said about the role of the major oil companies and car manu-

facturers in helping to produce and popularize ways of living based on very high levels of energy consumption. This is not a question of balancing the history of oil production and distribution with an analysis of its consumption, so much as understanding that production involved both producing energy and producing forms of life that were increasingly dependent on that energy.

35. On the peak of British coal production in the 1920s, see Bardi (2007) citing Kirby (1977) and Neuman (1934).

36. Jevons's son, H. Stanley Jevons, returned to the question of the exhaustion of coal reserves in Jevons (1915). He revised his father's estimate of the date of the possible exhaustion of British coal mines from one hundred years to "less than two hundred years" (ibid., pp. 756–57).

37. Originally a lecture delivered on April 21, 1936. The coal question is quoted on p. 517.

38. "Fifty years ago this summer [in 1955] a little book was published in London. It was entitled *The Free Convertibility of Sterling* and was authored by an experienced financial journalist George Winder. In the front, Antony Fisher wrote as director of the Institute of Economic Affairs, 'It [the book] is of vital concern to all those who are interested in their own freedom and the freedom of their country'. Henry Hazlitt gave the book a brilliant review in *Newsweek* on July 25th 1955 and all 2,000 sold out. One can make a very good case for saying this was the start of the free market public policy institute movement that today encircles the world. That little book did so well that Fisher was emboldened to approach a young economist named Ralph Harris. Harris in turn saw a chance to do good by challenging the post-World War II Keynesian consensus" (Blundell, 2005, p. 6).

39. The price of oil fell from $31 a barrel in 1920 to $9 in 1970 (in 2006 prices). The average price per decade also declined, from $18 per barrel in the 1920s, to $15 per barrel in the 1930s and 1940s, $14 per barrel in the 1950s, and $12 per barrel in the 1960s (BP, 2007).

40. Block (1977, pp. 164–202) makes no mention of the oil dimension of the crisis.

41. In 1961 Iraq had reduced the concession area of the foreign-owned Iraq Petroleum Company to the fields currently in production, in the Kirkuk region in the north. In 1969 Iraq had signed an agreement with the Soviet Union to help develop oil production in the south and to build a pipeline to a new refinery on the Persian Gulf. When production from the new field began in April 1972, IPC cut its production at Kirkuk by 50%. The government nationalized IPC in June (Bamberg, 2000, pp. 163–71; Tripp, 2007, p. 200). Algeria had taken 51% control of its French-owned oil industry in February 1971, and Libya began to nationalize foreign-owned oil production in December 1971. Syria had nationalized its small oil industry in 1964.

42. Podobnik (2006) discusses this question of the differing expertise relating to coal and oil.

43. In the 1950s about 95% of US arms exports were financed by government aid; by the 1990s the figure was about 30% (Nitzan and Bichler, 2002, p. 216).

44. Subsequent developments are discussed in Mitchell (2002a).

References

Aleklett, K., and Campbell, C. J. (2003). The peak and decline of world oil and gas production. *Minerals and Energy*, 18(1), 5–20.

Al-Khafaji, I. (2004). *Tormented births: Passages to modernity in Europe and the Middle East.* London: I. B. Tauris.

Arrow, K. J., and Debreu, G. (1954). Existence of an equilibrium for a competitive economy. *Econometrica*, 22(3), 265–90.

Bamberg, J. (2000). *The history of British Petroleum, Vol. 3, British Petroleum and global oil, 1950–1975: The challenge of nationalism.* Cambridge: Cambridge University Press.

Bardi, U. (2007). Peak oil's ancestor: The peak of British coal production in the 1920s. *Newsletter of the Association for the Study of Peak Oil and Gas*, 73 (January), 5–7.

Barry, A. (2006). Technological zones. *European Journal of Social Theory*, 9(2), 239–53.

Beblawi, H., and Luciani, G. (eds.) (1987). *The rentier state.* New York: Croom Helm.

Beinin, J., and Lockman, Z. (1987). *Workers on the Nile: Nationalism, Communism, Islam, and the Egyptian working class, 1882–1954.* Princeton, NJ: Princeton University Press.

Blair, J. M. (1976). *The control of oil.* New York: Pantheon Books.

Blewett, N. (1965). The franchise in the United Kingdom 1885–1918. *Past and Present*, 32 (December), 27–56.

Block, F. (1977). *The origins of international economic disorder: A study of United States international monetary policy from World War II to the present.* Berkeley, CA: University of California Press.

Blundell, J. (2005). *IEA turns 50: Celebrating Fisher meeting Hayek.* Atlas Investor Report.

Boal, I., and Martinez, D. (2006). Feast and famine: A conversation with Iain Boal on scarcity and catastrophe. January 3, 2006. Available at http://www.wordpower. co.uk/platform/Feast-and-Famine-A-Conversation.

Bowden, G. (1985). The social construction of validity in estimates of U.S. crude oil reserves. *Social Studies of Science*, 15(2), 207–40.

BP (2007). *BP statistical review of world energy 2007.* Available at www.bp.com.

Bromley, S. (1991). *American hegemony and world oil.* University Park, PA: Pennsylvania State University Press.

Bromley, S. (2005). The United States and the control of world oil. *Government and Opposition*, 40(2), 225–55.

Cabinet Office (1967). *The Torrey Canyon.* London: HMSO.

Callon, M. (1986). Some elements of a sociology of translation: Domestication of the scallops and the fishermen of St Brieuc Bay. In J. Law (ed.), *Power, action and belief: A new sociology of knowledge?* London: Routledge.

Canning, K. (1996). *Languages of labor and gender: Female factory work in Germany, 1850–1914.* Ithaca, NY: Cornell University Press.

Chernow, R. (1998). *Titan: The life of John D. Rockefeller, Sr.* New York: Random House.

Church, R. A., Outram, Q., and Smith, D. N. (1991). The militancy of British miners, 1893–1986: Interdisciplinary problems and perspectives. *Journal of Interdisciplinary History*, 22(1), 49–66.

Citino, N. J. (2000). Defending the "postwar petroleum order": The US, Britain and the 1954 Saudi-Onassis tanker deal. *Diplomacy and Statecraft*, 11(2), 137–60.

Citino, N. J. (2006). The rise of consumer society: Postwar American oil policies and the modernization of the Middle East. Paper presented at the 14th International Economic History Congress, Helsinki.

Corbin, D. (1981). *Life, work, and rebellion in the coal fields: The southern West Virginia miners, 1880–1922*. Champaign, IL: University of Illinois Press.

Coronil, F. (1997). *The magical state: Nature, money and modernity in Venezuela*. Chicago, IL: University of Chicago Press.

Daly, H. E. (1991). *Steady-state economics* (2nd ed.). Washington, DC: Island Press.

Deffeyes, K. S. (2005). *Beyond oil: The view from Hubbert's Peak*. New York: Farrar, Straus & Giroux.

Dennis, M. A. (1985). Drilling for dollars: The making of US petroleum reserve estimates, 1921–25. *Social Studies of Science*, 15(2), 241–65.

Denord, F. (2001). Aux origines du néolibéralisme en France: Louis Rougier et le Colloque Walter Lippmann de 1938. *Le mouvement social*, 195, 9–34.

Dix, K. (1988). *What's a coal miner to do? The mechanization of coal mining*. Pittsburgh, PA: University of Pittsburgh Press.

Dukes, J. S. (2003). Burning buried sunshine: Human consumption of ancient solar energy. *Climatic Change*, 61(1–2), 33–41.

Edwards, P. K. (1981). *Strikes in the United States, 1881–1974*. New York: St. Martin's Press.

Eley, G. (2002). *Forging democracy: The history of the Left in Europe, 1850–2000*. New York: Oxford University Press.

Engels, F. (1939 [1873]). The Bakunists at work. In K. Marx and F. Engels (eds.), *Revolution in Spain*. London: Lawrence and Wishart (originally published in *Der Volksstaat*, October 31, November 2 and 5, 1873).

Fagge, R. (1996). *Power, culture, and conflict in the coalfields: West Virginia and South Wales, 1900–1922*. Manchester: Manchester University Press.

Forrestal, J. (1941–49). *Papers*. Princeton, NJ: Public Policy Papers Collection, Seeley G. Mudd Manuscript Library.

Forrestal, J. (1951). *The Forrestal diaries* (W. Millis and E. S. Duffield, eds). New York: Viking Press.

Frank, A. F. (2007). *Oil empire: Visions of prosperity in Austrian Galicia*. Cambridge, MA: Harvard University Press.

Goldberg, E., Wibbels, E., and Mvukiyehe, E. (2008). Lessons from strange cases: Democracy, development, and the resource curse in the U.S. states. *Comparative Political Studies*, 41(4–5), 477–514.

Goodrich, C. (1925). *The miner's freedom: A study of the working life in a changing industry*. Boston: Marshall Jones.

Haberl, H. (2006). The global socioeconomic energetic metabolism as a sustainability problem. *Energy*, 31(1), 87–99.

Hansen, J., Sato, M., Kharecha, P., Russell, G., Lea, D. W., and Siddall, M. (2007). Climate change and trace gases. *Philosophical Transactions of the Royal Society A*, 365, 1925–54.

Hansen, J., Sato, M., Kharecha, P., Beerling, D., Berner, R., Masson-Delmotte, V., Pagani, M., Raymo, M., Royer, D.L., and Zachos, J.C. (2008). Target atmospheric CO_2: Where should humanity aim? *Open Atmospheric Science Journal*.

Harrison, R. (ed.) (1978). *Independent collier: The coal miner as archetypal proletarian reconsidered*. New York: St. Martin's.

Heal, G. M., and Dasgupta, P. (1979). *Economic theory and exhaustible resources*. Cambridge: Cambridge University Press.

Hicks, J. (1939). *Value and capital*. Oxford: Oxford University Press.

Hobsbawm, E. (1989). *The age of empire, 1875–1914*. New York: Vintage.

Hubbert, M. K. (1956). *Nuclear energy and fossil fuels*. Publication no. 95. Exploration and Production Research Division, Shell Development.

IEA (2005). *Coal in world in 2005*. International Energy Agency. Available at http://www.iea .org/Textbase/stats/index.asp.

IPCC (2007). *Fourth assessment report*. Geneva: Intergovernmental Panel on Climate Change.

Jaggers, K., and Gurr, T. R. (1995). Tracking democracy's third wave with the Polity III data. *Journal of Peace Research*, 32(4), 469–82.

Jevons, H. S. (1915). *The British coal trade*. London: E. P. Dutton.

Jevons, W. S. (1865). *The coal question: An inquiry concerning the progress of the nation and the probable exhaustion of our coal-mines*. London: Macmillan.

Karl, T. L. (1997). *The paradox of plenty: Oil booms and petro-states*. Berkeley, CA: University of California Press.

Kerr, C., and Siegel, A. (1934). The interindustry propensity to strike: An international comparison. In A. Kornhauser, R. Dubin, and A. M. Ross (eds.), *Industrial conflict*. New York: McGraw-Hill.

Keynes, J. M. (1936a). *The general theory of employment, interest and money*. London: Macmillan.

Keynes, J. M. (1936b). William Stanley Jevons 1835–1882: A centenary allocation on his life and work as economist and statistician. *Journal of the Royal Statistical Society*, 99(3), 516–55.

King, W. L. M. (1918). *Industry and humanity: A study in the principles underlying industrial reconstruction*. Boston: Houghton Mifflin.

Kirby, M. W. (1977). *The British coal mining industry 1870–1946: A political and economic history*. Hamden, CT: Archon Books.

Klare, M. (2001). *Resource wars: The new landscape of global conflict*. New York: Henry Holt.

Klare, M. (2008). *Rising powers, shrinking planet: The new geopolitics of energy*. New York: Metropolitan Books.

Lippmann, W. (1925). *The phantom public*. New York: Harcourt, Brace.

Lippmann, W. (1938). *The good society*. Boston: Little, Brown.

Luxemburg, R. (1925 [1906]). *The mass strike, the political party, and the trade unions*. Detroit: Marxist Educational Society (originally published as *Massenstreik, Partei und Gewerkschaften*, 1906).

Mahdavy, H. (1970). The patterns and problems of economic development in rentier states: The case of Iran. In M. A. Cook (ed.), *Studies in economic history of the Middle East*. Oxford: Oxford University Press.

McKinzie, R. D. (1973). Oral history interview with Charles P. Kindleberger, economist with the Office of Strategic Services, 1942–44, 1945; chief, Division German and Austrian Economic Affairs, Department of State, Washington, 1945–48; and Intelligence Officer, 12th U.S. Army group, 1944–45. Independence, MO: Harry S. Truman Library.

Meadows, D. H., Meadows, D., Randers, J., and Behrens III, W. W. (1972). *The limits to growth: A report for the Club of Rome's project on the predicament of mankind.*

Mitchell, T. (1998). Fixing the economy. *Cultural Studies*, 12(1), 82–101.

Mitchell, T. (2002a). McJihad: Islam in the US global order. *Social Text*, 20(4), 1–18.

Mitchell, T. (2002b). *Rule of experts: Egypt, techno-politics, modernity.* Berkeley, CA: University of California Press.

Mitchell, T. (2005). Economists and the economy in the twentieth century. In G. Steinmetz (ed.), *The politics of method in the human sciences: Positivism and its epistemological other.* Durham, NC: Duke University Press.

Mitchell, T. (2008). Culture and economy. In T. Bennett and J. Frow (eds.), *The Sage handbook of cultural analysis.* Thousand Oaks, CA: Sage.

Mumford, L. (1934). *Technics and civilization.* New York: Harcourt, Brace.

Neuman, A. M. (1934). *The economic organization of the British coal industry.* London: George Routledge.

Neville, R. G. (1978). The Courrières colliery disaster, 1906. *Journal of Contemporary History*, 13(1), 33–52.

Nitzan, J., and Bichler, S. (2002). The weapondollar–petrodollar coalition. In *The global political economy of Israel.* London: Pluto Press.

Nowell, G. (1994). *Mercantile states and the world oil cartel, 1900–1939.* Ithaca, NY: Cornell University Press.

Nye, D. E. (1999). *Consuming power: A social history of American energies.* Cambridge, MA: MIT Press.

Oppenheim, V. H. (1976–77). Why oil prices go up (1): The past: We pushed them. *Foreign Policy*, 25(Winter), 24–57.

Painter, D. S. (1984). Oil and the Marshall Plan. *Business History Review*, 58(3), 359–83.

Parker, R. B. (2001). *The October war: A retrospective.* Gainesville, FL: University Press of Florida.

Podobnik, B. (2006). *Global energy shifts: Fostering sustainability in a turbulent age.* Philadelphia, PA: Temple University Press.

Pollard, S. (1981). *Peaceful conquest: The industrialization of Europe, 1760–1970.* Oxford: Oxford University Press.

Pomeranz, K. (2000). *The great divergence: China, Europe, and the making of the modern world economy.* Princeton, NJ: Princeton University Press.

Quataert, D. (2006). *Miners and the state in the Ottoman Empire: The Zonguldak Coalfield, 1822–1920.* New York: Berghahn.

Reifer, T. E. (2004). Labor, race & empire: Transport workers and transnational empires of trade, production, and finance. In G. G. Gonzalez, R. Fernandez, V. Price, D. Smith, and L. T. Vo (eds.), *Labor versus empire: Race, gender, and migration.* London: Routledge.

Retort (Iain Boal, T. J. Clark, Joseph Matthews, and Michael Watts) (2005). *Afflicted powers: Capital and spectacle in a new age of war.* New York: Verso.

Robelius, F. (2007). "Giant oil fields—the highway to oil: Giant oil fields and their importance for future oil production." Unpublished PhD dissertation, Uppsala University.

Rodgers, D. T. (1998). *Atlantic crossings: Social politics in a progressive age.* Cambridge, MA: Belknap Press.

Ross, M. L. (2001). Does oil hinder democracy? *World Politics*, 53(3), 325–61.

Rosser, A. (2007). Escaping the resource curse: The case of Indonesia. *Journal of Contemporary Asia*, 37(1).

Sachs, J. D., and Warner, A. M. (1995). Natural resource abundance and economic growth. In *Development Discussion Paper No 517a*. Cambridge, MA: Harvard Institute for International Development.

Salamé, G. (ed.) (1994). *Democracy without democrats? The renewal of politics in the Muslim world*. New York: I. B. Tauris.

Samuelson, P. A. (1947). *Foundations of economic analysis*. Cambridge, MA: Harvard University Press.

Sartre, J.-P. (1977). *Critique of dialectical reason, Vol. 1, Theory of practical ensembles*. London: Verso.

Schumacher, E. F. (1973). *Small is beautiful: Economics as if people mattered*. New York: Harper & Row.

Sheail, J. (2007). Torrey Canyon: The political dimension. *Journal of Contemporary History*, 42(3), 485–504.

Sieferle, R. P. (2001). *The subterranean forest: Energy systems and the industrial revolution*. Cambridge: White Horse Press.

Silver, B. J. (2003). *Forces of labor: Workers' movements and globalization since 1870*. Cambridge: Cambridge University Press.

SIPRI (n.d.). *Arms transfers database*. Available at http://armstrade.sipri.org.

Sorel, G. (1914 [1908]). *Reflections on violence* (T. E. Hulme, trans.). New York: B. W. Huebsch (originally published as *Réflexions sur la violence*, 1908).

Stewart, R. (2006). *Occupational hazards: My time governing in Iraq*. London: Picador.

Suny, R. G. (1972). A journeyman for the revolution: Stalin and the labour movement in Baku, June 1907–May 1908. *Soviet Studies*, 23(3), 373–94.

Tilly, C., and Tilly, C. (1998). *Work under capitalism*. Boulder, CO: Westview Press.

Tripp, C. (2007). *A history of Iraq*. Cambridge: Cambridge University Press.

UNCTAD (2007). *Review of maritime transport 2007*. Geneva: United Nations Commission on Trade and Development.

Vitalis, R. (2006). *America's kingdom: Mythmaking on the Saudi oil frontier*. Palo Alto, CA: Stanford University Press.

Wantchekon, L. (2002). Why do resource dependent countries have authoritarian governments? *Journal of African Finance and Economic Development*, 5(2), 17–56.

Yates, D. A. (1996). *The rentier state in Africa: Oil rent dependency and neocolonialism in the Republic of Gabon*. Trenton, NJ: Africa World Press.

Zittel, W., and Schindler, J. (2007). *Coal: Resources and future production*. EWG Paper no. 1/01, July 10, 2007. Available at http://www.energywatchgroup.org/files/ Coalreport.pdf.

))) Energopower: An Introduction

Dominic Boyer

Returning to the Anthropology of Energy

The articles in this special collection explore the intersection of energic forces and fuels with projects of governance and self-governance across the world today. To adopt our language here, we are studying the entanglement of "biopower" (the management of life and population) and "energopower" (the harnessing of electricity and fuel). Since biopower will undoubtedly be the more familiar term, I concentrate this introduction on mapping the origins and analytical method of "energopower." Since "energopower" is a new concept (Boyer 2011), a more extensive definition and discussion is obviously in order. But, first, it is important to position this intervention in the context of previous anthropological engagements with energy. Although a recent flurry of important publications in the "anthropology of energy" (e.g., Behrends et al. 2011; Crate and Nuttall 2009; McNeish and Logan 2012; Nader 2010; Strauss et al. 2013) underscores the field's contemporary vitality, the fact remains that this is not the first, but rather by our count the third, generation of anthropology's engagement of energy.

The first generation was defined principally by the work of Leslie White (1949, 1959), a maverick who granted energy a prominent place in his efforts to resurrect evolutionary theory in anthropology. For White, energy was not one research area among others; it was the conceptual key to understanding everything about human life and history. In what is arguably his most influential work, "Energy and the Evolution of Culture," White outlines a literally universal theory:

> Everything in the universe may be described in terms of energy. Galaxies, stars, molecules and atoms may be regarded as organizations of energy. Living organisms may be looked upon as engines which operate by means of energy derived directly or indirectly from the sun. The civilizations or cultures of mankind, also, may be regarded as a form or organization of energy. (1943:335)

White's key equation was that "cultural development varies directly as the amount of energy per capita per year harnessed and put to work" (1943:338).

At the beginning of cultural history then, with only the energy of its bodies with which to operate, humanity's cultural development remained at a very low level. To satisfy needs and to improve capabilities, both physical and intellectual, humanity sought to harness new sources of energy, first in the domestication of animals and plants, increasing "tremendously the amount of energy per capita

Anthropological Quarterly 87, no. 2 (2014): 309–34

available for culture-building" (White 1943:343). But after several thousand years of steady improvement, humanity once again plateaued until the advancements of the eighteenth and nineteenth centuries allowed for the widespread harnessing of fuel and the invention of engines. Fuel and engines were the technological-cum-sociological basis of modern civilization, White said, and the source of all its great cultural leaps forward from machinery to the arts.

Although in many respects a familiar teleological narrative, White also emphasized a dialectical materialism. Social systems optimized for a given energy regime typically resisted new technologies designed to unleash greater magnitudes of energy. So when "cultural advance" ceased under a given energy regime, "it can be renewed only by tapping some new source of energy and by harnessing it in sufficient magnitude to burst asunder the social system which binds it" (1943:348). Thus, the fuel regime exploded the social apparatus of the agricultural regime in the modernization of society, a story that political economy misrecognized as the struggle between two systems of human production (capitalism and feudalism). White here sows the seeds of an idea that had great, but mostly undeveloped, critical potential: the notion that modern capitalist society was a fuel society to its core; its achievements were fundamentally predicated on fuel consumption such that rampant consumption had become archetypal throughout its culture. Perhaps, White did not pursue the cultural critique because he felt the fuel regime was nearing its end anyway. He wrote of peak oil and peak coal, of dwindling reserves of fuel for a world demanding more and more energy.

A committed but closeted socialist, White doubtless found his energy theory a reassuring rationale for impending revolution. He invoked the second law of thermodynamics frequently in this and later writings to establish a Manichean struggle between the dissipation and concentration of energy in the universe, history, and culture (e.g., "All life is a struggle for free energy" [1987:118]). Life being, in his view, the fundamental struggle to concentrate energy against entropy, it could not cease with the dimming of fuel. Instead, new energies would eventually be harnessed and new social systems and cultural advancements would accrete around them. In the mid-1940s, White sensed an energic revolution in the making, and he was not alone. In the late 1930s, successful experiments in nuclear fission set off a great deal of scientific and popular speculation about what a possible several million-fold increase in the energy at humanity's disposal would mean for all aspects of social life (see, e.g., O'Neill 1940; Potter 1940). Then, just months before the publication of White's landmark article, Enrico Fermi and his Manhattan Project team accomplished, although in secret, the first self-sustaining nuclear chain reaction—the pioneering controlled operationalization of nuclear energy that would pave the way towards "the atomic age." White noted the terrific potentiality of nuclear energy in his famous article: "To be able to harness sub-atomic energy would, without a doubt, create a civilization surpassing sober imagination of today" (1943:351), although he ultimately seemed more attracted to the possibility that the sun would become "directly our chief source of power in the future" (1943:351).

Although prescient in these and other ways, White probably was as much an obstacle to future anthropological research on energy as an inspiration. His politics and personality won him a few friends but many more enemies, enemies who succeeded in isolating him and minimizing his work for many productive years (see Peace 2004). Not only did his universalist model stand in sharp contrast to the Boasian historicism and individualism that dominated American anthropology of the 1940s and 1950s, but White also compounded this contrast with a strident insistence that the evolutionist thinking of figures like Lewis Henry Morgan and Edward B. Tylor was more coherent than and theoretically superior to the work of Boas. It was many years before some rapprochement could be found between White and the Boasians, by which time other less controversial figures like Julian Steward, Marshall Sahlins, and Elman Service had become more central to the movement that had come to be known as "cultural materialism," pushing White's thermodynamic and energetic focus deep into the shadows of mainstream social-cultural anthropology.

There is thus less of a lineage than one might expect between the first generation of anthropological interest in energy and the second generation, which emerged in the 1970s and 1980s. Some Whitean influence remained, particularly through the networks of Michigan anthropology (see, e.g., Adams 1975; Rappaport 1975), but there were concerns as to how to develop White's energy theory further. In his 1977 Presidential Address to the American Anthropological Association, Richard Adams recalled his tutelage under White and the problems that arose from importing physical and chemical conceptualizations of energy into social analysis:

> Basically, we relied on the first and second laws of thermodynamics and followed the example of community ecologists. But the fact that the second law was applicable only to closed systems posed substantial barriers to its use in analyzing human societies. They were, after all, clearly open systems. I think the dilatory development of energy study in social anthropology is in part the result of an inadequate theoretical basis provided by physics and chemistry. (1978:298)

Adams did not view this problem as insurmountable, seeing great promise, for example, in the chemist Ilya Prigogine's "foundation for an understanding of thermodynamic open systems" (1978:302), but he noted that the road ahead required a shift to the "holistic study of larger societies" (1978:301) for which, in his judgment, anthropological theory was as yet poorly equipped. Adams' address ended with a rousing and once again prescient call for anthropologists to take a more active role in the critical investigation of the magnitudes and sources of energy flow with an ultimate aim "to develop more direct dependence on solar energy and to reduce both the use of nonrenewable energy forms and the derangement of the natural ecological processes that construct the energy forms we need for our own nutrition and that of other species" (1978:307).

Yet, by and large, this call was not taken up immediately. Second-generation anthropologists of energy typically did stand in a critical relationship to the energy forms and norms of the developed world. But they were significantly less interested in advancing energetic and thermodynamic cultural theory than they were interested in highlighting the cultural and social impacts of energy development for indigenous peoples (Nordstrom et al. 1977), especially in terms of nuclear power (Robbins 1980), uranium mining (Robbins 1984), and oil extraction (Kruse et al. 1982; Jorgensen 1990). The second generation thus drew energy into wider debates over the rights of indigenous communities, environmental impacts, and resource exploitation, debates that remain central features of the anthropology of energy today (e.g., Barker 1997; Love and Garwood 2011; Powell and Long 2010; Sawyer 2004; Sawyer and Gomez 2012). But, much as Adams feared, the mainstream discipline's interest in the topic of energy remained "dilatory." The AAA flagship journal, American Anthropologist, for example, published scarcely any energy-related research after the mid-1970s (see, however, Whitehead 1987). In part, this gap can be attributed to the "applied" character of many second-generation projects that were designed foremost to help improve relationships among indigenous groups, corporations, and governments as regarded energy development (see, e.g., Jorgensen et al. 1978; Jorgensen 1984). But it also expressed the sense of theoretical impasse to which Adams had alluded as well as the progressive retreat from anthropological holism that was occasioned by the growing subspecialization of anthropological research and the waning capacity of the subfields to communicate and collaborate effectively with each other.

An important exception to this trend was the work of Laura Nader (1980, 1981), whose participation in the late 1970s in the US National Academy of Science's Committee on Nuclear and Alternative Energy Systems (CONAES) served as an inspiration to Adams as to what anthropologists might be capable of accomplishing (Adams 1978:307). Through her research on behalf of CONAES, Nader became interested in the contribution of worldviews, both popular and scientific, to the rejection of ideas of energy conservation and energy transition (Nader and Beckerman 1978). As the first anthropologist to research the energy policy community and energy scientists, Nader became both fascinated and deeply disturbed by "the culture of energy experts" (2004:775). She saw energy policy as "grounded in fear of deleterious change in life-styles and options" despite considerable evidence of "the very wide range of choices of life-styles that is available in any plausible energy future" (2010:241). Energy science meanwhile was plagued by an "inevitability syndrome" (2004:775) that resisted models not predicated upon ever-increasing resource use and energy expenditure:

> Also striking was the omnipresent model of unilinear development (a concept that anthropologists had left in the dust decades earlier), with little general understanding of macro-processes. For example, the recognition that civilizations arise but that they also collapse was missing from the thinking about the present. Prevalent was the nineteenth century belief that technological progress was equivalent to social progress. In such a progressivist frame science too could only rise and not fall or wane. Furthermore, the possibility that experts might be part of the problem was novel to the expert who thought that he stood outside of the problem. (2004:776)

Nader ended up laying much blame at the foot of the specialization and hierarchization of scientific communities in which "standardization and conformity" ruled the roost and prevented much needed creative thinking about energy futures (2004:776). Among anthropologists, Nader remains the only scholar to have researched the expert imagination of energy futures in such depth. Her work also paved the way for more recent ethnographies of energy experts (Mason and Stoilkova 2012) and for political anthropologies of carbon (Coronil 1997) and nuclear (Gusterson 1998; Masco 2006) statecraft.

If there is a lesson to be drawn from the timing of the first two generations of anthropological attention to energy, it is that they have accompanied vulnerable or transitional moments in dominant regimes of energopower. In White's case, his positioning of energy as the key to understanding all human culture (and indeed all existence) accompanied the nuclear energy revolution and its new magnitudes of creative and destructive power. In the case of the second generation, the context was what Nader termed "the energy decade" of the 1970s. The oil shocks of 1973 signaled the end of a certain phase of northern imperial control over carbon fuel (discussed below in more detail). A short-lived political willingness to explore alternative energy sources followed, helping to generate energopolitical fissures and tremors that attracted anthropological attention. The political recommitment to carbon and nuclear energy across the industrialized world in the 1980s blunted the aspirations and urgency of energy research in anthropology as elsewhere in the human sciences. Indeed, between the mid-1980s and the mid-2000s, anthropological research on energy seemed to go into a kind of hiatus (see, however, Traweek 1988; Dawson 1992; Coronil 1997), before displaying signs of renaissance (Henning 2005; Love 2008; Mason 2007; Sawyer 2004; Strauss and Orlove 2003; Wilhite 2005). Over the past several years, a mounting body of intriguing case studies has begun to generate more profound theoretical challenges (e.g., Reyna and Behrends 2008; Sawyer 2007; Winther 2008), including, we modestly hope, the rethinking of political power through energic power that is the subject of this special collection.

But if earlier iterations of the anthropology of energy clustered around moments of energopolitical change, then it is worth reflecting further on what is occurring now that helps to explain the recent and quite rapid accumulation of disciplinary interest in energy. Some 70 years after White's landmark paper, it seems as though energy has at long last become a subject worthy of serious attention in social-cultural anthropology. So, why now?

Inspirations: The Anthropocene and the "Anti-anthropocentric" Turn

The reasons for the recent (re)turn to energy are doubtless several. We should take note, first of all, that the return is occurring across the human sciences. In the last decade, we have witnessed an intensification of research and conversation around many different aspects and consequences of energy use: for example, the ethical considerations of climate change (Chakrabarty 2009; Jamieson 2001, 2011), the formation of climatological expertise (Edwards 2010; Parker 2010), entanglements of carbon fuels and political power (Klieman 2008; Mitchell 2009; Kashi 2008), sustainable and low carbon urban designs (Davis 2010; Wheeler and Beatley 2004), the potentialities of ecological theory (Morton 2010; Taylor 2009), fuel's presence in literature and the arts (Pinkus 2008; PMLA 2011; Wenzel 2006), and critical investigations of "petroculture" in its broadest sense (Szeman 2007, 2013), among many other topics. The seemingly spontaneous and uncoordinated eruption of kindred problematics and analytics across the human sciences suggests that a deeper, perhaps even epochal transformation of commitments has begun. As Dipesh Chakrabarty has put it elegantly, "anthropogenic explanations of climate change spell the collapse of the age-old humanist distinction between natural history and human history" (2009:201). And this collapse has spread well beyond history to challenge the epistemologies of other humanities and humanistic social sciences as well.

One can readily agree with Chakrabarty that the sense of urgency surrounding energy research today connects closely to how anthropogenic climate change and the necessity of energy transitions have become increasingly potent features of scientific and political truth (e.g., Oreskes 2004; United Nations 1992). Over the past two decades, public cultural commentary on signs and implications of climate change has boomed. One should not underestimate, for example, the impact of films like *An Inconvenient Truth* (2006) in animating popular imagination of not only impending global calamity but also possibilities of change and remediation. Even taking into account the works of an active and well-financed industry of climate change skeptics (Boykoff 2011), the mediation of extreme weather and pollution events has become so commonplace that "facticity" of the Anthropocene is becoming an increasingly secure feature of everyday knowledge.

At the same time, one hopes it is obvious that the Anthropocene is more than a discursive phenomenon. However one stands on the truth and accuracy of news representations, evidence of the ecological effects of human use of energy is mounting from new patterns and intensities of temperature, drought, and rainfall across the world to the poison skies of Beijing and the toxic soils of Fukushima. It has been an eye-opening experience to hear rural farmers and ranchers in remote parts of southern Mexico speak of *cambio climatico* as though it were an obvious environmental condition. But, as we have seen in the excruciating serial failures of the UNCCC (United Nations Convention on Climate Change) to limit carbon emissions on a planetary scale, empirical obviousness is no guarantee of serious political attention, let alone action. A panel set up by the UN in 2012 to evaluate its Clean Development Mechanism and the carbon market meant to rein in global emissions concluded that the system had "essentially collapsed" (Clark 2013), with the right to pollute now being so cheap as to offer no disincentive whatsoever. To return to White's glimpse of modernity's carbon core, it may well be the case that trying to fight overheated consumption with consumption-oriented remedies like "carbon markets" risks reinforcing rather than rupturing problematic modes of thought and action.

The sense of urgency intensifies with each "super storm," record-setting heat wave, and endless drought. Apocalyptic imaginaries swirl in the wake of political impasse; the well-founded fear of nuclear winter that I grew up with in late Cold War America has now mutated into nightmares of flooding, burning anthropogenic summer, threatening equivalently to be humanity's last season. These visions are symptomatic of the third generation's energopolitical rupture, parallel to the birth of nuclear energy in the 1940s and to the carbon imperial crisis of the 1970s. Today, we try to navigate the rising waters of certainty that our current course of intensive carbon and nuclear energy use combined with exponential human population growth will lead to unprecedented miseries for human and nonhuman life and probable civilizational collapse. All the modern promises of endless growth, wealth, health, and productive control over "nature" now appear increasingly deluded and bankrupt, designs for Malthusian tragedy.

The crisis may be showing up on the public radar only now, but we have been feeling the ache for decades. Before energy burst onto the scene again, we were already sensing a powerful dis-ease within modernity. In academic life, that sensibility manifested, for example, in a series of conceptual turns in the human sciences that, for lack of a less ugly term, I will call "anti-anthropocentric."[1] First, just at the moment that the dominant energopolitics was recalibrating itself to the rise of OPEC, science and technology studies began to come of age, exploring the contingencies of the production of expert knowledge across space, society, and time. Figures such as Michel Callon (1986) and Bruno Latour emerged as early prophets of the bankruptcy of modern nature/culture oppositions (1993) and of the "actancy" of objects and materials (1988). The parallel rise of Foucauldian analysis of power/knowledge (e.g., 1979) further underscored a lost faith in modern expertise and the rejection of the technocratic imaginaries that had seemed so robust until the oil shocks changed everything.

Subsequently, posthumanism (e.g., Haraway 1991) challenged the human empire over other forms of life, especially human species-ism and the careless manipulation of companion species and companion materials. In the past decade, we have seen a marvelous array of new conceptual movements working through the implications of banners such as "new materialism," "objected oriented ontology," "new realism," "speculative realism," and so on (see, e.g., Bennett 2010; de Landa 2002; Harman 2002; Meillasoux 2008). Such thinking is far from homogeneous. But they have a family resemblance to one another as collaborators in extending non-anthropocentric reasoning in the human sciences. Although concept work typically believes itself undetermined by its socio-environmental circumstances of origin, one cannot help but find the timing of these movements uncanny. They are all taking shape in the deepening shadow of the Anthropocene and intensifying public discourse on environmental degradation and disaster. Whatever more specific intellectual agendas they are pursuing, all of them index the problematic legacies of human-centered thinking and action. And thus, in more or less remote fashion, I believe they offer commentaries on carbon modernity's accelerating death-bringing in the name of enfueling human life.

The anti-anthropocentric turn in the human sciences should not be underestimated in its inspiration and reinforcement of third-generation energy research in anthropology. But it has more and less constructive interventions in my opinion. Posthumanism, for example, is a laudable ethical project both inside and outside the academy. As Timothy Morton has put it, under our contemporary circumstances of ecological risk and decline, there is an obvious need to "change our view from anthropocentrism to ecocentrism" (2007:2). We must force ourselves to confront to what extent our contemporary understanding of "the human" has been achieved "by escaping or repressing not just its animal origins in nature, the biological, and the evolutionary, but more generally by transcending the bonds of materiality and embodiment altogether" (Wolfe 2010:xv). The message here is the necessity of constituting new worldviews and modes of action appropriate to the recognition of ecological interdependency and interresponsibility. In a way, posthumanist ethics complement well the aspiration toward macrostructural "sustainability" in contemporary modernity, the attempt to achieve a (as yet fantastic) modernity that can retain its pleasures and powers without a constant demand for increasing its intensities and magnitudes. Such a project seems entirely salutary.

However, I find much less useful the onslaught of criticism against Kantianism and the phenomenological tradition. There has been an unfortunate tendency in the anti-anthropocentric turn, stoked no doubt by its revolutionary fervor, to dismiss enduring inquiry into human reason and agency as though that inquiry were itself somehow part of the problem rather than a complementary project of truth-finding or, better still, part of the solution to our contemporary challenges. Ian Bogost writes, for example, that "the speculative realists share a common position less than they do a common enemy: the tradition of human access that seeps from the rot of Kant" (2012:4). This seems to me misdirected rancor, both forgetting Kant's own critique of anthropocentrism and overestimating philosophy's capacity to orchestrate ideas, culture, and behavior in the world around it. The Anthropocene is, anyway, a peculiar time in which to become animated by an ideological project of denigrating the significance of human understanding and agency relative to the actancy of objects and materials. It is a bit like the Hegelian master cat toying with its dying prey, trying to resuscitate a tortured object into a subject worthy of domination. In other words, we cannot forget that the postwar period has seen several quantum leaps in "human" intervention into "nature" (synthetic biology, nuclear weaponry, anthropogenic climate change, to name a few of the more obvious). We might wonder about the stakes of theoretically rebalancing the relative powers of agency and actancy under these conditions. Is the more pressing need not to acknowledge the new magnitudes of agency and demand responsibility for them?

The impatient dismissal of the Kantian tradition (which includes heterodox critics from Hegel to Marx, Freud, Nietzsche, and Foucault) of questioning categories and practices of knowledge is thus unhelpful in that it, whether intentionally or not, deflects a species-specific mediating responsibility for the current state of the planet into arguments over whether we "have ever been modern" (Latour 1993) from an ontological standpoint. But, the ultimate ontological status of human agency seems less relevant than the problem of *accountability* for the fact that we have been acting as though we have been modern for a long time. "Logically, then," Chakrabarty writes, "in the era of the Anthropocene, we need the Enlightenment (that is, reason) even more than in the past" (2009:211). I concur on the condition that whatever reason comes next must also incorporate the positive fruits of anti-anthropocentric thinking, namely its deep criticism of how humanity (and, in particular, northern humanity), through its modernist fantasies of command and control over something called "nature," generated new conditions of contingency and vulnerability for the planetary ecology. In other words, if Enlightenment can accelerate the process of taking responsibility for the Anthropocene, then I am all for it. If not, frankly, it will amount to little more than the various distractions and delay tactics already being exercised within carbon energopower.

In this respect, I surely reveal my disciplinary as well as personal sympathies. The anti-anthropocentric turn places anthropology in a somewhat awkward situation. Anthropology's craft (as a fieldwork-based social science focusing on humanity) has traditionally been highly anthropocentric even as it has certainly never been uninterested in matters of materiality, ontology, and the nonhuman (see, e.g., Ingold 2011). Even though it is factual that movements like science and technology studies have exercised a massive impact upon anthropological research over the past two decades, the epistemic core of anthropology remains stubbornly "Kantian" in its praxiological, semiological, and phenomenological attentions to human experience. This is, to repeat, a good thing—in my view, responding to the Anthropocene requires all of these analytical traditions for their expertise in modeling human understanding and behavior. Still, there is now more so than ever a generative discussion, also a good thing, about the limits of anthropocentrism in anthropology. Our latest ensemble of trends mirror those of the human sciences more generally: neomaterialism, multispecies inquiry, a deepening interest in ontology, to name a few. As Hoon Song presciently remarked to me several years ago, "everyone's going Deleuzian," gesturing to the new analytic sensibility taking shape in anthropology. Thinking with Spinoza, Deleuze became fascinated by a matrix of cryptoenergic forces and flows that Brian Massumi has termed "ontopower."[2] This ontopower resembles Foucault's sense of power in certain respects, but it moves beyond his still Kantian interest in discourse and truth. Ontopower is not an epistemic mediator; it is said to be a real force that flows through us, primes us, shocks us, composes us, relates us nominal human beings with the diverse elements of our equivalently ontopowered environments. So, even if we have maybe not all "gone Deleuzian," the discipline's current fascination with Deleuze-inspired topics like affect (see, e.g., Stewart 2008) suggests that to do anthropology today means to be attentive to matters of force, flow, matter, and charge alongside its more traditional coordinates. The competition of Kantian and post-Kantian impetus that typifies anthropological theory today colors our effort to develop an energopolitical analysis commensurate with dominant strategies of biopolitical analysis.

Surfacing Energopolitics in a Biopolitical Era

For the moment, the analytics of political power in anthropology today closely align with Foucault's concept of "biopower" (Foucault 1978, 2002).[3] But the concept itself is somewhat diffuse. In an important article by Rabinow and Rose (2006:199), we learn that Foucault's own conceptual work on biopower was both incomplete and historically specific, that is, a way of denoting the gradual conjoining of two force clusters during the eighteenth and nineteenth centuries in Europe. The first force cluster was the anatomo-politics of the human body, "seeking to maximize its forces and integrate it into efficient systems," while the second was "one of regulatory controls, a biopolitics of the population, focusing on the species body, the body imbued with the mechanisms of life: birth, morbidity, mortality, longevity" (2006:196). Rabinow and Rose themselves suggest a more precise and generalizable formulation of "biopolitics" as "the specific strategies and contestations over problematizations of collective human vitality, morbidity and mortality; over the forms of knowledge, regimes of authority, and practices of intervention that are desirable, legitimate and efficacious" (2006:197). One notes immediately that this formulation, like Foucault's original, contains a certain anthropocentrism, although by no means a programmatic one. Biopower and biopolitics mark a domain of power that specifically concerns the management and control of *human* vitality. Rabinow and Rose, and many other anthropologists besides (e.g., Briggs and Nichter 2009; Fassin 2001; Greenhalgh and Winckler 2005; Petryna 2002), have effectively retooled the biopower concept for twentieth- and twenty-first-century conditions by bringing together the sciences, politics, and economies of life (where "life" itself involves issues as far-ranging as sexuality, reproduction, genomics, population, care of the self, and so on).

Still, life in the Foucauldian analytical imagination (much as in the governmental biopolitics it is modeling) clearly centers on human life. This close anchorage to "the human," even as it denies the authoritative overtures of "humanism," is, I strongly suspect, one reason why the concept has proven so useful in the discipline of anthropology in a time of transition, as a way of modeling power in the nascent "posthuman" era. Yet, given the epistemic and experiential challenges raised by the Anthropocene, "biopower" is clearly conceptually ripe for further reexamination, specifically as to whether its anthropocentrism is adequate to the analytics of the contemporary. Foucault, I think, would approve in that his genealogical method was not designed to inquire into timeless conditions that endure throughout history, but rather to examine "the constitution of the subject across history" (Foucault 1993:202). That is to say, if biopower has become one of our most potent keywords for analyzing political power today, it seems appropriate in the original spirit of Foucault's articulation to subvert it through new genealogical exercises lest we come to believe that "biopower" denotes a transhistorical dimension of modern power and subjectivity.

Our exploration of "energopower" in this special collection is precisely such an exercise of respectful subversion. Biopower continues to capture many of the most salient features of political power today, especially interventions of expertise and authority concerning health, security, and population. But the Anthropocene is challenging contemporary biopower to think beyond narrowly anthropocentric models of intervention and remediation. At the same time, the shockwaves affecting carbon and nuclear energy (from peak oil hypotheses to very real environmental toxicities and nuclear tragedies) have shaken the foundations of the contemporary biopolitical regime in such a way that we find fissures opening and fuel, in some cases quite literally, flowing into the groundwater of bios.

The concepts of energopower and energopolitics are children of this rupture, ways of putting into words the increasing recognition that conditions of life today are increasingly and unstably intertwined with particular infrastructures, magnitudes, and habits of using electricity and fuel. Timothy Mitchell's (2009, 2011) "Carbon Democracy" project has been pathbreaking in this respect. Mitchell, no stranger to biopolitical analysis himself, digs deeply into the history of carbon energy to surface the dependency of modern democratic power upon carbon energy systems; first coal, later oil, and now perhaps we are witnessing a third carbon revolution looming with natural gas and "unconventional" hydrocarbons like tar sands. What Reza Negarestani (2008) imagines as a black or rotting sun within the earth has crucially supplied the intensities of power for modern life and governance and, through these dependencies, has subtly shaped the trajectory and forms of modern political power. Mitchell (2009:407) shows, for example, how the consolidation of social democracy in the late nineteenth century crucially depended on the materialities and infrastructures of coal that allowed miners to choke political power until it acceded to labor reforms. He explains how the biopolitical norms of twentieth-century Keynesian welfarism were authorized by a regime of expertise concerning oil as an inexhaustible resource capable of fueling the endless growth of national economies. But he also shows how that understanding of oil was enabled in turn by a geopolitics of neoimperial control over the Middle East and its subsoil resources (Mitchell 2011:173fn). Once that control was ruptured with the formation of OPEC and the oil shocks of the 1970s, the magic of Keynesian biopolitical thinking was disrupted. "Growth" declined radically across the Global North and different powers of life exploited this crisis to rise to dominance, the politics we normally gloss as "neoliberal."

Mitchell's is an excellent example of energopolitical analysis in action. My second example is drawn from current field research in Mexico (see also Howe 2014). Mexico's state of crisis has become a routine feature of international news media coverage over the past several years. Usually this crisis is presented, in essence, as the biopolitical crisis of a war on drugs gone bad, spawning war machines across the country, perversely devouring life in the name of preserving it. Yet, there is another less publicized crisis in Mexico, the energic crisis of a steep decline in petroleum production (over 25% in the past 7 years) by the giant parastatal Pemex. Mexico is a petrostate par excellence in that Pemex's profits have supplied as much as 40% of the operating revenue of the Mexican federal government in recent years, meaning that every aspect of Mexican biopower also depends critically on the now fading light of the black sun. In the state of Oaxaca, Cymene Howe and I have been studying the attempt to capture a powerful but elusive new energy form, the winds of the Isthmus of Tehuantepec.

Those winds are literally a force to be reckoned with. When *El Norte* blows strongest in the winter months, with routinely tropical storm–like intensity, its 110–20 km/h winds can easily blow over tractor trailers and mangle turbines designed for less fierce and turbulent air. Transnational energy companies are enchanted by the vision of harnessing this perfect storm of energy and by Mexico's high electricity tariffs, which guarantee profits as strong and steady as the wind itself. But, alas, these winds blow across land, much of it organized under the collective stewardship of indigenous *binnizá* (Zapotec) and *ikojts* (Huave) communities, where the political climate is no less fierce. For centuries, the Isthmus has prided itself on negating international, national, and regional projects of control over its people and resources. In response, the area has come to be regarded as a dangerous and murky margin to the exercise of legitimate political authority. The Istmeños are known in Oaxaca City and Mexico City for their ignorance and poverty, for their inclination toward violence, for their manipulation by corrupt political bosses who ritually practice a *liderazgo* (leadership) of impeding state development projects until blackmail demands are satisfied. The managing editor of one of Oaxaca's largest newspapers lamented to us, "we have the most blockades and occupations of any Mexican state but also the fewest schools and the most poverty."

Still, the lure of the wind, "the most perfect jewel" as one government official described it, is too great to give up. Numerous representatives of the federal and regional levels of the state have assured us that wind development is biopolitical, a project to jolt this poor and highly indigenous region into a state of modernizing "progress." And yet in the first several years of serious wind development, we have found that the installation and exercise of institutional biopower (schools, medicine, even factories and prisons) has been little more than an afterthought. Instead, the dominant politics are the politics of transnational investment, grid extension, and electricity provision, a politics that is being orchestrated by another parastatal CFE, the electricity utility, whose biopolitical imagination is rudimentary to say the least. In Ixtepec—known locally as *Tristetepec* for its lack of employment opportunities—there is a plan now to build the first community-owned wind park in Latin America. The Ixtepecan *comuneros* want, above all, funds for social development and sustainable progress, and to become a beacon for community-owned energy in the Western Hemisphere. But neither the government nor the utility supports them—renewable energy is costly and complicated in terms of current grid technology, and communities cannot be forced to pay for infrastructural improvements like new substations and grid extensions in the way that transnational corporations can. CFE, as a "para-state," has thus literally taken it upon itself to overwrite sections of the Mexican constitution and tender law to prevent the community park from happening. As one of the leaders of the community park project growled, "CFE is strong but they are also working against the interests of the Mexican people." In the overlapping of neoliberal and neocolonial modes of abandonment that Mexico knows all too well, biopower in southern Mexico is, for good or for ill, an often forgotten partner in the transactions between old and new regimes of energopower.

Defining Energopower

Another lesson learned from Foucault is that it is sometimes better to offer a provocative placeholder than a definite statement, some rolls of intriguing fabric rather than a dazzling corset. We wish to lure imaginative designers to our workbench. In this spirit, I would describe energopower as an alternative genealogy of modern power, as an analytic method that looks in the walls to find the wiring and ducts and insulation, that listens to the streets to hear the murmur of pipes and sewage, that regards discourse on energy security today as not simply about the management of population (e.g., "biosecurity") but also about the concern that our precious and invisible conduits of fuel and force stay brimming and humming. *Above all, energopower is a genealogy of modern power that rethinks political power through the twin analytics of electricity and fuel. Energeia*, for Aristotle, was being-at-work. In modern physics, power is the rate at which energy is transferred, used, or transformed. We thus regard energopower as a discourse and truth phenomenon to be sure, but as one that searches out signals of the energo-material transferences and transformations incorporated in all other sociopolitical phenomena.

I would reiterate that our intention is *not*, in the tradition of Leslie White, simply to import the truth propositions of physical science into anthropology. Anthropology's unique strengths as a discipline do not lie (solely) in ontology as I have argued above; we may search for truths of an ontological kind, but we cannot ignore the many pathways of mediation that are involved in such truths' processes of epistemic formation and sedimentation. Energopower is thus a concept designed to bridge discourse, materiality, and history—we feel that the concept, and the multi-attentional method (Boyer 2010) that informs the concept, will help undermine impasses among the analytics of modernity and power that come to us through the Marxian and Foucauldian traditions and through more recent iterations of the anti-anthropocentric turn.

But let us anticipate the objection that energopower is ultimately just another modality of biopower. In other words, could all this talk of grids, fuels, and forces simply be mistaking the instruments of biopolitics for the agents of energopolitics? Personally, I agree with Doug Rogers (2014) that the either/or character of that challenge is not particularly fruitful. Our intervention here is rather of the "both/and" variety. We are not proposing that energopower displace biopower in anthropological theory of power anymore than we believe that energopower is a footnote to biopower. Both energopower and biopower offer analytics of modern power lenses through which to comprehend the organization and dynamics of political forces across different scales. Although it is very tempting to make an ontological or historical argument for one kind of power exerting determinative causal force over the other, that is not our mission here. Energopower is not a "kind of power," after all, but rather the conceptual lens for an analytical method of understanding power. Neither Foucault's biopower nor energopower can pretend to model the absolute truth of power because whose truth would that be? As anthropologists, for better or for worse, the profound multiplicity of human languages, knowledges, institutions, and experiences remains the muse and medium of our intellectual practice.

This argument for recognizing energopower is therefore phenomenological, or if you will Hegelian–Marxian, rather than ontological, since it accepts the mediation of cultural-historical experience in the making of epistemic categories and analytic concepts. Foucault located the origins of "biopower" in early modern Europe, but the concept itself was deeply imprinted by the Keynesian welfarism of his time. Likewise, "energopower" is a concept that makes sense now because of a series of events that have drawn our attention to tensions, contradictions even, between governmental institutions and aspirations and energic forces and fuels. Events such as the oil shocks of the 1970s or the more recent recognition of the Anthropocene help us to see new dimensions of power. But that does not mean that they trivialize the dimensions of power we have already recognized nor that these new insights somehow complete our understanding of power.

Because "power" is itself a shifter, a category of volatile reference, the "power" in biopower is *pouvoir*—which in its modal form means only the ability to do something, enablement, forces that allow other forces to happen. Following this logic, it would be impossible to say where the power of energy ends and that of life begins. Put in more concrete terms, the instrumentalities and truth discourses of a modern hospital or school would have little extensional force without electrification, without discourses of endless safe clean power, without highly energy-intensive building materials like cement that literally provide the foundations for biopolitical edifices. In other words, there could have been no consolidation of any regime of modern biopower without a parallel securitization of energy provision and synchronization of energy discourse. In this respect, biopower has always plugged in.

But likewise energopower has always been shaped by particular forms and politics of life. Fuel and electricity, needless to say, are institutionalized with biopolitical missions like "development" (see, e.g., Winther 2008). Biopower and energopower should thus not be viewed as oppositional. What we are exploring in this special collection is the interdependency, or at least the entanglement, of energopolitical and biopolitical regimes across the contemporary world from petroculture in Russia (Rogers 2014) to renewable energy in Mexico (Howe 2014), to waste management (Alexander and Reno 2014) and low carbon (Knox 2014) projects in the United Kingdom, to the experimental urban future of Masdar City in the United Arab Emirates (Günel 2014). In these cases, we see how the promises of Keynesian and neoliberal biopolitics (ever more freedom, ever more luxury, ever more use valuable goods and consumption opportunities) are straining the planet's eco-environmental nexus in ways that rupture not only the image of neoliberal autology but also the image of a self-governing biopower. We encounter more signs of how the dominant carbon and nuclear energopolitical regime is increasingly disrupting and poisoning life across the world. But at the same time, we glimpse fascinating new mutations in that regime's discourse and techniques of governance with the appearance of new anthropocentric and ecocentric biopolitical imaginaries responding to climate change. Energopolitical crisis is generating biopolitical effects and vice versa.

Conclusion: Transitions and Futures

This is, true to our final theme, a time of transitions. If, as Mitchell teaches us, the postwar period and its promises of endless growth were defined above all by a remarkable integration of energic systems (transnational oil and nuclear energy) and biopolitical order (Keynesian welfarism), then since the 1970s the world has experienced an accelerating process of dis-integration in which the seams between bios and energos are increasingly taut and visible. What comes next is abundantly unclear. A potential for revolutionary transition is there, as the renewable energy visionary and German politician Hermann Scheer (2004, 2005) asserted forcefully. Scheer argued not only against carbon and nuclear fuel, but also against the long, inefficient supply chains materially intrinsic to carbon and nuclear energy systems. These lend political authority a centralized energic infrastructure through which to oppress and ignore communities and individuals. Instead, Scheer imagined a future of decentralized renewable energy supply, which would create an unprecedented transformation of modern life and power. Imagine, Scheer mused, if we could truly harness the power of the yellow sun, maintaining all the pleasures and potentialities of energy-intensive modernity without a grid, without pipelines, without carbon emissions. "[O]nly a solar global economy can satisfy the material needs of all mankind and grant us the freedom to re-establish our social and democratic ideals" (2004:32). The solar economy, he promised, would generate new political possibilities by freeing citizenship from centralized, grid-based authority.

Scheer's voice still sounds like a cry in the wilderness. Mainstream media and public culture tend to resist the idea of revolutionary change for good or ill. There, the black sun still oozes, consigning post-carbon energopower and biopower to the realm of heliocentric fantasy. Natural gas and shale oil are instead said to be our true saviors; the US will become the next Saudi Arabia; carbon capture and geoengineering will protect us from global warming; and so on. This is, to reiterate, not only a fascinating moment in which to return to the anthropology of energy, it is a moment of supreme political and cultural urgency and opportunity. Anthropology has long excelled in gathering and analyzing epistemic signals from elsewhere. What we advocate here is that that "elsewhere" be reconsidered not only as encompassing other places, cultures, and times but also the signals of force and fuel surrounding us in the here and now, the humming of enablement. And for those of us who wish not only to analyze the world but also to change it, we can take heart in one thing: alternatives to the anthropocentric status quo are emerging abundantly in the human imagination if not yet in human institutions. The anthropology of energy has much to say not only about the limits and dead-ends of thinking about energy today; it also recognizes the many minds and hands—whether in the dust and wind of Álvaro Obregón (Howe 2014) or in the labs of Manchester (Knox 2014) and Masdar (Günel 2014)—where new alignments of life and energy are being brought into focus and form.

Notes

1. There are a large series of glosses that capture more specific aspects of this turn such as, for example, "ontological," "neomaterialist," "post-constructivist," "post-Kantian," and "anti-correlationist." A major recent conference clustered several of the philosophical and theoretical trends I cite here as a "nonhuman turn" (Center for 21st Century Studies 2012; see http://www.c21uwm.com/nonhumanturn/). While all these adjectives capture certain elements of contemporary debate and discourse very well, I find "anti-anthropocentric" the more compellingly accurate term at the level of the human sciences. For one thing, these literatures share more strongly in a critical project than in any positive project. Many fall well short, for example, of articulating positive biocentric or ecocentric positions. Also, their conceptual and thematic stakes vary: some grapple with metaphysical questions of ontology and materiality, others concentrate on the rights and politics of biotic nonhumanity, still others explore the possibility of "eco-phenomenology." Where they intersect is in the rejection of intellectual traditions that manifestly or latently assume human superiority or centricity as a pillar of their epistemic practice.

2. Massumi has stated, for example, "How can 'we' master what forms us? And reforms us at each instant, before we know it? But that is not to say that we're impotent before onto-power. Quite the contrary, our lives are capacitated by it. We live it; the power of existence that we are expresses it" (as quoted in McKim 2009:11).

3. There will undoubtedly be those who wish to contest this characterization. I do not mean to diminish in any way the importance, for example, of Marxian models of power, which have been resilient in anthropology and which are clearly resurgent in the past few years as well. Rather, I mean to suggest that Foucault's work has served as an especially intuitive and generative theoretical resource for conceptualizing power in ethnographic contexts.

References

Adams, Richard Newbold. 1975. *Energy and Structure: A Theory of Social Power*. Austin: University of Texas Press.

———. 1978. "Man, Energy, and Anthropology: I Can Feel the Heat, But Where's the Light?" *American Anthropologist* 80:297–309.

Alexander, Catherine, and Joshua O. Reno. 2014. "From Biopower to Energopolitics in England's Modern Waste Technology." *Anthropological Quarterly* 87 (2): 335–58.

Barker, Holly. 1997. "Fighting Back: Justice, the Marshall Islands and Neglected Radiation Communities." In Barbara Rose Johnston, ed., *Life and Death Matters: Human Rights and the Environment at the End of the Millennium*, 290–306. London: Alta Mira.

Behrends, Andrea, Stephen Reyna, and Guenther Schlee, eds. 2011. *Crude Domination: The Anthropology of Oil*. New York: Berghahn.

Bennett, Jane. 2010. *Vibrant Matter: A Political Ecology of Things.* Durham, NC: Duke University Press.

Bogost, Ian. 2012. *Alien Phenomenology, or What It's Like to Be a Thing.* Minneapolis: University of Minnesota Press.

Boyer, Dominic. 2010. "On the Ethics and Practice of Contemporary Social Theory: From Crisis Talk to Multiattentional Method." *Dialectical Anthropology* 34 (3): 305–24.

———. 2011. "Energopolitics and the Anthropology of Energy." *Anthropology Newsletter,* May, 5, 7.

Boykoff, Maxwell T. 2011. *Who Speaks for the Climate? Making Sense of Media Reporting on Climate Change.* Cambridge: Cambridge University Press.

Briggs, Charles L., and Mark Nichter. 2009. "Biocommunicability and the Biopolitics of Pandemic Threats." *Medical Anthropology* 28 (3): 189–98.

Callon, Michel. 1986. "Some Elements of a Sociology of Translation: Domestication of the Scallops and the Fishermen of St. Brieuc Bay." In John Law, ed., *Power, Action and Belief: A New Sociology of Knowledge,* 196–233. London: Routledge.

Chakrabarty, Dipesh. 2009. "The Climate of History: Four Theses." *Critical Inquiry* 35 (2): 197–221.

Clark, Pilita. 2013. "EU Emissions Trading Faces Crisis." *Financial Times,* January 21. Accessed from http://www.ft.com/cms/s/0/42e719c0-63f0-11e2-84d8-00144feab49a.html?ftcamp=published_links%2Frss%2Fworld_europe%2Ffeed%2F%2Fproduct#axzz2uRWfZhwt.

Coronil, Fernando. 1997. *The Magical State: Nature, Money, and Modernity in Venezuela.* Chicago: University of Chicago Press.

Crate, Susan, and Mark Nuttall, eds. 2009. *Anthropology and Climate Change.* Walnut Creek, CA: Left Coast Press.

Davis, Mike. 2010. "Who Will Build the Ark?" *New Left Review* 61 (January/February): 29–46.

Dawson, Susan E. 1992. "Navajo Uranium Mining Workers and the Effects of Occupational Illnesses: A Case Study." *Human Organization* 51 (4): 389–97.

de Landa, Manuel. 2002. *Intensive Science and Virtual Philosophy.* New York: Continuum.

Edwards, Paul N. 2010. *A Vast Machine: Computer Models, Climate Data and the Politics of Global Warming.* Cambridge, MA: MIT Press.

Fassin, Didier. 2001. "The Biopolitics of Otherness: Undocumented Immigrants and Racial Discrimination in the French Public Debate." *Anthropology Today* 17 (1): 3–7.

Foucault, Michel. 1978. *The History of Sexuality, Vol. 1: The Will to Knowledge.* London: Penguin.

———. 1979. *Discipline and Punish.* New York: Pantheon.

———. 1993. "About the Beginnings of the Hermenuetics of the Self: Two Lectures at Dartmouth." *Political Theory* 21 (2): 198–227.

———. 2002. *Society Must Be Defended: Lectures at the College de France, 1975–76.* New York: Picador.

Greenhalgh, Susan, and Edwin A. Winckler. 2005. *Governing China's Population: From Leninist to Neoliberal Biopolitics.* Stanford: Stanford University Press.

Guggenheim, David, dir. *An Inconvenient Truth.* Film, 100 min. Lawrence Bender Productions.

Günel, Gökçe. 2014. "Ergos: A New Energy Currency." *Anthropological Quarterly* 87 (2): 359–79.

Gusterson, Hugh. 1998. *Nuclear Rites: A Weapons Laboratory at the End of the Cold War.* Berkeley: University of California Press.

Haraway, Donna. 1991. "A Cyborg Manifesto: Science, Technology, and Socialist-Feminism in the Late Twentieth Century." In Donna Haraway, ed., *Simians, Cyborgs, and Women,* 149–81. London: Free Association Books.

Harman, Graham. 2002. *Tool-Being: Heidegger and the Metaphysics of Objects.* Chicago: Open Court.

Henning, Annette. 2005. "Climate Change and Energy Use." *Anthropology Today* 21 (3): 8–12.

Howe, Cymene. 2014. "Anthropocenic Ecoauthority: The Winds of Oaxaca." *Anthropological Quarterly* 87 (2): 381–404.

Ingold, Tim. 2011. *Being Alive: Essays on Movement, Knowledge and Description.* New York: Routledge.

Jamieson, Dale. 2001. "Climate Change and Global Environmental Justice." In Clark Miller and Paul Edwards, eds., *Changing the Atmosphere: Expert Knowledge and Global Environmental Governance,* 287–307. Cambridge, MA: MIT Press.

———. 2011. "Energy, Ethics and the Transformation of Nature." In Denis Arnold, ed., *The Ethics of Global Climate Change,* 16–37. London: Cambridge University Press.

Jorgensen, Joseph G. 1990. *Oil Age Eskimos.* Berkeley: University of California Press.

———, ed. 1984. *Native Americans and Energy Development, II.* Cambridge, MA: Anthropology Resource Center.

Jorgensen, Joseph G., Richard O. Clemmer, Ronald L. Little, Nancy J. Owens, and Lynn A. Robbins, eds. 1978. *Native Americans and Energy Development.* Cambridge, MA: Anthropology Resource Center.

Kashi, Ed. 2008. *The Curse of the Black Gold.* Michael Watts, ed. New York: Powerhouse Press.

Klieman, Kairn. 2008. "Oil, Politics, and Development in the Formation of a State: The Congolese Petroleum Wars, 1963–68." *International Journal of African Historical Studies* 41 (2): 169–202.

Knox, Hannah. 2014. "Footprints in the City: Models, Materiality, and the Cultural Politics of Climate Change." *Anthropological Quarterly* 87 (2): 405–29.

Kruse, John, Judith Kleinfeld, and Robert Travis. 1982. "Energy Development on Alaska's North Slope: Effects on the Inupiat Population." *Human Organization* 41 (2): 95, 97–106.

Latour, Bruno. 1988. *The Pasteurization of France.* Alan Sheridan and John Law, trans. Cambridge, MA: Harvard University Press.

———. 1993. *We Have Never Been Modern.* Catherine Porter, trans. Cambridge, MA: Harvard University Press.

Love, Thomas. 2008. "Anthropology and the Fossil Fuel Era." *Anthropology Today* 24 (2): 3–4.

Love, Thomas, and Anna Garwood. 2011. "Wind, Sun and Water: Complexities of Alternative Energy Development in Rural Northern Peru." *Rural Society* 20:294–307.

Masco, Joseph. 2006. *Nuclear Borderlands: The Manhattan Project in Post-Cold War New Mexico.* Princeton, NJ: Princeton University Press.

Mason, Arthur. 2007. "The Rise of Consultant Forecasting in Liberalized Natural Gas Markets." *Public Culture* 19 (2): 367–79.

Mason, Arthur, and Maria Stoilkova. 2012. "Corporeality of Consultant Expertise in Arctic Natural Gas Development." *Journal of Northern Studies* 6 (2): 83–96.

McKim, Joel. 2009. "Of Microperception and Micropolitics: An Interview with Brian Massumi, 15 August 2008." *Inflexions* 3:1–19. Accessed from www.inflexions.org.

McNeish, John-Andrew, and Owen Logan, eds. 2012. *Flammable Societies: Studies on the Socio-economics of Oil and Gas.* London: Pluto.

Meillasoux, Quentin. 2008. *After Finitude: An Essay on the Necessity of Contingency.* Ray Brassier, trans. London: Continuum.

Mitchell, Timothy. 2009. "Carbon Democracy." *Economy and Society* 38 (3): 399–432.

———. 2011. *Carbon Democracy: Political Power in the Age of Oil.* New York: Verso.

Morton, Timothy. 2007. *Ecology without Nature: Rethinking Environmental Aesthetics.* Cambridge, MA: Harvard University Press.

———. 2010. *The Ecological Thought.* Cambridge, MA: Harvard University Press.

Nader, Laura. 1980. *Energy Choices in a Democratic Society. A Resource Group Study for the Synthesis Panel of the Committee on Nuclear Alternative Energy Systems for the U.S. National Academy of Sciences.* Washington, DC: National Academy of Sciences.

———. 1981. "Barriers to Thinking New About Energy." *Physics Today* 34 (9): 99–104.

———. 2004. "The Harder Path—Shifting Gears." *Anthropological Quarterly* 77 (4): 771–91.

———, ed. 2010. *The Energy Reader.* Oxford: Wiley-Blackwell.

Nader, Laura, and Stephen Beckerman. 1978. "Energy as It Relates to the Quality and Style of Life." *Annual Review of Energy* 3:1–28.

Negarestani, Reza. 2008. *Cyclonopedia: Complicity with Anonymous Materials.* Melbourne: Re.Press.

Nordstrom, Jean Maxwell, James P. Boggs, Nancy J. Owens, and JoAnn Sootkis. 1977. *The Northern Cheyenne Tribe and Energy Developments in Southeastern Montana. Vol 1: Social and Cultural Investigations.* Lame Deer: Northern Cheyenne Research Project.

O'Neill, John J. 1940. "Enter Atomic Power." *Harper's* 181 (June): 1–10.

Oreskes, Naomi. 2004. "The Scientific Consensus on Climate Change." *Science* 306 (5702): 1686.

Parker, Wendy S. 2010. "Predicting Weather and Climate: Uncertainty, Ensembles and Probability." *Studies in History and Philosophy of Modern Physics* 41:263–72.

Peace, William J. 2004. *Leslie A. White: Evolution and Revolution in Anthropology.* Lincoln: University of Nebraska Press.

Petryna, Adriana. 2002. *Life Exposed: Biological Citizens after Chernobyl.* Princeton, NJ: Princeton University Press.

Pinkus, Karen. 2008. "On Climate, Cars, and Literary Theory." *Technology and Culture* 49 (4): 1002–9.

PMLA (Publications of the Modern Language Association of America). 2011. "Editor's Column: Literature in the Ages of Wood, Tallow, Coal, Whale Oil, Gasoline, Atomic Power, and Other Energy Sources." *PMLA* 126 (2): 305–26.

Potter, Robert D. 1940. "Is Atomic Power at Hand?" *Scientific Monthly* 50 (6): 571–74.

Powell, Dana E., and D. J. Long. 2010. "Landscapes of Power: Renewable Energy Activism in Diné Bikéyah." In Sherry Smith and Brian Frehner, eds., *Indians & Energy: Exploitation and Opportunity in the American Southwest*, 231–62. Santa Fe: SAR Press.

Rabinow, Paul, and Nikolas Rose. 2006. "Biopower Today." *BioSocieties* 1 (2): 195–217.

Rappaport, Roy. 1975. "The Flow of Energy in Agricultural Society." In Solomon Katz, ed., *Biological Anthropology: Readings from Scientific American*, 371–87. San Francisco: W. H. Freeman.

Reyna, Stephen, and Andrea Behrends. 2008. "The Crazy Curse and Crude Domination: Toward an Anthropology of Oil." *Focaal* 52 (1): 3–17.

Robbins, Lynn. 1980. *The Socioeconomic Impacts of the Proposed Skagit Nuclear Power Plant on the Skagit System Cooperative Tribes*. Bellingham: Lord & Associates.

———. 1984. "Energy Developments and the Navajo Nation: An Update." In Joseph Jorgensen, ed., *Native Americans and Energy Development*, 35–48. Cambridge, MA: Anthropology Resource Center.

Rogers, Douglas. 2014. "Energopolitical Russia: Corporation, State and the Rise of Social and Cultural Projects." *Anthropological Quarterly* 87 (2): 431–51.

Sawyer, Suzana. 2004. *Crude Chronicles: Indigenous Politics, Multinational Oil, and Neoliberalism in Ecuador*. Durham, NC: Duke University Press.

———. 2007. "Empire/Multitude—State/Civil Society: Rethinking Topographies of Power through Transnational Connectivity in Ecuador and Beyond." *Social Analysis* 51 (2): 64–85.

Sawyer, Suzana, and Terence Gomez, eds. 2012. *The Politics of Resource Extraction: Indigenous Peoples, Corporations and the State*. London: Palgrave Macmillan.

Scheer, Hermann. 2004. *The Solar Economy: Renewable Energy for a Sustainable Global Future*. London: Earthscan.

———. 2005. *A Solar Manifesto*. London: Earthscan.

Stewart, Kathleen. 2008. *Ordinary Affects*. Durham, NC: Duke University Press.

Strauss, Sarah, and Ben Orlove, eds. 2003. *Weather, Climate, Culture*. Oxford: Berg.

Strauss, Sarah, Thomas Love, and Stephanie Rupp, eds. 2013. *Cultures of Energy*. Walnut Creek, CA: Left Coast Press.

Szeman, Imre. 2007. "System Failure: Oil, Futurity and the Anticipation of Disaster." *South Atlantic Quarterly* 106 (4): 805–23.

———. 2013. "What the Frack? Combustile Water and Other Late Capitalist Novelties." *Radical Philosophy* 177 (January/February). Accessed from http://www.radicalphilosophy.com/commentary/what-the-frack.

Taylor, Bron. 2009. *Dark Green Religion: Nature Spirituality and the Planetary Future*. Berkeley: University of California Press.

Traweek, Sharon. 1988. *Beamtimes and Lifetimes: The World of High Energy Physicists*. Cambridge, MA: Harvard University Press.

United Nations. 1992. *United Nations Framework Convention on Climate Change*. Accessed from http://unfccc.int/resource/docs/convkp/conveng.pdf.

Wenzel, Jennifer. 2006. "Petro-Magic-Realism: Toward a Political Ecology of Nigerian Literature." *Postcolonial Studies* 9 (4): 449–64.

Wheeler, Stephen M., and Timothy Beatley, eds. 2004. *The Sustainable Urban Development Reader*. New York: Routledge.

White, Leslie. 1943. "Energy and the Evolution of Culture." *American Anthropologist* 45 (3): 335–56.

——. 1949. *The Science of Culture: A Study of Man and Civilization.* New York: Farrar, Straus & Giroux.

——. 1959. *The Evolution of Culture: The Development of Civilization to the Fall of Rome.* New York: McGraw-Hill.

——. 1987. *Ethnological Essays.* Beth Dillingham and Robert L. Carniero, eds. Albuquerque: University of New Mexico Press.

Whitehead, John A. 1987. "The Partition of Energy by Social Systems: A Possible Anthropological Tool." *American Anthropologist* 89:686–700.

Wilhite, Harold. 2005. "Why Energy Needs Anthropology." *Anthropology Today* 21 (3): 1–3.

Winther, Tanja. 2008. *The Impact of Electricity: Development, Desires and Dilemmas.* Oxford: Berg.

Wolfe, Cary. 2010. *What Is Posthumanism?* Minneapolis: University of Minnesota Press.

))) Past Connections and Present Similarities in Slave Ownership and Fossil Fuel Usage

Jean-François Mouhot

In order to minimise the risk of the reappearance of slavery on a large scale (a possible consequence of an energy shortage caused by our relentless consumption) or the danger of unbridled climate change (also caused by our extravagant burning of fossil fuels), perhaps we do not need technological fixes or "superhero" politicians but a change in general attitudes. After all, the overwhelming majority of industrialised countries are democracies and elected representatives are supposed to do what their name implies, i.e., represent their citizens. A general change in attitude will only happen if people are convinced, emotionally as well as intellectually, that our relentless use of fossil fuels has become dangerous and morally wrong. This section will attempt to do exactly that.

We have seen above that, since Aristotle, people have anticipated that automated engines could replace human labour and slaves. The idea, far from fading away, is currently making a powerful comeback, at the very time when western societies are contemplating a possible end of this era. Suddenly, the services given by machines, which had been for a time taken for granted, become visible again. There are nowadays an ever growing number of people who argue, again, that modern technology is now replicating the services once provided, in rich families, by slaves and servants.

Fossil fuel–powered machines and slaves play(ed) similar economic and social

Climatic Change 105 (2011): 329–55 [excerpt, 339–50]

roles in the societies in which they operate(d). Both slave societies and developed countries externalise(d) labour. In the first case, labour came from slaves; in the other, "work" is provided by machines. Consequently, there are many similarities between our dependence on fossil fuels and slave societies' dependence on bonded labour. In addition, both slaves and modern machines free(d) their owners from daily chores. They gave and continue to give individuals the leisure to read and write, perform arts, get informed, and participate in politics. If we all wanted to benefit from our current lifestyle without any fossil fuels, we would need to employ several dozen people working full time for us. The human exploitation and suffering resulting (directly) from slavery and (indirectly) from the excessive burning of fossil fuels are now morally comparable, even though they operate in a different way. We now know that when we burn oil or gas above what the ecosystem can absorb or when we are depleting non-renewable resources for leisure, we are indirectly causing suffering to other human beings, today and in the future. Similarly, cheap fossil fuels facilitate imports of goods produced in countries with little or no social protection and hence help externalise labour and perpetuate slave-like conditions.

Definitions and Limits to the Argument

Before going any further, it is important to start by defining slavery, usually understood as "someone who is legally owned by another person and works for them for no money" (*Longman Dictionary of Contemporary English*). However, such a definition is inadequate for at least two sets of reasons. First, historically a large number of what most people would consider as genuine slaves were not "legally owned" at all. That is why Patterson has argued that defining humans as property is not an essential constituent of slavery. For him slavery is "the permanent, violent, and personal domination of natally alienated and generally dishonoured persons" (quoted in Davis 2006: 30). This definition still applies today for most slaves. Second, slaves were occasionally given money as a reward for their work, or they could own, sometimes, considerable properties or their own business.

As Davis points out, it is extremely difficult to find a workable definition of slavery:

> One can imagine a spectrum of states of freedom and dependency or powerlessness, with various types of serfdom and peonage shading off into actual slavery. Within the category of slavery itself, we can also imagine a spectrum of slave systems beginning with those that accord slaves a variety of protections and rights. . . . It is clear that some forms of contract and prison labor have been harsher and more lethal than most examples of slavery. . . . If the labouring prisoners in the Nazi death camps and in Russia's gulag were not legally defined as owned chattel property, they were thereby completely made expendable and could be starved or frozen to death or simply shot, without any

recognizable loss. In terms of material standards of living, the slaves in the nineteenth century American South were clearly far better off than most slaves and forced labor in history; yet they were victims of one of the most oppressive slave systems ever known. (Davis 2006: 36)

For the purpose of my comparison, and because most people associate slavery not only to a form of legal ownership and free work, but also to the absence of liberty, harsh-treatment, family separation, and fear, I will adopt the following definition, based on Patterson's definition: slavery is "the permanent and violent domination of alienated and generally dishonoured persons who are performing work or services for the benefit of their owner." From this perspective, slavery is an end point on a spectrum of exploitation rather than a practice qualitatively different from any other form of exploitation.

Readers may find this definition of slavery problematic. For many, nineteenth-century African slavery is the archetype of all forms of slavery, whose consequences are felt most profoundly in the US. The idea of ownership was a key component in both the justification for and ultimate demise of chattel slavery. Thus, the comparison between fossil fuel use and slavery does not fully work if one applies it specifically to the Americas. Yet chattel slavery was not necessarily the worst form of bonded labour: today some forms of slavery, which do not involve formal ownership, might be even more cruel; and I am comparing our use of fossil fuels to all forms of slavery, not "just" nineteenth-century American slavery.

Also, when the word "machine" or the phrases "energy slaves" or "virtual slaves" are used in this text, I imply that they refer to machines powered by fossil fuels. This is both for the sake of brevity and because most machines today are directly or indirectly powered by oil, gas, or coal. Of course, some machines are powered by hydro or nuclear electricity, which is (nearly) carbon free. However, a vast majority of electricity worldwide is produced by coal or gas-fired power stations. According to the most recent estimates, thermal electricity generation (coal, gas, and oil burning) accounted for 66% of the total, while hydro-electricity accounted for 16%, nuclear 14%, and geothermal, solar, wind, wood, and waste only 2.2% (AER 2006). In theory, fossil fuels could be replaced by nuclear or wind and solar power. However, most commentators doubt that these energies could be scaled up to the extent needed to replace fossil fuels in the near future (Monbiot 2006). It is impossible to discuss nuclear power here in detail. Nuclear power—if all goes well—again would not stand the second part of my comparison with slavery. However, if things go wrong, its potential to create harm and suffering is very powerful indeed and has been demonstrated in the past (Power 2008). There is also the huge moral problem relating to the considerable amount of hazardous waste we will leave future generations and the potential for the spread of nuclear weapons.

The comparison between slavery and the excessive consumption of fossil fuel which I suggest in the following pages is by no means a perfect match. My argument is not that slavery and the excessive "luxury consumption" of fossil fuels are equivalent, but rather that they present striking similarities, with some notable differences.

The first distinction lies in the different ways by which suffering caused by slavery, on the one hand, and by the burning of fossil fuels, on the other, operate. In the case of slavery, oppression operates more or less directly. Slaves usually have names, faces, personalities, and their owners can directly make them suffer and immediately see the results. In the fossil fuel economy, however, the sufferings engendered by the burning of fossil fuels are indirect and often imperceptible by those who are causing it. It is difficult to recognize the potential connection between a coal-fired power plant in Europe and a refugee camp in Africa today and even more complicated to grasp the effects of climate change for future generations. The comparison thus ignores the direct human experience that characterised slavery. We cannot see the consequences of our burning of fossil fuels in the same way as slaveholders could see the suffering inflicted on their slaves; realisations of the consequences are delayed chronologically and removed geographically (Hulme 2009: 200–201). (However, many people—for example, consumers of slave goods—also benefited from slavery without maintaining direct connections to it. These people can certainly be said to have committed a morally equivalent sort of human transgression to people who benefit from fossil fuel use.)

A second, crucial difference is that there is no willingness to cause harm or dehumanize others by burning fossil fuels. By contrast, motives for enslavement were, and are, by no means limited to economic needs and frequently included a willingness to control others, even sometimes including sadism. As Aristotle himself recognised, in contrast to his vision of machines doing the work of slaves, there are strong psychological benefits from totally dominating other human beings. Adam Smith wrote in 1776 that the main motive for slavery was to "dominate, degrade, humiliate and control—often in order to confirm their own sense of pride and superiority." As Davis puts it, "chattel slavery is the most extreme example we have not only of domination and oppression but of human attempts to dehumanize other people" (Davis 2006: 2, 3, 29).

Machines and Slaves Play(ed) Similar Economic and Social Roles

In a recent article, Marc Davidson, drawing on two earlier comparisons (Orr 2000; Azar 2007), convincingly argued that there are similarities "between the rationalisation of slavery in the abolition debates and the rationalisation of ongoing emissions of greenhouse gases in the US congressional debates on the Kyoto Protocol." For example, Davidson argues that "in both debates US congressmen and Southern congressmen, respectively, represent an electorate with substantial interests in maintaining the status quo, costs are shifted to people who are not part of the electorate, and Congress rejects proposals for change" (Davidson 2008: 67–68). Yet it is possible to go much further and find even deeper connections. My comparison starts with an hypothesis that it is a feature of human nature that whenever societies have had the opportunity to find someone *or something* else to work for them for free, or for a small cost, they have almost always taken advantage of it, whatever the moral cost. The way this operates, irrespective of gender, class, religion, or ethnicity, is amply demonstrated by the fact that there were a number of slaveholders, in the American South or elsewhere, who had themselves been slaves. If slavery reminds us of "our [slaves and slave-owners alike] shared humanity, not only our triumphal possibilities but also our profound limitation" (Davis 2006: 180), the same could be said of fossil fuel usage.

Both slave owners and inhabitants of developed countries relied, and still rely, on work generated from an external source of energy to enjoy their particular lifestyle. In the former case, labour came from slaves; in the latter, it is derived from "work"—in the sense physicists use the word—which is mostly provided by energy from fossil fuels. As Davidson puts it: "today the United States is as dependent on fossil fuels for its patterns of consumption and production as its South was on slavery in the mid-nineteenth century."

Of course, Davidson's comparison cannot be measured exactly. Quantifying the value in today's money of all slaves in the Southern States in a meaningful way is extremely difficult, as it is to determine the value of fossil fuels in today's world. Moreover, the dependence on fuels cannot be accounted for by quantitative economic analysis alone (Debeir et al. 1991: 124).

Even though it is impossible to make a robust, like-for-like comparison, as forms of energy, there are clear similarities between our current economic dependence on fossil fuels and the nineteenth-century economy's dependence on slaves. The United States and, to a large extent, the rest of the world are far more reliant on fossil fuels today than the US economy—or any other economy—has ever been on slavery. Several scholars have long claimed that this dependency jeopardises our very survival: "proper alternative sources of energy that can substitute for fossil fuels must be found to prevent mankind from reverting to an agricultural level of activity which would mean a dramatic and painful reduction of both mankind's size and its level of living" (Cipolla 1978: 63; many comparable claims are made in the literature on "peak oil").

In fact, the US Southern State reliance on slaves was partially an illusion, as demonstrated by the rise in cotton production after the abolition of slavery. Likewise, our current dependency on fossil fuels is also somewhat of an illusion; we could consume far less energy than our present levels, while being as healthy and happy (perhaps even more so) as our current high-energy consuming world (Illich 1974). Yet, if we want to maintain current standards of living, our dependency on fossil fuel is excessive. This is especially true when new structural constraints, which were irrelevant or of peripheral concern to previous societies—such as the substantial increase in the global population, urban-sprawl, or the hyper speciali-sation of contemporary economies—are taken into account.

Another way of evaluating current dependencies on "virtual slaves" is to con-sider how much people-effort would be required to replace work done by machines. At the end of the nineteenth century, for example, human labour still accounted for about 95% of all industrial work in the US—and this was generally true of all traditional societies—whereas, today, it constitutes only 8% (Berry et al. 1993). A century ago, Oscar Wilde considered the steam engine to be the functional equiv-alent of servants or slaves. However, the term "energy slave" (and subsequent ef-forts to quantify the average number of such slaves per person in the world) seems to have first been coined by American philosopher R. B. Fuller in the early 1950s. His estimate was that in 1950, each individual on earth had approximately 38 "energy slaves" at his disposal (Fuller 1969; Marks 1964). This concept has since been used widely: I have found over twenty references to "energy slaves" in publi-cations, although these are sometimes phrased slightly differently, for example as "mechanical slaves."

The implication is clear; if we wanted to maintain the same lifestyle without petroleum, coal, natural gas, we would need to employ several dozen persons, or more, on a full-time basis (the exact number of "virtual slaves" depends not only on the period and the country considered, but also whether one takes the average amount of mechanical work a healthy human could do in a year [as Fuller does] or the average amount "energy slaves" working 24 h a day, 365 days could do in a year [as McNeill does]). This astoundingly high figure comes from the fact that a single litre of petrol contains the equivalent of about 9 kWh of energy, while the output of an average human being is about 3 kWh in the course of a 40-h working week. Even if human labour involves more than just an output of kWh, it is reasonable to suggest that we pay little for our oil when compared to the amount of "work" fossil fuel can provide. It is no wonder that people in past centuries enthusiasti-cally adopted new energies, or that the majority of us want to continue to enjoy the numerous positive aspects of fuel-powered machines.

Both Fossil Fuel Usage and Slavery Cause(d) Harm to Others

Besides the similarities between the convenience brought to us by fossil fuel–powered machines and the convenient life slaves brought to slave owners, another parallel exists between the harm caused to human beings by slavery and the harm caused by the current large scale burning of fossil fuels. I do not want to imply that all fossil fuel burning is bad. A useful distinction can be made between "luxury emissions" and "survival emissions." "One unit of carbon dioxide emitted by an Indian peasant farmer, essential for subsistence, carrie[s] a different moral weight to a unit of carbon dioxide emitted by an American tourist flying to the Bahamas" (Hulme 2009: 159; Agarwal and Narain 1991).

Some might argue that it is not possible to compare pain triggered by the use of slaves and pain caused by the use of oil, gas, or coal, as in the latter case we are dealing with inanimate objects that cannot suffer. However, when we burn oil or gas above what the eco-system can absorb, we are indirectly causing suffering to other human beings. The reports from the Intergovernmental Panel on Climate Change (IPCC) make it clear that the release of carbon dioxide is already causing harm by increasing droughts and flooding, threatening crop yields, and displacing large numbers of people, and this damage and suffering is set to rise in years to come. The future looks grim for the world, whether the IPCC looks at freshwater resources and their management, ecosystems, food, or health. The predictions for Africa, the continent where the slave trade involved more people and lasted longer than in any other place in the world, are even more worrisome: "By 2020, between 75 million and 250 million people [in Africa] are projected to be exposed to increased water stress due to climate change. . . . Agricultural production, including access to food, in many African countries and regions is projected to be severely compromised by climate variability and change" (IPCC 2007).

If we accept the conclusion of the IPCC, we must recognize that we are now fully aware of both the causes of climate change and its consequences. It is no longer possible to argue that our use of oil is morally neutral. Driving cars or flying *does*—however indirectly and unwittingly—hurt people now. And because emissions accumulate in the atmosphere, they will increasingly continue to do so in the future, unless the trends reverse, somehow.

For those who are still unpersuaded that climate change is hurting people, or do not believe that the climate is being significantly altered by human activity, there is still the moral, as well as practical, problem that by using fossil fuels, we are depleting very valuable resources that are not renewable. Scholars like Jared Diamond (2006) have vividly told cautionary tales of societies disappearing because they relied on a staple source of food or energy that they subsequently exhausted. Britain had oil in the North Sea, much of which is now gone forever. The nation has used a resource that its children will not be able to benefit from. Because the oil and gas have been burnt in large part for "luxury emissions" rather than "survival emissions" and with little concern beyond the "carpe diem" mantra, it has not replaced it by anything of similar value. The next generation will inherit the worst consequences of this cheap energy lifestyle (if we leave aside anthropogenic climate change, we could add pollution, obesity, the spread of concrete over arable land and greenbelt, and the consequent development of highly unsustainable suburbia to the litany).

One could of course rightly argue that "human history since the dawn of agriculture is replete with unsustainable societies, some of which vanished but many of which changed their ways and survived" (McNeill 2000: 358). History is useful as it can remind us both of the ingenuity of the human species to solve problems in the past and, at the same time, of the dangers involved in depleting resources too quickly. It is also true that previous generations have often left huge debts to their children. However, climate change is a new problem taking place on an unprecedented scale. In any case what previous generations did cannot justify our wrongdoings now.

Similarly, how should we respond to the moral problem that in a world where poor people struggle to find enough food to feed their families, we are burning *food* to run our cars or heat our homes? This had direct consequences for the world's poor, as bio-fuel production tends to drive upwards the price of cereals worldwide, directly affecting the poor. This led, for example, to riots occurring in Haiti in 2008 over the price of corn. Some remorseless companies go as far as to encourage people to burn corn instead of wood pellets: "heat your home for a bushel of corn per day" claims a Canadian company (Caneco 2008). Filling up a large car's fuel tank with ethanol uses enough maize to feed a person for a year (Economist 2007). Put starkly, the rich are buying up food to run cars at the expense of the world's poor (Douthwaite 2007). We are also clearly putting our societies and our children at risk by relying so heavily on fossil fuels. In the same way as slave owners frequently worried about slaves escaping or revolting, we also regularly worry about our suppliers of oil or gas stopping deliveries of this precious liquid. The US, UK, and Netherlands, for example, refused to sell oil to Japan in 1941, with the consequences we know; in the 1970s, the OPEC reduced their production of crude oil, triggering a worldwide crisis; and more recently still, Russia has cut off natural gas shipments to Ukraine. Industrial countries have also become involved in an increasingly violent politics of oil in countries such as Iraq, the Sudan, and Nigeria to mention a few.

This situation could—once more—be compared to the attitude of slave owners who benefited for a limited period of time from free labour, but then left the task of dealing with the dire consequences of slavery to their children. For example, African Americans as a group continue to be disadvantaged economically and socially in many areas compared to other US citizens. It is often difficult to separate the legacy of the segregation era from the legacy of slavery (and other factors), but it has long been suggested that the separation of slave men from their families might be an ongoing contributory factor to the fragility of black families in contemporary American society (Elkins 1959; Moynihan Report 1965). Black families are less likely to form and more likely to break up than White, Hispanic, or Asian families, and, to take a trivial example, a greater proportion of children from single families commit certain crimes than children from two-parent families. This, combined with some rampant racism (another legacy of the slavery era), makes it easier for Blacks to get arrested: "although blacks account for only 12 percent of the U.S. population, 44 percent of all prisoners in the United States are black. . . . Indeed, nearly five percent of all black men, compared to 0.6 percent of white men, are incarcerated" (HRW 2003; USCB 2006). This comes at a high cost for US society as a whole, which still has to be paid nearly 150 years after the Thirteenth Amendment to the United States Constitution abolishing slavery.

One might at this point object that the comparison still does not stand, because slavery is against the law, while using fossil fuel is not. This is easily dismissed, however, as slavery *was* perfectly legal at one point and was only prohibited after the culmination of several contributing factors (such as anti-slavery campaigns, the changing socio-economic climate, and the shift in moral attitudes, as discussed above). Yet, even when slavery was legal, numerous alibis were invented to justify it. A similar phenomenon can be observed today, as people increasingly feel the need to justify their high energy use. Is it completely unthinkable that one day driving highly inefficient vehicles might be outlawed when much "cleaner" alternatives exist?

Another objection to the present argument might be that the harm caused by climate change is unintentional. On this view, as the damage is not deliberate and the consequences are not always tangible, we cannot be blamed for this damage, or at least not as severely as if it were deliberate and the consequences were known. For example, since many people were unaware that burning fossil fuels could cause harm until very recently, previous generations cannot be blamed for it. This is a perfectly valid argument. The context in which machines were introduced is important. It would be unfair to blame American farmers who purchased tractors and fertilizers and saw these as incontestable help in feeding their families and the rest of the world; or women who bought washing machines to be freed from family chores. It would not seem just either to say that people who were emitting carbon dioxide 100, or perhaps even 20, years ago are responsible in the same way as we are now with our knowledge of the climate consequences (a similar case, with hindsight, could be made against them for the rate at which they were burning valuable resources, destroying eco-systems, or damaging people's health by creating smog, but we could not blame them for climate change for they did not know, or at the very least did not have a clear idea of, the likely consequences). If previous generations made a mistake by burning fossil fuel carelessly, they mostly did this with a clear conscience and in the good faith that they were trying to improve their own and other people's lives.

It is also true that it is more difficult for us to grasp the consequences of our actions than it was for slave owners, as the consequences of greenhouse gas emissions are indirect and usually mediated through a vapid media. Some people remain genuinely perplexed about the causes of climate change, a responsibility for which rests, to a considerable extent, on the fossil fuel lobbies as well as recalcitrant governments unwilling to clearly impart scientific findings to their citizens (Gore and Guggenheim 2006; Oreskes 2007).

The "unintentional damages" argument only stands for previous generations who had no idea of the likely consequences of burning fossil fuels. For people living today, it only stands as long as we are ignorant of the fact that the way we live is having devastating consequences for others. For most reasonably informed people now, it is hard to ignore the warnings of scientists. Even for those who are not fully convinced of the causes of climate change, should we not be using precautionary measures if there are reasonable grounds to believe that we might be changing the climate? As many already pointed out in the 1990s, "if the models were faulty, future climate change could be worse than predicted, not better" (Weart 2003: 180). Davidson makes a similar point when he compares the inconsistent attitude of US congressmen who decided to approve the war in Iraq but constantly refuse to act on climate change on the basis that there is, they argue, insufficient evidence that man-made emissions are creating harm (Davidson 2008).

If the lobbies referred to above have been so easily able to spread doubts among many intelligent people, it is perhaps because many of us prefer to avoid the conclusions of the IPCC and the present comparison helps us to understand why. If we disagree about climate change, it is perhaps not only because of "our different attitudes to risk, technology and well-being; our different ethical, ideological and political beliefs; our different interpretations of the past and our competing visions of the future" (Hulme 2009: xxvi), but because, at root, we also all have strong vested interests in not believing the climate science. Rich countries are mostly democracies where large sections of the general public do not want to change their lifestyles, something politicians know very well. In Eastern Europe, when parliamentary regimes were re-established after communism collapsed, environmental problems "caused by foreigners, the military, or specific factories were often addressed and sometimes resolved. Those caused by the consumption patterns of ordinary citizens often got worse under democracy." This seems to be a general pattern in parliamentary regimes (McNeill 2000: 348, 353).

Even then polities and publics alike continue to clutch at straws. For example, IPCC predictions that a 1–3°C temperature rise will see some potential advantages for global food production and "bring some benefits, such as fewer deaths from cold exposure," in temperate regions (predominately in industrialised countries) are sometimes treated as grounds for extolling the fossil fuel economy, including for future generations. And this in spite of the IPCC's overall estimation that "in the aggregate, however, net effects will tend to be more negative the larger the change in climate" (IPCC 2007).

In the effort to claim that climate change might be a good thing, however, once again we can hear the same kinds of self-serving justifications used by slave-owners. Nineteenth-century slave-owners, for example, claimed that the work of slaves would benefit future generations or that slaves were actually better off "working" in the US than they would be "working" in England's factories. For instance, US vice president John C. Calhoun argued on the senate floor in 1837: "The Central African race . . . had never existed in so comfortable, so respectable, or so civilized a condition as that which it now enjoyed in the Southern States." Slavery was not "an evil. Not at all. It was a good—a great good" (quoted in Davidson 2008: 72). Some also thought that freed slaves would be unable to feed themselves or be responsible for their own fate. The same kind of arguments had been used by apologists of slavery since antiquity.

It is also true that it is practically impossible in our contemporary world, even for the most virtuous, to live without relying on some sort of energy of the fossil variety. We are subjected, as individuals, to constant invitation (from corporations, governments, or peers) to consume ever more. We are perhaps, like drug addicts, as much victims as culprits of this consumer or technological society (Illich 1974). It could even be argued that we ourselves are slaves of our way of life: it is a well-known paradox that slave owners tend to become dependent on their slaves, or as Emerson put it (1863: 98), "If you put a chain around the neck of a slave, the other end fastens itself around your own." Our dependency on fossil fuels for energy extends to the very machines we rely on in our daily lives—the computer I am using to write this article confirms this general rule. If we cannot do otherwise, one could argue that we cannot be blamed nor can we prevent ourselves from hurting others. Slave owners were therefore more to blame than we are, because they could at least emancipate their slaves and choose a virtuous life.

Yet, one should not underestimate the struggle that most slave owners experienced in the decision to free their slaves. In the nineteenth-century American South, state laws "restricted or in effect prohibited manumission" (Davis 2006: 193). By the same token, we should not overestimate our own difficulties in reducing carbon dioxide emissions. After all, while it is fairly easy to install low-consumption light bulbs or to switch to a provider of renewable electricity, still very few of us do this. If we were able to distribute an equitable share of carbon dioxide allowance per person, and to keep the overall international emissions under the threshold of what worldwide carbon sinks can absorb each year, our (reduced) emissions would gradually slow down the rate of climate change. This is exactly what—perhaps over simplistic—schemes such as "Contraction and Convergence" propose (Bows and Anderson 2008).

There remains, however, a final objection to the slavery vs. fossil fuel analogy. As we have seen above, most definitions of slavery emphasise the idea of complete legal ownership and control by a master over a person who has to work for them for free. By contrast, we would claim that we do not compel anybody to work for us for free as consumers of fossil energy. Phrased differently, if the result of the work of a slave can be compared to the result of the action of machines through the combustion of fossil fuel, the process of the exploitation itself is different. This is again true to some degree, even though I have already partly addressed this point and shown that slave owners did not necessarily own their slaves legally nor compel them to work for free. Besides, this objection can be challenged on two grounds:

First, our "global economy" rests on Ricardo's famous comparative advantage concept, which is based on the assumption of negligible transport costs. The availability of comparatively cheap energy is a required condition for the transportation of foreign goods on a massive scale over large distances. One of the main reasons why rich countries import inexpensive products is due primarily to the fact that manpower, in the so-called "developing world," is much less expensive than in more affluent countries. Yet inexpensive energy has, in a way, enabled us to delocalise sub-standard working conditions. Workers in "developing countries" on which we rely to sew our jeans and trainers often have little or no liberty of movement or choice of employer and often experience ill-treatment, when they are not simply inmates in state prisons (conditions which sadly fit my definition of slavery only too well). Their conditions are sometimes perhaps harsher than those of slaves in traditional Caribbean or American economies: "chattel slaves represented a valuable investment, an investment of rising value in much of the new world, but that slightly protective aspect of chattel slavery . . . does not apply to the many millions of bonded and coerced workers in today's so-called developing world" (Davis 2006: 329–30). Reports of these appalling working conditions frequently appear in Western newspapers, and so we cannot claim that we do not know. Cheap transport, relying on fossil fuels and, as a consequence, contributing massively to climate change, is what makes this deplorable reality possible. It is true, however, that even if fossil fuel use makes this direct human exploitation possible, fossil fuel use cannot be equated to human exploitation itself. Secondly, the harm of climate change often amounts to violence or force against a large number of people, and will increasingly do so. Through a series of steps, starvation or the destruction of eco-systems amounts to denying people the freedom to make decisions about their lives. Floods, droughts, and rises in sea level will force millions of people to become refugees. According to estimates, "when global warming takes hold, there could be as many as two hundred million people overtaken by sea-level rise and coastal flooding, by disruptions of monsoon systems and other rainfall regimes, and by droughts of unprecedented severity and duration" (Myers 2002). Many of these refugees may end up in camps, where they will not be much better off than prisoners. Forced from their homes and families, their land may be taken away and some may end up having to work for unscrupulous masters or in prostitution rings (another form of slavery in which refugees are already over-represented).

Even those fortunate enough to stay on their land, such as the vast number of peasants in the "developing world," may find themselves, in their struggle to feed their families, victims of debt bondage, a condition which "can hardly be distinguished from traditional slavery" (OHCHR 1991). As crop failure is a common trigger to debt bondage, it is not unreasonable to link, however indirectly, our climate change–inducing emissions to mechanisms that are reducing people to a form of slavery. This is not a new phenomenon: ecological factors such as droughts (triggering dearth and/or debts) have long been an important contributing factor in reducing people to slavery during the Atlantic slave trade era (Pétré-Grenouilleau 2004: 118).

Lessons from History and Conclusion

Comparing the attitude of slave owners and our own attitude to petroleum is both adequate and useful.

By emphasising the human tendency and desire for convenience, which drove (and continues to drive) the use of slaves (real or virtual), our collective inaction in the face of climate change and our addiction to cheap energy can be explained. It enables us to take stock of our resistance to change and legislation against ever increasing fossil fuel usage. This explanation also shows that the problem of climate change is deeply rooted. The human tendency to "externalise labour" and "dominate, degrade, humiliate and control," as evidenced by the almost universal practice of slavery, will percolate in any and every system.

However, climate change and fossil fuel depletion could be the galvanising forces that enable us to address many of the excesses and injustices so visible in the world today. Climate change can thus be seen as an opportunity, rather than as a problem. "Investigations into the causes of climate change have shown us—in case we preferred to forget it—that our world is a very unequal one." Similarly, climate change "teaches us—in case we were complacent—that our current energy portfolio is not sustainable" and that "we should do what we can to conserve carbon-based fuels and that we should accelerate the search for new, non-carbon based energy sources" (Hulme 2008).

The comparison is also useful because virtually everybody now agrees that slavery is wrong. If we accept the comparison, it follows that we are better enabled to see and *feel* the iniquity of continuing to burn fossil fuels inconsiderably, and emotions are often more capable of mobilising us than logical reasoning (Weber 2006: 104). That is, if we are convinced that we are behaving like slave-owners (whom we morally condemn), we are more likely to want to act differently.

References

AER (2006) Energy information administration, "World Net Electricity Generation by Type, Most Recent Annual Estimates, 2006," http://www.eia.doe.gov/emeu/international/RecentElectricityGenerationByType.xls.

Agarwal A, Narain S (1991) *Global warming in an unequal world*. Centre for Science and Environment, Dehli.

Azar C (2007) Bury the chains and the carbon dioxide. *Climatic Change* 85:473–75.

Berry BJL, Conkling EC, Ray DM (1993) *The global economy: Resource use, locational choice, and international trade*. Englewood Cliffs, NJ: Prentice Hall.

Bows A, Anderson K (2008) Contraction and convergence: An assessment of the CCOptions model. *Climatic Change* 91:275–90.

Caneco (2008) Advertisement for company. http://www.ecobusinesslinks.com/corn-stoves .htm.

Cipolla CM (1978) *The economic history of world population*. Penguin.

Davidson MD (2008) Parallels in reactionary argumentation in the US congressional debates on the abolition of slavery and the Kyoto Protocol. *Climatic Change* 86:67–82.

Davis DB (2006) *Inhuman bondage, the rise and fall of slavery in the new world*. Oxford University Press, Oxford.

Debeir J-C, Deléage J-P, Hémery D (1991) *In the servitude of power: Energy and civilisation through the ages*. Zed Books, London.

Diamond J (2006) *Collapse: How civilisations choose to fail or survive*. Penguin, London.

Douthwaite R (2007) Sharing out the rations. *Irish Times*, January 13, 2007.

Economist (2007) The end of cheap food. *Economist*, December 6, 2007. Available at http://www.economist.com/opinion/displaystory.cfm?story_id=10252015.

Elkins SM (1959) *Slavery: A problem in American institutional and intellectual life*. University of Chicago Press, Chicago.

Emerson RW (1863) *Essays, first series*. Ticknor and Fields, Boston.

Fuller RB (1969) *Utopia or oblivion: The prospects for humanity*. Allen Lane, Penguin Press, London.

Gore A, Guggenheim D (2006) *An inconvenient truth*. DVD, Paramount.

HRWIncarceratedAmerica (2003) *Human rights watch backgrounder*. Available at http://www.hrw.org/backgrounder/usa/incarceration.

Hulme M (2008) Five Lessons of Climate Change: A personal statement. March 2008. Available at http://www.mikehulme.org/wp-content/uploads/the-five-lessons-of-climate -change.pdf.

Hulme M (2009) *Why we disagree about climate change: Understanding controversy, inaction and opportunity*. Cambridge University Press, Cambridge.

Illich I (1974) *Energy and equity*. Calder & Boyars, London.

IPCC (2007) Summary for policymakers. In: Parry ML, Canziani OF, Palutikof JP, van der Linden PJ, Hanson CE (eds.), *Climate change 2007: Impacts, adaptation and vulnerability*. Contribution of Working Group II to the Fourth Assessment Report of the Intergovernmental Panel on Climate Change. Cambridge University Press, Cambridge, pp. 7–22.

Marks RW (1964) *Space, time, and the new mathematics*. Bantam Books.

McNeill JR (2000) *Something new under the sun: An environmental history of the twentieth century*. Allen Lane, London.

Monbiot G (2006) *Heat: How to stop the planet burning*. London, Penguin.

Moynihan Report (1965) *The Negro family: The case for national action*. Office of Policy Planning and Research, United States Department of Labor, March 1965. Available at http://www.dol.gov/oasam/programs/history/webid-meynihan.htm.

Myers N (2002) Environmental refugees: A growing phenomenon of the 21st century. *Philos Trans R Soc B: Biol Sci* 357:609–13. Available at http://www.ncbi.nlm.nih.gov/pmc/articles/PMC1692964/.

OHCHR Office of the High Commissioner for Human Rights (1991) Fact Sheet No. 14, Contemporary Forms of Slavery. Available at http://www.ohchr.org/Documents/Publications/FactSheet14en.pdf.

Oreskes N (2007) The American denial of global warming. Online lecture at http://www.uctv.tv/search-details.asp?showID=13459.

Orr D (2000) 2020: A proposal. *Conserv Biol* 14:338–41.

Pétré-Grenouilleau O (2004) *Les traites négrières: essai d'histoire globale*. Gallimard.

Power MS (2008) *America's nuclear wastelands: Politics, accountability, and cleanup*. Washington State University Press, Pulman.

USCB U.S. Census Bureau (2006) American community survey. Available at http://factfinder.census.gov/servlet/ADPTable?_bm=y&-geo_id=01000US&-qr_name=ACS_2006_EST_G00_DP5.

Weart SR (2003) *The discovery of global warming*. Harvard University Press, Cambridge, MA.

Weber EU (2006) Experience-based and description-based perceptions of long-term risk: Why global warming does not scare us (yet). *Climatic Change* 77:103–20.

))) Imperial Oil: The Anatomy of a Nigerian Oil Insurgency

Michael Watts

Blood may be thicker than water, but oil is thicker than both.

—Perry Anderson (2001, 30)

Introduction

In his 2006 State of the Union address, George Bush finally put into words what previous presidents could not bring themselves to utter in public: addiction. The US, he conceded, is "addicted" to oil—which is to say addicted to the car—and as a consequence unhealthily dependent upon Middle East suppliers.

* * *

It is no surprise, then, that alternative sources of oil-supply should be very much on the Bush radar screen (since conservation strategies or increased gas taxes are conspicuously absent). [Vice president Dick] Cheney's National Energy Strategy Report in 2001 bemoaned the US oil habit—"a dependency on foreign powers that

Erdkunde 62 (2008): 27–39

do not have America's interests at heart"—long before the State of the Union address. A recent report in the *Financial Times* headline (March 1, 2006) makes the new agenda crystal clear. If Africa is not as well endowed in hydrocarbons (both oil and gas) as the Gulf states, nevertheless the continent "is all set to balance power," and as a consequence is "the subject of fierce competition by energy companies" (ibid., 1). IHS Energy—one of the oil industry's major consulting companies—expects African oil production, especially along the Atlantic littoral, to attract "huge exploration investment" contributing over 30% of world liquid hydrocarbon production by 2010. Over the last 5 years when new oil-field discoveries were a scarce commodity, Africa contributed one in every four barrels of new petroleum discovered outside of Northern America. As the Catholic Relief Services put it in their exemplary study of oil geopolitics, "The Bottom of the Barrel" (Gary and Karl 2003), a new scramble is in the making. The battleground consists of the rich African oilfields—the continent's "copious reserves of natural gas and its sweet light oil" (ibid., 1).

Energy security is the name of the game. No surprise, then, that the Council of Foreign Relation's call for a different US approach to Africa in its new report "More than Humanitarianism" (CFR 2005) turns on Africa's "growing strategic importance" for US policy (2005, xiv). It is the West African Gulf of Guinea, encompassing the rich on- and offshore fields stretching from Nigeria to Angola, that represents a key plank in Bush's alternative to the increasingly volatile and unpredictable oil-states of the Persian Gulf. Nigeria and Angola account for nearly 4 million bpd (almost half of Africa's output), and US oil companies alone have invested more than $40 billion in the region over the last decade (and another $30 billion expected between 2005 and 2010). Oil investment now represents over 50% of all foreign direct investment (FDI) in the continent (and over 60% of all FDI in the top four FDI recipient countries), and almost 90% of all cross-border mergers and acquisition activity since 2003 has been in the mining and petroleum sector (WIR 2005, 43). Strategic interest for the US certainly means cheap and reliable low-sulphur ("sweet") oil imports but also keeping the Chinese (for example, in Sudan) and South Koreans (for example, in Nigeria)—aggressive new actors in the African oil business—and Islamic terror at bay (Africa is, according to the intelligence community, the "new frontier" in the fight against revolutionary Islam).

In short, the geo-strategic importance of Nigerian and the wider Gulf of Guinea oil turns on not only the failure of the US global oil acquisition strategy of the post-war period but also the additional ingredient of Global War on Terror (GWOT). In the face of support by neoconservative promoters and opportunistic Washington lobbyists, strategists at the Pentagon have invented a new security threat to increase funding for European Command's (EUCOM's) footprint in Africa (Lubeck et. al. 2007). Recently, Deputy Assistant Secretary of Defense for African Affairs

Teresa Whelan announced the discovery of a "new threat paradigm"—the threat of "ungoverned spaces" in Northwest and West Africa (http://www.jhuapl.edu/POW/rethinking/video.cfm#whelan). In practice all four of the military services—including an Africa Clearing House on security information, supported by a Pentagon think tank, the Africa Center for Strategic Studies housed at the National Defense University—are involved and implicated in the new scramble for the continent. Against a backdrop of spiraling militancy across the Delta, US interests have met up with European strategic concerns in the Gulf in the establishment of the "Gulf of Guinea Energy Security Strategy" (GGESS). By December 2005, the American ambassador and the Managing Director of Nigerian National Petroleum Corporation (NNPC) agreed "to establish four special committees to co-ordinate action against trafficking in small arms in the Niger Delta, bolster maritime and coastal security in the region, promote community development and poverty reduction, and combat money laundering and other financial crimes" (*This Day*, December 9, 2005). The establishment of a new African command (AFRICOM) in 2007 is the final capstone in the militarization of American energy security policy in Africa. Energy security, it turns out, is a terrifying hybrid, a perplexing doubleness, containing the old and the new: primitive accumulation and American militarism coupled to the war on terror (Harvey 2003; Retort 2005; Barnes 2005).

This perfect storm of forces linking dispossession, war, and energy provides the broad context in which one can grasp the descent into violence and ungovernability that has characterized the political dynamics of the Nigerian oilfields across the Niger delta, arguably one of the most strategic centers of oil supply for the United States, currently providing over 12% of all US imports. It is the broad dynamics of what is in effect now an insurgency in the oil-producing Niger delta that I shall address here. The simultaneous growth of militancy and oil disruption commenced in the late 1990s. By 2003 oil supply had been compromised by 750,000 barrels per day as a result of militant attacks on oil installations across the region. In April 2004, another wave of violence erupted around oil installations—at the end of April, Shell lost production of up to 370,000 barrels per day, largely in the western Delta—this time amid the presence of armed insurgencies. Two so-called "ethnic militias" led by Ateke Tom (of the Niger Delta Vigilante [NDV]) and Alhaji Asari Dokubo (of the Niger Delta People's Volunteer Force [NDPVF]), each driven and partly funded by oil monies and actively deployed (and paid) by high-ranking politicians as political thugs during the 2003 elections, have transformed the operating environment in the Niger delta oilfields. Since late 2005, the situation in the Delta has only worsened. Following attacks on oil installations and the taking of hostages in late December 2005 and early 2006, a hitherto unknown group of insurgents from the Warri region, the Movement for the Emancipation of the Niger Delta (MEND), began calling for the international community to evacuate from the Niger Delta by February 12, or "face violent attacks." Two weeks later, the group claimed responsibility for attacking a Federal naval vessel and for the kidnapping of nine workers employed by the oil servicing company Willbros, allegedly in retaliation for an attack by the Nigerian military on a community in the Western Delta. More than fifteen Nigerian soldiers were killed, and between May and August 2006 there were at least three kidnappings per month (typically the hostages have all been released following the payment of substantial ransoms by the government).[1] In the last 9 months the escalation of attacks—including electronically detonated car bombings, attacks on government buildings, and massive disruption of oil installations deploying sophisticated military equipment, and the kidnapping of workers sometimes from platforms 40–60 kilometers offshore—have spiraled out of control. In a deteriorating environment in which many oil companies have withdrawn personnel and cut back production—currently (May 2007) there is a 900,000 barrel per day shut-in—Julius Berger, the largest construction company operating in the country, announced its withdrawal from the Niger Delta in the middle of 2006. President Obasanjo has sent in additional troops to bolster the Joint Military Task Force (JMTF) in the Delta, but it is clear that they are incapable of operating effectively within the riverine creeks. The violence has continued—indeed deepened; at least 60 militants were reported killed and another 100 arrested in two days of brutal fighting in Bayelsa State late August 2006. According to the Center for Strategic and international Studies (CSIS 2007) 123 expatriate hostages have been taken since January 2006 (until early March 2007), and there have been 42 attacks on oil installations over the same period. As I write, the residence of the new vice president elect (the governor of Bayelsa State) has been bombed, Chevron has temporarily shut down its operations, and following a massive pipeline explosion at Bomu, a total of 900,000 barrels of oil per day are currently shut in (30% of official production). It is quite unclear, when located on this larger canvas, what Petroleum Minister Edmund Daukoru could possibly have meant when he announced to OPEC in February 2007 in Greece that "the worst is over," that "it is a very, very temporary thing" (United Press International, January 28, 2007, http://www.upi.com/Energy/analysis_nigeria_hopeful_for_oil_future).

The costs of the oil insurgency are vast. A report prepared for the Nigerian National Petroleum Company (NNPC) published in 2003 entitled *Back from the Brink* paints a very gloomy "risk audit" for the Delta. NNPC estimated that between 1998 and 2003, there were four hundred "vandalizations" on company facilities each year (and 581 between January and September 2004); oil losses amounted to over $1 billion annually. In early 2006 MEND claimed a goal of cutting Nigerian output by 30% and has apparently succeeded. Within the first 6 months of 2006, there were nineteen attacks on foreign oil operations and over $2.187 billion lost in oil revenues; the Department of Petroleum Resources claims this figure represents 32% of the revenue the country generated this year. The Nigerian government claims that between 1999 and 2005 oil losses amounted to $6.8 billion, but in November 2006 the managing director of Shell Nigeria reported that the loss of revenues due to "unrest and violence" was $61 million per day (a shut-in of about 800,000 barrels per day), amounting to a staggering $9 billion since January 2006. Against a backdrop of escalating attacks on oil facilities and a proliferation of kidnappings, the Joint Revolutionary Council (apparently an umbrella group for insurgents) threatened "black November" as an "all out attack on oil operating companies" (*Observer*, November 5, 2006).

The Road to Serfdom

The backdrop to the new scramble is the calamity of African poverty—in the language of *Our Common Interest: The Report of The Commission on Africa* (2005), assembled by Tony Blair and Gordon Brown, "the greatest tragedy of our time." They dubbed 2005 the "Year of Africa." In June of that year the Live 8 concerts drew a global audience of 2 billion, and a week later the G8 pledged to double aid to Africa ($25 billion by 2010) and forgive the debts of fourteen African states. African poverty had forced itself into the international limelight aided and abetted by a motley crew of humanitarians from Bono to Jeffrey Sachs to the Pope. The milestones in the growing international visibility of the African crisis include the *United Nations Millennium Declaration* in 2000; the Millennium Challenge Account; the President's Emergency Plan for AIDS Relief (PEPFAR); and the African Growth And Opportunity Act, all launched by President Bush; and now, the new World Bank African Action Plan. Collectively these palliatives were belated responses to the unacceptable face over two decades of globalization, reform, and the search for the Holy Grail of good governance. On the continent itself, the New Economic Partnership for African Development (NEPAD) (2001) and the revamped African Union (formerly the Organization for African Unity) offered the prospect that poor leadership (the pathologies of the African postcolonial state variously described as patrimonialism, prebendalism, predation, quasi-statehood, the postcolony, and politics of the belly) was to be taken seriously by an African political class that purportedly represented a new sort of democratic dispensation unleashed by a raft of the political transitions during the 1990s.

To see the African crisis, however, as a moral or ethical failure on the part of the "international community" (not least in its failure to meet the pledges promised by the Millennium Development Goals of reducing poverty by half by 2015) is only a partial truth. The real crisis of Africa is that after 25 years of brutal neoliberal reform, and savage World Bank structural adjustment and IMF stabilization, African development has failed catastrophically. William Easterly (2006), former high-ranking World Bank *apparatchik*, in his new lacerating demolition of structural adjustment—"a quarter century of economic failure and political chaos" —boldly states that the entire unaccountable enterprise of planned reform is "absurd" (http://www.nyu.edu/fas/institute/dri/Easterly/). It was Africa after all that was the testing ground for the Hayekian counter-revolution that swept through development economics in the 1970s. It began with the Berg Report in 1980, which was the first systematic attempt to take the Chicago Boys experience in post-Allende Chile—on some readings the birth of neoliberalism (Harvey 2005)—and impose it on an entire continent. The ideas of Elliot Berg and his fellow travelers marked the triumph of a long march by the likes of Peter Bauer, H. G. Johnson, and Deepak Lal (ably supported by the monetarist think tanks such as the Institute of Economic Affairs and the Mont Pelerin Society, and the astonishing rise to power from the early presence of Leo Strauss and Friedrich Hayek of the "Chicago School") through development institutions like the World Bank. Long before shock therapy in Eastern Europe or even the debt-driven "adjustments" in Latin America, it was sub-Saharan Africa that was the playground for neoliberalism's assault on the "over extended public sector," "excessive physical capital formations," and the "proliferation of market distortions" (Toye 1987, 48–49). According to the UN, twenty-six of thirty-two sub-Saharan states had a "liberal" economic regime by 1998.

If the 1980s were Africa's Lost Decade—collapsing commodity prices, deteriorating terms of trade, and the first crashing waves of IMF austerity—then how might one characterize the last 15 years (a long decade in which the benefits of reform were to be finally felt) in which life expectancy across sub-Saharan Africa steadily fell and per capita income has at best stagnated? A "Lost Generation"? And all of this during a period in which net official ODA fell by 40% (from $18.7 to $10 billion). If it is lucky, Africa will achieve its Millennium Goals of universal primary education and a 50% reduction in poverty by the middle of the next century (2150).

In Africa, the court of neoliberalism has been concluded, and the verdict is in. The picture is not pretty. Over the last 30 years there has been no growth in income for the average African. Life expectancy is 46 years. Twenty-three of forty-seven sub-Saharan states have currently a GDP of less than $3 billion (ExxonMobil's net profit in the first quarter of 2006 was $8 billion). By 2005, thirty-eight of the top fifty-nine priority countries that failed to make headway toward the Millennium Goals were sub-Saharan states, and according to the *Chronic Poverty Report 2004–05*, all sixteen of the most "desperately deprived" countries are located in sub-Saharan Africa. Over 300 million people live on less than $2 per day—and this is expected to rise to 400 million by 2015. One-third of the population of the continent is undernourished (Benson 2004); stunting rates run at almost 40%. According to a United Nations Food and Agriculture Organization assessment in January 2006, twenty-seven countries are in need of emergency food relief.

The neoliberal tsunami broke with a dreadful ferocity on African cities, and the African slum world in particular. Reform—the privatization of public utilities creating massive corporate profits and a decline in service provision, the slashing of urban services, the immiseration of many sectors of the public workforce, the collapse of manufactures and real wages, and often the disappearance of the middle class—was, as Mike Davis (2006) notes, remorselessly anti-urban in its effects. As a consequence, African cities confronted the horrifying realities of an economic contraction of 2%–5% per year combined with sustained population growth of up to 10% per annum (Zimbabwe's urban labor market grew by 300,000 per year in the 1990s while urban employment grew by just 3% of that figure). In Dar es Salaam public service expenditures per capita fell by 10% a year in the 1980s; in Khartoum adjustment created one million "new poor," and urban poverty in Nigeria almost tripled between 1980 and the mid-1990s. No wonder that 85% of urban growth in Nairobi, Kinshasha, and Nouakchott in the 1980s and 1990s was accommodated in the slums barracks of sprawling and ungovernable cities. Lagos, everyone's worst urban apocalypse (Packer 2006), grew from 300,000 to 13 million in over 50 years and is expected to become part of a vast Gulf of Guinea slum expected to accommodate 60 million poor along a littoral corridor 600 kilometers stretching from Benin City to Accra by 2020. Black Africa will contain 332 million slum dwellers by 2015, a figure expected to double every 15 years (Davis 2006, 19). The pillaging and privatization of the state and the African commons is the most extraordinary spectacle of accumulation by dispossession, all made in the name of foreign assistance (Harvey 2005). The involution of the African city has as its corollary not an insurgent lumpenproletariat but, as Davis himself laconically admits, the rise of "Mohammed and the Holy Ghost," a vast political universe of Islamism and Pentecostalism. It is this occult world of invisible powers—whether populist Islam in Kano or witchcraft in Soweto (Ashforth 2005)—that represents the most compelling ideological legacy of neoliberal utopianism in Africa.

* * *

Of course, there are those within the development business for whom the failure of secular nationalist development is a result not of too much neoliberalism, but not enough. The complaint here, typically from those within the free-market establishment, is that adjustment and stabilization has never really been implemented (a right-wing version of the left-wing claim that adjustment was asking African ruling classes to commit political suicide). SAP's were simply "vetriloquism" in which, as van de Walle (cited in Easterly 2005, 146) sees it, the IMF/IBRD had given up trying to get African governments to do things but simply want to be told what governments might do to get a loan. There is, of course, some truth to this, but the cry of any planned failure will always be "we were defeated by not going far enough." David Harvey (2005) has described the radically uneven geographical patterns of neoliberal governance and rule. Yet he makes the point that within this complexity is a "universal tendency" to increase social inequality and expose the poor to austerity and marginalization. And the reality is that in Africa World Bank reforms, and the pressures imposed by the WTO from the mid-1990s onwards, did have drastic consequences for trade and investment—the litmus test of neoliberal development—seen in the widespread dismantling of state marketing boards and of trade protections. And here the picture is devastating. In absolute terms African exports grew quite rapidly from 1963 to 2000, but at a much slower rate than world trade generally. Africa's share of world exports fell from almost 6% in 1962 to 2% in 2000. In non-oil products (food and manufactures) growth rates of exports between 1980 and 1998 were miserable. Rodrik (1999) argues that given African conditions (income, geography, and socioeconomic conditions), the performance is "average." Yet it is incontestable that African exports are characterized overall by a "disintegration from Northern markets" and "isolation from more dynamic developments in the composition of international trade" (Gibbon and Ponte 2005, 44). UNCTAD showed that of the exports from twenty-six African states, the average concentration on primary exports has remained basically unchanged (roughly 85%) since 1980. In all categories, sub-Saharan Africa has failed to move up the value-added chain away from primary commodities.

The African accumulation crisis and the dynamics of capital and trade flows are in practice complex and uneven (Hart 2003). In addition to oil (and the very few cases of manufacturing growth in places like Mauritius which are simply national export-processing zones), the other source of economic dynamism is the (uneven) emergence of global value chains especially around high-value agricultures such as fresh fruits and vegetables (Friedberg 2004; Gibbon and Ponte 2005) in South Africa, flowers in Kenya, green beans in Senegal. These forms of contract production, typically buyer-driven commodity chains in which retailers exert enormous power, have created islands of agrarian capitalism, but as Gibbon and Ponte show, they contribute to and deepen patterns of existing inequality across Africa and rest on "concentrations of private economic power" (2005, 160), which typically means non-African business elites. The deepening of commodification—driven in part by the dismantling of the marketing boards under neoliberalism—in tandem with demographic pressures (caused as much by civil war and displacement as high fertility regimes) has made land struggles a vivid part of the new landscape of African development (Peters 2004; Woodhouse et. al. 2000).

In reality what is on offer is a bleak world of military neoliberalism. At one pole are enclaves of often militarily fortified accumulation (of which the oil complex is the paradigmatic case; see Ferguson 2006) and the violent, sometimes chaotic, markets so graphically depicted in the powerful documentary film by Hubert Sauper *Darwin's Nightmare*. On the other are the black holes of recession, withdrawal, and uneven commodification (Bernstein 2004). These complex trajectories of accumulation are dominated at this moment by the centrality of extraction and a return to primary commodity production.

The New African "Gulf States"

One aspect of the doubleness that I referred to earlier is that on this bleak canvas of development failure in Africa is the undisputed fact that currently the continent stands at the centre of a major oil boom. To the extent there is any economic dynamism, in other words, it resides in the primary commodities sector—oil and gas especially—which is the most important source of capitalist accumulation on the continent. Over the period 1981–1985, FDI inflow into Africa was running at $1.7 billion per annum; by 1991–1995 it had grown to $3.8 billion (Asiedu 2005; WIR 2005). Yet as a percentage of all developing country FDI inflow, the figure represented a secular decline from 9% to less than 5% (all-in-all miniscule compared to South and East Asia and Latin America).

Between 1995 and 2001, FDI inflow amounted to $7 billion per year, but almost two-thirds of the portfolio was destined for three countries (Angola, Nigeria, and South Africa, in which oil FDI accounted for 90% of all FDI inflow). Half of Africa's states had effectively none. Two-thirds of FDI was derived from the same three countries (UK, Germany, and the US) that had dominated FDI supply in 1980. According to the World Investment Report (2005), FDI into Africa is currently $18 billion; four oil-producing countries account for 50%, and the top ten almost three-quarters. To put the matter starkly, the vast bulk of private transnational investment—the hallmark of success for the neoliberal project—was monopolized by a quartet of mining-energy economies.

The continent accounts for roughly 10% of world oil output and 9.3% of known reserves. Though oil fields in Africa are generally smaller and deeper than the Middle East—and production costs are accordingly 3–4 times higher—African crude is generally low in sulfur and attractive to US importers. As a commercial producer of petroleum, Africa arrived, however, rather late to the hydrocarbon age. Oil production in Africa began in Egypt in 1910 and only in earnest in Libya and Algeria (under French and Italian auspices) in the 1930s and 1940s. Now there are twelve major oil producers in Africa—members of the African Petroleum Producers Association—dominated, in rank order of output, by Nigeria, Algeria, Libya, and Angola, which collectively account for 85% of African output. Up until the 1970s North Africa dominated production of oil and gas on the continent, but in the last three decades it has moved decisively to the Gulf of Guinea encompassing the rich on- and offshore fields stretching from Nigeria to Angola (Hodges 2001; Frynas 2004; Yates 1996). The Gulf—constituted by the so-called West African "Gulf States"—has emerged as the predominant African supplier to an increasingly tight and volatile world oil market. All of the major Gulf oil producers are highly oil-dependent: for the top six African oil states, petroleum accounts for between 75% and 95% of all oil export revenues, between 30% and 40% of GDP, and between 50% and 80% of all government revenues.

All of these petro-states are marked by staggering corruption, authoritarian rule, and miserable economic performance (the so-called "resource curse"; Ross 2001). The deadly operations of the alliance between corporate oil and autocratic oil states have partially helped force the question of transparency of oil operations onto the international agenda. Tony Blair's Extractive Industries Transparency Initiative, the IMF's Oil Diagnostic program, and the Soros Foundation's Revenue Watch are all "voluntary" regulatory efforts to provide a veneer of respectability to a rank and turbulent industry (Zalik 2004; Gary and Karl 2003; Watts 2005).

Nigeria: The Rise and Fall of an Oil State

Nigeria is the jewel in the Gulf of Guinea crown. One of every five Africans is a Nigerian—the country's population is currently estimated to be 137 million—and it is the world's seventh largest exporter of petroleum, providing the US market with roughly 12% of its imports. A longtime member of OPEC, Nigeria is an archetypical "oil nation." With reserves estimated at close to 40 billion barrels, oil accounted in 2004 for 80% of government revenues, 90% of foreign exchange earnings, 96% of export revenues according to the IMF, and almost half of GDP. Crude oil production runs currently at more than 2.1 million barrels per day valued at more than $45 billion per year in oil revenue. Mostly lifted onshore from about 250 fields dotted across the Niger Delta, Nigeria's oil sector now represents a vast domestic industrial infrastructure: more than three hundred oil fields, 5,284 wells, 7,000 kilometers of pipelines, ten export terminals, 275 flow stations, ten gas plants, four refineries (Warri, Port Hartcourt I and II, and Kaduna), and a massive liquefied natural gas (LNG) project (in Bonny and Brass).

The rise of Nigeria as a strategic player in the world of oil geopolitics has been dramatic and has occurred largely in the wake of the civil war that ended in 1970. In the late 1950s petroleum products were insignificant, amounting to less than 2% of total exports. Between 1960 and 1973 oil output exploded from just over 5 million to over 600 million barrels. Government oil revenues in turn accelerated from 66 million naira in 1970 to over 10 billion in 1980. A multi-billion-dollar oil industry has, however, proved to be a little more than a nightmare. To inventory the achievements of Nigerian oil development is a salutary exercise: 85% of oil revenues accrue to 1% of the population; perhaps $100 billion of $400 billion in revenues since 1970 have simply gone "missing" (the anti-corruption chief Nuhu Ribadu claimed that in 2003, 70% of the country's oil wealth was stolen or wasted; by 2005 it was "only" 40%). Over the period 1965–2004, the per capita income fell from $250 to $212; income distribution deteriorated markedly over the same period. Between 1970 and 2000 in Nigeria, the number of people subsisting on less than one dollar a day grew from 36% to more than 70%, from 19 million to a staggering 90 million. According to the IMF, oil "did not seem to add to the standard of living" and "could have contributed to a decline in the standard of living"

(Sala-i-Martin and Subramanian 2003, 4). Over the last decade GDP per capita and life expectancy have, according to World Bank estimates, both fallen.

What is on offer in the name of petro-development is the terrifying and cata- strophic failure of secular nationalist development. It is sometimes hard to gasp the full consequences and depth of such a claim. From the vantage point of the Niger Delta—but no less from the vast slum worlds of Kano or Lagos—development and oil wealth is a cruel joke. These paradoxes and contradictions of oil are nowhere greater than on the oilfields of the Niger Delta. In the oil rich states of Bayelsa and Delta the UN human development index fell between 1995 and 2005 (UNDP 2006). Oil has wrought only poverty, state violence, and a dying ecosystem, says Ike Okonta (2005). The government's presence, Okonta notes, "is only felt in the form of the machine gun and jackboots" (2005, 206). It is no great surprise that a half-century of neglect in the shadow of black gold has made for a combustible politics. All the while the democratic project initiated in 1999 appears ever more hollow. The May 2007 elections in which the ruling party was returned to power were widely seen to be massively fraudulent, perhaps the worst in Nigerian history.

The nightmarish legacy of oil politics must be traced back to the heady boom days of the 1970s. The boom detonated a huge influx of petro-dollars and launched an ambitious (and largely autocratic) state-led modernization program. Central to the operations of the new oil economy was the emergence of an "oil complex" that overlaps with, but is not identical to, the "petro-state" (Watts 2005). The latter is comprised of several key institutional elements: (1) a statutory monopoly over min- eral exploitation, (2) a nationalized (state) oil company that operates through joint ventures with oil majors who are granted territorial concessions (blocs), (3) the security apparatuses of the state (often working in a complementary fashion with the private security forces of the companies) who ensure that costly investments are secured, (4) the oil-producing communities themselves within whose custom- ary jurisdiction the wells are located, and (5) a politico-financial mechanism by which oil revenues are distributed to the federation (states, local governments, and central government).

The oil revenue distribution question—whether in a federal system like Ni- geria or in an autocratic monarchy like Saudi Arabia—is an indispensable part of understanding the combustible politics of imperial oil. In Nigeria there are four key distribution mechanisms: the federal account (rents appropriated directly by the federal state); a state derivation principle (the right of each state to a propor- tion of the taxes that its inhabitants are assumed to have contributed to the federal exchequer); the Federation Account (or States Joint Account), which allocates revenue to the states on the basis of need, population, and other criteria; and a Special Grants Account (which includes monies designated directly for the Niger Delta, for example, through the notoriously corrupt Niger Delta Development Com- mission). Over time the derivation revenues have fallen (and thereby revenues directly controlled by the oil-rich Niger Delta states have shriveled) and the States Joint Account has grown vastly. In short, there has been a process of radical fiscal centralism in which the oil-producing states (composed of ethnic minorities) have lost and the non-oil-producing states (composed of ethnic minorities) have lost and the non-oil-producing ethnic majorities have gained—by fair means or foul. Since the return to civilian rule however in 1999 the derivation principle has increased to 13%, and as a result the oil-producing states in the Niger delta—especially Bayelsa, Delta, and Rivers—have been awash in oil monies, a process that has simultaneously expanded the political powers of the state governors and vastly increased the opportunities for corruption (in effect there has been a decentralization of oil corruption away from the federal center). Neither of the processes of decentralization—of political power and of oil revenues—has had any development consequence for the impoverished oil fields and for the alienated ethnic minorities that constitute the heart of the oil region.

Overlaid upon the Nigerian petro-state is, in turn, a volatile mix of forces that give shape to the "oil complex." First, the geo-strategic interest in oil means that military and other forces are part of the local oil complex. Second, local and global civil society enters into the oil complex either through transnational advocacy groups concerned with human rights and the transparency of the entire oil sector, or through local social movements and NGOs fighting over the consequences of the oil industry and the accountability of the petro-state. Third, the transnational oil business—the majors, the independents, and the vast service industry—are actively involved in the process of local development through community development, corporate social responsibility, and stakeholder inclusion. Fourth, the inevitable struggle over oil wealth—who controls and owns it, who has rights over it, and how the wealth is to be deployed and used—inserts a panoply of local political forces (ethnic militias, paramilitaries, separatist movements, and so on) into the operations of the oil complex (the conditions in Colombia are an exemplary case). In some circumstances oil operations are the object of civil wars. Fifth, multilateral development agencies (the IMF and the IBRD) and financial corporations like the export credit agencies appear as key "brokers" in the construction and expansion of the energy sectors in oil-producing states (and latterly the multilaterals are pressured to become the enforcers of transparency among governments and oil companies). And not least, there is the relationship between oil and the shady world of drugs, illicit wealth (oil theft for example), mercenaries, and the black economy. It is out of this volatile and combustible mix of forces that an ungovernable and insurgent space called the Niger delta has emerged.

The Anatomy of an Oil Insurgency

How, then, can one grasp the transformation of the Niger delta into a space of insurgency? I cannot provide a full accounting here but rather want to identify a key number of processes generated from within the heart of the oil complex. Each is an expression of a long and deeper geography of exclusion and marginalization by which the oil-producing delta came to suffer all of the social and environmental harms of the oil industry and yet receives in return (until recently) very little of the oil revenues. It is from the geo-political contradiction of oil without wealth—a bequest of the oil complex—that the insurgency has drawn sustenance. In this sense, the insurgency does not appear to be a shining example of the influential predation theory of rebellion proposed by Paul Collier and his World Bank associates (2003). In this view insurgency is less about grievance than greed and rebellion is a form of organized crime. While the Niger delta has its fair share of predation and greed, to see the insurgency as a product of youth crime is to misconstrue its geopolitical and historical origins.

What were the forces that emerged from this geopolitical contradiction? The first, not surprisingly in a region of sixty of more ethnic groups and a powerful set of institutions of customary rule, was ethnonationalism. This was central, of course, to the Ogoni movement, but the banner has been taken up in the last decade or so by the Ijaw, the largest ethnic or so-called "oil" minority in the Delta. Their exclusion from the oil wealth (and the federal revenue allocation process), to say nothing of bearing the environmental costs of oil operations across the oilfields, became central to the emergence of a new sort of youth politics. The establishment of the Ijaw Youth Congress in 1998 marked a watershed in this regard, and it became the vehicle through which a new generation of youth leaders took up the struggle. Many were mobilized in and around youth movements and came to assume local positions of power, including a number who took up an explicitly militant anti-state insurgent stance especially in the wake of the hanging of Ken Saro-Wiwa when Gandhian tactics were, in some quarters, seen to have failed catastrophically.

The second force was the inability and unwillingness of the Nigerian state in its military and civilian guises to address this political mobilization in the Delta without resorting to state-imposed violence by an undisciplined military, police, and security forces. In this sense the history of the Ogoni struggle was a watershed too insofar as it bequeathed a generation of militants for whom MOSOP represented a failure of non-violent politics. The return to civilian rule in 1999 saw a further militarization of the region in which communities were violated and experienced the undisciplined violence of state security forces. The destruction of Odi (1999) and Odiama (2005) by military forces and the violence meted out by the Joint Military Task Force based in Warri were the most dramatic instances of state intimidation. This unrelenting militarization of the region to secure "national oil assets" further propelled the frustrations of a generation of youth who, in the period since the 1980s, had grown in their organizational capacities.

Third, the militant groups themselves represented the intersection of two important forces: on the one hand the rise of youth politics in which a younger generation, whose economic and political prospects were stymied, began to challenge both customary forms of chiefly power, and on the other the vast corruption of the petro-state (whether military or civilian). These twin processes have a long history dating back at least to the famous Twelve Day Republic in which, in 1965, a group of young Ijaw men proclaimed, against a backdrop of expanding oil output, an independent Ijaw state. But the political mobilization of the youth turned from a sort of peaceful civic nationalism increasingly toward militancy, and this in turn, as I have suggested, was driven by the violence of the Nigerian military forces. But in addition the politicians, especially the increasingly powerful governors who in the period since 1999 have assumed direct control over huge flows of oil monies through the federal allocation process and derivation, sought to make use of the youth movements for their own electoral purposes (that is to say political thugs to intimidate voters). Paradoxically a number of the militias often got their start by being bankrolled by the state and politicians and indeed the NDF and NPDVF were both fuelled by machine politicians during the notoriously corrupt 1999 and 2003 elections. It has been the radical decentralization of power and corruption downward and the escalation of youth politics upwards that has produced a fertile soil in which popular grievances could take on a militant cast—aided and abetted of course by the ease with which weapons could be obtained (through the privatization of the arms trade and the corruption of local military). The means of violence were "democratized."

Fourth, the existence and proliferation of oil theft, known locally as "oil bunkering," provided a financial mechanism through which militants could (after being abandoned by their political patrons) finance their operations and attract recruits. The organization of the oil theft trade, which by 2004 was a multi-billion-dollar industry involving high-ranking military, government officials, and merchants, drew upon the local militia to organize and protect the tapping of pipelines and the movement of barges through the creeks and ultimately offshore to large tankers. This is, on its face, a case of the sort of organized crime that Collier invokes in his account of the economics of rebellion—and indeed there are explicitly criminal elements and syndicates at work in the operations of a vast bunkering business in Nigeria—yet the theft of oil provided a lubricant for a ready existing set of grievances. Furthermore, it throws into question the sharp distinction between state and insurgent upon which the entire Collier edifice rests: oil bunkering precisely links the insurgent with state and military officials. Rebel organizations and insurgents were, in this sense, not merely criminal gangs.

And finally, the operations of the oil companies—in their funding of youth groups as security forces, in their willingness to use military and security forces against protestors and militants alike, and in their corrupt practices of distributing rents to local community elites—all contributed to an environment in which military activity was in effect encouraged and facilitated. A number of companies used violent youth groups to protect their facilities (see WAC 2003). Corporate practice, and community development in particular, had the net effect of inserting millions of dollars of so-called "cash payments" into the local economy by paying corrupt chiefs, violent youth groups, or corrupt local officials in the hope that the oil would keep flowing. In practice the uneven record of community development projects and the illicit forms in which cash payments were made produced a growing hostility (expressed in the growth of oil platform occupations, attacks on pipelines, and more recently hostage taking) to the companies. Directly and indirectly corporate practice were essential to the dynamics of local violence and the escalation of insurgent activity.

The emergence of MEND in 2005 represents the almost inevitable end-point of the operations of this quarter of forces, powered by a process of marginalization and alienation that assumed a growing militancy during the 1990s. MEND has grown from an earlier history of increasingly militant youth embracing, for example, the Egbesu Boys of Africa, the Meinbutu Boys, and others in the Warri region dating back to the early 1990s. It is now something like a "franchise" insofar as it operates in a tense and complex way with other shady militant groups such as the Martyrs Brigade and the Committee on Militant Action. What is important to grasp is that MEND cannot be understood outside of the operation of the quartet of forces that I briefly outlined, and yet at the same time MEND is inextricably linked to local politics: struggles among and between two key Ijaw clans (Gbaramantu and Egbema) over access to oil monies, struggles with Chevron over the lack of a Memorandum of Understanding for so-called "host communities" in their clan territory, control of oil bunkering territories, and not least the complex politics of Warri city, the large oil town to the north. Here is a multi-ethnic city that has imploded since the 1990s as warring ethnic groups (fuelled by machine politics) have fought for the establishment of new local government authorities as a basis for laying claim to federal oil monies. Into this mix was the catalytic effect of the Nigerian special military task force (Operation Hope) that came to quell the growing militancy across the region in which the Gbarmantu clan territory was repeatedly attacked and bombed (Courson 2007). The social geography of clan territory was in this way converted into a space of insurgency.

Reflections

The insurgency across the Niger delta, involving a welter of differing groups and interests it needs to be said, is inextricably wrapped up with the intersection of generational politics, a corrupt and violent petro-state, irresponsible oil company practice, and the existence of a vast oil bunkering network. As Kalyvas (2001, 113) suggests, viewed from the microlevel these sorts of insurgencies—an oil insurgency in this case—resemble "welters of complex struggles" in which the notion that the rebels are criminals who operate against law-abiding states fails to capture the dynamics at work. Group interests are often "localistic and region-specific" (Kalyvas 2001, 112), yet, as I have tried to argue, their specificity emerges from the structured totality of the oil complex. It all makes for an enormously unstable and volatile mix of political, economic, and social forces, now located on a larger, and more intimidating, canvas of global oil instability and the Global War on Terror.

The operations of the oil complex and the violent and unstable spaces it creates seem to endorse Harvey's (2005) notion of accumulation by dispossession. The oil complex is a vast forcing house of primitive accumulation, repeating the original sin of robbery. It operates as if through a chain of enclosures, violent economies that dispossess at a variety of levels and through a raft of modalities. The rise of the resource control movement over the last 15 years, the rise of the oil minority, and the complex mix of ethnonationalism and insurgent politics across the Delta are reactive to—or drawing from Polanyi (1947) one might say a double movement against—Imperial Oil. What it has produced of course is a fragmented polity in which we have forms of parcellized and turbulent sovereignty (see Mbembe 2001), including insurgent spaces, rather than a robust modern oil nation.

Notes

1. The companies and government have typically denied the payments of ransoms to militants, but there have been reports in the press, by activists and others of payments in excess of $250,000. In fact the decline in oil bunkering since 2004 has seen militias turning to kidnapping and extortion as sources of revenues as bunkering income has fallen.

References

Anderson, P. (2001): Scurrying toward Bethlehem. In: *New Left Review* 10, 5–31.

Ashforth, A. (2005): *Witchcraft, violence and democracy in South Africa*. Chicago.

Asiedu, E. (2005): Foreign direct investment in Africa. Research Paper 24. WIDER, United Nations University, Helsinki.

Barnes, S. (2005): Global flows: terror, oil and strategic philanthropy. In: *African Studies Review* 48 (1), 1–23.

Benson, T. (2004): Africa's food and nutrition security situation. IFPRI Discussion Paper 37. Washington, DC.

Bernstein, H. (2004): Considering Africa's agrarian questions. In: *Historical Materialism* 12/4 (3–5), 115–44.

CFR (Council on Foreign Relations) (2005): More than humanitarianism. Task Force Report 56. New York.

Collier, P.; Elliott, V. L.; Hegre, H; Hoeffler, A.; Reynal-Querol, M.; and Sambanis, N. (2003): Breaking the conflict trap, civil war and development policy. World Bank Policy Research Report. New York.

Courson, E. (2007): The burden of oil: social deprivation and political militancy in Gbaramatu clan, Warri southwest LGA, Delta State, Nigeria. Niger Delta: Economies of Violence Project. Working Paper 15. Berkeley. http://globetrotter.berkeley.edu/NigerDelta/.

CSIS (Centre for Strategic and International Studies) (2007): Briefing on the Niger Delta. (March 14). Washington.

Davis, M. (2006): *Planet of the slums*. London.

Easterly, W. (2006): *The white man's burden*. London.

Ferguson, J. (2006): *Global shadows: Africa in the neoliberal world order*. Durham, NC.

Friedberg, S. (2004): *French beans and food scares*. Oxford.

Frynas, G. (2004): The oil boom in equatorial Guinea. In: *African Affairs* 103, 527–46.

Gary, I.; and Karl, T. L. (2003): *Bottom of the barrel: Africa's oil boom and the poor*. Baltimore, MD.

Gibbon, P.; and Ponte, S. (2005): *Trading down*. Philadelphia.

Hart, G. (2003): *Disabling globalization*. Berkeley.

Harvey, D. (2003): *The new imperialism*. London.

———. (2005): *A brief history of neoliberalism*. London.

Hodges, T. (2001): *Angola: From afro-stalinism to petro-diamond capitalism*. Oxford.

Kalyvas, S. (2001): New and old civil wars. In: *World Politics* 54, 99–118.

Lubeck, P.; Watts, M.; and Lipschitz, R. (2007): Convergent interests: US energy security and the securing of democracy in Nigeria. International Policy Report. Washington, DC.

Mbembe, A. (2001): At the edge of the world. In: *Public Culture* 12 (1), 259–84.

Okonta, I. (2005): Nigeria: Chronicle of a dying state. In: *Current History*, May, 203–8.

Packer, G. (2006): The Megacity: Decoding the chaos of Lagos. In: *New Yorker*, November 13. New York.

Peters, P. (2004): Inequality and social conflict over land. In: *Journal of Agrarian Change* 4/3, 269–314.

Polanyi, K. (1947): *The great transformation*. Boston.

Retort (2005): *Afflicted powers: Capital and spectacle in a new age of war*. London.

Rodrik, D. (1999): *The new global economy and developing countries*. London.

Ross, M. (2001): Does oil hinder democracy? In: *World Politics* 53, 325–61.

Sala-i-Martin, X.; and Subramanian, A. (2003): Addressing the resource curse: An illustration from Nigeria. International Monetary Fund (IMF) Working Paper 03/139. Washington, DC.

Toye, J. (1987): *Dilemmas of development*. Oxford.

UNDP (2006): The Niger Delta Human Development Report. Abuja.

WAC Global Services (2003): Peace and security in the Niger Delta. Port Harcourt.

Watts, M. (2005): Righteous oil? Human rights, the oil complex and corporate social responsibility. In: *Annual Review of Environment and Resources* 30, 373–407.

WIR (World Investment Report) (2005): United Nations. New York.

Woodhouse, P.; Bernstein, H.; and Hulme, D. (eds). (2000): *African enclosures*. London.

Yates, D. (1996): *The rentier state in Africa: Oil rent dependency and neocolonialism in the Republic of Gabon*. Trenton, NJ.

Zalik, A. (2004): The peace of the graveyard: The voluntary principles on security and Human Rights in the Niger Delta. In: Van der Pijl, K.; Assassi, W.; and Wigan, D. (eds.): *Global regulation: Managing crisis after the imperial turn*. London.

››› Excerpt from *The Cheviot, the Stag and the Black, Black Oil*
John McGrath

Music: Grannie's Hielan' Hame on accordion.

Enter TEXAS JIM, *in 10-gallon hat. He greets the audience fulsomely, shakes hands with the front row, etc.*

TEXAS JIM (*to the backing of the accordion*) In those far-off days of yore, my great-great grand-pappy Angus left these calm untroubled shores to seek his fortune in that great continent across the Atlantic Ocean. Well, he went North, and he struck cold and ice, and he went West, and he struck bad times on the great rolling plains, so he went South, and he struck oil; and here am I, a free-booting oil-man from Texas, name of Elmer Y. MacAlpine the Fourth, and I'm proud to say my trade has brought me back to these shores once more, and the tears well in my eyes as I see the Scottish Sun Sink Slowly in the West behind. . . . (*Sings.*)

My Grannie's Hielan' Hame.

Blue grass guitar in, country style. He changes from nostalgia to a more aggressive approach.

> For these are my mountains
> And this is my glen
> Yes, these are my mountains
> I'll tell you again—
> No land's ever claimed me
> Though far I did roam
> Yes these are my mountains
> And I—have come home.

Guitar continues: he fires pistol as oil rigs appear on the mountains.

Fiddle in for hoe-down. Company line up and begin to dance hoedown.

JIM *shakes hands with audience, then back to mike and begins square dance calls:*

The Cheviot, the Stag and the Black, Black Oil (London: Bloomsbury, 2015 [originally West Highland Publishing, 1974]), 145–60

TEXAS JIM Take your oil-rigs by the score,
    Drill a little well just a little off-shore,

    Pipe that oil in from the sea,
    Pipe those profits—home to me.

    I'll bring work that's hard and good—
    A little oil costs a lot of blood.

    Your union men just cut no ice
    You work for me—I name the price.

    So leave your fishing, and leave your soil,
    Come work for me, I want your oil.

    Screw your landscape, screw your bays
    I'll screw you in a hundred ways—

    Take your partner by the hand
    Tiptoe through the oily sand

    Honour your partner, bow real low
    You'll be honouring me in a year or so

    I'm going to grab a pile of dough
    When that there oil begins to flow

    I got millions, I want more
    I don't give a damn for your fancy shore

    1 2 3 4 5 6 7
    All good oil men go to heaven

    8 9 10 11 12
    Billions of dollars all to myself

    13 14 15 16
    All your government needs is fixing

    17 18 19 20
    You'll get nothing, I'll get plenty

21 22 23 24
Billion billion dollars more

25 26 27 28
Watch my cash accumulate

As he gets more and more frenzied, the dancers stop and look at him.

27 28 29 30
You play dumb and I'll play dirty

All you folks are off your head
I'm getting rich from your sea bed

I'll go home when I see fit
All I'll leave is a heap of shit

You poor dumb fools I'm rooking you
You'll find out in a year or two.

He stops, freaked out. The dancers back away from him.

He gets himself under control and speaks to the audience.

Our story begins way way back in 1962. Your wonderful government went
looking for gas in the North Sea, and they struck oil.

Guitar.

Well, they didn't know what to do about it, and they didn't believe in all
these pesky godless government controls like they do in Norway and Algeria
and Libya, oh, my God—no, you have a democracy here like we do—so your
government gave a little chance to honest God-fearing, anti-socialist
businessmen like myself—

Guitar.

Two Company members stand in their places to speak.

M.C.1 Shell-Esso of America, Transworld of America, Sedco of America,
Occidental of America—and of Lord Thomson.

M.C.2 Conoco, Amoco, Mobil, Signal.

TEXAS JIM All of America.

M.C.1 And British Petroleum—

TEXAS JIM A hell of a lot of American money, honey.

Guitar.

Enter WHITEHALL, *a worried senior Civil Servant.*

WHITEHALL You see, we just didn't have the money to squander on this sort of thing.

TEXAS JIM That's my boy—

WHITEHALL And we don't believe in fettering private enterprise: after all this is a free country.

TEXAS JIM Never known a freer one.

WHITEHALL These chaps have the know-how, and we don't.

TEXAS JIM Yes sir, and we certainly move fast.

M.C.1 By 1963 the North Sea was divided into blocks.

M.C.2 By 1964 100,000 square miles of sea-bed had been handed out for exploration.

WHITEHALL We didn't charge these chaps a lot of money, we didn't want to put them off.

TEXAS JIM Good thinking, good thinking. Your wonderful labourite government was real nice: thank God they weren't socialists.

M.C.1 The Norwegian Government took over 50% of the shares in exploration of their sector.

M.C.2 The Algerian Government control 80% of the oil industry in Algeria.

M.C.1 The Libyan Government are fighting to control 100% of the oil industry in Libya.

Guitar.

WHITEHALL Our allies in NATO were pressing us to get the oil flowing. There were Reds under the Med. Revolutions in the middle-east.

TEXAS JIM Yeah, Britain is a stable country and we can make sure you stay that way. (*Fingers pistol.*)

WHITEHALL There is a certain amount of disagreement about exactly how much oil there actually is out there. Some say 100 million tons a year, others as much as 600 million. I find myself awfully confused.

TEXAS JIM Good thinking. Good thinking.

WHITEHALL Besides, if we produce our own oil, it'll be cheaper, and we won't have to import it—will we?

M.C.1 As in all Third World countries exploited by American business, the raw material will be processed under the control of American capital—and sold back to us at three or four times the price—

M.C.2 To the detriment of our balance of payments, our cost of living and our way of life.

TEXAS JIM And to the greater glory of the economy of the US of A.

Intro. to song. Tune: souped-up version of "Bonnie Dundee."

TEXAS JIM *and* WHITEHALL *sing as an echo of* LOCH *and* SELLAR.

TEXAS JIM AND WHITEHALL
    As the rain on the hillside comes in from the sea
    All the blessings of life fall in showers from me
    So if you'd abandon your old misery
    Then you'll open your doors to the oil industry—

GIRLS (*as backing group*) Conoco, Amoco, Shell-Esso, Texaco, British Petroleum, yum, yum, yum. (*Twice.*)

TEXAS JIM There's many a barrel of oil in the sea
  All waiting for drilling and piping to me
  I'll refine it in Texas, you'll get it, you'll see
  At four times the price that you sold it to me.

TEXAS JIM AND WHITEHALL As the rain on the hillside, etc. (*Chorus.*)

GIRLS Conoco, Amoco, etc. (*Four times.*)

WHITEHALL
  There's jobs and there's prospects so please have no fears,
  There's building of oil rigs and houses and piers,
  There's a boom-time a-coming, let's celebrate—cheers—

TEXAS JIM *pours drinks of oil.*

TEXAS JIM For the Highlands will be my lands in three or four years.

No oil in can.

Enter ABERDONIAN RIGGER.

A.R. When it comes to the jobs all the big boys are American. All the technicians are American. Only about half the riggers are local. The American
  companies'll no take Union men, and some of the fellows recruiting for
  the Union have been beaten up. The fellows who get taken on as roust-
  abouts are on a contract; eighty-four hours a week in twelve hour shifts,
  two weeks on and "Conoco, Amoco, Shell-Esso, Texaco . . ." one week off.
  They have to do overtime when they're tell't. No accommodation, no leave,
  no sick-pay, and the company can sack them whenever they want to. And
  all that for twenty-seven pounds a week basic before tax. It's not what I'd
  cry a steady job for a family man. Of course, there's building jobs going
  but in a few years that'll be over, and by then we'll not be able to afford
  to live here. Some English property company has just sold eighty acres
  of Aberdeenshire for one million pounds. Even a stairhead tenement with
  a shared lavatory will cost you four thousand pounds in Aberdeen. At
  the first sniff of oil, there was a crowd of sharp operators jumping all
  over the place buying the land cheap. Now they're selling it at a hell of
  a profit.

Drum. Company step on stage again, speak to the audience.

M.C.1 In the House of Commons, Willie Hamilton, MP, said he was not laying charges at the door of any particular individual who had *quote:* moved in sharply to cash-in on the prospect of making a quick buck. There is a great danger of the local people being outwitted and out-manoeuvred by the Mafia from Edinburgh and Texas . . . end quote.

M.C.2 The people must own the land.

M.C.3 The people must control the land.

M.C.1 They must control what goes on it, and what gets taken out of it.

M.C.3 Listen to this. Farmers in Easter Ross have had their land bought by Cromarty Firth Development Company.

M.C.2 Crofters in Shetland have had their land bought by Nordport.

M.C.1 Farmers in Aberdeenshire have had their land bought by Peterhead and Fraserburgh Estates.

M.C.3 All three companies are owned by Onshore Investments "of Edinburgh."

M.C.2 Onshore Investments, however, was owned by Mount St Bernard Trust of London and Preston, Lancashire.

M.C.3 A man named John Foulerton manages this empire. But whose money is he handling? Who now owns this land in Easter Ross, Shetland, and Aberdeenshire? Whose money is waiting to buy *you* out?

Drum roll.

M.C.1 Marathon Oil?

M.C.2 Trafalgar House Investments?

M.C.3 Dearbourne Storm of Chicago?

M.C.4 Apco of Oklahoma?

M.C.5 Chicago Bridge and Iron of Chicago?

M.C.2 P&O Shipping?

M.C.3 Taylor-Woodrow?

M.C.1 Mowlems?

M.C.2 Costains?

M.C.5 Cementation?

M.C.4 Bovis?

M.C.3 Cleveland Bridge and Engineering?

M.C.2 These people have been buying up the North of Scotland.

TEXAS JIM A-a, a-a. With the help of your very own Scottish companies:
Ivory & Sime of Edinburgh; Edward Bates & Son of Edinburgh;
Noble Grossart, of Edinburgh; and the Bank of Scotland, of—er—
Scotland.

M.C.4 And the Shiek of Abu Dhabi's cousin who owns a large slice of the
Cromarty Firth . . .

M.C.2 Mrs Cowan, of the Strathy Inn, was offered a lot of money by a small
group of Japanese.

TEXAS JIM What can you little Scottish people do about it?

Silence. Exit TEXAS JIM.

M.C.2 Mr Gordon Campbell, in whose hands the future of Scotland rested at
this crucial period, said:

WHITEHALL *gets up, does nothing, sits down.*

M.C.2 Scottish capitalists are showing themselves to be, in the best tradition of
Loch and Sellar—ruthless exploiters.

Enter S.N.P. EMPLOYER.

S.N.P. EMPLOYER Not at all, no no, quit these Bolshevik haverings. Many of us captains of Scottish industry are joining the Nationalist party. We have the best interest of the Scottish people at heart. And with interest running at 16%, who can blame us?

M.C.2 Nationalism is not enough. The enemy of the Scottish people is Scottish capital, as much as the foreign exploiter.

Drum roll.

Actor who played SELLAR *and* WHITEHALL *comes on.*

ACTOR (*as* SELLAR) I'm not the cruel man you say I am. (*As* WHITEHALL.) I'm a Government spokesman and not responsible for my actions . . .

TEXAS JIM I am perfectly satisfied that no persons will suffer hardship or injury as a result of these improvements.

Drum roll.

Short burst on fiddle. JIM *and* WHITEHALL *go to shake hands. Enter between them, in black coat and bowler hat,* POLWARTH—*not unlike* SELKIRK's *entrance.*

POLWARTH I am Lord Polwarth, and I have a plan. The present government seems to have no control over the hooligans of the American oil companies and their overpaid government servants, so the government has appointed me to be a knot-cutter, a troubleshooter, a clearer of blockages, and a broad forum to cover the whole spectrum. However, I am not a supremo. In this way, the people of Scotland—or at least the Bank of Scotland—will benefit from the destruction of their country.

M.C.2 Before becoming Minister of State, Lord Polwarth was Governor of the Bank of Scotland, Chairman of the Save and Prosper Unit Trust, a Director of ICI and was heavily involved in British Assets Trust, Second British Assets Trust and Atlantic Assets Trust, which at that time owned 50% of our old friend, Mount St Bernard Trust.

Musical intro. Tune: Lord of the Dance.

TEXAS JIM *and* WHITEHALL *turn* LORD POLWARTH *into a puppet by taking out and holding up strings attached to his wrists and back. They sing:*

ALL Oil, oil, underneath the sea,
    I am the Lord of the Oil said he,
    And my friends in the Banks and the trusts all agree,
    I am the Lord of the Oil—Tee Hee.

POLWARTH I came up from London with amazing speed
    To save the Scottish Tories in their hour of need:
    The people up in Scotland were making such a noise,
    That Teddy sent for me, cos I'm a Teddy-boy . . .

WHITEHALL: "I am a Government spokesman and not responsible for my
    actions"

ALL Oil, oil, etc.

POLWARTH Now all you Scotties need have no fear,
    Your oil's quite safe now the trouble-shooter's here,
    So I'll trust you, if you'll trust me,
    Cos I'm the ex-director of a trust company.

ALL Oil, oil, etc.

POLWARTH Now I am a man of high integrity,
    Renowned for my complete impartiality,
    But if you think I'm doing this for you,
    You'd better think again cos I'm a businessman too—

ALL Oil, oil, etc.

At the end of the song, LORD POLWARTH *freezes.* TEXAS JIM *and* WHITEHALL *let
go of his strings, and he collapses.* M.C.2. *catches him on her shoulder and carries
him off.* JIM *and* WHITEHALL *congratulate each other, then turn to the audience.*

TEXAS JIM And the West is next in line.

LORD POLWARTH: "I am the Lord of the Oil said he"

WHITEHALL And the West is next in line.

GAELIC SINGER And the West is next in line. Even now exploration is going on
    between the Butt of Lewis and the coast of Sutherland.

WHITEHALL Don't worry, it will take at least five years.

GAELIC SINGER They've started buying land already.

WHITEHALL We can't interfere with the free play of the market.

TEXAS JIM Leave it to me, I'll take it out as quick as I can and leave you just as I found you.

GAELIC SINGER Worse, by all accounts.

WHITEHALL Now look here, we don't want you people interfering and disturbing the peace—what do you know about it?

GAELIC SINGER We'd like to know a hell of a lot more . . . (*Exit.*)

WHITEHALL As our own Mr. Fanshaw of the HIDB said:
"These oil rigs are quite spectacular. I hear they actually attract the tourists—"

Enter CROFTER *and his* WIFE.

WHITEHALL You can give them bed and breakfast.

Doorbell rings.

WIFE Get your shoes on, that'll be the tourists from Rotherham, Yorks, and put some peats on top of that coal—they'll think we're no better than theirselves.

CROFTER Aye, aye, aye—go you and let them in . . .

WIFE Put off that television and hunt for Jimmy Shand on the wireless.

CROFTER *mimes this action.*

Oh God, there's the Marvel milk out on the table, and I told them we had our own cows—

Bell rings again.

CROFTER Aye, aye, aye, they'll be looking like snowmen stuck out there in this blizzard—

WIFE Och, it's terrible weather for July—

CROFTER It's not been the same since they struck oil in Loch Duich.

WIFE Now is everything right?

She wraps a shawl round her head; he rolls up his trouser leg and throws a blanket round himself to look like a kilt, and puts on a tammy.

WIFE Get out your chanter and play them a quick failte.

CROFTER How many would you like?

WIFE Just the one—

He plays a blast of Amazing Grace. She takes a deep breath and opens the door. The visitors are mimed.

WIFE Dear heart step forward, come in, come in. (*Clicks fingers to* CROFTER.)

CROFTER (*brightly*) Och aye!

WIFE You'll have come to see the oil-rigs—oh, they're a grand sight, right enough. You'll no see them now for the stour, but on a clear day you'll get a grand view if you stand just here—

CROFTER Aye, you'll get a much better view now the excavators digging for the minerals have cleared away two and a half of the Five Sisters of Kintail.

WIFE You'll see them standing fine and dandy, just to the west of the wee labour camp there—

CROFTER And you'll see all the bonnie big tankers come steaming up the loch without moving from your chair—

WIFE You'll take a dram? Get a wee drammie for the visitors—

CROFTER A what?

WIFE A *drink*. I doubt you'll have anything like this down in Rotherham, Yorks. All the people from England are flocking up to see the oil-rigs. It'll be a change for them.

CROFTER Here, drink that now, it'll make the hairs on your chest stick out like rhubarb stalks.

WIFE When the weather clears up, you'll be wanting down to the shore to see the pollution—it's a grand sight, right enough.

CROFTER Aye, it's a big draw for the tourists: they're clicking away at it with their wee cameras all day long.

WIFE Or you can get Donnie MacKinnon to take you in his boat out to the point there, to watch the rockets whooshing off down the range—but he'll no go too far, for fear of the torpedoes. Himself here would take you but he gave up the fishing a while back.

CROFTER It's no safe any more with the aerial bombs they're testing in the Sound. Anyway, all the fish is buggered off to Iceland.

WIFE What does he do now? Oh, well, he had to get a job on the oil-rigs.

CROFTER Oh, aye, it was a good job, plenty money . . .

WIFE He fell down and shattered his spine from carelessness. (*Clicks her fingers at him.*)

CROFTER (*brightly*) Och aye!

WIFE And now he can't move out of his chair. But he has a grand view of the oil-rig to give him something to look at, and helping me with the visitors to occupy him.

⟩⟩⟩ Nuclear Ontologies

Gabrielle Hecht

I

In October 2002, George W. Bush claimed that Saddam Hussein had "recently sought significant quantities of uranium from Africa." Pressed for details, his administration cited CIA intelligence that Iraq had tried to purchase 500 tons of uranium from Niger. Skepticism mounted, with the CIA itself expressing doubts about the intelligence. By the 2003 State of the Union address, senior officials instead credited the information to British intelligence. But the source mattered less, according to Bush, than the inescapable conclusion: Iraq planned to build nuclear weapons.

Since then, both the uranium and the weapons claims have been decisively refuted. In February 2002, the CIA sent former diplomat Joseph Wilson to investigate whether Niger had indeed concluded a deal with Iraq. Wilson found no trace of the alleged sale. When he heard Bush's statement a few months later, he initially assumed that the president meant some other uranium-producing African nation. Upon realizing that Bush really did mean Niger, an appalled Wilson went public in the *New York Times*. In an attempt to discredit him, an "anonymous source" from the Bush administration outed Wilson's wife as a CIA operative. Meanwhile, it turned out that the proof that Saddam sought uranium "from Africa" consisted of forged documents peddled by an Italian businessman. Most analysts now agree that there simply was no Iraq–Niger uranium deal.[1]

The re-emergence of this story in the media produced a cacophony of opinion, mostly pitting Bush stalwarts against defenders of Wilson. To the administration's obvious delight, bickering over personal credibility drowned out the pivotal issue of whether Bush misled the nation into war. It also bypassed the real significance of the Niger episode for contemporary global nuclear relations.

Consider the political and technical parameters of the administration's claims. Bush officials repeatedly stated that Iraq had sought uranium "from Africa." Had Saddam been suspected of approaching Kazakhstan, would they have asserted that he'd sought uranium "from Asia"? Highly unlikely. Africa remains the "dark continent," mysterious and politically corrupt—highly plausible qualifications for a nuclear supplier. And when details were required, what better candidate for shady dealings than Niger, a nation most Americans couldn't distinguish from Nigeria? Consider also the assumption that acquisition of "uranium" would constitute prima

Constellations 13, no. 3 (2006): 320–31

facie evidence of a bomb program's existence. In public discourse, "uranium" seems inseparably linked to nuclear weapons. But before "uranium" becomes weapons-usable, it must be mined as ore, processed into yellowcake, converted into uranium hexafluoride, enriched, and pressed into bomb fuel. "Uranium" is therefore as underspecified technologically as "Africa" is underspecified politically.

The Niger uranium episode draws attention to the ambiguities of the nuclear state, and to the state of being nuclear. What exactly is a "nuclear state," and how do we know? Are the criteria scientific, technical, political, systemic? The ambiguities underlying the episode cannot merely be dismissed as Bush administration doublespeak. On the contrary: they lie at the heart of today's global nuclear order —or dis-order, as the case may be.

From 1945, both cold warriors and their activist opponents cultivated the notion of nuclear exceptionalism. Atomic weapons were portrayed as fundamentally different from any other human creation. "The bomb" became the ultimate trump card; geopolitical status seemed directly proportional to the number of nukes a nation possessed. And nuclear exceptionalism—along with an accompanying rhetoric of rupture—went well beyond geopolitics. "Nuclear" scientists and engineers gained prestige, power, and funding far beyond their colleagues in "conventional" research. Anti-nuclear activists argued that nuclear technologies posed qualitatively and quantitatively distinct, never-before-encountered dangers. The fact that "going nuclear" involved splitting atoms—creating rupture in nature's very building blocks—only strengthened this exceptionalism. Asserting the ontological distinctiveness of "the nuclear" carried political, cultural, and economic stakes amplified by morality-talk, which tended to boil down to a simple duality: nuclear technology represented either salvation or depravity.

In Europe and North America, the simplifications of nuclear exceptionalism did not end with the Cold War. They merely shifted terrain, as the "clash of civilizations" replaced the "superpower struggle."[2] Once again, the discourse of exceptionalism crossed political boundaries. In 1990, left-wing intellectual Régis Debray opined that "broadly speaking, green [i.e., Islam] has replaced red as the rising force." This was especially frightening because "the nuclear and rational North deters the nuclear and rational North, not the conventional and mystical South."[3] On the right, Bush's "axis of evil" formulation escalated fears that nuclearity might escape the control of that "rational North."

Yet the "nuclear" has never been the exclusive province of "the North"; its tentacles have always wrapped around the globe. Consider uranium production. Uranium for the Hiroshima bomb came from the Belgian Congo. Britain's weapons program exploited imperial ties to uranium-supplying regions in Africa. Uranium reserves gave Australia and South Africa a material role in the "defense of the West." France's nuclear program depended on uranium in its African colonies.

Such a list could continue for pages. But many policy wonks would immediately raise an objection: by itself, they would say, uranium mining does not make a nation "nuclear." In 1995, for example, a major report by the Office of Technology Assessment with a summary of the "nuclear activities" of 172 nations did not list Gabon, Niger, or Namibia—which together accounted for nearly 25% of the world's uranium production that year—as having any "nuclear activities."[4] Does this mean that uranium mines are somehow *not* nuclear? Can nuclearity, like radioactivity, be measured with a Geiger counter? Working in uranium mines exposes miners to radiation. Usually (though not always) they face lower exposure levels than workers in uranium enrichment facilities (whose nuclearity no one doubts). In the 1950s, Tandroy men worked small open-cast mines in Madagascar with hand tools. The ore they dug underwent only the most rudimentary treatment before shipment to France for refining and conversion. Tandroy miners themselves knew nothing of atom bombs, nuclear power plants, or radioactivity. They certainly received some radiation exposure, but no one will ever know how much.[5] Were the Malagasy mines somehow less "nuclear" than the Ranger mine in Australia, which has been subject to intense political, social, and environmental scrutiny since 1975? Were all of these mines, in turn, less "nuclear" than the French conversion plant at Malvési, the European enrichment facility in Almelo, or the US weapons labs at Livermore?

Such questions are problems of ontology, not merely of rhetoric. The *nuclearity* of a nation, a program, a technology, or a material—that is, the degree to which any of these things counts as "nuclear"—can never be defined in simple, clear-cut, scientific terms. Rather, nuclearity is a technopolitical spectrum that shifts in time and space. It is a historical and geographical condition, *as well as* a scientific and technological one. And nuclearity, in turn, has significant consequences for politics, culture, and health. Degrees of nuclearity structure global control over the flow of radioactive materials; they constitute the conceptual bedrock of anti-nuclear movements; they affect regulatory frameworks for occupational health and compensation for work-related illnesses. This matter of degree is crucial; nuclearity is not an on-off condition. Nuclear ontologies have a history, and a geography.

We cannot understand the geography of nuclearity without taking into account another kind of geopolitical rupture-talk from the Cold War period: the discourse of decolonization. Less than 3 months after the US bombed Hiroshima, the United Nations charter became the first document of international law to refer to "the principle of equal rights and self-determination of peoples." In principle (though certainly not in practice), a new world order had emerged built upon a foundation of equality for all. Independence would free Africans and Asians from the shackles of white rule. Formerly colonized people could choose their leaders, pursue economic prosperity, educate their children, and join the global community as peers. New nation-states would serve the interests of their people, who for the first time would be citizens rather than subjects. Like those of nuclearity, these ruptures too were matters of morality: the 1948 Universal Declaration of Human Rights was construed as a moral leap forward for humankind.

Political leaders blended nuclear and postcolonial discourses about rupture and morality in a variety of ways. Postwar French and British leaders not only hoped that the atom bomb would substitute for colonialism as an instrument of global power; they also saw in it a means of preventing their own colonization by the superpowers. Consider this remark by Churchill's chief scientific advisor, Lord Cherwell, in 1951: "If we have to rely entirely on the United States army for this vital weapon, we shall sink to the rank of a second-class nation, only permitted to supply auxiliary troops, like the native levies who were allowed small arms but no artillery." Or French parliamentary deputy Félix Gaillard that same year: "those nations which [do] not follow a clear path of atomic development [will] be, 25 years hence, as backward relative to the nuclear nations of that time as the primitive peoples of Africa [are] to the industrialized nations of today." Nuclear = colonizer. Non-nuclear = colonized. Africa remained the eternal site of, and metonym for, backwardness.[6]

For Europeans this act of technopolitical mapping had deep roots, extending the assumptions and practices of the "new imperialism" to the nuclear state and to the state of being nuclear. Colonial warfare rested on the assumption that different moral structures underlay the rules of war for battles between "civilized" nations and conflicts with "savages." Aerial bombing followed machine guns as tools of extermination. Its first victims lived in oases outside Tripoli (1911) and villages in Morocco (1913). Even as ecstatic prophets in Europe and America proclaimed the airplane's ability to ensure world peace, the RAF experimented with strategic bombing in Baghdad (1923) and the French bombarded Damascus (1925). For prescient science fiction writers, it was only a matter of time before atomic energy would follow suit. And in a Pacific war with virulent racial overtones, it did. Several hundred thousand Japanese became the first victims of the "white race's superweapon."[7] As the Atomic Bomb Casualty Commission industriously erected colonial scientific structures to study the explosions' aftermath,[8] the US and Britain had already begun to scour African colonies in a desperate bid to monopolize the magic new stuff of geopolitical power: uranium.

II

In 1953, "Atoms for Peace" signaled an emerging new form of nuclearity: atomic power plants. As the program morphed into the International Atomic Energy Agency, the atomic creed left space for postcolonial leaders to challenge the technopolitical geography of nuclearity asserted by the West. The Indians stepped in first, as Jawaharlal Nehru proclaimed nuclear development a fundamental building block of Indian national identity. During negotiations over the IAEA statute, Indian delegates raised a challenge. If representation on the IAEA Board of Governors relied solely on technical achievement and a Cold War East-West balance, they charged, the agency would only reproduce immoral global imbalances. Instead, qualification for Board membership should combine nuclear "advancement" with regional distribution. In the end, India succeeded. A complex formula allocated five permanent IAEA Board seats to member states deemed the "most advanced in the technology of atomic energy including the production of source materials" globally, and another five according to geographical region.[9] Uranium producers in Eastern and Western bloc nations would rotate through another two seats, and "suppliers of technical assistance" would rotate through one seat. The ten final spots would be electively distributed among the eight IAEA regions. The resulting emphasis on "advancement" made the Cold War obsession with technological rankings a structural feature of the IAEA. But geography and national history also mattered. The regional framework accommodated—even encouraged— postcolonial fantasies of nuclear nationalism.

But what would make a nation count as "most advanced in the technology of atomic energy including the production of source materials"? What were "source materials," and how significant a manifestation of nuclearity were they? In the half-century since these phrases laid the foundation for the global nuclear order, their meanings have been negotiated and renegotiated in treaties, contracts, and practices. A few examples of these ontological shifts are enough to glimpse the high stakes of nuclear exceptionalism.

Consider the role of apartheid South Africa, whose delegate was responsible for including "source materials" as an indicator of nuclear technological "advancement" in the IAEA statute. By 1956, contracts signed with the US and Britain had made uranium production vital to South Africa's economy.[10] Anticipating that the IAEA would play a central role in shaping the emerging uranium market, South Africa was determined to obtain a statutory seat on the Board. Because the apartheid state represented the very antithesis of the postcolonial settlement pursued by India, only by asserting a depoliticized, technical vision of nuclearity could South Africa hope to secure its seat.

When IAEA statute discussions took place in 1954–56, South African nuclearity was limited to uranium production, underwritten by a very small research program. This was an increasingly tenuous basis for a claim to superior "advancement," for in the mid-1950s the nuclearity of uranium was in flux. Prior to that, the uranium narrative went something like this:

• Uranium was the only naturally occurring radioactive material that could fuel atomic bombs. These, in turn, were a fundamentally new kind of weapon, capable of rupturing not only global order but the globe itself.

- Uranium was a rare ore. If the West could monopolize its supply, it could keep the Communist ogre at bay and make the world safe for democracy. The West therefore had to secure all sources of uranium around the world. Nothing mattered more.

- Uranium's crucial importance made it imperative to proceed as secretly as feasible. Geological surveys, actual and potential reserves, means of production, terms of sales contracts: state secrets one and all. And if uranium's nuclearity imposed secrecy, that secrecy in turn reinforced the ore's nuclearity. Uranium was the only ore ever subject to legislation specifically targeted at ensuring the secrecy of its conditions of production.

By the mid-1950s, however, it had become clear that while pitchblende (very high grade uranium) was rare, lower grades of ore were not. The stuff was everywhere. Meanwhile, the Soviets had plainly found their own sources. So Western monopoly of "source material"—if defined simply as raw ore—would be impossible. The challenge lay not in *finding* uranium ore, but in *processing* it to weapons-grade quality.

In IAEA statute discussions, one sign that the nuclearity of uranium ore had eroded was that nations whose primary claim to nuclearity lay in uranium production would have to rotate seats on the IAEA Board. Indeed, India tried to relegate South Africa and Australia to mere "producers" (rather than "most advanced" in their regions). Prevailing on their powerful customers (the US and Britain) for support, South African delegates insisted that "source materials" should count as an indicator of "advancement."[11]

In terms of technological practice, South Africa was no more "advanced" in 1957 than, say, Portugal (which also mined uranium). The difference lay in technopolitical geography. Portugal was in Western Europe, a region at the pinnacle of nuclear "advancement." South Africa was in the IAEA's Africa / Middle East region, where its competitors for nuclearity—Israel and Egypt—carried political baggage even heavier than its own. In a technopolitical geography where the Cold War trumped racial injustice, South Africa's uranium production could serve as the pinnacle of African nuclearity.

III

Matters of ontology also drove discussions about how to safeguard the nuclear order: how to control the flow of materials and technologies to ensure that instruments of planetary destruction didn't fall into the wrong hands. States not only had to agree on *how* to regulate, they also had to agree on *whom* and *what* to regulate. Who could be trusted with which systems? Which materials, knowledges, and systems were specifically nuclear? Of these, which were unique to atomic weapons and which were dual-use? It seemed understood that strongly nuclear materials or systems should be subject to stricter controls than weakly nuclear ones, but what did this mean in practice?

For example, "fissionable materials" clearly needed safeguarding. But what *was* a "fissionable material"? Uranium ore had to undergo milling, refinement into yellowcake, conversion to uranium hexafluoride, and enrichment before it could become fuel for nuclear reactors (or bombs). At exactly what point did uranium stop being "source material" and become "fissionable material"? The difference mattered enormously, because the two categories would be subject to different controls. In the words of one South African scientist, "the definitions would have to be essentially practical, rather than 'textbook' in nature, . . . legally watertight, and must take account of certain political implications." In the end, the IAEA definitions committee abandoned the more ambiguous term "fissionable material" (preferred by Indian delegates) in favor of three other categories: "source materials," "special fissionable materials," and "uranium enriched in the isotope 235 or 233."[12]

Exactly what would "safeguards" mean? Definitions alone didn't determine a method of control. The US promoted a pledge system in which purchasers of nuclear technologies and materials would agree not to use their purchases toward military ends, and accept international inspections verifying compliance. Most other nations *selling* nuclear systems paid lip service, at least, to controls on the flow of nuclear materials and technologies. Buyers, however, were underwhelmed by the prospect of controls. India, in particular, argued that regulating access would perpetuate colonial inequalities and undermine national sovereignty.

Neither side was completely disingenuous, but nor did these arguments tell the whole story. They obscured more mundane political and commercial issues. South Africa and Canada, for example, wanted to avoid controls on uranium end-uses which might put their product at a commercial disadvantage. Within India, experts and institutions disagreed over whether to build an atomic bomb, but wanted to keep their options open by minimizing international controls. Meanwhile, the US, the UK, and the Soviet Union refused to accept inspections of their nuclear installations. Western European nations accused the US and the UK of seeking competitive advantage, arguing that they too should be exempt from IAEA inspections, and be subject only to Euratom safeguards. Many "Third World" nations saw such arrangements as straightforward moves by the North to dominate the global South by writing the rule book in its own favor.[13]

The 1970 Treaty on the Non-Proliferation of Nuclear Weapons (NPT) expressed all of these tensions. Under the NPT, "nuclear weapons states" pledged not to transfer atomic weapons or explosive devices to "non-nuclear weapons states." The latter, in turn, renounced atomic weapons and agreed to accept IAEA safeguards and compliance measures. Strikingly, the NPT invoked human rights language and the rhetoric of development:

1. Nothing in this Treaty shall be interpreted as affecting the *inalienable right* of all the Parties to the Treaty to develop research, production and use of nuclear energy for peaceful purposes. . . .

2. All the Parties to the Treaty undertake to facilitate, and *have the right to participate in*, the fullest possible exchange of equipment, materials and scientific and technological information for the peaceful uses of nuclear energy. Parties to the Treaty in a position to do so shall also cooperate in contributing alone or together with other States or international organizations to the further development of the applications of nuclear energy for peaceful purposes, especially in the territories of non-nuclear-weapon States Party to the Treaty, *with due consideration for the needs of the developing areas of the world.*[14]

In an effort to accommodate postcolonial morality and palliate the ascendancy of the Cold War paradigm, the NPT essentially declared that nuclearity—of the "peaceful" persuasion—was a fundamental right. It thus codified nuclear exceptionalism: no other international agreements referred to any scientific or technological activity as an "*inalienable* right" of special importance to "the developing areas of the world." And what the NPT codified, the IAEA implemented, via its program of "technical assistance" to developing nations.[15] This, incidentally, is the historical basis for Iran's current invocation of its "inalienable right" to nuclearity—and one reason why European negotiators found it legally impossible to deny that claim outright.

The NPT made nuclearity into a global "right," but it left the ontology of safeguards to the IAEA. Between 1961 and 1972, the IAEA produced five different documents, each with a somewhat different solution to the ontological problem of which materials and technologies were sufficiently nuclear to demand safeguards and inspections. By now South Africa had cemented the technological justification for its Board seat with an extensive nuclear R&D program, and wanted to minimize external oversight of its uranium industry. The nation led other uranium producers in continually, and successfully, pushing to exclude mines and ore-processing plants from official definitions. The 1968 safeguards document, for example, defined a "principal nuclear facility" as "a reactor, a plant for processing nuclear material, irradiated in a reactor, a plant for separating the isotopes of a nuclear material, a plant for processing or fabricating nuclear material *(excepting a mine or ore-processing plant)*. . . ."[16] Uranium mines and mills were thus specifically excluded from being nuclear in any way—even from the residual category of "other types" of "principal nuclear facilities." In 1972, uranium ore was specifically excluded, as well, from the category of nuclear "source material."[17]

By 1972, then, the nuclearity of uranium ore and yellowcake had plummeted. This fall from grace had concrete consequences. NPT signatories did have to inform the IAEA of yellowcake exports, but this did not trigger safeguards actions.

No one was required to track yellowcake shipments or their fate, and uranium mines and mills were exempt from international inspection.

IV

The technopolitics of nuclear ontologies had fraught consequences in many other ways. For example, until the late 1960s the US Atomic Energy Commission refused to monitor working conditions in uranium mines and mills. In essence, the agency argued that mines were not really nuclear workplaces because, among other things, they did not contain significant radiological hazards. A 1957 Public Health Service report categorically disagreed, finding high levels of radon gas in several Colorado Plateau mines.[18] Nonetheless, limits on radiological exposure in mines were not set until 1967, after extensive bureaucratic wrangling.[19] Subsequent developments reflected a peculiarly American admixture of nuclear exceptionalism and colonial politics. The 1990 Radiation Exposure Compensation Act (RECA) followed a decade-long struggle to compensate Navajo uranium miners. RECA payments—which ultimately were not limited to Navajos—were framed as "compassionate payments" to honor miners (and downwinders) who had sacrificed their health to the Cold War.

In South Africa, meanwhile, uranium did not count as a nuclear material for health regulatory purposes until 1999. Before then, South African mines implemented no special protections against radiological hazards, despite studies conducted in the 1950s, '60s, and '70s documenting extremely high radon levels in some shafts. The battle to bring mines under the purview of the National Nuclear Regulator lasted over 10 years, and saw fierce resistance from the mining industry to the very principle of classifying any mine or plant as "nuclear."[20] Drawing on the appearance of good recordkeeping offered by apartheid-era systems of influx and labor control, mine operators argued that the workforce of black miners was so transient that no single individual could have received dangerous levels of radon exposure. In contrast, one well-placed South African radiation protection expert I interviewed estimates that thousands of miners multiplied their risk of cancer from two-fold to over ten-fold through radon exposure in the mines.

Until very recently, the principal ontological border of nuclearity for health purposes was the notion of the "permissible dose": the idea that below a certain threshold, health effects of radiation exposure are statistically negligible. Always controversial, this notion now seems conclusively refuted. A National Research Council report released in June 2005 concludes that there is "no threshold of exposure below which low levels of ionizing radiation can be demonstrated to be harmless or beneficial."[21] The regulatory consequences of this challenge to an established nuclear ontology remain to be seen.

Yet whatever they may be, they will certainly make waves in the virulent contemporary debate over the nuclearity of depleted uranium (DU) munitions, first deployed against Iraqi tanks in the 1991 Persian Gulf war and used again in the Balkans, Afghanistan, and the current Iraq war. Some scientists, activists, and former military personnel question DU munitions' legality as well as their morality. They emphasize the nuclearity of DU weapons, calling them "dirty bombs," "weapons of mass destruction"—or, at the extreme, "nuclear weapons." For them, DU's nuclearity has to do with its radioactivity, its persistent toxicity, and its effect on civilians living near places where DU munitions were used. Some activists have even coined a new term—"atomicity"—to designate the number of radioactive particles released; hence the much-quoted phrase "800 tons of DU is the atomicity equivalent of 83,000 Nagasaki bombs."[22] Nuclear nightmare narratives have resurfaced, ranging from reports of increased childhood leukemia rates in Basra to lurid accounts of horrifying birth defects among Afghani children and the babies of DU-exposed US veterans. The US and UK governments—along with the IAEA and the WHO—downplay the nuclearity of DU weaponry and deny any proven causality between birth defects and DU exposure. At least until the recent NRC report, they asserted that the radiation levels released by DU munitions fall within permissible dose ranges, and would not cause discernible health effects.

Between these two extremes, a few challengers are seeking middle ground. "Atomicity" is meaningless and misleading as a scientific term, they argue, because it has "no regard for the type of radiation present, its relative biological impact, method of dispersal, etc."[23] At the same time they insist on the need for substantive data, maintaining that the Pentagon has failed to conduct appropriate research into the health effects of DU weapons, and even at times suppressed evidence.[24] Clearly the outcome of this debate—whatever it turns out to be—will have major effects on those caught in (nuclear?) battle in the future.

During the Cold War the amazing political flexibility of nuclear exceptionalism served to entrench it on the left and the right, in the North and the South. But the ontologies of nuclearity are ever-shifting. Global warming, Western fears about the impact of the alleged "clash of civilizations" on the world's oil supply (why we're really in Iraq), and Bush's bedroom relationship with the "nucular" industry are combining to transform nuclear power from ecological Satan to planetary savior—even James Lovelock, author of the Gaia hypothesis, is on board. South Africa is once again on the frontlines, with plans to design and build the first so-called "inherently safe" pebble bed reactor. This time its main partner is China, which intends to increase its nuclear power capacity five-fold by 2020. Iran, meanwhile, insists that its uranium enrichment program merely aims to guarantee access to nuclear electricity and ensure independence from the West. If nuclear power experiences the renaissance its promoters so deeply desire, what scheme will shape who gets to develop it, who only gets to buy it, and who doesn't get it at all? What will determine which technologies and materials are "safe"? And safe for or from whom? Anti-DU activists may not have the right answers, but they're asking the right questions. From Iran to North Korea, from the battlegrounds of Iraq to the suburbs of America—and yes, even in Niger—the stakes of nuclear exceptionalism remain high.

Notes

My deepest gratitude extends to Paul Edwards and to Bruce Baird Struminger for their significant and wide-ranging input into this essay. Thanks also to Susan Lindee, John Krige, Catherine Kudlick, Itty Abraham, Charlie Bright, Juan Cole, Geoff Eley, Tony Judt, and audiences at the University of Michigan and Princeton University for their comments and suggestions.

1. Iraq had acquired uranium from Niger, Portugal, and Brazil in the 1970s, when launching its nuclear program, but had stopped these purchases in the 1980s. Bush's claim did not refer to these earlier purchases. See Joseph Cirincione, "Niger Uranium: Still a False Claim," *Carnegie Proliferation Brief* 7, no. 12.

2. Samuel Huntington, *The Clash of Civilizations and the Remaking of the World Order* (New York: Simon & Schuster, 1996).

3. Régis Debray, *Tous azimuts* (Paris: Odile Jacob, 1990).

4. Office of Technology Assessment, *Nuclear Safeguards and the International Atomic Energy Agency*, OTA-ISS-615, April 1995, Appendix B.

5. Gabrielle Hecht, "Rupture-Talk in the Nuclear Age: Conjugating Colonial Power in Africa," *Social Studies of Science* 32, nos. 5–6 (October–December 2002): 691–728.

6. Quotes in Alice Cawte, *Atomic Australia, 1944–1990* (Sydney: New South Wales, 1992), 41; and Gabrielle Hecht, *The Radiance of France: Nuclear Power and National Identity after World War II* (Cambridge, MA: MIT Press, 1998), 62.

7. Sven Lindqvist, *A History of Bombing* (New York: New Press, 2001); John Dower, *War without Mercy: Race and Power in the Pacific War* (New York: Pantheon, 1986).

8. M. Susan Lindee, *Suffering Made Real: New York: American Science and the Survivors at Hiroshima* (Chicago: University of Chicago Press, 1994).

9. In 1956, members of the first category were the US, the USSR, the UK, France, and Canada; members of the second were South Africa, Brazil, Japan, India, and Australia. See David Fischer, *History of the International Atomic Energy Agency: The First Forty Years* (IAEA, 1997).

10. South Africa's uranium was located in the same mines that produced its gold. In the decade following WWII, supplying uranium to the US and Britain saved many of these mines from economic collapse and served as conduits for massive foreign investment in the nation's industrial infrastructure. See Thomas Borstelmann, *Apartheid's Reluctant Uncle: The United States and Southern Africa in the Early Cold War* (New York: Oxford University Press, 1993); and Jonathan E. Helmreich, *Gathering Rare Ores: The Diplomacy of Uranium Acquisition, 1943–1954* (Princeton, NJ: Princeton University Press, 1986).

11. "International Atomic Energy Agency," Annex to South Africa minute no. 79/2, 28/7/56, pp. 10–11. National Archives of South Africa (hereafter NASA): BLO 349 ref. PS 17/109/3, vol. 2. The position of South Africa vis-à-vis the IAEA is thoroughly documented in the BLO 349, BVV84, and BPA 25 series of these archives.

12. Ibid. Plutonium fell into the category of "special fissionable materials." As a highly radioactive, extremely explosive, human-made material, it represented (and continues to represent) the pinnacle of nuclearity, the most exceptional of all things nuclear.

13. Lawrence Scheinman, *The International Atomic Energy Agency and World Nuclear Order* (Washington: Resources for the Future, 1987); Itty Abraham, *The Making of the Indian Atomic Bomb: Science, Secrecy and the Postcolonial State* (New York: St. Martin's Press, 1998); George Perkovich, *India's Nuclear Bomb: The Impact on Global Proliferation* (Berkeley: University of California Press, 1999); Astrid Forland, "Negotiating Supranational Rules: The Genesis of the International Atomic Energy Agency Safeguards System" (Dr. Art., University of Bergen, 1997).

14. Article IV of *The Treaty on the Non-Proliferation of Nuclear Weapons* (signed at Washington, London, and Moscow July 1, 1968). Emphasis mine. For the full text of the treaty, and the US State Department's triumphalist version of its history, see https://www.iaea.org/publications/documents/treaties/npt.

15. Between 1958 and 1993, the IAEA gave out $617.5 million in "technical assistance." The top 10 recipients were Egypt, Brazil, Thailand, Indonesia, Peru, Pakistan, Philippines, Bangladesh, South Korea, and Yugoslavia. OTA, *Nuclear Safeguards and the International Atomic Energy Agency*, 53.

16. IAEA, INFCIRC/66/Rev. 2, September 16, 1968.

17. IAEA, INFCIRC/153 (Corrected), June 1972.

18. Duncan A. Holaday et al., *Control of Radon and Daughters in Uranium Mines and Calculations on Biologic Effects*, Public Health Service Publication No. 494 (Washington: US Government Printing Office, 1957).

19. Eric W. Mogren, *Warm Sands: Uranium Mill Tailings Policy in the Atomic West* (Albuquerque: New Mexico University Press, 2002); J. Samuel Walker, *Containing the Atom: Nuclear Regulation in a Changing Environment, 1963–1971* (Berkeley: University of California Press, 1992).

20. This battle is extensively documented in archival material from the South African Chamber of Mines and the Council for Nuclear Safety.

21. See the press release on the National Academy of Sciences website: http://www8.nationalacademies.org/onpinews/newsitem.aspx?RecordID=11340.

22. Leuren Moret, "A Death Sentence Here and Abroad. Depleted Uranium: Dirty Bombs, Dirty Missiles, Dirty Bullets," Centre for Research on Globalisation, August 21, 2004. http://globalresearch.ca/articles/MOR408A.html. Accessed September 29, 2004.

23. Jack Cohen-Joppa, "Disinformation about Depleted Uranium," https://www.highbeam.com/doc/1P3-739697791.html.

24. Dan Fahey, "The Emergence and Decline of the Debate over Depleted Uranium Munitions, 1991–2004," June 20, 2004, http://www.wise-uranium.org/pdf/duemdec.pdf.

))) A Dark Art: Field Notes on Carbon Capture and Storage Policy Negotiations at COP17

Gökçe Günel

Introduction

I started learning about the controversies surrounding carbon capture and storage (CCS) negotiations during my ethnographic fieldwork on the development of a clean technology and renewable energy sector in Abu Dhabi, United Arab Emirates, between September 2010 and June 2011. The environmental consultants I worked with had been preparing a policy submission to the United Nations Framework Convention for Climate Change (UNFCCC) regarding the inclusion of CCS technology under the Clean Development Mechanism (CDM).[1] My involvement in the project as an anthropologist and an intern allowed me to develop an understanding of how the CDM operated, as well as what CCS technologies comprised. I became further interested in how the CCS issue in the CDM debate would be resolved. In this essay, I trace the unfolding and resolution of the CCS in the CDM negotiations in Durban, South Africa, during the COP17. In this way, I hope to present a critique of climate change policy infrastructures, underlining the various incongruities that characterized the negotiations.

CDM is a market-based "flexibility mechanism"[2] that was initiated under the Kyoto Protocol with the intention of encouraging industrialized countries to invest in greenhouse gas emission reduction programs in developing countries, such as hydropower, wind energy, or solar energy projects.[3] This way, the environmental consultants explained to me, industrialized countries could meet their own emission reduction commitments, while fostering sustainable development within host countries. Most importantly, they stressed, CDM projects had to satisfy the so-called "additionality" requirement. In other words, the project proponents had to prove that the given project would not have been initiated without the additional CDM incentive from the UNFCCC. As such, the first step for starting a CDM application to the UNFCCC constituted proving how the project would not have happened without this additional push. The environmental consultants that I worked with produced baselines, estimating future greenhouse gas emissions in the absence of the proposed projects. They suggested that they needed to act like attorneys and defend the proposal as if it were a legal case.

These project proposals would then be evaluated by third-party Designated Operational Entities (DOEs) to guarantee that the project would instigate valid emission reductions. If the DOE gave approval to the project, the proposal would be submitted to the CDM Executive Board within the UNFCCC, waiting to be

ephemera 12, nos. 1/2 (2012): 33–41

registered. "But the registration of hundreds of Clean Development Mechanism (CDM) projects at the United Nations Framework Convention for Climate Change (UNFCCC) only shows how successful the consultants that work within these procedures are, rather than proving the success of CDM as a program," a senior environmental consultant that I worked with told me, thereby questioning the legitimacy of the policy infrastructures that they worked with. Upon registration at the UNFCCC, the project would start to produce carbon credits for the involved entities, based on the supposed emissions reductions gained from its implementation.

In this framework, if China decided to build a solar power station, with technology or expertise from a German company, rather than relying on lower-cost energy from coal plants, the reduced carbon emissions attributed to this investment could be credited towards the German company's emission reduction commitment, set by the Kyoto Protocol. The development of a solar power station would also contribute to sustainable development in China, or at least this is what CDM proposed.[4]

However, if CCS were to be included under the CDM, the environmental policy consultants explained to me, China could build a coal-powered plant, provided that it is equipped with CCS technologies, and still receive carbon credits for it. CCS technology, as my interlocutors outlined, operated by obtaining carbon dioxide from large industrial compounds, such as coal plants, carrying it in solid, liquid, or gas form to storage sites, and injecting it into geological formations such as deep saline aquifers, unmineable coal seams, or maturing oilfields, kilometers below the ground. Accordingly, the inclusion of CCS in the CDM would mean that carbon credits would be issued for carbon dioxide sequestered through future CCS projects undertaken in so-called developing countries, providing incentives for further investments in this technology.

CCS Controversies

Yet, ever since its inception as a climate change mitigation technique, my interlocutors reminded me, CCS had been a controversial technology. "Issues such as site feasibility, high operational costs, future safety, and unresolved legal liability make carbon capture and storage projects challenging to initiate, implement, and operate," the environmental consultants summarized. In addition, parties who were critical of CCS projects often suggested that including CCS in the CDM could incur a crowding out effect, leading investment away from other climate change mitigation strategies, such as renewable energy or energy efficiency projects. So, rather than building a solar power plant and reducing carbon emissions, developing countries could proceed with coal-powered plants and attempt to use CCS technologies to later bury the emissions resulting from such operations. Surely, this development could negatively influence the flourishing of renewable energy projects around the world. Accordingly, the opponents of CCS in CDM argued,

CCS projects do not necessarily reduce dependence on coal or oil, thereby failing to promote the transition from coal- or oil-based power sources to renewable energy. In this way, it was underlined, CCS is not in line with the main principles of the CDM.

Secondly, the environmental consultants noted, when implementing CCS projects, oil-producing countries could use maturing oilfields as storage locations for the carbon dioxide that they obtained, as these oilfields are considered naturally sealed reserves. And yet, injecting gas into oil reservoirs leads to increased oil production as well, a process commonly known in the industry as enhanced oil recovery (EOR). By injecting carbon dioxide into ageing fields and pumping oil out, oil producers may increase the lifetime of the fields by up to 30%, while freeing up the natural gas more commonly used in such processes. The inclusion of CCS as an eligible technology for decreasing carbon emissions then becomes a perverse incentive for further oil production. The entities that earn carbon credits from CCS activities in turn become oil-producing countries.

Regardless of these controversies, the environmental consultants I worked with believed that the 17th Conference of the Parties to the United Nations Framework Convention on Climate Change (UNFCCC), or COP17, in Durban, South Africa, would be a milestone for CCS negotiations, allowing this controversial climate change mitigation technology to be included under the CDM. While Durban negotiations did prove to be a victory for the proponents of CCS in the CDM, in this essay I would like to show that they also highlighted the numerous inequalities that are part and parcel of the production and implementation of climate change policies.

Constructive Ambiguity

The long-winded CCS deliberations, which officially started at the COP11 in Montreal in 2005, had reached a breaking point last year at the COP16 in Cancun when the decision was made to include CCS in CDM, with the provision that safety and liability protocols could be resolved. In February 2011, parties submitted proposals regarding modalities and procedures guidelines on CCS projects. After collecting the submissions, the UNFCCC secretariat put together a synthesis report. Next, a technical workshop was organized in Abu Dhabi, in September 2011, inviting parties to learn more about the current status of CCS technologies. Following the technical workshop, the secretariat published a workshop report and a modalities and procedures draft, which was opened for negotiation in the Durban meeting.[5] After two long contact group discussions, modalities and procedures guidelines were finally accepted on December 3, 2011, with liability protocols remaining as the only outstanding issue. The parties had not been able to agree upon whether host countries or carbon credit holders should be liable for the stored carbon dioxide, or if the liability should be shared between the two stakeholders.

During the second week of COP17, the liability provisions were settled as well, requiring that countries hosting projects develop thorough regulations for carbon dioxide storage and liability. It was stipulated that project developers place 5% of the carbon credits earned from CCS projects in a reserve fund. The carbon credits in this reserve fund would be awarded to the project proponents only after 20 years of monitoring, provided that no carbon dioxide leaks from the underground storage site. It was also decided that in case a project participant was unable to go on with the project, liability would automatically be transferred to the host country. Such provisions were expected to mitigate concerns for the uncertainties of CCS technologies, especially in regards to long-term liability.

However, there remained certain inconclusive issues as well. For instance, what did it mean to defer the liability for CCS projects completely to host countries? This type of provision evidenced the current inability to put together an international treaty on the issue, while making it more difficult for future provisions to be produced, as they could potentially contradict host country rulings. Resolutions on transboundary movement of carbon dioxide, which involves capturing carbon dioxide in one nation state, transporting it, and storing it in another, were also postponed to the COP18 in Doha, Qatar, as it would require defining project boundaries, characterization of carbon dioxide as toxic or non-toxic material, its legitimacy under other international treaties, and administering the participation of multiple project proponents. However, as significant as this issue may be for the future implementation of CCS projects, it did not hinder the process for including CCS projects in the CDM. As such, at the end of the Durban negotiations, many parties argued, CCS did manage to receive the legitimacy that it sought. While waiting for the Subsidiary Body for Scientific and Technological Advice (SBSTA) meeting where the CCS in CDM decision would be announced, I chatted with three CCS experts, at times working with the secretariat. "How big is the damage done, you think?" one of them asked another. "The monitoring criteria were supposed to be *stringent*," he replied, quoting the initial policy document. When I asked what adjective he would use instead, he laughingly proposed that "wishy-washy" would be a good alternative. "What we are trying to achieve in putting together this document is constructive ambiguity," one of them later told me. Here, "constructive ambiguity" implied a quick resolution of the debates, without producing further controversy amongst the delegates. He understood the production of constructive ambiguity as an aesthetic challenge as well, created step by step through highlighting the document in different colors, bracketing unresolved sentences and finally cleansing the text of colors and brackets.[6] The application of such constructive ambiguity could eventually result in "wishy-washy" protocols as well, wherein the goal-oriented nature of the negotiations could at times curtail a rigorous analysis of the final policy decisions. Finally, they argued that the inclusion of CCS in CDM was symbolic, more than anything else. "It will be technically complicated to implement CCS projects and acquire carbon credits in the next few years, with the given state of technology. So even when CCS is included in the CDM, it's not like we're going to have an upsurge of CCS projects," they summarized with much relief.

Bargaining Devices

"One of the West African countries says they don't want their country to be used as a video game," Lisa,[7] a Greenpeace campaigner, reported after concluding her meetings with various delegates participating in the CCS in CDM policy-making sessions. "They say that including CCS in CDM will pave the way for developed countries to test unverified technologies within developing countries." Through the video game analogy, the West African delegate pointed to how the decision-makers were detached from the actual space and time in which the results of their actions would be experienced. He showed disbelief in the functional purpose of the practices of implementing this specific technology. It was more like a game, where unproven technologies would be experimented with and perhaps later discarded. And then the nation state in which this game had been played would have to attend to the possibly dire consequences.

But Lisa doubted that the West African delegate would state this position during the debates. "There must be other countries with opposing views," she sighed, "what about Panama or Jamaica, or maybe Uganda?" She sat down to write an email to one of these delegates, whom she had briefly interacted with after a contact group meeting, when he asked a question regarding the current state of CCS technologies. "A lot of countries don't have the resources or the time to pay attention to different issues, so they may not know anything about CCS in particular," Lisa reminded me. If a delegate were unsure or uninformed, then Greenpeace would provide infrastructure and give information on policies. "In the past year, there have been twelve failed CCS projects in Europe," she said, "I don't understand why they want to export a failed technology to developing countries." She added this argument to the end of her email and wondered if the delegates she had been in touch with would be attending the meetings during the next few days.

Overall, CCS negotiations have been characterized by low levels of participation, with the major stakeholders being Saudi Arabia, Brazil, Norway, the European Union, Australia, the United Arab Emirates, Kuwait, and the Alliance of Small Island States (AOSIS) countries. When I asked why this is the case, Michael, a member of the UNFCCC secretariat, who has been following the debates, told me that many countries do not have the technical expertise to participate in the debates on an emergent technology such as CCS. Countries that already have full-fledged oil industries, and thereby first-class geologists and reservoir engineers, were able to negotiate better, given their access to a more thorough understanding of the subsurface. They could rely upon their oil experts in presenting arguments for and against CCS in CDM. As such, it was not surprising that Brazil's CCS delegate was an executive at Petrobras, the state-owned oil company, while the Saudi Arabian delegate worked with Saudi Aramco. In this sense, expertise seemed to be highly permeable in the climate change debates, allowing a Petrobras or Aramco representative to temporarily give up his affiliations and to serve as a delegate for his country. While this enabled countries to have stronger and more reliable perspectives on technical issues, it may also raise questions on whose interest becomes represented in the debates.

Many participants to the Durban meetings were curious about why AOSIS countries refrained from engaging with the CCS in CDM negotiations, especially after being strong opponents for many years. When I asked a senior negotiator about the absence of AOSIS countries in the Durban negotiations, he suggested, "One of the AOSIS members seemed like it was opposing CCS in CDM but then again, the delegate is not well prepared, does not really know what he's saying, so his interventions do not make much sense. It's not like they've studied the issue before," he underlined, "and I mean, they just needed to read eight policy papers in preparing for the meeting here." The senior negotiator later suggested that AOSIS does not want to invest more time on CCS in the CDM, especially because they are more concerned about general Kyoto Protocol issues and added how they do not have a common understanding of CCS, which prohibits them from intervening further.

When I asked Michael why AOSIS delegations were no longer active, he explained, "They cannot really oppose this issue anymore. This is all that Saudi Arabia wants. If it doesn't get it, then it will put sand in all other negotiations. AOSIS have so many more stakes in the climate debates—they would like to have the support of Saudi Arabia and the other oil-producers," he argued, and emphasized how the politics behind these debates made him very frustrated. "In an ideal world, every issue would be thought through separately, so when producing arguments regarding CCS in CDM, parties would not think about how this decision would impact other climate issues, or see this as a bargaining device. But this is not an ideal world and this is all we have," he concluded. According to him, participants did not prioritize studying the various problems associated with the issue, such as the reliability of technology or its potential environmental impact, but rather focused on the political power that they would accrue by bargaining in a specific manner. Likewise, another delegate, who had been actively following the negotiations, suggested how Brazil had been against the inclusion of CCS in CDM for many years, especially because they did not want to divert attention away from unavoided deforestation projects, known as REDD+, which constituted another battleground for inclusion under the CDM. After years of opposition, when finally giving support to CCS in CDM, Brazil also expected that Saudi Arabia would be favourable to the inclusion of REDD+ under the CDM in Durban and next year at the COP18 in Doha, Qatar.

But why was Saudi Arabia so dedicated to CCS in the CDM? The Saudi Arabian delegate explained to me how his country does not have any CDM projects. "If CCS is included in the CDM, then Saudi Arabia can also start to play its part in contributing to climate change mitigation," he added. When I asked him about the role of EOR and whether EOR projects would be included in the CDM, he explained that such projects should be considered on a case-by-case basis. According to him, it did not make sense to produce an international treaty on this issue. EOR, which had been a significant subject matter in the debates on CCS in the CDM, was not mentioned in the final modalities and procedures draft. A secretariat member that I spoke with explained how "no one brought up EOR in the debates," finally leading to the omission of the whole issue from the documents. As the Saudi Arabian delegate told me, it would be considered on a case-by-case basis.

Positions of Criticality

"Do you know what civil society organizations think about the decision," some secretariat members asked me on the day when the results of the negotiations were going to be publicly announced, "how did Greenpeace or CDM Watch react?" Having had the opportunity to spend time with Greenpeace and CDM Watch campaigners throughout the negotiations, at times helping them with their campaigning work, I explained how these NGOs did not believe that big oil should also earn carbon credits, in addition to the extra oil that they procure through enhanced oil recovery. "Well, they are right," they responded, "We really have nothing to say."

Greenpeace and CDM Watch members worked long hours, developing arguments and communication strategies to oppose to the inclusion of CCS in CDM. Every morning, they picked up recently printed copies of the new policy draft along with the daily program and went through them to underline the changes that had been made during the previous day's contact group meetings. They tried to identify resisting parties, consulted legal and technical experts inside and outside the conference to find loopholes in the policy documents, and looked for ways of manipulating the decision-making process. Pointing out the inequality of resources among different delegations, and showing how certain countries do not have enough staff to follow each climate change issue, they produced material on CCS for delegates to use and rely upon and provided both big picture information and small details. They produced press releases, organized press conferences where they could express their understandings of the context, and briefed individual journalists. Overall, Greenpeace and CDM Watch members had managed to develop a vast network of contacts and a clear understanding of how the COP works, thereby serving a position of criticality throughout the negotiations.

Besides NGOs like Greenpeace and CDM Watch, other lobbying organizations such as Global CCS Institute, CCS Association, or Bellona occupied prominent positions during the Durban CCS debates. Organizing many side events with oil industry representatives, energy ministers, corporate figures, or geologists, they managed to give shape to the predominant discourse on CCS during the meeting, framing it as a critical climate change mitigation strategy. "We need every bit of energy we can get and therefore CCS is vital. It allows us to consume coal or oil, without worrying about the carbon emissions they produce," a Shell representative, who had presented at one of the side events, told me later during a short interview, "I imagine that if I came back to the world in 100 years, maybe then I could see a place which is fuelled by renewable energy sources, but not before then," he added. Karen, a geologist from a research university in the United States, who had also participated in the technical workshop in Abu Dhabi in September 2011, concurred and suggested that she does not understand why people are so afraid of CCS. "CCS is not a dark art," another CCS lobbyist added.

Yet, most importantly, many underlined, CCS would help development continue in countries such as China and India, which still relied upon coal plants. During one of these side events, when a representative from a German NGO got up to explain how and why civil society organizations in Germany were resisting the implementation of CCS technologies within the country's boundaries and proposed that the capital invested in CCS should actually be utilized to improve renewable energy infrastructures, a delegate from a West African country adamantly stated, "We can't improve our industry on solar power. We need to uplift our people and we will need coal for that. Germany has educated its people and now it's time for it to clean up." It was time for developed countries to give up their coal plants, but the developing countries would need them for longer, so as to create industrial infrastructures that match countries like Germany. Maybe CCS in the CDM could be helpful in such cases, the representative said. As much as CCS was criticized as a way in which the fossil fuel industry was reinventing itself or testing unverified technologies within developing countries, in this case it was perceived as a desirable means of development.

A New Definition of Justice

Overall, the CCS negotiations in Durban disclosed the many inequalities that parties suffer from in both bargaining for and implementing climate change mitigation techniques. The resources that parties can spare for specific issues, their levels of preparedness, negotiating powers, and existing industrial infrastructures all constituted factors influencing decision making.

In discussing such incongruities, one researcher I spoke with proposed that we should come up with a new definition of justice, wherein vulnerability would be prioritized, more than anything else. In this framework, the most vulnerable countries' interests would be served first, making climate policy relatively simpler. "Islands, for instance," he reminded me, "they will be suffering from fresh water problems very soon." In fabricating climate policy, this understanding of justice would perhaps serve as a useful principle to keep in mind.

Notes

1. The environmental consultants that I worked with have advanced engineering degrees. They come from different countries around the world, and mostly were in the UAE for temporary periods. The individuals who informed this essay—through meetings, interviews, or informal conversations—originate specifically from Algeria, Germany, India, Iran, Lebanon, the United Arab Emirates, and the United Kingdom.

2. For a summary of carbon trading and flexibility mechanisms under the Kyoto Protocol, see C. Hepburn, C. "Carbon Trading: A Review of the Kyoto Mechanisms," *Annual Review of Environment and Resources* 32 (2007): 375–93. Also see L. Lohmann, "Carbon Trading: A Critical Conversation on Climate Change, Privatisation and Power," *Development Dialogue*, no. 48 (2006), special issue, http://www.dhf.uu.se/Publications/dd.html.

3. A helpful journalistic account of the workings of the CDM and carbon trading can also be found at M. Schapiro, "Conning the Climate: Inside the Carbon-Trading Shell Game," *Harper's*, February 2010, 31–39.

4. For some critiques of CDM mechanisms, see C. Fogel, "The Local, the Global, and the Kyoto Protocol," in *Earthly Politics: Local and Global in Environmental Governance*, ed. S. Jasanoff and M. L. Martello (Cambridge, MA: MIT Press, 2004); and Y. Schreuder, *The Corporate Greenhouse: Climate Change Policy in a Globalizing World* (London: Zed, 2009).

5. The modalities and procedures draft text that started the discussions in Durban is available at http://unfccc.int/resource/docs/2011/sbsta/eng/04.pdf.

6. See A. Riles, *The Network Inside Out* (Ann Arbor: University of Michigan Press, 2000).

7. The names provided here are pseudonyms.

))) Gendering Oil: Tracing Western Petrosexual Relations

Sheena Wilson

By employing a feminist lens to "follow the oil" and trace "the webs of relations and cultural meanings through which oil is imagined as a 'vital' and 'strategic' resource,"[1] I wish to interrogate the relationship between human rights and gender and racial equality and the petro-discourses that are newly oriented around ecology in our contemporary moment.[2] As with many cultural transformations and their associated ideological turns, women's relationship to oil, to the environment, and to the petrocultures of the twentieth and twenty-first centuries in the West

Oil Culture, ed. Ross Barrett and Daniel Worden (Minneapolis: University of Minnesota Press, 2014), 244–66.

is portrayed in the mainstream media in a limited number of largely superficial ways that include but are not limited to the use of pseudo (embedded) feminist discourses that invoke women's rights as a justification for colonial extraction politics, and the recuperation of the female body as a medium for spectacularized politics—largely with explicit consumer aims.

The latter half of this article focuses on representations that either neutralize or trivialize women's political and economic relationship to oil, through an analysis of the Ethical Oil media campaign that began in 2010 in Canada. This campaign demonstrates the way women's identities have been intentionally constructed to naturalize a particular relationship between Western women and oil. The narratives and gender constructs that inform many advertising and media images also function to undermine contemporary women's more serious engagements with the environment, and as such contribute to a broader array of cultural, rhetorical, and legal efforts currently under way to criminalize environmentalists and activists in legislation as well as in the popular media.[3] The first half of this essay discusses the media reception of two examples of contemporary women's environmental activism—the Idle No More protests in Canada and Chief Theresa Spence's subsequent hunger strike in December 2012 and January 2013—in order to establish the cultural and ideological context in which recent representations of women and oil intervene. By examining these forms of activism, I will show how Indigenous women's political activism in the context of mainstream media is reconstructed to the detriment of the resistance movements and to other community members—namely, Indigenous men and youth, who risk being constructed as terrorists in discourses of petro-violence.

Chief Theresa Spence and Idle No More: Petropolitics and the Construction of the Terrorist

In the current dominant discourses circulating in Canada and the Western world, environmental messages are acceptable when they are controlled and shaped by petro-invested governments, industry organizations, and corporations. These environmental messages articulate concerns about health and safety, environmental stewardship, and performance, all within the context of a neoliberal discourse of increasing expansion and exploitation of resources. Within this paradigm, environmentalism, environmentalists, environmental science, and scientists—especially women and minority citizens acting on behalf of environmental agendas—become the targets of media attack, perceived not only as potential obstacles to oil extraction but also as threats to the proliferation of capitalism itself, since oil and capitalism are imagined as symbiotic.

In our contemporary moment, environmental activists are increasingly constructed as environmental terrorists, through both counterterrorism units and associated mainstream discourses. In Canada, on February 9, 2012, the federal government responded to left-leaning environmental movements resisting oil sands extraction through a sliding semantics that identified these groups and their actors as terrorists. The government-authored report, "Building Resilience against Terrorism: Canada's Counter-terrorism Strategy," contends that "although not of the same scope and scale faced by other countries, low-level violence by domestic issue-based groups remains a reality in Canada. Such extremism tends to be based on grievances—real or perceived—revolving around the promotion of various causes such as animal rights, white supremacy, environmentalism, and anti-capitalism."[4] By May 1, 2012, Canadian environment minister Peter Kent had also used criminalizing rhetoric to characterize the activities of Canadian environmental groups, claiming that these organizations had been used "to launder offshore foreign funds for inappropriate use against Canadian interest."[5] And, as Indigenous communities have begun to mount more organized protests against the infringement of their treaty rights that are linked to issues of environmental protection and that include barricades and other forms of demonstration, it is important to remain cognizant of what feminist scholar Heather M. Turcotte has already identified as the slippages within "academic and state representations of petro-terrorists, petro-gangs, and victims of gender violence . . . that produce the figure of the petro-terrorist-gang-member for public consumption and foreign policy."[6] It is imperative to consider how these discourses use conflicting representations of women and of feminism in the West that have ramifications not only for women in Western petrocultures but also for ethno-cultural communities and other marginalized groups.

Media and advertising tactics performed to minimize women's relationship to the environment are in direct contrast to the very serious involvements of women actively engaged in environmental movements in Canada and around the world. Robert R. M. Verchick, for example, points out that many of the "most visible and effective environmental justice organizations are led by and consist mainly of women. . . . Thus, while 'environmental justice' describes an environmental movement and a civil rights movement, it also describes a women's movement . . . a feminist movement."[7] In Canada, as around the world, a significant percentage of female environmental activists are women of color or women from minority communities—especially Indigenous women who are disproportionately impacted by environmental changes.

In her work on the Niger Delta, for example, Heather Turcotte has found that women in petropolitics are typically invoked only as objects of law who struggle against violence in their communities. She theorizes that activist women are naturalized as mothers and grandmothers in mainstream discourses—maternal protectors of the environment "rationalized as unpolitical and external to the political economy."[8] She argues that women's protests are rearticulated "in ways that omit deeper histories of interconnected state violence and women's anti-imperialist engagements with state power."[9] By contrast, racialized men are constructed as "terrorists" when active in these same petro-resistance movements that are sometimes initiated or led by women.[10]

The Idle No More movement went public at a November 10, 2012, Saskatoon teach-in organized by four women activists—Nina Wilson, Sheelah McLean, Sylvia McAdam, and Jessica Gordon. These activists were concerned about the effects of a Harper government omnibus bill C-45 that infringed on Aboriginal rights.[11] Nonviolent political events then proliferated across the country and around the world. Chief Theresa Spence of the Attawapiskat band in northern Ontario intensified public attention when she endured a 44-day hunger strike between December 11, 2012, and January 24, 2013. These two actions, often conflated, led to broad public discussion of environmental and First Nations, Inuit, and Métis issues. Idle No More continues to gain momentum. Chief Spence has made headlines fighting for better educational opportunities and living standards for her community since she became a chief in 2006. In spite of this, the media configures her in a formulaic manner—as an incompetent and possibly corrupt politician. Alternatively, as part of the recent petro-political resistance, she is for the most part a symbolic figure. Media attention tends to confine her principled political activism to health-related issues related to dietary regimes rather than to the long traditions of nonviolent civil disobedience and the hunger strike in particular. Twelve of the thirteen demands that she made to the Canadian government have been largely overwritten. Only one demand became highly visible: her desire to meet with Prime Minister Harper and Governor General David Johnston. This became the subject of much public debate and deflected focus from the issues at the heart of Spence's protest, positioning her as somewhat obstinate. Like the Idle No More movement, the remainder of Spence's concerns were centered on the omnibus bills C-38 and C-45, which undermine aspects of Section 35 of the Canadian Constitution, and treaty rights and First Nations, Métis, and Inuit (FNMI) ways of life in Canada. However, they were rarely, if ever, fully explained by the media.

Meanwhile, the mainstream media provided space for male FNMI leaders to declare that "It's time for the men to step up."[12] And the political discord within the movement and its leadership was given significant attention. This media focus rhetorically constructs Indigenous struggles and debates as signs of the inability to organize and an ethno-cultural group turning in on itself—as opposed to the very common leadership conflicts any group faces when dealing with political issues. The Idle No More youth movement that Chief Spence inspired is currently being undermined through reports that its leaders "appear to have little control over the direction of the movement," and as of the end of January 2013 the movement itself is being increasingly linked to violence:[13] violence initiated by "aggressive elements within the existing [A]boriginal leadership structure" and violence against FNMI peoples in Canada that rhetorically blames the victims for stirring up racist reaction.[14] Newspaper headlines read, "PM Harper believes Idle No More movement creating 'negative public reaction,' say confidential notes."[15] The gang-rape of an Indigenous woman in Thunder Bay, Ontario, on December 27, 2012, which initially received minimal attention from mainstream media, was finally reported on *The Current*, a major Canadian radio broadcast on January 25, but it was contextualized by an introduction that places the onus on the movement and not on the history of colonialism and racism: "Idle No More is inflaming long-standing tensions between FNMI and non-FNMI communities. In Thunder Bay, police investigate a possible hate crime and the mayor regrets that his plan to keep people safe has failed."[16] This rape is relevant to the Idle No More movement because, as it was reported, "During the attack the men allegedly told the victim it wasn't the first time they had committed this type of crime and 'it wouldn't be the last.' She [the victim] told police they [the attackers] also told her, 'You Indians deserve to lose your treaty rights,' making reference to the recent Idle No More events in Thunder Bay."[17] This attack and another alleged "starlight tour" reported in the CBC broadcast,[18] while part of a long history of violence against Indigenous women and men, are now inspiring fear in Indigenous communities and being used to suggest that youth might be safest distancing themselves from the activism of the Idle No More movement. The violated female body, in association with the Idle No More movement, functions to spectacularize and market petro-violence to the media-consuming public as a reaction to the Idle No More resistance movement rather than as endemic to the legacies of colonial logic, human rights abuses, and gender-sexual violence on which Western petrostates are founded. This is but one manifestation of how the appropriation of women's political power can create dangerous outcomes for entire communities. The foundations for these practices are outlined in the following section, which analyzes how the frameworks of consumerism and embedded feminism have defined women's relationships to oil in such a way as to naturalize imperial and neoliberal agendas.

Western Petrocultures, Women, and Minorities

The histories of feminism and oil are intertwined.[19] In the century and a half since oil's potential as a major energy source was first discovered and harnessed, a world oil industry has emerged to transform the distribution of world power between nations, the everyday lived reality of all citizens of the Western world, and increasingly the daily experience of people across the globe. The age of oil in the West is virtually synchronous with the women's rights movement:[20] after similar periods of development, both the oil industry and the Western women's rights movements had gained significant momentum by the early twentieth century. In this same moment, photography began to be used as a tool to construct new feminine identities for the public imaginary that have promoted certain concepts of beauty, domesticity and housewifery, motherhood, personal and family hygiene, and women's autonomy. Alternately affirming and contesting these concepts, women's rights movements and women's lives have transformed over the last century of the age of oil; many of these transformations can be attributed directly to industrialization and the petroleum-related innovations that came to define gender dynamics and gender roles in Western petroculture(s). The age of oil is rife with ironies that have resulted in both feminist advances as well as the reinforcement of long-standing patriarchal conceptualizations of woman as object and as property, popularized through the pervasiveness of the female image as it has been recuperated by capitalist, consumerist, neoliberal discourses of the late twentieth and twenty-first centuries.

There are a number of visual and rhetorical tropes that position women's relationships to oil—particularly Western, white, middle- and upper-class women's relationships—in consumer terms. These tropes build on historical practices that have linked women to social and cultural developments by targeting them as consumer and as commodity—the "consumer consumed." Women have long been identified as major consumers of petroleum products—fashion products being one example. "Ecofeminism" in mainstream popular media has come to be signified by reductive and trivial issues. Magazine and new-media headlines ask questions such as "Are you a Green Beauty?" "What's your clothing's footprint?" and "Are you an eco-fashionista?" Advice is given about "How to Go Green: Fashion and Beauty."[21] In the Western cultural practice of reducing women's social engagement to consumerism and the neoliberal practice of Starbucks logic, as Slavoj Žižek has called it, consumerism and social justice are collapsed into one act through "products that contain the claim of being politically progressive acts in and of themselves . . . [and in which] political action and consumption become fully merged."[22] And this logic—Starbucks logic—expresses itself in gendered ways. Specifically in relationship to petroleum and oil, this logic reinforces patriarchal social, political, and economic norms.

This consumerist logic is neither new nor restricted to a surface resistance of oil and the petroleum industry. Similar instances might include anti-animal testing labels on cosmetic products, or the grunge movement's ironic or distorted cultural refraction of the global antipoverty movements of the last several decades, whereby designers sell a spectacle of scarcity. These trends disguise consumption as political awareness or even activism and elide discussion of the more systemic and infrastructural processes that imbricate oil, our capitalist economy, and our culture in ways that are much more complex to untangle than simply purchasing petroleum-free mascara.

This practice of both objectifying women, particularly middle- and upper-class white women, as consumable products and supposedly empowering them as consumers themselves, through fashion and beauty products, translates into a performance of political engagement both as the eco-fashionista with a green wardrobe and green cosmetic bag and as the often dehumanized object onto which social resistance is draped, projected, and performed as a commercial strategy in the guise of political resistance. Western women's relationships to petroleum have been constructed in the social imaginary as a site of spectacle through which resistance to petroculture is signified yet undercut.

Canada and the Ethical Oil Billboard Campaign: Gender, Sexual Relations, and Colonialism as Petro-Discourses

Ezra Levant, a conservative Canadian media personality, published the book *Ethical Oil: The Case for Canada's Oil Sands* in 2010.[23] This book promotes a right-wing, pro-Canadian-oil-sands vision for the country. In fact, Levant was at one time, around the new millennium, a political candidate in the same party as Stephen Harper—current prime minister of Canada.[24] Furthermore, Levant uses polemical and prejudiced vocabulary to characterize foreign interests and the nationals of multiple countries. In fact, Levant has a history of publishing pejorative material.[25] Furthermore, the book targets certain ethnic groups within Canada, including FNMI people. The rhetoric first promoted by this book has been sustained through a website, EthicalOil.org, started by Alykhan Velshi. In 2011, Ethical Oil also ran a media campaign that included billboards and television commercials (also made available online). The campaign explicitly rebranded Canadian oil/tar sands oil as an ethical source of oil.[26] This is relevant at this particular historical moment because it, at least temporarily, reoriented the public debate in Canada away from a discussion of the environment and toward a discussion of oil's foundational and integral role in Canadian national identity, by fetishizing oil as a national resource linked to Canadian pioneerism and innovation, ingenuity, and integrity, as well as Canada's international reputation as a liberal democratic peace-keeping nation with an excellent human rights record.[27]

The campaign links what are apparently disparate issues in ways that rely on the preexisting sociocultural fetishization of oil as a thing, with powers that make it capable of changing the nation. In this case, the Ethical Oil campaign imbues the oil sands with the potential to redirect power toward good and ethical ends— the propagation of multiculturalism, for example, or the advancement of the Canadian dream, a ubiquitous myth linking economic expansion with personal and financial fulfillment. Playing on historical identity tropes and the myth of the Canadian nation, Levant simply affirms what Canadians wish to believe about themselves, and he does so by weaving oil into the social, political, and economic fabric of the nation.

These popular beliefs about multiculturalism and racial and gender equality became the focal point of the Ethical Oil billboard and television campaign of 2011, which employs images meant to represent foreign female nationals, as well as white and nonwhite Canadian women, as a strategy to reinscribe oil into the nation myth. The concept of saving foreign women in this campaign invokes a power-discourse that rhetorically positions and visually reinforces for all Canadians the narrative myth of nation whereby Canadians are benevolent, tolerant, democratic protectors of human rights. One of the billboards juxtaposes the images of two women. The image on the left of the billboard appears under a red-banner heading that reads "Conflict Oil" with a secondary red-banner message, "Conflict Oil Countries: Women Stoned to Death," superimposed over the black-and-white image of a burka-clad woman who is being buried alive in preparation for stoning. The image on the right half of the billboard is a color photo of the graduation-gown-clad female mayor of Fort McMurray, the largest urban center in the oil sands region of Northern Alberta.[28] Her picture appears under the green-banner message "Ethical Oil" and the second green-banner message superimposed over the middle of the image reads "Canada's Oil Sands: Woman Elected Mayor." Her identity is emphasized in the fine print in the bottom right-hand corner of the graduation photo that reads "Mayor Melissa Blake." By contrast, the woman in the left-hand photo is not similarly identified. According to various sources, this image dates back to either the 1970s or 1980s, and the female subject is an Iranian woman. However, the billboard itself provides no such context.[29] This burka-clad woman simply stands in as a synecdoche for the perceived oppression of women in Muslim areas of the world. The visual rhetoric of this billboard not only situates foreign women in a position of victimization whereby they must be rescued,[30] a form of embedded feminism whereby women's rights are used to justify foreign policy, often in the form of political or military intervention by Western nations into the affairs of Eastern nations, but it also validates the status of Western women. In this paradigm, women's liberation from traditional private-sphere roles becomes the only evidence required or necessary to demonstrate the superior status and civilization of Western nations, despite the ongoing feminist struggles of women in the twenty-first-century West.

This billboard promotes a variation on the themes of embedded feminism and consumerism as social justice. At the bottom of the billboard, another green-banner message reads, "Ethical Oil. A Choice We Have to Make." The color codes of red and green make evident that viewers are being encouraged to choose—or rather that they have no choice but to choose—Canadian oil-sands oil for ethical reasons that link Canadian Oil, by proxy, to women's rights. The billboard is not a direct call for war or invasion in the name of women's rights, although one could easily argue that it is a tentative step toward that, as political-media attention has increasingly highlighted tensions with Iran over its nuclear program and "the threat posed by Iran to Middle Eastern oil supplies";[31] however, as a reaction to women's oppression, it calls for embargo, even if only at the level of individual consumer choice. The use of the Iranian woman's image is an example of what Turcotte theorizes as the "uncritical representation of gender violence in other geopolitical locales [that suppresses] the state's simultaneous and daily consumption and violation of women for its own nation-building practices."[32] The use of the foreign burka-clad woman is part of a larger rhetorical practice that fails to read and understand exceptional moments of gender violence in Other contexts as such, and instead invokes these images of violence against women because they sustain the Western narratives of foreign-woman-as-victim. As Turcotte argues in regards to the United States, locating violence against women in other places "obscures and denies the dismal histories of gender and sexual violence endemic to the United States"—or, as I would argue, in the West in general—and the "consequent 'rescue narratives' demand victimized 'third world women' . . . must be saved from 'ethnic' perpetrators."[33] In the Canadian context, the Ethical Oil campaign is written into an accepted history of petropolitics that justified the suppression of Other peoples (domestically and internationally) and the invasion of foreign nations—for the purpose of gaining control of petroleum resources—by invoking the national myth of Canada as a defender of human rights. This billboard implies, through its false logic, that Canadians can choose continued security and freedom not only for Canadian women but also for foreign women by supporting Canadian oil. Resisting the oil sands industry, within this visual rhetorical frame, becomes akin to supporting female repression in other regimes, to supporting terrorism as it has been loosely defined in the post-9/11 era, and to forsaking the advances of Western feminist movements. The consumer "choice" provided by the artificially constructed parameters of this campaign justifies the perpetuation of Western oil-consuming lifestyles driven by current petro-economies as a strategy to shift the discussion away from environmentally focused critiques of the oil sands industry and its planned expansion.

Considering this billboard campaign within the context established by the book *Ethical Oil: The Case for Canada's Oil Sands*, with its blatantly racist, imperialist, and sexist language, adds another layer of meaning to the visuals. Take, for example, the billboard that juxtaposes the image of a beautiful and smiling Indigenous woman dressed in oil-rig garb: her head tilted and looking beatifically upward to the left from under an Esso hardhat, protective industrial eyewear, earmuffs, and overalls. The green-banner messages read "Ethical Oil" and "Canada's Oil Sands: Aboriginals Employed." This is juxtaposed with the red-banner messages of "Conflict Oil" and "Sudan's Oil Fields: Indigenous Peoples Killed" superimposed on the image of militiamen walking through the desert with a human skull in the forefront of the image. The Indigenous woman's presence in the billboard campaign appropriates her identity at two levels. Her sexuality and ethnicity are employed simultaneously as a tool of erasure in at least two ways: by preempting potential criticism of the rampant gender inequality in the oil industry by creating a simplified visual claim that women do benefit financially from oil-field employment; and by undermining the ongoing collective criticism and resistance against oil industry expansionism coming from FNMI communities in Northern Alberta and elsewhere in Canada. This image visually reenforces Levant's scathing criticism of any FNMI resistance to the oil sands as hypocritical and unjustified because he claims that the oil industry provides jobs and fiscal return for these communities.[34] This image of the young Indigenous woman reinvents the Canadian oil industry, and by proxy Canada itself, as a site of gender equality, in contrast to the alleged misogyny and sexism found in other countries. The image also aims to create doubts about the pervasiveness of anti-oil sands attitudes among Indigenous peoples, and to raise questions around the legitimacy of these communities to resist treaty breaches that disrupt traditional practices on Indigenous lands. However, by this same economic logic, all citizens of the global West, if not the entire world, can be silenced into complicity due to the way petroleum has shaped our daily lives and the global economy. This is not the case, nor should it be, in light of increasing evidence of environmental impact that has inspired a demand for change from across the political spectrum. Nevertheless, this oversimplified visual and rhetorical strategy attempts to weaken public sympathy for Indigenous resistance against the environmental impacts of the oil industry.

Prior to and during the period of the Ethical Oil billboard campaign, a number of FNMI communities publicly resisted industrial encroachment on their land and exposed the impacts of air, water, and land pollution in the area of Northern Alberta. There were also a number of environmental groups and activists making headlines with their anti-tar sands protests both in Canada and around the world at the time of the campaign. Rather than address these issues that involve both human rights and environmental concerns within Canada, the Ethical Oil campaign resituates petro-violence and human rights abuses outside the Canadian context. Situating the Aboriginal body—both the beautiful female in the "Aboriginals Employed" billboard as well as the ethnic male, potentially Indigenous, in the "Good Jobs" billboard discussed below—within the neoliberal infrastructure of the petroleum industry serves to erase the invisibilized Indigenous body that resists the oil sands and the oil industry. In doing so, petro-protests that demand social and eco-justice can be redefined as petro-violence. The activists in these movements are also easily reinscribed as terrorists.[35]

The racist underpinnings of the appropriation of the Indigenous female figure and Indigenous identities in general become even more explicit within the context of the other stereotypical and reductive billboard messages in the campaign. First is a billboard that, on the left, declares "Conflict Oil" and "Dictatorship" over a collage of the flags of Saudi Arabia and Iran, with faint superimposed images of Ayatollah Khomeini and President Ahmadinejad overlaid on the flag of Iran, all contrasted against the message on the right side of the billboard that features the Canadian flag flying behind green-banner messages of "Ethical Oil" and "Democracy." Another billboard represents the dichotomy of the red-banner messages,

"Conflict Oil" and "Forced Labour," with an image of Hugo Chavez, versus the green-banner messages of "Ethical Oil" and "Good Jobs" visually supported by the image of a young pleasant-looking man, quite possibly Indigenous. Yet another billboard presents the idea of "Conflict Oil" and "Degradation" illustrated with an image of a black man walking through a muddy or oily field gesturing toward a large fire burning in the background, likely in the Niger Delta, which is then contrasted with the green-banner messages of "Ethical Oil" and "Reforestation" and the image of someone standing in a wooded area—possibly meant to represent one of the people responsible for the Syncrude reforestation project in Northern Alberta. Yet another set of red-banner messages reads "Conflict Oil" and "Funds Terrorism," juxtaposing a photograph of the back of someone's head, clad in a Saudi-Arabian keffiyeh and looking into the distance at an oil rig, with the adjacent green-banner messages "Ethical Oil" and "Funds Peacekeeping" superimposed on the image of a Canadian peace-keeping monument. Photographed from below so as to heighten its majestic appearance, this sculpture is silhouetted against a blue sky and accompanied by a Canadian flag blowing in the wind. A final billboard pairs "Conflict Oil" and "Persecution" substantiated by the image of two blindfolded men being prepared for hanging, on the left, paralleled by the green-banner message on the right-hand side that reads "Ethical Oil" and "Pride" visually substantiated by two interlocking male hands, clasped together, each with a rainbow bracelet at the wrist.

In these simplistic binary arguments, the Ethical Oil campaign reveals the gendered and racialized messages that have been naturalized as part of Canadian, and even Western, petrocultural narratives. Read together, the various billboard images situate in bas-relief the identity-based fantasies of the entire Ethical Oil campaign, whereby foreign women of color are figured as victims of horrific violence, Canadian FNMI women are recuperated as symbols of Western gender and ethnic equality and representatives of the progressive employment practices of the oil industry, and white Canadian women are celebrated as civic leaders and symbols of democracy. These images obfuscate the historical, ongoing systemic and cultural racism against First Nations peoples in Canada and overwrite the very low percentage of female politicians.[36] By focusing on gender-sexual violence and inequality elsewhere, the campaign accomplishes what Elizabeth Swanson Goldberg has identified as the practice of creating markets of violence for consumption that erases these issues within the nation-state.[37]

Conclusion

Understanding the role of gendered petro-relations as they have been historically constructed and as they continue to be perpetuated by Ethical Oil and mainstream media representations of resistance movements such as Idle No More reveals the degree to which Western neoliberal petro-discourses are invested in promoting specific female identities and definitions of Western feminism. These iterations justify our current oil-consuming lifestyles as an issue of women's rights through basic rhetorical strategies that reinforce women's relationship to petroleum products in consumer terms and that recuperate the female body as a canvas on which to spectacularize and perform politics. Furthermore, the export of these specific female identities brings with it the promise of new female consumer markets. Therefore, empowered feminist identities outside those sanctioned by the mainstream neoliberal petro-discourses are depoliticized and renegotiated in the public sphere. Women activists and women activists of color who refuse to conform to the consumerist identities marketed to them, and whose agendas to protect the environment will potentially disrupt industrial development and business as usual, pose a particular threat to established forms of hegemonic power and therefore risk becoming the target of violence,[38] just as the men that align themselves with these women also risk being framed by discourses of petro-violence and marked as terrorists. It is not merely the threat of alternative energies that is being resisted. Reliance on wind and solar energy will not, in and of themselves, reconfigure power relations, nor eliminate racism, sexism, and class disparity, any more than oil in and of itself causes wealth, poverty, war, or militarism.[39] Therefore, as we make the necessary moves toward alternative energy sources in the twenty-first century, it is valuable to consider how we might be simply fetishizing these energies, as we have done with oil, in ways that perpetuate the capitalist status quo, and the web of associated relationships and power dynamics of gender, race, ethnicity, and class. Rather than succumb to the fears of what the end of cheap oil might mean for the future in general, or for feminism in particular,[40] it may be more productive to reevaluate current discourses around oil and alternative energy sources for their potential to disrupt rather than to simply reproduce intersectional social inequalities.[41]

Notes

1. Matthew Huber's use of "following the oil" and tracing the "webs of relations" builds on the ideas of David Harvey, Timothy Mitchell, and others. For more details see Matthew T. Huber, "Oil, Life, and the Fetishism of Geopolitics," *Capitalism Nature Socialism* 22, no. 3 (2011): 32–48.

2. Ibid., 36.

3. In 2012, a counterterrorism unit was established in Alberta to protect the energy sector. See "Counter-terrorism Unit to Protect Alberta Energy Endustry," *Canadian Broadcasting Corporation*, June 6, 2012, http://www.cbc.ca/news/business/story/2012/06/06/rcmp-counter-terrorism-oil.html. And over the summer of 2012, the media reported on concerns of environmental terrorism. An article published on July 29, 2012, in the *Canadian Press* cites a "newly declassified intelligence report" stating that there is a "growing radicalized environmentalist faction" in Canada that is opposed to the country's energy-sector policies. See "Rad-

ical Environmentalism on the Rise, RCMP Report Says," *Canadian Broadcasting Association*, July 29, 2012, http://www.cbc.ca/news/politics/story/2012/07/29/pol-radical-environmental ism-growing-intelligence-report-warns.html.

4. "Building Resilience against Terrorism: Canada's Counter-terrorism Strategy," *Public Safety Canada*, last modified December 4, 2012, http://www.publicsafety.gc.ca/cnt/rsrcs/ pblctns/rslnc-gnst-trrrsm/index-en.aspx.

5. Quoted in "Environmental Charities 'Laundering' Foreign Funds, Kent Says," *Canadian Broadcasting Corporation*, May 1, 2012, http://www.cbc.ca/news/politics/story/2012/ 05/01/pol-peter-kent-environmental-charities-laundering.html.

6. Heather M. Turcotte, *Petro-Sexual Politics: Global Oil, Legitimate Violence, and Transnational Justice* (Charleston, SC: BiblioBazaar, 2011), 213.

7. Robert R. M. Verchick, "Feminist Theory and Environmental Justice," in *New Perspectives on Environmental Justice: Gender, Sexuality, and Activism*, ed. Rachel Stein (New Brunswick, NJ: Rutgers University Press, 2004), 63.

8. Turcotte, *Petro-Sexual Politics*, 208.

9. Ibid.

10. Ibid.

11. "Idle No More: How It Began," *Province*, January 7, 2013, http://www.theprovince .com/news/bc/Idle+More+began/7780924/story.html.

12. Bruce Campion-Smith, "Idle No More: Spence Urged by Fellow Chiefs to Abandon Her Fast," *Toronto Star*, January 18, 2013, https://www.thestar.com/news/canada/2013/01/18/ idle_no_more_spence_urged_by_fellow_chiefs_to_abandon_her_fast.html.

13. Deveryn Ross, "Idle No More's Real Challenge," *Winnipeg Free Press*, January 24, 2013, http://www.winnipegfreepress.com/opinion/analysis/idle-no-mores-real-challenge -188173011.html.

14. Ross, "Idle No More's Real Challenge," *Winnipeg Free Press*, January 24, 2013.

15. Jorge Barrera, "PM Harper Believes Idle No More Movement Creating 'Negative Public Reaction,' Say Confidential Notes," *APTN National Notes*, January 25, 2013, http:// aptn.ca/news/2013/01/25/pm-harper-believes-idle-no-more-movement-creating-negative -public-reaction-say-confidential-notes/.

16. Duncan McCue, "Idle No More and Tensions in Thunder Bay," Special edition of *The Current*, CBC Radio, January 25, 2013, http://www.cbc.ca/player/News/Canada/Audio/ ID/2329105939/.

17. Valerie Taliman, "Rape, Kidnapping Being Investigated as Hate Crime in Thunder Bay," Indian Country Today Media Network, January 7, 2013, http://indiancountrytodayme dianetwork.com/article/rape-kidnapping-being-investigated-hate-crime-thunder-bay -146797.

18. "Starlight tours" refers to the unofficial police practice of dropping people, often Indigenous or other marginalized individuals, outside of city limits in what are often intemperate and life-threatening Canadian weather conditions, forcing them to walk back to the city or home. See "Starlight Tours," *Superior Morning*, CBC Radio, January 7, 2013, http:// www.cbc.ca/superiormorning/episodes/2013/01/07/starlight-tours/.

19. The Seneca Falls women's rights convention was held in New York in 1848. See "Women's Rights Convention in Seneca Falls, NY," *The Susan B. Anthony Center for Women's*

Leadership, n.d., http://www.rochester.edu/sba/suffrage-history/womens-rights-convention -in-seneca-falls-ny/. Ten years later, in 1858, the first commercial oil well was established in Oil Springs, Ontario. See "Black Gold: Canada's Oil Heritage," *County of Lambton Libraries Museums Galleries*, n.d., http://www.lclmg.org/lclmg/Museums/OilMuseumofCanada/Black Gold2/OilHeritage/OilSprings/tabid/208/Default.aspx. A year later, an oil rig in Pennsylvania struck oil. See "The Story of Oil in Pennsylvania," *Petroleum Education: History of Oil*, The Paleontological Research Institution, n.d. http://www.priweb.org/ed/pgws/history/ pennsylvania/pennsylvania.html.

20. It is important to note that there have been other class and race/ethnic struggles during the petroleum era, which this essay also references, and cannot escape referencing, because of the manner in which popularly naturalized discourses around oil and new-energies are situated in relationship to women, women of color, and other marginalized groups.

21. "Quizzes: Fashion & Beauty" and "How to Go Green: Fashion & Beauty," Planet-green.com, n.d. These quizzes are no longer available on the Planetgreen website.

22. Slavoj Žižek, "Censorship Today: Violence, or Ecology as a New Opium for the Masses, part 1," *Lacan.com*, n.d., http://www.lacan.com/zizecology1.htm.

23. This section borrows and builds on ideas previously published in Sheena Wilson, "Ethical Oil: The Case for Canada's Oil Sands, Review," *American Book Review* 33, no. 3 (2012): 8–9.

24. In 2002, Levant was nominated as the Canadian Alliance candidate in the Calgary Southwest riding, but he stepped down so that Stephen Harper, who had been elected party leader, could run in that riding. This led to the eventual election of Harper, leader of the Canadian Conservative Party, as Prime Minister of Canada. See Jane Taber, "Meet Harper's Oil-Sands Muse," *Globe and Mail*, September 10, 2012, http://m.theglobeandmail.com/news/ politics/ottawa-notebook/meet-harpers-oil-sands-muse/article1871340/.

25. Levant has had various complaints brought against him: both human rights complaints for the publication of the Danish cartoons of the Prophet Muhammad in 2006 and for making false claims about George Soros, to name two examples. And again in March 2014, yet another lawsuit was filed against Levant, this time for $100,000, related to blog posts he made that allegedly damaged the reputation of a Saskatchewan lawyer by labeling him a "jihadist and a liar." http://www.advocatedaily.com/sun-news-host-ezra-levant-sued -for-libel-lawyer-seeking-100k-in-damages.html.

26. The Ethical Oil organization misappropriates terms and ideas, such as "grassroots" and "fair trade," that are typically associated with left-leaning social justice agendas, for the purpose of supporting a right-wing neoliberal agenda focused on promoting the expansion of the oil industry in Alberta. For more details on the book, see Wilson, "Ethical Oil," 8–9.

27. This redirection in public debate was certainly evident in Alberta and in Canada starting at the time of the book's release in the fall of 2010, and for the subsequent 2 years, 2011 through 2012. The book and its author received significant attention at the time of the book's release, and the ethical oil message was further perpetuated by the associated Ethical Oil.org website. The site's "About" page explains that "EthicalOil.org began as a blog created by Alykhan Velshi to promote the ideas in Ezra Levant's bestselling book *Ethical Oil: The Case for Canada's Oil Sands*."

28. Fort McMurray is not officially a city but part of the Regional Municipality of Wood Buffalo in Northern Alberta.

29. According to a July 20, 2008, news article published by Radio Free Europe / Radio Liberty online, the image is from the 1980s. "Nine Iranians Sentenced to Death by Stoning," *Radio Free Europe / Radio Liberty*, July 20, 2008, http://www.rferl.org/content/Iranians _Sentenced_Death_Stoning/1184938.html. However, a major Canadian newspaper, *Globe and Mail*, claimed the image was from the 1970s. See "IN PICTURES: Ethical Oil Ad Campaign," *Globe and Mail*, July 28, 2011.

30. For a more specific discussion and definition of "embedded feminism," albeit in a different context, see the following: Helmut W. Ganser, "'Embedded Feminism': Women's Rights as Justification for Military Intervention?" (presentation at "Coping with Crises, Ending Armed Conflict: Peace Promoting Strategies of Women and Men," the international conference presented by the Gunda Werner Institute of the Heinrich Böll Foundation, Berlin, November 15, 2011), Gunda Werner Institute, n.d., http://www.gwi-boell.de/web/ un-resolutions-helmut-ganser-embedded-feminism-presentation-2953.html.

31. Oliver Wright, "Britain Meets Gulf Allies over Growing Tensions in Iran," *Independent*, January 24, 2013, http://www.independent.co.uk/news/uk/home-news/britain-meets -gulf-allies-over-growing-tensions-in-iran-8278101.html.

32. Heather M. Turcotte, "Contextualizing Petro-Sexual Politics," *Alternatives: Global, Local, Political* 36, no. 3 (2011): 204.

33. Ibid.

34. Levant provides an inflected critique of Chief Al Lameman, and by proxy all other Aboriginal people who contest the oil sands, in the book *Ethical Oil*. Levant writes: "Of course, oil sands firms are careful to make sure they stay on good terms with Aboriginals in the area, even if a few noisy malcontents like Beaver Lake Cree chief Al Lameman figure they'd rather sue the industry. . . . They've unilaterally declared themselves to be the sole 'keepers' of an enormous swath of land that crosses the border between Saskatchewan and Alberta. But Lameman and his reserve are not keepers of that massive swath of land in any meaningful way; they don't tend to it; they don't look after it or protect it; they don't improve it or develop it, and they certainly don't work it. They just claim it for themselves. Keepers is right. If Al Lameman has his way, he'll end up keeping thousands of Aboriginals from improving their lives, getting an education, escaping poverty. All of those jobs, all that education, all those opportunities ended, all for one band" (213–15).

35. Turcotte discusses at length how, in the context of the Niger Delta, the reporting of and even the academic analysis of "ethnoracial, national, and gender-sexual inequalities within and between Nigeria–U.K.–U.S. relations" have misidentified petro-protests as terrorist activities, in part because there has been a failure to recognize the degree to which these movements "developed out of strategies of community justice within women's organizations. These were neither 'terrorist acts' nor the community 'turning in on itself' but, rather, expressions of justice often supported by male youth groups, which created spaces through which community engagement could address inequalities." "Contextualizing Petro-Sexual Politics," 208.

36. According to the database for "Women in national parliaments" compiled by the Inter-Parliamentary Union based on information from 190 countries, last updated October

31, 2012, Canada ranks 47th in the world for female participation in the "lower or single house" with a percentage of only 24.7. Afghanistan, by contrast, ranks significantly higher in 37th place, with 27.7% participation. *Inter-Parliamentary Union*, http://www.ipu.org/wmn-e/classif.htm.

37. See Elizabeth Swanson Goldberg, *Beyond Terror: Gender, Narrative, Human Rights* (New Brunswick, NJ: Rutgers University Press, 2007).

38. "Women Human Rights Defenders," *Association for Women's Rights in Development*, n.d., http://www.awid.org/Our-Initiatives/Women-Human-Rights-Defenders; Inmaculada Barcia and Analía Penchaszadeh, "Ten Insights to Strengthen Responses for Women Human Rights Defenders at Risk," *Association for Women's Rights in Development*, 2012, http://www.awid.org/sites/default/files/atoms/files/ten_insights_to_strengthen_responses_for_women_human_rights_defenders_at_risk.pdf.

39. Huber, "Oil, Life, and the Fetishism of Geopolitics."

40. Sharon Astyk has written and blogged about how "the women's movement has never fully acknowledged the degree to which women's social roles have changed not just due to activism, but due to energy resources. This comparative blind spot means that we have also failed to grasp how vulnerable those gains are." While Astyk rightly indicates that women's lives have been transformed by the petroleum-derived energy sources, the over-simplified cause-and-effect relationship drawn between oil and feminism fetishizes the power of oil and fails to acknowledge the many other socioeconomic and political power relations that have played a role in the feminist advances of the last two centuries. See "Peak Oil Is Still a Women's Issue and Other Reflections on Sex, Gender, and the Long Emergency" (blog post), *Casaubon's Book*, Science-blogs.com, January 31, 2010, http://scienceblogs.com/casaubonsbook/2010/01/31/peak-oil-is-still-a-womens-iss/.

41. For more information on the notion of intersectionality, see Sheena Wilson, "Gender," in *Fueling Culture*, ed. Imre Szeman, Jennifer Wenzel, and Patricia Yaeger (New York: Fordham University Press, forthcoming).

))) Anthropocenic Ecoauthority: The Winds of Oaxaca

Cymene Howe

Lagoons

I would like to be able to say that I have a great recording of the conversation that Jesús and I had out by the lighthouse, at the end of a tiny peninsula that slowly disappeared into water around us. But microphone technology being what it is, and the wind being what it is in the Isthmus of Tehuantepec, there is nothing but a loud a rush of white noise blowing through the digital spaces where our conversation should have been. These are no average winds and their value has been carefully metered, in terms of both their profit making potential and their greater

Anthropological Quarterly 87, no. 2 (2014): 381–404

ethical possibilities in the global reduction of greenhouse gases. If you ask anyone in the global wind energy industry, they will likely tell you that the Isthmus of Tehuantepec is one of the best places on the planet to generate renewable electricity. In San Dionisio del Mar, where Jesús was born and raised, the largest single-phase wind park in Latin America has been slated for construction; it is also where Jesús and others have been leading a movement against the park's installation on their collectively held land. One of the reasons that Jesús and other *comuneros*[1] have opposed the Mareña Renovables wind park is that they are convinced that the construction of the park will endanger local fish and shrimp populations. San Dionisians live by the sea in both senses: the town is located on a lagoonal and maritime peninsula, and many residents are dependent upon the surrounding waters for both income and subsistence. San Dionisians and others in the *resistencia* are also wary about renewable energy development in the region because, thus far, it seems to have recapitulated the old habits of capitalism, *caciquismo*,[2] and corruption. Many Istmeños believe they have been enrolled in programs of renewable energy development without being fully informed or included in the process. Or, to use the language of the *resistencia*, they have been tricked and forced to bear the consequences of climate change mitigation and green capitalist aspirations in ways that are all too familiar.

Events that will determine the future of wind energy in the Isthmus and, in turn, set precedents for renewable energy development in Mexico, have been unfolding on an almost daily basis in the Isthmus: most of them in the form of confrontations, *bloqueos* (blockades), and *barricadas* (barricades) on the ground and in the courts. This article is based on 16 months of collaborative field research[3] and hundreds of conversations with Isthmus residents as well as activists, wind industry lobbyists, investment bankers, journalists, and government and industry representatives in Oaxaca City and Mexico City. For governmental officials and renewable energy company executives, wind power echoes with opportunity in all directions: from local biopolitical development to climatological aspirations to enhance Mexico's laudable international reputation in carbon reduction. For those challenging wind energy development, or actively asserting an "anti-eolic" position, protecting lifeways and ecological spaces is a more fundamental and immediate concern. Following the development of wind parks across the Isthmus, this article analyzes how a politics of resistance and local perceptions of environmental peril have challenged renewable energy transitions. In the debates and stand-offs that have transpired in reaction to the Mareña Renovables project planned in San Dionisio, a tension has emerged between local perceptions of ecological conditions and environmental knowledge that is gauged to global climate remediation. These divergences indicate distinct ways of imagining and articulating what I call "anthropocenic ecoauthority." Anthropocenic ecoauthority is predicated on a series of experiential, scientific, and managerial truth claims regarding ecological knowledge and future forecasting in an era of global anthropogenic change. Whether enunciated by resident communities, state officials, corporate representatives, or environmental experts, ecoauthority gains its particular traction by asserting ethical claims on behalf of, and in regards to, the anthropogenically altered future of the biosphere, human and nonhuman.

An Ecologics of Transition

The transition to sustainable forms of energy demands attention to multiple scales of engagement: from the places where energy production, distribution, and consumption physically occur to the logics and ethics that guide energy and climate policies. Anna Tsing (2004) has, among others (Comaroff and Comaroff 2003; Marcus 1995), demonstrated how anthropology's expertise is tested by competing scalar engagements, whether these are specifically located within the rubric of "the environment" (Choy 2011), "energy" (Wilhite 2005; Winther 2008), "globalization" (Appadurai 1996), or something else altogether. In order to create what she calls an ethnography of "global connections," Tsing (2004) has been interested in finding the points of contention, as well as cohesion, in multiple registers of discourse and interaction. She calls these nodal encounters "zones of awkward engagement . . . where words mean something different across a divide even as people agree to speak" (2004:xi). Following how environmental movements developed in parts of Indonesia, Tsing finds that self-determination and a codification of indigenous ecological knowledge were combined to engender novel forms of ecoconciousness and political movements in response to extractive practices. In a similar move to delineate how relationships between state agencies and local populations either facilitate or foreclose the growth of particular environmental identities, Arun Agrawal (2005) describes how "nature" is made available for individual, subjective management and identity formation. As in Tsing's and Agrawal's case studies, very specific imaginaries of natural environments and their utilitarian, spiritual, and climatological value echo across the Isthmus of Tehuantepec. In the case of forest management in Oaxaca, as Andrew Mathews (2011) has described, specific forms of environmental and scientific knowledge are produced in tandem, between experts and publics. State authority is accomplished and official knowledge coproduced through alliances between the state and powerful local actors, rendering very particular "stable representations" of knowledge, leaving others diminished (Mathews 2008:485). The ways that environmental logics shape social relationships among and between agencies, individuals, and subjective experience provide an important optic for understanding the cultural and political contingencies of energy transitions in Oaxaca. Equally relevant are the ways in which ecological discourses and practices underpin the reciprocal relationships between state authority, knowledge regimes, and shifting perceptions of "the environment."[4]

Environmental knowledge and renewable energy transitions are evolving in an era that is increasingly being called the Anthropocene—a time of unprecedented human-generated deviations from our geological, climatological, and biological past (Chakrabarty 2009). Our anthropocenic conditions would seem to compel us to speak across very specific registers of local environmental awareness (such as concerns about local ecosystems and aquaspheric and terrestrial damage) and those that are reckoned climatologically and globally (such as worries about greenhouse gas emissions and carbon contamination) (e.g., Crate and Nuttall 2009). As will be clear in what follows, local environmentally informed responses and those that purport to speak on behalf of a global scale are often conflicted, and their sources of knowledge disparate. What will also be readily apparent is that the politics of renewable energy in the Isthmus are steeped in neoliberal development logics, persuading government agencies and functionaries to align with the profit seeking interests of renewable energy corporations (Gledhill 1995; McDonald 1999; Ochoa 2001; Schwegler 2008). Rather than focusing attention on the political economy of energy transition, however, I want to signal the ways in which energy futures are profoundly shaped by discourses and practices that assert an ecological and environmental authority; these epistemological and ethical exercises suggest a symbiotic awareness, fundamentally founded in moral claims to protect the biosphere, including humans and other biotic life, now and in the future.[5]

Opportunities

It has become a well-known fact that Mexico's national oil company, PEMEX (*Petróleos Mexicanos*), is in crisis. The company's oil production capacity has declined dramatically over the past several years. Given that PEMEX's profits have provided up to 40% of the country's federal operating budget, drops in oil production also mean a significant loss of revenue for the state and nation. Mexico has, however, made significant commitments to renewable energy, with the Ministry of Energy aiming to generate 35% of electricity from non-carbon sources by the year 2024. Think tanks and environmentalists alike have heralded Mexico's achievements as one of the major success stories for climate change mitigation in recent years. While the Kyoto Protocol did not demand emissions reductions for the country, Felipe Calderón's administration instituted some of the most ambitious and comprehensive climate change legislation[6] in the world. However, the advent of renewables and carbon reduction policies have also raised questions about how these projects may disenfranchise local populations and limit their autonomy regarding how ecological spaces and environmental resources are to be used and managed (Howe 2011; Lifshitz-Goldberg 2010; Love and Garwood 2011; Krauss 2010; Oceransky 2009; Pasqualetti 2011). The ways in which state functionaries, private corporations, and local residents imagine the fate of the land, and the wind that gusts above it, are often widely divergent. Even as each position resonates with certain ethical truth claims of protection, the ways that ecoauthority is being formulated and asserted illustrate an often incommensurable rift.

The Mexican government has made substantial investments to determine the future of wind development and renewable energy implementation in Oaxaca. The Mareña Renovables project, for one, articulates well with the ecological and energy development aspirations of the Mexican state. In its National Infrastructure Program, the Mexican Ministry of Energy (SENER, *Secretaria de Energía*) has lauded the growth of renewable energy resources. For SENER, wind energy, specifically, has several vectors of potential, in terms of both local development and national contributions. The Ministry sees the winds of Oaxaca as

> an opportunity to reduce emissions without compromising national economic development; an opportunity to contribute to climate change mitigation; an opportunity to attract investment to Mexico; an opportunity to develop local capabilities; an opportunity for technological development; an opportunity to increase the nation's global competitiveness. (SENER 2007:32–33)

Opportunity is an important trope for the Ministry as it envisions a path ever upward toward growth and development, the presumed zenith of social prosperity. One of the reasons that "opportunity" resounds so easily in state discourses is that the Isthmus of Tehuantepec is one of the best locations for the production of wind power in the world. It is also located in one of the poorest states in the country where economic development is often prioritized in government programs and politicians' ambitions. The former director of Sustainable Energy for the State of Oaxaca explained to us, for example, "if it weren't for the wind, there would be no development [in the Isthmus]." From the vantage of state officials, the authority and right to promote economic development (in this case, renewable energy projects) is, in part, an ethical calling that will, putatively, bring further degrees of prosperity to the region. The future of the Isthmus is, in this way, married with the development plans of private capital in a tidy moral tale that is often described as "win-win." Opportunity is articulated across all dimensions of social life and social "health"—from national economic metrics, to local capabilities enhancement, to global climate change mitigation. Federal governmental offices, like SENER among others, are able to denote a series of environmental and economic futures as the proper way forward. Each of these claims for ecological, economic, and technological possibilities is also framed as an ethical exercise intended to benefit the "greater good," whether this is scaled to the Mexican nation or on behalf of the planetary bios. Charting development through future economic and environmental interrelationships is one way in which the state is able to constitute its own ecoauthority.

Like the federal government, the state government of Oaxaca has underscored the value of wind energy for Isthmus communities, state development, and carbon reduction aspirations. The governor is confident of the benefits to be reaped by *energía eólica* (wind energy), at all levels of society. He describes that

> the initiative and confidence that the business sector has placed in our state [Oaxaca] has made it possible, within the first two years of this government, to see a total of 15.852 billion pesos (about $1.22 billion USD) invested in the wind sector, making our state a leader both nationally and internationally in the production of clean energy, directly combatting the effects of global warming. ("Mareña Renovables" 2012)

The state government is keen to take advantage of the wind's potential, developing both its economic and ethical possibilities. Ecoauthoritative claims converge here to foretell future prospects for the region while also promising relief from contaminative energy production. In these narratives, both global and local populations are poised to benefit from the dual environmental "goods" of wind power and liberal development policies. However, the production and consumption of renewable energy draw attention to the uneven ways in which the benefits of climate mitigation and renewable energy development are being distributed.

Wind power in the Isthmus, once it is made electric, is largely dedicated to a system called *autoabastecimiento* (industrial self-supply). Electricity generated in relatively remote locations in *parques eólicos* (wind parks) is channeled through a series of substations and high-tension wires to offset electricity prices and create a portfolio of green energy consumption for the companies who have contracted to purchase this power from the *Comisión Federal de Electricidad* (CFE). The CFE is a parastatal monopoly that manages all electricity distribution in Mexico, controlling the entire grid. Corporate consumption and privately managed electricity generation projects (like wind parks) are channeled through the CFE's electrical infrastructure. Currently, in the Isthmus of Tehuantepec, 14 *parques eólicos* are in operation; all of them (except for a CFE pilot park) are operated by transnational renewable energy corporations and multinational consortiums. Electricity production in the Isthmus has now reached more than one gigawatt, or enough to power close to one million Oaxacan homes. However, the electricity is specifically not going to homes or municipalities in the Isthmus; it goes further afield and is purchased by corporations such as Coca Cola, Heineken, Walmart, and the baked goods manufacturer Bimbo to offset their manufacturing costs and environmental impact. Electricity is being produced, but not for local populations, who complain vociferously that the cost of domestic and commercial electricity is inordinately high. To redouble this extractive ethos, companies have not built factories or industries in the region, nor have wind parks been able to provide more than a handful of permanent jobs for local people. For average Istmeños, the wind industry gives them few perceptible benefits.[7] Although wind parks are predicted to "guarantee development," this rings hollow for many people in the region who have seen many a megaproject come and go (Howe, Boyer, and Barrera in press). Nevertheless, the ecoauthoritative discourses surrounding wind parks provide ethical traction to the expectation that they will both remedy the future of the climate and ameliorate regional socioeconomic deficits.

The Land and Sea beneath the Wind

The Mareña Renovables wind park in San Dionisio was slated to begin construction in early 2012. It has not. The blades, towers, and docking materials are instead moldering in warehouses awaiting an increasingly uncertain future. The 396 megawatt park, if it were to be constructed, would be the largest single-phase wind power installation not only in Mexico, but in all of Latin America. The project is financed by an Australian consortium (the Maquarie Group), Mitsubishi, and the Dutch pension fund PGGM. It has also been funded by loans from several banks, including the Inter-American Development Bank, which prides itself on its rigorous environmental and human rights standards. The majority of the 132 turbines proposed for the park were to be located on a sliver of sand bar, the Barra of Santa Teresa, which is also very near indigenous peoples' ceremonial sites.[8] Equally concerning to local populations is that this is an area where many residents earn their livelihood fishing and harvesting shrimp. The communities of Santa Maria del Mar and San Dionisio del Mar, where the development is proposed, each operate under a communal property system in Mexico called *bienes comunales*.[9] Since the wind park contract was signed in 2004, many *comuneros* who initially agreed to the development are now in opposition. Their argument is that the community was inadequately informed about the scale and impact of the park from the outset. For instance, many believed that there would be only 40 turbines, rather than 132. In June of 2012, the validity of the San Dionisio contract was put to a judicial review. While the fate of the contract continues to be the object of legal interpretation, the local *resistencia* has, with pro bono legal assistance, successfully lobbied for an *amparo* (injunction or staying order) on the Mareña project. As of this writing, the park's installation has been halted and its future development appears unlikely.

Although wind park projects have the enthusiastic support of federal and state governments, rental contracts for them were crafted under somewhat suspicious conditions. Following a USAID study that deemed the area rich in wind resources, and an aperture in Mexican law that allowed international investors to develop Mexican energy resources,[10] the Isthmus became a bonanza for wind power. Putting it in more critical terms, as many in the opposition do, it became the site of a *nueva conquista*.[11] In the 1990s and early 2000s, several companies, many of them Spanish, working in conjunction with local agents on the ground, acquired contracts and agreements from individuals and communities. Essentially, the Isthmus was divided into corporate districts controlled by investment interests and development companies; this arrangement also prevented landholders from seeking competitive counteroffers on contracts.[12] People in San Dionisio and other communities in the area that would be affected by the Mareña project often speak of a *despojo de nuestra tierra*, being robbed, or stripped, of their land. They are concerned about losing control of the rights to their land; with 20- to 30-year contracts for turbine and road easements, which are renewable up to 60 years, this seems eminently possible.

Rental payment structures often depend not only on the quantity of land being occupied by turbines and roads, but according to when and with which company a land owner signed on. Compensation rates and rental agreements in the Isthmus vary widely. However, they are far less than rental payments for similar installations in the US or Europe.[13] In Unión Hidalgo, a woman with four hectares of land under contract said that she receives only 90 pesos (or less than $7 USD) per month. She explained that she wants to revisit the contract in order to have it coincide with current rates in the region. But, as she said, the legal counsel required to do so will be very costly. In addition to landholders, communities surrounding the wind parks may also receive a collective payment—from between 0.97% to 1.5% of profits from electricity generation—to be put toward social development programs. In the community of Santa Maria del Mar, where Mareña has planned to put 30 of the total 132 turbines, *comuneros* are said to be receiving 1,000 pesos (or about $77 USD) a year for their consent to the park's construction, channeled through the local municipal authority. Mareña Renovables is quick to note that it has invested huge sums of money in the region: a payment of 20,500,000 pesos (about $1.5 million USD) to San Dionisio del Mar for the *Licencia de Construcción*. In addition, the company has agreed to pay a fee of 1,866,623 pesos (over $140,000 USD) to the community once the project has begun generating electricity. From the perspective of the company, each of these sums can be put toward social works (such as schools, health clinics, and road paving) that will help to develop the community and its infrastructure. For many Istmeños, the promised payments are, as they put it, *una miseria* (a pittance) in comparison to the profits the companies are making. Nonetheless, these sums are purported to be local, biopolitical investments derived from ethically correct resource use and carbon mitigation.

If the Mareña project were to be built in San Dionisio del Mar, it would be the first wind park in Mexico to be built on actively managed *bienes comunales* lands. Given the collective nature of decision making and the open forum required of voting on *comuna* matters, the park has the potential to be a referendum on community control over renewable energy development, rather than one determined by corporate and state development interests (Cohen 1999). As Mike Hulme (2009) and Hermann Scheer (2004) have argued, renewable energy transitions and climate change mitigation measures have the potential to foster "new political, economic and cultural freedom" (Scheer 2004:67). While a wind park located on communal lands could offer a new model for renewable energy implementation, unfortunately, as of yet, it has not.[14] Instead, corruption, manipulation, and threats of violence have been the operative logic. In January 2011, the *resistencia* against the Mareña *megaproyecto eólico* (wind megaproject) occupied the municipal palace in San Dionisio del Mar and unseated the municipal president, claiming that he accepted bribes from the company totaling about six million pesos (or about $460,000 USD). Across the lagoon in Santa Maria del Mar, one resident explained that the people of Santa Maria have agreed to the park only because "the people in Santa Maria are repressed (*reprimida*) and silenced (*callada*)." Concerns about intimidation, threats, and manipulation by local leaders have led to profound suspicions and antagonisms in the region. Accusations of bribes and purchased consent have been central to nearly every conversation that we have had about wind development in the Isthmus; for many, these are symptoms of "traitorous" behavior.

A fundamental critique of the Mareña project is that the Mexican government has failed to fulfill its responsibility, mandated by national and international law, to create a process that ensures free, prior, and informed consent for indigenous communities regarding projects that will directly affect them. The development desires of the Mexican state, along with those of green capital, each resound with the global chorus for climate change mitigation. But these calls become dubious when they are predicated on an apparent willingness to trample constitutional rights as well as more recent international human rights protocols enshrined in Mexican law. The entire contracting process for wind park development, according to the *resistencia*, was undertaken with a "dynamic of dispossession, abuse, lies, and contempt toward indigenous peoples." Or, as one organization has described it, the development is a "simulation of legality in order to cover up land grabs and environmental damage" (AMAP 2012:2). The gigantic wind park has been tainted by corruption. At worst, it exists as an infamous account of leaders being massively enriched at the expense of the communities they represent. At best, it is an example of green capitalism's failure to adequately communicate with and compensate the people who live in the economically marginalized spaces where, invariably, the wind seems to blow the strongest.

Damage

If green developers and state agencies have used ecoauthoritative registers to substantiate the benefits of renewable energy—for the Isthmus region and for the world—the *resistencia* has also taken up an ecoauthoritative voice of its own to argue the reverse. Rather than local economic benefit, they maintain, renewable energy is predicated on local environmental damage that will impact both human and nonhuman inhabitants. The *resistencia* asserts that corporate and state developers have failed to present adequate information about the negative environmental consequences of wind parks and how they will affect livelihoods in the communities where they are located. They question whether the government is actually invested in local communities' right to determine the fate of their land and other elements of the natural environment, including wind and water. Istmeños' confidence in the apparatus of the state is limited, and they often voice suspicions about the government's ability and willingness to provide credible reports of environmental risks and ecological damage.

The heart of the resistance—as Antonio L., one of the key spokespersons, explained—is made up of people who "live by the sea" and live "*por la pesca*" (by fishing). Three communities situated on the lagoon and oceanfront have been at the forefront of the anti-Mareña battle, believing that their livelihoods and the environmental future of the region will be endangered by the park. Fishermen from Juchitán de Zaragoza, the municipal capital and epicenter of the southern Isthmus, have joined forces with the opposition because they too are worried about *la pesca*. They are concerned about the effects that the construction phase may have upon marine and lagoonal life and they are troubled about the more ambiguous and uncharted effects of the turbines when they are in operation. Vibration, noise, and light, as every fisherman with whom we have spoken has emphasized, will scare away their catch and, thus, their ability to survive. Both ikojts and binnizá fishermen[15] believe that the noise generated by 132 turbines will likely affect the fish population, driving them to migrate to other areas. Lights atop the towers are a source of concern, brightening the waters in ways that, again, may make fish depart. Fishermen also described that the turbines will create vibrations that will rattle the sandbar and emanate across the bottom of the ocean and lagoon. Their worries center on the potential environmental damage that might occur if the project is brought to fruition. But their environmental concerns are also directly related to their livelihoods and their ability—and that of future generations—to work with, in, and around the waters surrounding their communities. State and corporate promises of social development and infrastructural improvements to these communities seem, to those in the resistance, both improbable and malformed. In Álvaro Obregón, the hamlet that is the gateway to the Mareña construction site, one fisherman put it this way:

Maybe this company or the government will come in here and pave these roads. Though I doubt it; we've been waiting 30 years and they never have. Look around you, it is dirt and dust everywhere. But even if they do pave these roads it won't matter. If I can't fish I can't live and if the people of Álvaro Obregón can't fish, they can't live and so we will have to leave here anyway. There will be no one left to enjoy those paved streets.

The boons of development, even if they are made real, appear to be an illogical remedy if environmental harm will spell the end of an important form of human subsistence. As they critique the ancillary effects of noise, light, and vibration,

Isthmus fishermen exercise their own ecoauthority to challenge whether wind parks, humans, and non-human species can cohabitate in the maritime spaces of the Isthmus.

Descriptions of environmental damage, such as toxic oil seeping from turbine mechanisms, circulate freely in the Isthmus; however, there are significant differences in opinion as to how much harm, permanent or temporary, has occurred with the rise of the wind parks. The unique geographic conditions of the Mareña park make its environmental impacts difficult to fully estimate. It is, apparently, the only wind park in the world slated to occupy a sandbar. The Inter-American Development Bank's (IDB) Environmental and Social Management Report acknowledged the possibility of short-term "economic displacement" from the disruption of fishing during the construction phase of the park. It also detailed the environmental consequences the park was likely to have on marine turtles, jackrabbits, and bats. However, the report did not analyze the long-term impacts of the park's presence on local fish populations. The IDB's assessment notes that the proposed Mareña sites "have been exposed to intense human activities in the past decades which have led to a deterioration of the 'natural' character of the area" (IDB 2011:10), and that "both sites [of the proposed project] have been severely affected by anthropogenic activity" (2011:6). Local fishermen, for their part, also confide that the local fish stock has been on the decline, in part due to pollution and in part due to overfishing. Placing massive turbines atop the narrow sand bar of Santa Teresa, fishermen believe, will worsen these conditions and hinder their ability to fish for both subsistence and a modest income.

Fishermen's concerns and the ecoauthoritative statements generated by the *resistencia* are not without their ecological and environmental "impact." Their worries about the region's ecological viability are derived from an awareness of aquaspheric limits; or put another way, they are environmental assessments about how much an anthropogenically injured lagoon and sea can be expected to yield. These concerns are ecologically oriented and they have local traction. As reasoned evaluations made by those who live by and from the sea—as well as for others in the region who find truth in these concerns—these are credible doubts about the potential environmental changes that will follow the park's installation. In a context of somewhat sketchy scientific reporting, the environmental assertions of fishermen and others may garner more credibility than they otherwise might. Wind industry professionals, bankers, and state officials in Oaxaca City and Mexico City are, probably unsurprisingly, quick to scoff at fishermen's claims. While the question of whether noise, vibration, and light will result in the deleterious outcomes that fishermen predict is unknown, it does uncover an important contingency. It indexes the difficulty of producing credible ecological knowledge when state and corporate interests appear to be compromised by financial incentives rather than attentive to environmental protection.

The Treasure of the Isthmus

The Mareña Renovables website is a deep digital pool filled with ecologically authoritative proclamations, reports, and results.[16] It has lofty aspirations for the wind of the Isthmus, claiming that "the wind is the treasure of the Isthmus that will bring Mexico into the future of sustainable electric energy generation." With an uncanny doubling, the website material manages to focus simultaneously on the narrow strip of the Barra de Santa Teresa and its surrounding waters as well as on the greater planetary biosphere, captured under the rubric of climate change mitigation. Echoing neoliberal development impulses as well as biopolitical growth, the wind park is, first and foremost, described as "a comprehensive project investing in renewable energy that promotes economic growth for the state and enhances the well-being of the communities in the Isthmus of Tehuantepec." "Compared with power generation using fossil fuels," the website rightly maintains, "wind energy generates minimal impact on the environment, making the construction of wind parks a great option to ensure the future of electricity using sustainable means." Guarantees of local development coupled with greenhouse gas reductions are laudable to be sure, and the company seems keenly aware of the value of this anthropocenic moral authority. Indeed, the very origins of the Mareña consortium itself, its genesis, are framed in terms of climatological goodwill:

> The [Mareña] consortium was formed in order to construct a wind park that would generate clean energy and mitigate climate change and it has worked very conscientiously to ensure that the park that will be built on the Barra de Santa Teresa respects both the natural and cultural heritage of the region and to preserve the fishing activities of the lagoon. This is a world class project that will . . . generate the equivalent of electricity consumption for half a million homes. Additionally it will avert 879,000 tons of greenhouse gas emissions each year. The park will give the state of Oaxaca the opportunity to be an example of sustainable development, promoting clean energy projects that will, at the same time, promote the economic growth of communities. The project follows the highest environmental standards and has obtained local and federal construction permits.

Adherence to state regulations and observance of "the highest environmental standards" are key elements of the company's ecological self-portrait. These distinctions are also a means to leverage ecoauthority and bolster the company's credibility and correctness. Very similar claims were made during a press conference held in Juchitán in December 2012. In the presentation, the social communication team hired by Mareña Renovables detailed the environmental conservation practices the company had begun or would be undertaking. These included waste management, bird monitoring, mangrove protection, reforestation and transplantation, marine turtle preservation, animal rescue, and safeguarding the endangered

Tehuantepec jackrabbit (Chaca 2012). In addition, as one of Mareña's social development team members explained in our interview, there are projects to enhance human "capacity" in the region. Human development and skill enhancement, such as artisanal workshops where participants learn to weave plant fibers, are counted among the ways that the company is seeking to ensure human well-being and regional development.

The company's plans to promote human social development and protect various forms of life, from mangroves to jackrabbits, offer a comprehensive set of ecological and social remedies. Indeed, balancing local concerns of damage against the global gains of climatological cleansing has been a critical element in positioning the wind park and the company itself as environmentally upstanding. In an effort to assuage local concerns about fish and shrimp stock, Mareña Renovables developed an interior web page on its site specifically dedicated to fishing cooperatives. "The conclusion [of] studies and international experience," it states, "is that there are no effects on fishing caused by the operation of a wind park." While it goes on to note that the construction phase will entail a great deal of movement and added turbulence (and turbidity) in the water, the company has sought to make the construction process as rapid as possible, to avoid any unnecessary disruptions. The sentence centered at the foot of the page, in bold, conveys the central message behind these locally focused environmental statements: "The wind park project respects the culture of fisherman." The information presented is calibrated to diminish the worries of fishermen and others in the area, using globally circulated discourses of environmental protection and scientific authority regarding natural processes. From waste management to the grander claims of "environmental benefit," the corporation is invested in establishing its ecologically ethical credentials (Jamieson 2011).[17] Positioning itself as an entity that can, and will, foment these sorts of cures, the company marshals a powerful ecoauthoritative voice and position. However, it is also a position that presumes that local communities have faith in the process of permits, impact reports, and the unqualified good of "international experience." Given the history of the Isthmus and the suspicions that surround the interventions of both the state and transnational capital, ecoauthoritative assertions such as these have not appealed to local communities in the ways that the company has hoped for (Campbell et al. 1993).

The Mareña consortium has carefully employed its ecological authority as it has attempted to maintain equilibrium across different scales of environmental and human health. In one dimension, the park is said to have a "minimal (negative) impact on the environment." At the same time, a series of biopolitically productive metaphors claim that a "magnificent opportunity to guarantee sustainability" is likewise possible. The company's portrayal of the wind park emphasizes its positive impact upon ecological and economic conditions both local and global; this claim is possible only through establishing an ecoauthoritative register of environmental impact reports, "international experience," and greater ethical claims to protect the global biosphere. The question, however, is not so much whether the many environmental impact reports that were required to license the wind park are correct or not, nor is it a matter of challenging the veracity of wind power's positive contribution to climatological equilibrium. The Mareña project may in fact portend an empirical good especially over and against the use of fossil fuels. However, it is also important to draw attention to the ways in which anthropocenic ecoauthority is being used to ensure the park's development and, in turn, to limit local decision making about the environment's status, potential, and future. While local ecological interpretations and those of corporate and state functionaries often appear to be counternarratives, I would argue that both positions' concerns about fish and culture, as well as translocal worries regarding climatological corruption, are mutually codependent. They each emerge from similar ethical uncertainties and moral possibilities regarding the ecological shape of the future.

The competing scales of ecological remedy appear incommensurate because each of them sounds out a different audience and each of them focuses on slowing different sorts of environmental distress and harm. Local ecointerpretations are often reduced to naïve estimations, limited by a lack of education or at best predicated on "indigenous knowledge." Even if local ecoauthoritative perspectives and concerns are not entirely disregarded by state agencies or company representatives, local experiential claims are nonetheless limited in their scope and impact. They can only ever speak for the future of generations in the Isthmus and the continuation of fishing and shrimping as a mode of life. The grander claims of the company and the state calibrate their environmental remediation to a global good, not "merely" a regional concern, and thus can leverage somewhat greater ecological moral authority.

Biopolitics in the Anthropocene

The politics of energy transition and climate change mitigation, as they become articulated in their local and global dimensions, are, to put it euphemistically, a challenge. They have often proven resistant to the mechanisms of global governance and protocols, resulting in failed agreements and missed targets. The Isthmus of Tehuantepec is a case that illustrates several tensions that condition energy transitions, from sovereignty claims to dreams of sustainability. It raises critical questions about the future of renewability. For example, will energy transition be determined by the pipeline politics and grid formations of our carbon past or not (Coroníl 1997; Mitchell 2009, 2011; Sawyer 2004)? And how does a particular collective will or voice—either those of local communities or those of state institutions and private companies who purport to speak for national interests and global benefits—come to take precedence? Oaxaca is one of the poorest and least energy intensive states in the country of Mexico, a country which itself has not produced carbon contamination on the scale of the Global North. Yet, both government functionaries and renewable energy companies are fully committed to making the region a sustainable energy powerhouse. There is no inherent incommensurability between renewable energy implementation and economic and social equity. However, as we see in the Oaxacan case, these processes are profoundly shaped by the ways in which ecoauthoritative positions are generated, legitimated, and implemented or, alternately, refused and diminished.

Among residents of the Isthmus, there is remarkably little talk of climate change remediation or a commitment to providing the means for clean electricity generation. Instead, people voice concerns about land, fish, work, and culture. The Mexican state and transnational renewable energy corporations investing in Oaxaca are, conversely, speaking the language of climate change and greenhouse gas reduction. An anthropogenically altered world would seem to be a condition that demands new mobilizations of biopower.[18] Instead, however, the politics of climate change and the related process of renewable energy transition appear to depend on familiar biopolitical frameworks: improving schools and clinics, providing jobs and roads, and managing populations in the service of development, investment, and growth. These are biopolitical concessions spoken in the idiom of energopolitical shifts,[19] suggesting that indigenous peasants are state subjects whose precarity can be solved through publicly funded material improvements. Those in opposition to the way that renewable energy projects are being implemented in the Isthmus rely on these same biopolitical frameworks. The *resistencia* too calls for the protection of livelihoods to ensure that poor and marginalized populations are not dispossessed of the few resources they have been granted: land and access to ocean and lagoonal fishing. These sorts of biopolitically motivated solutions parallel the logics that have driven development policies for decades. They also reiterate many of the climate change discourses and policies of the last several years, or what Anthony Giddens (2009:8) has called the "convergence" of political and economic stimuli to facilitate the use and distribution of sustainable energy.

Central to the process and politics of transition, as I have argued here, are narratives, policies, and actions that utilize ecoauthoritative claims on behalf of a larger, if not always well-defined, "environment." For local protestors, their ecological objects are fish, shrimp, and lagoonal spaces. For global investors and state bureaucrats, the greater global climate is positioned as the ultimate ecological objective. Though the targets of ecological harm reduction and environmental remediation are distinct, the ecoauthoritative registers used to advocate for them ring surprisingly similar in their future orientation. As Mike Hulme (2013) has argued in the case of climate modeling science, these discourses become, effectively, a particular way of "anticipating the future" and they "become a prosthetic-to-human moral and ethical deliberation about long-term decision-making" (Hulme 2013:50;

see also Edwards 2010; Hastrup and Skrydstrup 2013). As a series of knowledge claims, anthropocenic ecoauthority animates policies and protests; these are couched in ethical terms as works and deeds intended to benefit either a planetary bios or the interests of the region's human population. These are questions, ultimately, of how life will be managed, or not, in a new energic era.

The biopolitical and energopolitical strategies being deployed in Mexico are confounding, in part because they cannot seem to simultaneously address the management of life at the local level (in this case, the people and environmental "resources" of the Isthmus of Tehuantepec) and life at the planetary level (which requires decreased emissions and transitions to renewable energy). Familiar forms of biopower, as we have understood them—from their colonial manifestations, to those based in Keynesian welfare models, to post-industrial neoliberal constellations—appear inadequately equipped to address the biotic demands of the Anthropocene. The remediations available seem to only reiterate well-worn models of endless growth and biopolitical liberalism. If Foucault (2009) originally formulated biopower as a way to enact and exercise power over and through the basic biology of the human species, the anthropocenic era demands that we ask how biopower, as an analytic, can make the leap to a much greater dimension of engagement. The original sites of biopolitical intervention have been population and the human species, but the energic, atmospheric, aquaspheric, and lithospheric changes that have been dubbed the Anthropocene demand that our focus extend to life beyond the human.[20] Shifts in the planetary climate call for a new ecologics of the present and the future.

Notes

1. Joint holders/managers of collective lands.

2. System of dominance by local political authorities often associated with nepotism, corruption, and petty tyranny.

3. Field research was conducted in collaboration with Dominic Boyer and was funded by a grant from the National Science Foundation, Cultural Anthropology Program (NSF 1127246). Initial research in 2009 and 2011 was followed by extended fieldwork from May 2012 to August 2013.

4. In a similar way, the congruences between human subjectivity and agentive nonhuman objectivity (e.g., Bennett 2010; Latour 2004; McKibben 2006) have shown how technicians of governance encounter and evoke nature in their expert management of larger biotic bodies or "ecosystems" (e.g., Darier 1999; Luke 1999; Malette 2009).

5. By "symbiotic awareness," I refer to the increased recognition that human practices (such as fossil fuel consumption) are having climatological effects that impact the entire biosphere. Whether there is a growing human consciousness of our mutuality and interdependent relationships with other biotic life or abiotic materials is an open question, but "mutualisms" (Gilbert, Sapp, and Tauber 2012) in biology, "ecological thought" (Morton

2012), and object oriented ontology (Harman 2011) in the humanities suggest new imaginaries of the human condition that question the boundaries of biological or species-specific individuality.

6. In May 2007, President Felipe Calderón announced the National Climate Change Strategy, instituting climate change mitigation as a central part of national development policy. The Renewable Energy and Energetic Transition law was passed in 2008, requiring that 35% of electricity come from non-fossil fuels by 2024, 40% by 2030, and 50% by 2050. In June 2009, the federal government formally committed to a detailed long-term plan for emission reductions, *Programa Especial para Cambio Climático*, that monitors improvements and establishes reduction guidelines, sector by sector. Wind power accounts for the majority of Mexico's clean development mechanisms (almost 2,300 kilotons of CO_2 reduction, compared to 900 kilotons through methane recovery). In 2012, before leaving office, Calderón signed the General Climate Change Law, which proposed to formalize targets in previous legislation, inaugurate the National Institute of Ecology and Climate Change, and coordinate federal offices to develop holistic mitigation and accommodation planning.

7. In the community of La Ventosa, a small number of local residents work at the neighboring wind parks as managers, engineers, and technicians. However, most direct employment thus far in the region has been in the form of temporary jobs during the construction phase and benefits for a few property owners and restauranteurs who cater to the handful of Europeans employed by the wind parks.

8. The small island of Tileme, for example, is considered a sacred site by many ikojts people.

9. *Ejido* (cooperative land tenure) and *bienes comunales* land tenure models are legacies of the Mexican Revolution that confer management decisions among members who collectively decide the fate of their land (see McDonald 1999; Nugent and Alonso 1994; Castellanos 2010).

10. NAFTA, the Agrarian Law Reform, and PROCEDE—coupled with the 1992 Electric Energy Public Service Law—allowed local landholders to more easily sell and contract their land to private interests and gave private sector companies the ability to participate in electric power generation.

11. The sentiment that wind parks may be signs of a *nueva conquista* is related to the fact that the majority of transnational companies involved are Spanish. Likewise, there is a long history of megaprojects and infrastructural works that have often been associated with colonial or extractive motives, including (long-standing) plans for a trans-isthmus canal, railroads, highways, logging, and hydroelectric dam projects (Barabas and Bartolomé 1973).

12. In November 2004, Grupo Preneal obtained the usufruct rights to 1,643 hectares (approximately 16.5 million meters squared) of *bienes comunales* land. However, *comuneros* aver that they were misinformed about the project and that signatures were obtained through obfuscation and manipulation. Further, they argue that the constitutional provision afforded to indigenous communities—demanding consultation and prior consent be provided—was not adequately met.

13. In the US, concession agreements signed between 2005 and 2008 document lease rates from $3,000 to over $8,000 per year per turbine across several different US states.

14. There is a proposal afoot to establish a community wind park outside the city of

Ixtepec that would offer much higher returns to *comuneros* and fund community development projects. The community-owned wind park has been stalled due to restrictions put in place by the Federal Electricity Commission (see Boyer 2014).

15. Ikojts (or Huave) and binnizá (Zapotec) indigenous populations have inhabited the area for many centuries.

16. One particularly odd fact presented is that cats are responsible for more bird deaths in the Isthmus than are wind turbines. However, this calculation does not account for the effect of an additional 132 turbines, if and when they are made operational. One also has to wonder exactly how this calculation was made; or how does one count the number of killer cats (or birds eaten) in Juchitán? The website was originally accessible at http://marena-renovables.com.mx/ but has been deactivated, likely sometime in December 2013.

17. While the website speaks to local concerns, its content is apparently gauged to a global audience. No fishermen we encountered had ever actually seen the Mareña website.

18. The task of Michel Foucault's biopower is the care of all aspects of human life, including religion, morals, health, infrastructure, safety, arts, trade, industry, poverty, and so on (see Foucault 2000).

19. Energopolitics refer to the ways in which energic forces and fuels shape and compel political power in particular directions. For further discussion related to "energopower," see Boyer's chapter in this volume (Boyer 2014), as well as Barry and Born (2013) on social and material interrelationships, Mitchell (2009, 2011) on the ways in which hydrocarbons have shaped contemporary forms of governance and power, and Ferry and Limbert (2008) on the transformation of energic materials into national resource making projects.

20. Or, as Dipesh Chakrabarty has described it, climate change "requires us to bring together intellectual formations that are somewhat in tension with each other: the planetary and the global; deep and recorded histories; species thinking and critiques of capital" (2009:213).

References

Agrawal, Arun. 2005. *Environmentality: Technologies of Government and the Making of Subjects*. Durham, NC: Duke University Press.

AMAP, Alianza Mexicana por la Autodeterminación de los Pueblos. 2012. "Comunicado de prensa: gobernador de Oaxaca incumple sus compromisos y desconoce la realidad de las comunidades istmeñas." October 9.

Appadurai, Arjun. 1996. *Modernity at Large: Cultural Dimensions of Globalization*. Minneapolis: University of Minnesota Press.

Barabas, Alicia M., and Miguel A. Bartolomé. 1973. *Hydraulic Development and Ethnocide: The Mazatec and Chinantec People of Oaxaca, Mexico*. Mexico City: International Work Group for Indigenous Affairs.

Barry, Andrew, and Georgina Born, eds. 2013. *Interdisciplinarity: Reconfigurations of the Social and Natural Sciences*. London: Routledge.

Bennett, Jane. 2010. *Vibrant Matter: A Political Ecology of Things*. Durham, NC: Duke University Press.

Boyer, Dominic. 2014. "Energopower: An Introduction." *Anthropological Quarterly* 87 (2): 309–34.

Campbell, Howard, Leigh Binford, Miguel Bartolomé, and Alicia Barabas, eds. 1993. *Zapotec Struggles: Histories, Politics, and Representations from Juchitán, Oaxaca*. Washington, DC: Smithsonian Institution Press.

Castellanos, M. Bianet. 2010. "Don Teo's Expulsion: Property Regimes, Moral Economies, and Ejido Reform." *Journal of Latin American and Caribbean Anthropology* 15 (1): 144–69.

Chaca, Roselia. 2012. "Presenta Mareña Renovable proyecto eólico en Juchitán." *Noticias: Voz e Imagen*, December 7. Accessed from http://www.noticiasnet.mx/portal/oaxaca/general/gruposvulnerables/128457-presenta-marena-renovable-proyecto-eolico-juchitan.

Chakrabarty, Dipesh. 2009. "The Climate of History: Four Theses." *Critical Inquiry* 35 (Winter): 197–221.

Choy, Timothy. 2011. *Ecologies of Comparison: An Ethnography of Endangerment*. Durham, NC: Duke University Press.

Cohen, Jeffrey H. 1999. *Cooperation and Community: Economy and Society in Oaxaca*. Austin: University of Texas Press.

Comaroff, Jean, and John Comaroff. 2003. "Ethnography on an Awkward Scale: Postcolonial Anthropology and the Violence of Abstraction." *Ethnography* 4 (2): 147–79.

Coroníl, Fernando. 1997. *The Magical State: Nature, Money, and Modernity in Venezuela*. Chicago: University of Chicago Press.

Crate, Susan, and Mark Nuttall, eds. 2009. *Anthropology and Climate Change*. Walnut Creek, CA: Left Coast Press.

Darier, Eric. 1999. "Foucault and the Environment: An Introduction." In *Discourses of the Environment*, ed. Eric Darier, 1–34. Malden, MA: Blackwell.

Edwards, Paul N. 2010. *A Vast Machine: Computer Models, Climate Data and the Politics of Global Warming*. Cambridge, MA: MIT Press.

Ferry, Elizabeth Emma, and Mandana E. Limbert, eds. 2008. *Timely Assets: The Politics of Resources and Their Temporalities*. Santa Fe: School for Advanced Research Press.

Foucault, Michel. 2000. " '*Omnes et Singulatim*': Toward a Critique of Political Reason." In *Power/Michel Foucault*, ed. James D. Faubion, trans. Robert Hurley, 317–18. New York: New Press.

———. 2009. *Security, Territory, Population: Lectures at the Collège de France 1977–1978*. New York: Picador.

Giddens, Anthony. 2009. *The Politics of Climate Change*. Cambridge: Polity Press.

Gilbert, Scott F., Jan Sapp, and Alfred I. Tauber. 2012. "A Symbiotic View of Life: We Have Never Been Individuals." *Quarterly Review of Biology* 87 (4): 325–41.

Gledhill, John. 1995. *Neoliberalism, Transnationalization and Rural Poverty: A Case Study of Michoacán, Mexico*. Boulder, CO: Westview.

Harman, Graham. 2011. *The Quadruple Object*. Alresford, UK: Zero Books.

Hastrup, Kirsten, and Martin Skrydstrup, eds. 2013. *The Social Life of Climate Change Models: Anticipating Nature*. New York: Routledge.

Howe, Cymene. 2011. "Logics of the Wind: Development Desires over Oaxaca." *Anthropology News* 52 (5): 8.

Howe, Cymene, Dominic Boyer, and Edith Barrera. In press. "Los márgenes del Estado al viento: autonomía y desarrollo de energías renovables en el sur de México." *Journal of*

Latin American and Caribbean, Special Issue, "Energy, Transition and Climate Change in Latin America."

Hulme, Mike. 2009. *Why We Disagree about Climate Change: Understanding Controversy, Inopportunity and Inaction.* Cambridge: Cambridge University Press.

———. 2013. "How Climate Models Gain and Exercise Authority." In *The Social Life of Climate Change Models: Anticipating Nature*, ed. Kirsten Hastrup and Martin Skrydstrup, 3–44. New York: Routledge.

Inter-American Development Bank (IDB); Mexico. 2011. "Marena Renovables Wind Power Project." (MEL1107) Environmental and Social Management Report, November 21.

Jamieson, Dale. 2011. "Energy, Ethics and the Transformation of Nature." In *The Ethics of Global Climate Change*, ed. Denis Arnold, 16–37. London: Cambridge University Press.

Krauss, Werner. 2010. "The 'Dingpolitik' of Wind Energy in Northern German Landscapes: An Ethnographic Case Study." *Landscape Research* 35 (2): 195–208.

Latour, Bruno. 2004. *Politics of Nature: How to Bring the Sciences into Society.* Cambridge, MA: Harvard University Press.

Lifshitz-Goldberg, Yael. 2010. "Gone with the Wind? The Potential Tragedy of the Common Wind." *Journal of Environmental Law and Policy* 28 (2): 435–71.

Love, Thomas, and Anna Garwood. 2011. "Wind, Sun and Water: Complexities of Alternative Energy Development in Rural Northern Peru." *Rural Society* 20:294–307.

Luke, Timothy W. 1999. "Environmentality as Green Governmentality." In *Discourses of the Environment*, ed. Eric Darier, 121–51. Malden, MA: Blackwell.

Malette, Sebastien. 2009. "Foucault for the Next Century: Eco-Governmentality." In *A Foucault for the 21st Century: Governmentality, Biopolitics and Discipline in the New Millennium*, ed. Sam Binkley and Jorge Capetillo, 221–39. Cambridge: Cambridge Scholars.

Marcus, George E. 1995. "Ethnography in/of the World System: The Emergence of Multi-Sited Ethnography." *Annual Review of Anthropology* 24:95–117.

"Mareña Renovables Impulsa el desarrollo sustentable en beneficio de las comunidades del Istmo." 2012. Press release, November 19.

Mathews, Andrew. 2008. "Statemaking, Knowledge and Ignorance: Translation and Concealment in Mexican Forestry Institutions." *American Anthropologist* 110 (4): 484–94.

———. 2011. *Instituting Nature: Authority, Expertise and Power in Mexican Forests.* Cambridge, MA: MIT Press.

McDonald, James H. 1999. "The Neoliberal Project and Governmentality in Rural Mexico: Emergent Farmer Organization in the Michoacán Highlands." *Human Organization* 58 (3): 274–84.

McKibben, Bill. 2006. *The End of Nature.* New York: Random House.

Mitchell, Timothy. 2009. "Carbon Democracy." *Economy and Society* 38 (3): 399–432.

———. 2011. *Carbon Democracy: Political Power in the Age of Oil.* New York: Verso.

Morton, Timothy. 2012. *The Ecological Thought.* Cambridge, MA: Harvard University Press.

Nugent, Daniel, and Ana María Alonso. 1994. "Multiple Selective Traditions in Agrarian Reform and Agrarian Struggle: Popular Culture and State Formation in the Ejido of Namiquipa, Chihuahua." In *Everyday Forms of State Formation: Revolution and the Negotiation of Rule in Modern Mexico*, ed. Gilbert M. Joseph and Daniel Nugent, 209–46. Durham, NC: Duke University Press.

Oceransky, Sergio. 2009. "Wind Conflicts in the Isthmus of Tehuantepec: The Role of Ownership and Decision-Making Models in Indigenous Resistance to Wind Projects in
Southern Mexico." *commoner* 13 (Winter): 203–22.

Ochoa, Enrique C. 2001. "Neoliberalism, Disorder, and Militarization in Mexico." *Latin
American Perspectives* 28 (4): 148–59.

Pasqualetti, Martin J. 2011. "Social Barriers to Renewable Energy Landscapes." *Geographical
Review* 101 (2): 201–23.

Sawyer, Suzana. 2004. *Crude Chronicles: Indigenous Politics, Multinational Oil, and Neoliberalism in Ecuador*. Durham, NC: Duke University Press.

Scheer, Hermann. 2004. *The Solar Economy: Renewable Energy for a Sustainable Global Future*. London: Earthscan.

Schwegler, Tara A. 2008. "Take It from the Top (Down)? Rethinking Neoliberalism and
Political Hierarchy in Mexico." *American Ethnologist* 35 (4): 682–700.

SENER. 2007. "Energía eólica y la política energética mexicana." Ing. Alma Santa Rita
Feregrino Subdirectora de Energía y Medio Ambiente, SENER. Monterrey, México,
October.

Tsing, Anna Lowenhaupt. 2004. *Friction: An Ethnography of Global Connection*. Princeton,
NJ: Princeton University Press.

Wilhite, Harold. 2005. "Why Energy Needs Anthropology." *Anthropology Today* 21 (3): 1–3.

Winther, Tanja. 2008. *The Impact of Electricity: Development, Desires and Dilemmas*. Oxford:
Berg.

❯❯❯ Excerpt from *Encyclical on Climate Change & Inequality: On Care for Our Common Home*

Pope Francis

Global Inequality

48. The human environment and the natural environment deteriorate together;
we cannot adequately combat environmental degradation unless we attend to
causes related to human and social degradation. In fact, the deterioration of the
environment and of society affects the most vulnerable people on the planet: "Both
everyday experience and scientific research show that the gravest effects of all
attacks on the environment are suffered by the poorest."[1] For example, the depletion of fishing reserves especially hurts small fishing communities without the
means to replace those resources; water pollution particularly affects the poor
who cannot buy bottled water; and rises in the sea level mainly affect impoverished coastal populations who have nowhere else to go. The impact of present
imbalances is also seen in the premature death of many of the poor, in conflicts

Encyclical on Climate Change & Inequality: On Care for Our Common Home (Brooklyn: Mel-
ville House, 2015), 28–38

sparked by the shortage of resources, and in any number of other problems which are insufficiently represented on global agendas.[2]

49. It needs to be said that, generally speaking, there is little in the way of clear awareness of problems which especially affect the excluded. Yet they are the majority of the planet's population, billions of people. These days, they are mentioned in international political and economic discussions, but one often has the impression that their problems are brought up as an afterthought, a question which gets added almost out of duty or in a tangential way, if not treated merely as collateral damage. Indeed, when all is said and done, they frequently remain at the bottom of the pile. This is due partly to the fact that many professionals, opinion makers, communications media, and centres of power, being located in affluent urban areas, are far removed from the poor, with little direct contact with their problems. They live and reason from the comfortable position of a high level of development and a quality of life well beyond the reach of the majority of the world's population. This lack of physical contact and encounter, encouraged at times by the disintegration of our cities, can lead to a numbing of conscience and to tendentious analyses which neglect parts of reality. At times this attitude exists side by side with a "green" rhetoric. Today, however, we have to realize that a true ecological approach *always* becomes a social approach; it must integrate questions of justice in debates on the environment, so as to hear *both the cry of the earth and the cry of the poor.*

50. Instead of resolving the problems of the poor and thinking of how the world can be different, some can only propose a reduction in the birth rate. At times, developing countries face forms of international pressure which make economic assistance contingent on certain policies of "reproductive health." Yet "while it is true that an unequal distribution of the population and of available resources creates obstacles to development and a sustainable use of the environment, it must nonetheless be recognized that demographic growth is fully compatible with an integral and shared development."[3] To blame population growth instead of extreme and selective consumerism on the part of some, is one way of refusing to face the issues. It is an attempt to legitimize the present model of distribution, where a minority believes that it has the right to consume in a way which can never be universalized, since the planet could not even contain the waste products of such consumption. Besides, we know that approximately a third of all food produced is discarded, and "whenever food is thrown out it is as if it were stolen from the table of the poor."[4] Still, attention needs to be paid to imbalances in population density, on both national and global levels, since a rise in consumption would lead to complex regional situations, as a result of the interplay between problems linked to environmental pollution, transport, waste treatment, loss of resources, and quality of life.

51. Inequity affects not only individuals but entire countries; it compels us to

consider an ethics of international relations. A true "ecological debt" exists, particularly between the global north and south, connected to commercial imbalances with effects on the environment, and the disproportionate use of natural resources by certain countries over long periods of time. The export of raw materials to satisfy markets in the industrialized north has caused harm locally, as for example in mercury pollution in gold mining or sulphur dioxide pollution in copper mining. There is a pressing need to calculate the use of environmental space throughout the world for depositing gas residues which have been accumulating for two centuries and have created a situation which currently affects all the countries of the world. The warming caused by huge consumption on the part of some rich countries has repercussions on the poorest areas of the world, especially Africa, where a rise in temperature, together with drought, has proved devastating for farming. There is also the damage caused by the export of solid waste and toxic liquids to developing countries, and by the pollution produced by companies which operate in less developed countries in ways they could never do at home, in the countries in which they raise their capital: "We note that often the businesses which operate this way are multinationals. They do here what they would never do in developed countries or the so-called first world. Generally, after ceasing their activity and withdrawing, they leave behind great human and environmental liabilities such as unemployment, abandoned towns, the depletion of natural reserves, deforestation, the impoverishment of agriculture and local stock breeding, open pits, riven hills, polluted rivers and a handful of social works which are no longer sustainable."[5]

52. The foreign debt of poor countries has become a way of controlling them, yet this is not the case where ecological debt is concerned. In different ways, developing countries, where the most important reserves of the biosphere are found, continue to fuel the development of richer countries at the cost of their own present and future. The land of the southern poor is rich and mostly unpolluted, yet access to ownership of goods and resources for meeting vital needs is inhibited by a system of commercial relations and ownership which is structurally perverse. The developed countries ought to help pay this debt by significantly limiting their consumption of non-renewable energy and by assisting poorer countries to support policies and programmes of sustainable development. The poorest areas and countries are less capable of adopting new models for reducing environmental impact because they lack the wherewithal to develop the necessary processes and to cover their costs. We must continue to be aware that, regarding climate change, there are *differentiated responsibilities*. As the United States bishops have said, greater attention must be given to "the needs of the poor, the weak and the vulnerable, in a debate often dominated by more powerful interests."[6] We need to strengthen the conviction that we are one single human family. There are no frontiers or barriers,

political or social, behind which we can hide, still less is there room for the globalization of indifference.

Weak Responses

53. These situations have caused sister earth, along with all the abandoned of our world, to cry out, pleading that we take another course. Never have we so hurt and mistreated our common home as we have in the last two hundred years. Yet we are called to be instruments of God our Father, so that our planet might be what he desired when he created it and correspond with his plan for peace, beauty, and fullness. The problem is that we still lack the culture needed to confront this crisis. We lack leadership capable of striking out on new paths and meeting the needs of the present with concern for all and without prejudice towards coming genera-tions. The establishment of a legal framework which can set clear boundaries and ensure the protection of ecosystems has become indispensable, otherwise the new power structures based on the techno-economic paradigm may overwhelm not only our politics but also freedom and justice.

54. It is remarkable how weak international political responses have been. The failure of global summits on the environment makes it plain that our politics are subject to technology and finance. There are too many special interests, and eco-nomic interests easily end up trumping the common good and manipulating information so that their own plans will not be affected. The *Aparecida Document* urges that "the interests of economic groups which irrationally demolish sources of life should not prevail in dealing with natural resources."[7] The alliance between the economy and technology ends up sidelining anything unrelated to its imme-diate interests. Consequently the most one can expect is superficial rhetoric, spo-radic acts of philanthropy, and perfunctory expressions of concern for the envi-ronment, whereas any genuine attempt by groups within society to introduce change is viewed as a nuisance based on romantic illusions or an obstacle to be circumvented.

55. Some countries are gradually making significant progress, developing more effective controls and working to combat corruption. People may well have a growing ecological sensitivity, but it has not succeeded in changing their harmful habits of consumption, which, rather than decreasing, appear to be growing all the more. A simple example is the increasing use and power of air-conditioning. The markets, which immediately benefit from sales, stimulate ever greater demand. An outsider looking at our world would be amazed at such behaviour, which at times appears self-destructive.

56. In the meantime, economic powers continue to justify the current global system where priority tends to be given to speculation and the pursuit of financial gain, which fail to take the context into account, let alone the effects on human

dignity and the natural environment. Here we see how environmental deterioration and human and ethical degradation are closely linked. Many people will deny doing anything wrong because distractions constantly dull our consciousness of just how limited and finite our world really is. As a result, "whatever is fragile, like the environment, is defenceless before the interests of a deified market, which become the only rule."[8]

57. It is foreseeable that, once certain resources have been depleted, the scene will be set for new wars, albeit under the guise of noble claims. War always does grave harm to the environment and to the cultural riches of peoples, risks which are magnified when one considers nuclear arms and biological weapons. "Despite the international agreements which prohibit chemical, bacteriological and biological warfare, the fact is that laboratory research continues to develop new offensive weapons capable of altering the balance of nature."[9] Politics must pay greater attention to foreseeing new conflicts and addressing the causes which can lead to them. But powerful financial interests prove most resistant to this effort, and political planning tends to lack breadth of vision. What would induce anyone, at this stage, to hold on to power only to be remembered for their inability to take action when it was urgent and necessary to do so?

58. In some countries, there are positive examples of environmental improvement: rivers, polluted for decades, have been cleaned up; native woodlands have been restored; landscapes have been beautified thanks to environmental renewal projects; beautiful buildings have been erected; advances have been made in the production of non-polluting energy and in the improvement of public transportation. These achievements do not solve global problems, but they do show that men and women are still capable of intervening positively. For all our limitations, gestures of generosity, solidarity, and care cannot but well up within us, since we were made for love.

59. At the same time we can note the rise of a false or superficial ecology which bolsters complacency and a cheerful recklessness. As often occurs in periods of deep crisis which require bold decisions, we are tempted to think that what is happening is not entirely clear. Superficially, apart from a few obvious signs of pollution and deterioration, things do not look that serious, and the planet could continue as it is for some time. Such evasiveness serves as a licence to carrying on with our present lifestyles and models of production and consumption. This is the way human beings contrive to feed their self-destructive vices: trying not to see them, trying not to acknowledge them, delaying the important decisions, and pretending that nothing will happen.

A Variety of Opinions

60. Finally, we need to acknowledge that different approaches and lines of thought have emerged regarding this situation and its possible solutions. At one extreme,

we find those who doggedly uphold the myth of progress and tell us that ecological problems will solve themselves simply with the application of new technology and without any need for ethical considerations or deep change. At the other extreme are those who view men and women and all their interventions as no more than a threat, jeopardizing the global ecosystem, and consequently the presence of human beings on the planet should be reduced and all forms of intervention prohibited. Viable future scenarios will have to be generated between these extremes, since there is no one path to a solution. This makes a variety of proposals possible, all capable of entering into dialogue with a view to developing comprehensive solutions.

61. On many concrete questions, the Church has no reason to offer a definitive opinion; she knows that honest debate must be encouraged among experts, while respecting divergent views. But we need only take a frank look at the facts to see that our common home is falling into serious disrepair. Hope would have us recognize that there is always a way out, that we can always redirect our steps, that we can always do something to solve our problems. Still, we can see signs that things are now reaching a breaking point, due to the rapid pace of change and degradation; these are evident in large-scale natural disasters as well as social and even financial crises, for the world's problems cannot be analyzed or explained in isolation. There are regions now at high risk and, aside from all doomsday predictions, the present world system is certainly unsustainable from a number of points of view, for we have stopped thinking about the goals of human activity. "If we scan the regions of our planet, we immediately see that humanity has disappointed God's expectations."[10]

Notes

1. Bolivian Bishops' Conference, Pastoral Letter on the Environment and Human Development in Bolivia *El universo, don de Dios para la vida* (March 23, 2012), 17.

2. Cf. German Bishops' Conference, Commission for Social Issues, *Der Klimawandel: Brennpunkt globaler, intergenerationeller und ökologischer Gerechtigkeit* (September 2006), 28–30.

3. Pontifical Council for Justice and Peace, *Compendium of the Social Doctrine of the Church*, 483.

4. *Catechesis* (June 5, 2013): *Insegnamenti* 1/1 (2013), 280.

5. Bishops of the Patagonia-Comahue Region (Argentina), *Christmas Message* (December 2009), 2.

6. United States Conference of Catholic Bishops, *Global Climate Change: A Plea for Dialogue, Prudence and the Common Good* (June 15, 2001).

7. Fifth General Conference of the Latin American and Caribbean Bishops, *Aparecida Document* (June 29, 2007), 471.

8. Apostolic Exhortation *Evangelii Gaudium* (November 24, 2013), 56: *AAS* 105 (2013), 1043.

9. John Paul II, *Message for the 1990 World Day of Peace*, 12: AAS 82 (1990), 154.43.
10. Id., *Catechesis* (January 17, 2001), 3: *Insegnamenti* 24/1 (2001), 178.

❱❱❱ "Night Ride"

Ken Saro-Wiwa

The car sped on, its engine roaring into the all-pervading silence of dusk. A light rain had fallen earlier. Puddles of water lay on the dirt road which led through bushes, over a little dilapidated wooden bridge, newly planted farmlands and past several mud villages towards town.

The man and the woman sat at the back of the car. Her son sat beside the driver, asleep. The man held the fingers of her left hand. She, unthinking, let him stroke her fingers gently.

Now and again, a woman or two returning from market with little baskets on their heads flipped past like memories from bye-gone days. Once they passed a little group carrying a man on a bed. The man was covered entirely in a mat. He was dead.

"It's the cholera," said she, and shivered.

Her husband had also died during the war. After a long trek in drenching rain, he had arrived home (an open mud classroom somewhere in the bush) that sombre evening, coughing, his legs bloated. A look at his handsome face and she knew something was wrong. They would have to go to hospital. But she knew it was futile. There were no drugs in the hospital. The night planes mostly brought ammunition. The few drugs which came through disappeared on the black market. The trickle which got into hospital was meant mostly for officers and officials— the ones who sat behind the stone-deaf desks. He lay in hospital for three days. Then the world ended. Her child would have to be born posthumously. There was not even time to bury him. The planes arrived on a bombing mission and smashed the mortuary. She had to move on to the next village.

He too had been in the war. But on the opposite side. His role had been to bring education, food, drugs, and succour to the war-weary, battered communities living on the flat scrubby plains where oil wells gushed night and day; the wells which were the main argument of, and fuel for the war. He also had several anecdotes from his terrible experience among the soldiers, among the despairing and the deprived for whom he had toiled untiringly for many agonising months. And

A Forest of Flowers: Short Stories (Harlow: Longman, 1995), 109–16. Pearson Education Limited.

now the war was over, it was still his task to rehabilitate all those who had fled at the first unaccustomed sound of either mortar fire or bombs. He was busy urging the adults to return to their farms and their occupations and to send the children back to school. It was a hard, frustrating task.

The newly planted farms flitted past them into the darkness behind.

These farms will be the death of you, my brothers and sisters, he thought. Only that morning he had stood in the village square in Dukana asking them to send the girls to school. But who will help us tend the farm, the women had asked? Who will baby-sit for us while we plant the yams? He was in despair. Each planting season you buy seed yams. You toil from January through December, then you eat the yield. Come next season and you have to buy seed yams again. You don't have a bank account, perhaps a new piece of cloth from the market at Egwanga, and your thatch roof is leaking disastrously. From year to year. And then you have to give your five-year-old daughter away in marriage—so you might buy seed yams to plant. Is that life? Was that why God created you? No, my sisters. These farms will be the death of us. They yield us nothing. Not until we can get more education. So send the kids to school. Send the girls to school. But who will baby-sit for us? Who will help us fetch water? Despair. An old woman had hobbled up to him. My son, they arrived this morning and dug up my entire farm, my only farm. They mowed down the toil of my brows, the pride of the waiting months. They say they will pay me compensation. Can they compensate me for my labours? The joy I receive when I see the vegetables sprouting? God's revelation to me in my old age? Oh my son, what can I do? What answer now could he give her? I'll look into it later, he had replied tamely.

Look into it later. He could almost hate himself for telling that lie. He cursed the earth for spouting oil, black gold, they called it. And he cursed the gods for not drying the oil wells. What did it matter that millions of barrels of oil were mined and exported daily; so long as this poor woman wept those tears of despair? What could he look into later? Could he make alternate land available? And would the lawmakers revise the laws just to bring a bit more happiness to these unhappy wretches whom the search for oil had reduced to an animal existence? They ought to send the oil royalties to the men whose farms and land were despoiled and ruined. But the lawyers were in the pay of the oil companies and the government people in the pay of the lawyers and the companies. So how could he look into it later? He should have told the woman to despair. To die. Not live in death. That would have been more honest and respectable. He tapped his foot on the floor, agitated.

You could always tell when the car was drawing close to Bori. The water tank stood against the skyline, and then the corrugated iron sheets in the police station and barracks, the brick walls of the government quarters, the poles bearing the telegraph wires and when it had not broken down, the generator in the hospital. The main road split the town into two. Night was almost falling. He ordered the driver to stop.

"I'll buy some aspirin here, while the driver goes further down the road to buy petrol. Won't you get down? We might have a soft drink or two."

She only heard his last words. She opened her door mechanically, a far away look in her eyes, and got out. They stepped into the shop. It was virtually empty. A bottle of aspirin cost five shillings, double its price before the war. He paid for it, mumbling words about inflation and profiteering. There was no soft drink. The shopboy shuffled to the beat of music from a transistor radio placed on one of the empty shelves.

The car returned.

He let her move out of the shop first, and took a long critical look at her. The total picture of her did not please him. She was no longer beautiful. Plain, he would say. The war had wrought a great change in her. She was still in her twenties. Maybe if she had a man . . .

The driver moved into gear and switched on the headlamps. As the car picked up speed, the headlamps scoured the countryside, picking out the yam stakes rising from little mounds of earth. Far to the left and right, in the distance, gas flares from oil-fields illumined the sky. A frown crossed his brows.

"You are very silent," she said, trying to start a conversation. He sighed.

"Why do you sigh?"

"It's the gas flares."

"They bother you?"

"Yes."

"I can't possibly see how."

"Politics, politics," he said impatiently, and tapped his foot on the floor.

Then added: "Let's forget it."

"Oh, you know I won't forget it."

"But you must. Look behind you there. See the young moon? Remember a night like this when we stood by the hedge that ran between our houses in Bori, holding hands?"

She smiled sadly in the darkness; she was relieved that he could not see her face.

"That was a long time ago. I got married soon after."

"Tell me why," he replied harshly.

"That is no story to tell," she answered, smiling sadly in the darkness. "You need not torment me. It was not a happy marriage."

"You let me down," he said harshly.

"You do not mean to hurt me, I hope. I am a young defenceless widow."

"And I a bachelor."

"So you will find happiness some day."

"Someday, I will find happiness. Yes."

The sarcasm in the tone of his voice was unmistakeable. She peered into his face and thought she detected a deep hardening of his features. He seemed to have grown old of a sudden.

"You're angry with me? Perhaps I should not have asked for the ride in your car?"

"Don't be childish," he rebuked her mildly.

"I'm not a child," she said, chuckling.

"Tell me why you turned sarcastic when I spoke of happiness a moment ago."

"I admit I find that idea rather strange."

"Strange? And you are a young man. Still in your twenties. You live in a decent flat, earn a good salary, own a chauffeur driven car. You have the rest of your life before you. You could go over your life once again, if you so desired. What else could you want?"

He laughed a hollow, cynical laugh that sent shivers down her spine. She recalled now how they had grown up together; the joys they shared, the vows they had made to each other. And how, one Christmas holiday, she had, in her innocence, fallen prey to a young man who worked in the tax office and who took advantage of her innocence and put her in the family way. How she had cried when she knew what had happened! How angry her parents had been. And her childhood friend, he who now sat in the car with her, how hurt, distressed, and distraught he had been! He had taken it out on her, laughing at her, mocking her to disguise his hurt. He still remembered the mocking laugh with which he had greeted her after she moved out of her parents' house to live with the man who had defiled her.

The headlamps of the car picked out a mangy dog eating excrement by the roadside. And he recalled that dreary, rainy night during the war when he had found a solitary dog in the deserted village of Ogale, coolly picking over the skull of a dead man. He had cursed the futility of the war, then.

"What else could I want?" he repeated, slowly emphasizing each word. "What else could I want?" And he grunted.

"You sound strange tonight."

"Do I?"

"You should get married so you'll know happiness."

"Did you know happiness when you were married?"

"Mine was different. You will marry the girl of your choice."

"And then I'll find happiness?"

"I believe it." There followed a long silence.

"You're a cynic," she said.

They passed a group of men carrying a little bundle wrapped in a mat.

Cholera, she thought, and shivered.

Behind the crowd walked a disconsolate woman, her hands thrown up in despair.

Cholera, he thought, and shrugged.

"Not exactly a cynic," said he. "Only I cannot find joy in those little bundles. There is not even a hospital to which they can go. And not enough vaccines to go round either. Vaccines sent from headquarters are sold to these miserable wretches. Someone's converted into private use money meant for the purchase of drugs. But there's no one to dismiss."

"Well?"

"So there's absolutely nothing to be happy about."

"You're not going to carry the world on your shoulders. You're not Atlas."

"No. Nor was I meant to be."

A heavy lorry flashed past them at terrific speed. The scent of stockfish.

"That driver's in a hurry," she remarked.

"And should be. Someone's about to profiteer from the sufferings of the poor. These are gifts from Norway carried in second-hand vehicles, reconditioned and donated by the people of Great Britain. For Africans dying of hunger and malnutrition. What will you have?"

"You are cynical about the gifts and the donors?"

"I fear the Greeks and the gifts they bring. Perhaps it's as well they will all end up in private pockets."

Then it occurred to him that he was not being fair in making the statement. It would have been more than difficult for them had not these gifts arrived. He fell silent.

Now they came to a gas flare. A village basked beneath the flare. Some of the houses had no roofs. They had been removed by heavy shells during the war. The village, ugly, squat, and untidy, had gone to sleep.

"Lucky villagers. They have light all night. No need for electricity," she said.

"None." And he laughed a hard, merciless laugh.

A short distance beyond the village, some men flagged them down excitedly. He ordered the driver to stop. She thought they might be armed robbers and said so. He said no. They stopped when they were sufficiently close to the men. One of the men held a little, emaciated woman up to him.

"She's got the plague," he said.

"Right. Put her in the back seat. I'll take her to hospital. Someone come along with her."

They all drew back in terror.

"Where's her husband?"

"He died this morning."

"Any grown-up child?"

"Her only daughter died yesterday."

"Who will go with her."

The crowd shrank back. He heard someone mutter, "They're a cursed family."

"Right, I'll take her along. And look to yourselves. Boil your water properly before you drink. Don't put any 'izal' into it. And don't call the juju priest either. Just keep the surroundings clean. Good night."

They drove off, and continued the journey in silence. The sick woman, clad in rags, sat between them. Twice she vomited and twice they both rallied to make her a bit more comfortable. He used his handkerchiefs to dry up the mess. She had to use her headtie for a mop. And she saw resurface that elemental goodness in him, which she had always admired and loved. He was not a cynic: perhaps too much of an idealist. In a country where there was no room whatever for idealism. She had heard him speak in Dukana that morning; she had listened to the few words he had spoken in the car, wondered at the mocking laugh which she found so detestable. If he were left alone, he might destroy himself. A thought flashed through her mind, but she killed it as soon as it was born. She continued to mop vomit with her headtie.

He, too, was struck by the devoted way in which she mopped the old woman's vomit with her headtie. He felt she was doing it for him. And a thought flashed through his mind. He quickly stifled the thought and let his mind wander to the problems of society. How could you stop people stealing food and medicines meant for the public? How could you stop them from taking bribes? . . . A man was going from Jerusalem to Jericho and fell among a band of robbers. And they beat him. . . . If he did not hold himself, he would keel over. And no one would mourn him. Absolutely no one. He wound down the glass and threw his soaked handkerchief through the window.

The sick woman began to whimper, whispering inaudible words.

And he thought, I cannot get this out of my system. It's absolutely impossible. . . . Her words came to him. You're not Atlas! No, nor was meant to be. The world had always got on with its imperfections. So what? . . . She's a good girl. I ought to have married her. Now she's a widow. But has a son from her first husband. I cannot take him. . . . She's no longer as beautiful as when she was younger. He sighed.

She looked in his direction. There was only the darkness between them, but she could see shafts of light breaking in from the homes in the distance, illumining the car somewhat.

Now the town loomed into sight.

"Drive to the hospital," he ordered the driver.

The sick woman died before they arrived at the hospital. He sent her corpse into the mortuary. It was already full. But he left her there all the same. There was no one to make an entry of the event. The nurse on duty was overburdened with work and had been rude.

He banged the door of his car, and lit himself a cigarette. She sat by his side, weeping.

"You will see me home, won't you," she asked.

"Oh yes, I will."

They drove to her house in silence. The neighbours had already gone to sleep when they arrived. She roused her child who had slept soundly through the ride. He alighted from the car and ran indoors quickly.

She opened the car door, walked over to that side of the car where he stood waiting for her and extended her hand. She had regained composure. He ignored her hand and took her in his arms.

"You will have to teach me to live with death," he pleaded. She disengaged herself from his arms and walked briskly away.

PART THREE

Energy in Philosophy: Ethics, Politics, and Being

The ways we have named and conceptualized energy have given rise to a range of philosophical questions and inquiries concerning ethics, ontology, epistemology, time, and history. Indeed, many of the most important interventions in the energy humanities concern the compelling philosophical challenges generated by energy, from its unique ontological status to the environmental consequences of its use and abuse. The concept of energy can be used to name everything from a basic property of objects to an animating force of history. Each and every way in which we put energy to work generates conceptual and theoretical problems that only philosophical analysis and speculation can properly name and explain.

The diverse approaches represented in this section are linked by a common recognition that energy needs to become a more important aspect of philosophical thought. Until recently, energy and the consequences of its use have played very little role in the elaboration of theories and concepts within the history of philosophy. This may be due in part to the very nature of energy. According to the first law of thermodynamics, energy is something that can be neither created nor destroyed, yet is essential to change, movement, and transformation—a curious form of being that seems to have been all but invisible to philosophers. It also plays a central role in shaping the form and character of society. Energy makes the world go round, and different sources of energy have made it go around differently throughout history; even so, the social implications of energy have been as hidden as its strange ontology. Given the ways in which it has long evaded thought, we might well concur with Karen Pinkus's claim that "energy can be understood as a heterogeneous set of self-mystifying systems or machines that block access to thought even as they fascinate us."[1] The pieces in this section aim to undo these blocks on thought, so that we might more fully grapple with the significance that energy has for us.

If there can be anything that links the diverse range of philosophical approaches to energy represented here, it is the provocation that a confrontation with energy compels us to engage in a thorough-going reimagination of many of our fundamental concepts and understandings. In modern scientific and philosophical thought, energy has become homogeneous, named and numbered via static measures that capture the power to do work. As in Pinkus's "Air" and the other pieces brought together here, this apparent conceptual control over energy has in fact limited our ability to truly understand it. In "The Draukie's Tale," poet Laura Watts describes the limits that scientists reach when they try to understand how the Draukie creates the energy of waves. A mythic creature, the Draukie remains silent when scientists try to "imprison" her by figuring her movements in the language of scientific data. It is only when one young scientist befriends the Draukie by singing to her that the mysteries of wave energy—a potential source of renewable energy—are revealed. Our old methods of dealing with energy stand in the way, a point that Martin McQuillan's contribution also highlights. "No philosopher as a philosopher has ever taken seriously the question of oil," McQuillan writes. His aim is to address this gap by making sense of just how philosophy has been shaped by the carbon era, so that we might puzzle out how philosophy needs to be reconstituted in the post-carbon era. "Our present understanding of all exchange, debt, and faith"—in other words, the core principles of modern humanism and Enlightenment thought—"runs through oil," McQuillan argues. He pushes us to grapple with the deep links between oil and modern thought, so that we might take on the work of generating new concepts as our oil reserves run dry.

Allan Stoekl is one of the few contemporary philosophers who have taken oil and energy head-on. In his groundbreaking book *Bataille's Peak: Energy, Religion, and Postsustainability* (excerpted here), Stoekl investigates the contemporary significance of the work of French philosopher Georges Bataille (1897–1962). Energy is at the heart of Bataille's thinking as it is for no other thinker. One of the questions that Stoekl poses here (and returns to throughout his book) is whether Bataille's understanding of the social and psychological significance of energy expenditure makes sense in relationship to our use of fossil fuels. For Bataille, energy is unlimited: it comes to us from the sun in endless amounts, and so we are safe to expend it, since we will always have more. This profusion of energy underwrites Bataille's conception of how human communities work best; what is missed in Bataille's thinking, Stoekl believes, is the concentrated, expanded expenditure enabled by the discovery of fossil fuels, which has allowed humans to expend well beyond the carrying capacity of the environment. The social limits that the environment might have imposed on us had we never discovered fossil fuels have been suspended during modernity, because fossil fuels have rendered these limits unnecessary or invisible—until now. Stoekl intends to recover from Bataille's work a distinct and original relationship to expenditure. We imagine that the way to address our overexpenditure of energy and its environmental consequences is to reign in our use of it—to conserve our energies so that future use is guaranteed. But framing our relationship to energy in this way has an outcome that is very much the opposite of what we imagine: we actually *affirm* the utility of energy and our means–ends vision of an economy that is shaped by the principle of unlimited growth. The challenge is to produce a model and practice of expenditure that is something other than what it has come to be at present—consumption. Stoekl finds in Bataille's understanding of the operations of a "general economy" a possible way of getting beyond an ideal of unlimited growth without having to deny the tendency of human communities to expend. An encounter with Bataille's philosophy can't help but upend the guiding assumptions at the core of most environmental ethics: that conservation is the only or best way to manage our use of energy and to mitigate environmental damage and destruction.

Another shared starting point for philosophers who have turned their attention to energy is a recognition that—like it or not—the post-carbon world has already begun. The most prominent name for this post-carbon era is the "Anthropocene," a term coined by atmospheric scientist Paul Crutzen and ecologist Eugene Stoermer to capture the impact of human activities on the planet. It is the proposed name for the present geological epoch (following the Holocene) and highlights the degree to which humans have been able to reshape the earth's environment. While there is disagreement about the precise beginning date of the Anthropocene (though it is usually identified with James Watt's invention of the steam engine in 1763), its length is understood to be a few hundred years—not the thousands of years that geological epochs typically demarcate (e.g., the Holocene is 11,700 years long).

This section includes four essays that speak to the significance of the Anthropocene for philosophical thought, and which pay specific attention to the implications of this category for energy pasts, presents, and futures. Roy Scranton's influential article in the *New York Times* treats the Anthropocene as a sign of the death of our civilization. While this might sound ominous, Scranton suggests that the Anthropocene gives us an opportunity to learn how to die, "not as individuals, but as a civilization"; if we treat this moment as a collective opportunity, the passing away of modernity brings with it new ways of being in the world that redress the material and conceptual limits of the old ones. Timothy Morton's equally influential book *Hyperobjects: Philosophy and Ecology after the End of the World* (excerpted here) also views the Anthropocene as a collective, conceptual opportunity. For Morton, the recognition of human beings as an environmental and geological force on a planetary scale brings about what might at first seem to be a surprising conclusion: humanity is *not* at the center of history, epistemology, and ontology, nor are we the ones who provide significance and value to the events and objects of the world. For Morton, our post-carbon world initiates a posthuman one, as well as an upending of the caveats and self-certainties of modern thought. Morton's forceful rearrangement of the long-standing order of things is a consequence of his recognition that our existing systems of thought have led to global warming and environmental crisis. It is also necessitated by the explanatory limits of existing philosophy when it comes to phenomena such as climate and global warming—phenomena that Morton names "hyperobjects" because they are "real entities whose primordial reality is withdrawn from humans." Just because it is difficult to see or experience global warming—an absence or invisibility that has impeded climate change politics—it is no less real. The presence of hyperobjects, writes Morton, "seem[s] to force something on us, something that affects some core ideas of what it means to exist, what Earth is, what society is."

A concept like the Anthropocene is intended to shake up deeply held ideas about how we understand the world, including our understanding of nature and the environment. Many philosophers, including Morton, are deeply suspicious about the concept of "nature," seeing it as a key element of the modern network of concepts and ideas that brought about the Anthropocene. Others, however, argue that nature has a key role to play in bringing about a new world after the end of modernity. In his contribution, Dale Jamieson outlines an ethics appropriate to the Anthropocene that is anchored in a respect for nature. In the brief excerpt from his *Reason in a Dark Time: Why the Struggle against Climate Change Failed—and What It Means for Our Future*, Jamieson is careful to outline a nuanced, multi-faceted idea of respect, as well as of nature. If we are ever to adopt a set of values appropriate to the Anthropocene, Jamieson believes they will need to be con-nected to the ethics we currently have; and these ethics, especially as they relate to climate change, are connected to nature. Yet this ethical stance toward nature is not meant to reproduce the perennial gap between man and nature. Rather, as Jamieson writes, "respecting nature is respecting ourselves."

As a number of other critics have done recently (most notably Donna Haraway and Jason Moore),[2] in "We Have Always Been Post-Anthropocene" Claire Cole-brook interrogates the Anthropocene to tease out the conceptual presumptions that lie buried within it. The *anthropos* foregrounds man or the human as the sub-ject of the Anthropocene, identifying a masculinized universality connected to a planetary scale, and doing so in a way that evades the gender difference, among others, named by the so-called age of man. But the limits of the Anthropocence go far beyond its compression of distinct human experiences into a single experience, one that (among other things) assigns blame equally to everyone on the planet for the energy-intensive transformations of the atmosphere brought about by a rela-tively small segment of the earth's inhabitants. The Anthropocene carries within it the idea of a *good* Anthropocene—the capacity to transform the planet for the better with the same powers and energies that messed it up in the first place. For Colebrook, the Anthropocene is symptomatic of a logic that has underwritten both modern philosophy (e.g., humanism) and those forms of thought that have imag-ined themselves as struggling against it (e.g., posthumanism). Connecting the Anthropocene with social and conceptual systems and mechanisms that generate difference, Colebrook explores the fascinating implications for philosophy and politics of an alternative imagining of the present: "What might it mean to think a counterfactual scenario where humans had not inflicted the difference of the Anthropocene on the planet?"

These inquiries into the Anthropocene provide a glimpse into the complexities involved in conceptualizing the relationship we have developed with the planet, and of the even greater complexities of pushing ourselves into new relationships. To create new concepts and connections, we will need to develop not just a post-carbon philosophy but, as Joseph Masco's essay highlights, a post-nuclear philosophy as well. Masco ties our insatiable desire for energy and our use of it in warfare to embodied and geophysical mutations over the course of the twentieth century. The thousands of nuclear explosions since 1945 have ensured that the body of every person on the planet is suffused with their remnants, which have been deposited within our genomes. There has been a change, too, in how we envision our health and well-being. Focusing on the work on military strategist Hermann Kahn, Masco shows that health has come to be understood as "incipient death" in the post-nuclear era. Life has been renarrated as stages toward death—a stunning development whose full social and political impact is difficult to fully apprehend.

Despite the huge changes that we have wrought on the planet and on ourselves, it remains surprisingly hard to understand the role of energy in shaping the concepts we use and the lives we lead. Those philosophers who have started to work out the meaning of energy for not just how we live but how we think, act, and understand the world face the considerable challenge of unnerving the quotidian weight of carbon life and the carbon philosophies that go with it. It is left to works such as Reza Negarestani's one-of-a-kind theory-fiction *Cyclonopedia: Complicity with Anonymous Materials* (excerpted here) to get the message across. A rich, nuanced book, the short excerpts included here offer a sense of the compelling way in which Negarestani approaches oil and energy. In *Cyclonopedia*, oil possesses qualities well beyond those we normally assign to it. It is nothing less than a "satanic sentience" that "possesses tendencies for mass intoxication on pandemic scales (different from but corresponding to capitalism's voodoo economy and other types of global possession systems)." The living mass of oil just under the surface of the earth has been endlessly nurtured by the petropolitics conducted on its surface—prayers and sacrifices to a demon that thrives on war and conflict. "Oil is the undercurrent of all narrations, not only the political but also that of the ethics of life on Earth," writes Negarestani. Simultaneously theological treatise, political manifesto, demonology, and science fiction, works such as *Cyclonopedia* might well provide an opening for us to deal with the problems of the present without the weight of concepts and ways of being that have shown themselves to be part of the problem rather than an element of the solution.

Notes

1. Karen Pinkus, *Fuel: A Speculative Dictionary* (Minneapolis: University of Minnesota Press, 2016).

2. See Donna Haraway, "Anthropocene, Capitalocene, Plantationocene, Chthulucene: Making Kin," *Environmental Humanities* 6 (2015): 159–65; and Jason Moore, *Capitalism in the Web of Life: Ecology and the Accumulation of Capital* (New York: Verso, 2015).

))) Bataille's Ethics

Allan Stoekl

Given Bataille's antecedents—among them gnosticism, alchemy, Bruno, Sade—it should come as little surprise that one of his major works—*The Accursed Share* (1988), first published in 1949, focuses on the importance of energy use and expenditure in society and nature. What may come as a surprise is that, considering the excesses of works like "The Use Value of D. A. F. de Sade," he should publish a sober historical and scientific analysis of energy and excess in traditional and modern societies. But 20 years—and World War II—separate these two works— years in which senseless destruction passed from being a desideratum of avant-garde thought to an all-too-daily event. Throughout his postwar focus on energy, Bataille nevertheless remains faithful to his earlier vision: how to see community ultimately not as the affirmation of the primacy of a narrow conservation of energy, justified through a remote, indeed inconceivable higher ideal (God, Man), but as the joyous and anguished expenditure of energy (momentarily) concentrated in beings and things. Bataille is still concerned, in other words, with the violent "transmutation" of matter at the expense of static and exhausted identities and ideals. By 1949, however, he has passed from provocation to patient analysis.

For Bataille nature and society are one and the same because both are nothing more than instances of energy concentration and waste.[1] The refocus on energy production and use has profound implications: "Man" is not so much the author of his own narrative, or the subject that experiences and acts, as "he" is the focal point of the intensification or slackening of energy flows. For this reason human life on earth must be seen as just one instance of many energy events: moments in which energy is absorbed from the sun lead to growth and reproduction but, just as important, energy is also blown off. Humans in this sense are no different from any other animals, though their wastage of energy might be more intense through its very self-consciousness. All social productions—all cultural productions—are therefore seen as modes of energy appropriation and squandering; their value or lack of value must be seen in the context of their role as conduits in the flows of energy through humans outward to the void of the universe. These flows are gifts not necessarily to other humans but to the emptiness of the sky. Gifts, or put another way, destroyed things, things whose end lies in immediate consumption, not utility and deferred pleasure.

Bataille's work anticipates much recent analysis, which sees value—economic, cultural—deriving from energy inputs: humans may "produce," but their produc-

Bataille's Peak: Energy, Religion, and Postsustainability (Minneapolis: University of Minnesota Press, 2007), 32–42, 46–50.

tive activity is dependent on the quantities of energy that they are capable of harnessing. Human evolution—physical and cultural—in this view is a function of the channeling of energy: taking advantage of abundant energy (derived from agricultural inputs or, later, from fossil fuels), humans reproduce and populate the earth; suffering, on the other hand, from a lack of energy, their society contracts, and they find ways to cope with eternal shortage. Surplus and shortage are thus intimately linked; each is always present in the other, and each must be recognized in its fundamental role in the preservation, extension, intensification, and ruin of the community.[2]

There are, no doubt, many ways in which the centrality of energy for life can be read. In the nineteenth century a kind of cultural pessimism was all-pervasive: since the second law of thermodynamics postulated the entropy of any given field of energy, we could then infer that any society, any life-form, any planet would eventually lose the energy it had at its disposal and sink into quietude, feebleness, death. From the larger argument about energy, and the eventual fate of the sun and all other stars, commentators were quick to see a similar effect in society: the fadeout of energy led to weakness and cultural decadence. Society was on a death trip just like the sun; humans, presented in this reactionary mode, could brood over their fate but could do little to prevent it.[3]

Bataille consciously points in the opposite direction. In Bataille's view, rather than entropy, the magnificent expenditure of energy, characterized by the violence and brilliance of the sun, leads to the conclusion that energy is limitless and that the chief problem lies not in its hoarding and in the warding off of the inevitable decline, but in the glorious burn-off of the sun's surplus. In effect, the problem becomes how best to expend rather than how best to envision the consequences of shortage. For all that, Bataille is not an optimist in the conventional sense of the word because he does not link abundant energy and its glorious throughput with the placid satisfactions and order of a middle-class existence.[4]

In the 1950s there was a lot of talk about "energy too cheap to meter": the promise of the nuclear energy industry. That was good news, apparently, because it would allow us to live happy lives with a maximum number of appliances; we could always own more, always spend more, with the ultimate goal being human comfort. Growth was the name of the game—it still is—and growth in comfort was made possible when more energy was produced than needed. If energy is nothing more than the power to do work, then an unending surplus of energy meant nothing more than a continuous rise in productivity, a concomitant rise in the number of objects citizens could look forward to possessing, and the personal satisfaction associated with those objects.

Bataille too envisages a constant surplus of energy, but his energy is very different from the metered or unmetered kind. True, one can momentarily put some of Bataille's energy "to work." But there is always too much of it to be simply controlled; it always exceeds the limits of what one would be capable of devoting to some end. Bataille's energy is therefore inseparable from the wildly careening atoms of Sade or even the profoundly formless matter envisaged by Bruno. "Cursed matter," be it the charged matter studied by Durkheim, or the "base matter" of Bataille's gnosticism, or the mortal meat of Sade's "transmutation," is not only matter that is left over and so can contribute its energy to further growth; it is also matter that is burned off, which leads nowhere beyond itself, and so is dangerous, powerful, sacred.[5] Bataille's energy shoots through a charged matter that obtrudes in sacred ritual and erotic "wounds": the "share" of energy is not a resulting order but a base disorder. Such matter is in excess, not inert but virulent, threatening, turning as easily against the one who would wield the power as against a supposed victim. But along with this, the excessive, material world is "intimate," not a useful, classifiable thing, but a moment of matter that does not lead outside itself, can serve no useful purpose, is not anchored in time in such a way that it becomes a means rather than an end.

Of course no energy can be surplus in and of itself. The supposed surplus energy, too cheap to meter, of the 1950s was only surplus in relation to a power grid: there was to be so much of it that it would pulse through the power grid, illuminating backyard patios and electrically heating split-level homes for free. And the more split levels that would be built, the more available—domesticated—energy there would be to fuel the world—and so on, presumably, to infinity. Bataille's energy, however, is in surplus on another kind of grid—that of the semiotic categories of a comprehensible social system. It is what is left over when a system completes itself, when a system depends on energy in order to complete itself—but it only does so by excluding the very energy that makes its completion possible. Put another way, we can say that a social system needs to exclude a surplus of energy (hence matter) in order to constitute itself as coherent and complete. There are, in other words, limits to growth, be they external (as in an ecology) or internal (as in a social philosophy or ideology).[6] That surplus/energy, in Bataille's terminology, is "cursed," always already unusable, outside the categories of utility. It is thus not servile, not ordered or orderable. A banal example: if a rural region can produce only so much food, then its "carrying capacity" is limited; the excess human population it produces will have to be burned off in some way. A surplus of humans in a given locale will lead to contraception, warfare, celibacy, sacrifice.[7] A certain equilibrium, tentative and never truly stable, will result. Human energy, human population, will have to be lost: effort that could be spent in nonsustainable growth—producing more things that could not be absorbed—will be spent, spewed out, in other, nonproductive activities: again, war, the production of (left-hand and right-hand) sacred artifacts, "useless" art, and so on. The inevitable limit of the system—economic, ecological, intellectual—always entails a surplus that precisely *defeats* any practical appropriation. This uncontrollable and useless energy courses through the body, is the body, animating it, convulsing it: this is a threatening energy that promises death rather than any straightforward appropriation.

"Excess" matter will therefore be different in kind from its double, the "share" that can be reabsorbed into the system: the excess matter-energy will not be easily classifiable, knowable, within the parameters of the grid. It will always pose itself as a profound challenge. Against the coherent oppositions and reliable significations found operating within a given system of energy use, it constitutes a series of instances of energy in flux: never stable, never predictable, but a matrix of free energy-symbolization at the ready, to open but also to undermine the coherency of the system. Rendered docile, energy makes the system possible (society, philosophy, physics, technology); revealing itself as excessive, unconditioned, at the moment the edifice achieves its fragile summit, energy opens the abyss into which the system plunges.

Bataille's Version of Expenditure

The Accursed Share, first published in 1949, has had a colorful history on the margins of French intellectual inquiry. Largely ignored when first published, it has gone on to have an interesting and subtle influence on much contemporary thought. In the 1960s, fascination with Bataille's theory of economy tended to reconfigure it as a theory of writing: for Derrida, for example, general economy was a general writing. The very specific concerns Bataille shows in his work for various economic systems is largely ignored or dismissed as "muddled."[8] Other authors, such as Michel Foucault and Alphonso Lingis, writing in the wake of this version of Bataille, have nevertheless stressed, following more closely Bataille's lead, the importance of violence, expenditure, and spectacular transgression in social life.[9]

The basis for Bataille's approach can be found in the second chapter of the work "Laws of General Economy." The theory in itself is quite straightforward: living organisms always, eventually, produce more than they need for simple survival and reproduction. Up to a certain point, their excess energy is channeled into expansion: they fill all available space with versions of themselves. But inevitably, the expansion of a species comes against limits: pressure will be exerted against insurmountable barriers. At this point a species' explosive force will be limited, and excess members will die. Bataille's theory is an ecological one because he realizes that the limits are internal to a system: the expansion of a species will find its limit not only through a dearth of nourishment but also through the pressure brought to bear by other species.[10] As one moves up the food chain, each species destroys more to conserve itself. In other words, creatures higher on the food chain consume more concentrated energy. It takes more energy to produce a calorie consumed by a (carnivorous) tiger than one consumed by a (herbivorous) sheep. The ultimate consumers of energy are not so much ferocious carnivores as they are the ultimate consumers of other animals and themselves: human beings.

For Bataille, Man's primary function is to expend prodigious amounts of energy, not only through the consumption of other animals high on the food chain (including man himself) but in rituals that involve the very fundamental forces of useless expenditure: sex and death.[11] Man in that sense is in a doubly privileged position: he not only expends the most, but alone of all the animals he is able to expend *consciously*. He alone incarnates the principle by which excess energy is burned off: the universe, which is nothing other than the production of excess energy (solar brilliance), is doubled by man, who alone is aware of the sun's larger tendency and who therefore squanders consciously in order to be in accord with the overall tendency of the universe. This for Bataille is religion: not the individualistic concern with deliverance and personal salvation, but rather the collective and ritual identification with the cosmic tendency to lose.

Humans burn off not only the energy accumulated by other species but, just as important, their own energy, because humans themselves soon hit the limits to growth. Human society cannot indefinitely reproduce: soon enough what today is called the "carrying capacity" of an environment is reached.[12] Only so many babies can be born, homes built, forests harvested. Then limits are reached. Some excess can be used in the energy and population required for military expansion (the case, according to Bataille, with Islam [OC, 7:83–92; AS, 81–91]), but soon that too screeches to a halt. A steady state can be attained by devoting large numbers of people and huge quantities of wealth and labor to useless activity: thus the large numbers of unproductive Tibetan monks, nuns, and their lavish temples (OC, 7:93–108; AS, 93–110). Or most notably, one can waste wealth in military buildup and constant warfare: no doubt this solution kept populations stable in the past (one thinks of the endless battles between South American Indian tribes), but in the present (i.e., 1949) the huge amounts of wealth devoted to military armament, worldwide, can lead only to nuclear holocaust (OC, 7:159–60; AS, 169–71).

This final point leads to Bataille's version of a Hegelian "absolute knowledge," one based on the certainty of a higher destruction (hence an absolute knowledge that is also a non-knowledge). The imminence of nuclear holocaust makes it clear that expenditure, improperly conceived, can threaten the continued existence of society. Unrecuperable energy, if unrecognized or conceived as somehow useful, threatens to return as simple destruction. Bataille's theory, then, is a profoundly *ethical* one: we must somehow distinguish between versions of excess that are "on the scale of the universe," whose recognition-implementation guarantee the survival of society (and human expenditure), and other versions that entail blindness to the real role of expenditure, thereby threatening man's, not to mention the planet's, survival.

This, in very rough outline, is the main thrust of Bataille's book. By viewing man as a spender rather than a conserver, Bataille manages to invert the usual order of economics: the moral imperative, so to speak, is the furthering of a "good" expenditure, which we might lose sight of if we stress an inevitably selfish model of conservation or utility. For if conservation is put first, inevitably the bottled-up forces will break loose but in unforeseen, uncontrollable, and, so to speak, untheorized ways. We should focus our attention not on an illusory conservation, maintenance, and the steady state—which can lead only to mass destruction and the ultimate wasting of the world—but instead on the modes of expenditure in which we, as human animals, should engage.[13]

But how does one go about privileging willed loss in an era in which waste seems to be the root of all evil? Over 50 years after the publication of *The Accursed Share*, we live in an era in which nuclear holocaust no longer seems the main threat. But other dangers lurk, ones just as terrifying and definitive: global warming, deforestation, the depletion of resources—and above all energy resources: oil, coal, even uranium. How can we possibly talk about valorizing heedless excess when energy waste seems to be the principal evil threatening the continued existence of the biosphere on which we depend? Wouldn't it make more sense to stress conservation, sustainability, and downsizing rather than glorious excess?

What Appears to Be Wrong with Bataille's Theory?

To think about the use value of Bataille, we must first think about the nature of energy in his presentation. For Bataille, excessive energy on the earth is natural: it is first solar (as it comes to us from the sun), then biological (as it passes from the sun to plants and animals to us), then human (as it is spent in our monuments, artifacts, and social rituals). The movement from each stage to the next involves an ever-greater disposal: the sun spends its energy without being repaid; plants take the sun's energy, convert it, and throw off the excess in their wild proliferation; and animals burn off the energy conserved by plants (carnivores are much less efficient than herbivores), all the way up the food chain. Humans squander the energy they cannot put to use in religious rituals and war. "On the surface of the globe, *for living matter in general*, energy is always in excess, the question can always be posed in terms of extravagance [*luxe*], the choice is limited to how wealth is to be squandered [*le mode de la dilapidation des richesses*]" (*OC*, 7:31; *AS*, 23; italics Bataille's). There never is or will be a shortage of energy, it can never be used up by man or anything else, because it comes, in endless profusion, from the sun and stars.

Georges Ambrosino, a nuclear scientist and Bataille's friend, is credited in the introduction of *The Accursed Share* (*OC*, 7:23; *AS*, 191) as the inspiration for a number of the theses worked out in the book. In some unpublished "notes preliminary to the writing of *The Accursed Share*" (*OC*, 7:465–69), Ambrosino sets out very clearly some of the ideas underlying Bataille's work:

> We affirm that *the appropriated energies produced during a period are superior in quantity to the appropriated energies that are strictly necessary to their production.*

For the rigor of the thesis, it would be necessary to compare the appropriated energies of the same quality. The system produces all the appropriated energies that are
necessary to it, it produces them in greater quantities than are needed, and finally it
even produces appropriated energies that its maintenance at the given level does not
require.

In an elliptical form, but more striking, we can say that *the energy produced is superior
to the energy necessary for its production.* (OC, 7:469)[14]

Most striking here is the rather naive faith that, indeed, there always will be an
abundance of refinable, usable energy and that spending energy to get energy inevitably results in an enormous surplus of energy—so much that there will always
be a surplus, "greater quantities than are needed." Ambrosino, in other words, projects a perpetual surplus of energy return on energy investment (EROEI).[15] One
can perhaps imagine that a nuclear scientist, in the early days of speculation about
peaceful applications of atomic energy, might have put it this way. Or a petroleum
geologist might have thought the same way, reflecting on the productivity of the
earth shortly after the discovery of a giant oil field.[16] Over 50 years later it is much
harder to think along these lines.

Indeed, these assumptions are among those most contested by current energy
theorists and experts. First, we might question the supposition that since all energy in the biosphere ultimately derives from the sun, and the sun is an inexhaustible source of energy (at least in relation to the limited life spans of organisms),
there will always be a surplus of energy *for our use.* The correctness of this thesis
depends on the perspective from which we view the sun's energy. From the perspective of an ecosystem—say, a forest—the thesis is true: there will always be
more than enough solar energy so that plants can grow luxuriantly (provided
growing conditions are right: soil, rainfall, etc.) and in that way supply an abundance of biomass, the excess of which will support a plethora of animals and, ultimately, humans. All living creatures will in this way always absorb more energy
than is necessary for their strict survival and reproduction; the excess energy they
(re)produce will inevitably, somehow, have to be burned off. There will always be
too much life.

If we shift perspective slightly, however, we will see that an excess of the sun's
energy is not always available. It is (and will continue to be) extremely difficult to
achieve a positive energy return directly from solar energy.[17] As an energy form,
solar energy has proven to be accessible primarily through organic (and fossilized)
concentration: wood, coal, and oil. In human society, at least as it has developed
over the last few millennia, these energy sources have been tapped and have allowed the development of human culture and the proliferation of human population. It has often been argued that this development/proliferation is not due solely
to technological developments and the input of human labor; instead, it is the

ability to utilize highly concentrated energy sources that has made society's progress possible. Especially in the last 200 years, human population has expanded
mightily, as has the production of human wealth. This has been made possible by
the energy contributed to the production and consumption processes by the combustion of certain fuels in ever more sophisticated mechanical devices: first wood
and then coal in steam engines, and then oil and its derivatives (including hydrogen, via natural gas) in internal combustion engines or fuel cells. Wealth as it has
come to be known in the last 300 or so years, in other words, has its origins not
just in the productivity of human labor and its ever more sophisticated technological refinements, as both the bourgeois and Marxist traditions would argue, but
in the energy released from (primarily) fossil fuels through the use of innovative
devices. In the progress from wood to coal and from coal to oil, there is a constant
progression in the amount of quantifiable and storable energy produced from a
certain mass of material.[18] Always more energy, not necessarily efficiently used;
always more goods produced, consumers to consume them, and energy-based fertilizers to produce the food needed to feed them. The rise of civilization as we
know it is tied directly to and is inseparable from the type of fuels used to power
and feed it—and the quantities of energy derived from those fuels at various stages
of technology.[19]

Certainly Bataille, following Ambrosino, would see in this ever-increasing energy use a continuation—but on a much grander scale—of the tendency of animals
to expend energy conserved in plant matter. Indeed, burning wood is nothing more
than that. But the fact remains that by tapping into the concentrated energy of
fossil fuels, humans have at their disposal (ancient) solar energy—derived from
fossil plants (coal) and algae (oil)—in such a concentrated form that equivalent
amounts of energy could never be derived from solar energy alone.[20]

In a limited sense, then, Bataille and Ambrosino are right: all the energy we
use ultimately derives from the sun. There is always more of it than we can use.
Where they seem to be wrong is that they ignore the fact that for society as we
know it to function, with our attendant leisure made possible by "energy slaves,"
energy derived from fossil fuels, with their high EROEI, will be necessary for the
indefinite future.[21] There is simply no other equally rich source of energy available
to us; moreover, no other source will likely be available to us in the future.[22]
Bataille's theory, on the other hand, ultimately rests on the assumption that energy is completely renewable, there will always be a high EROEI, and we need not
worry about our dependence on finite (depletable) energy sources. *The Accursed
Share* for this reason presents us with a strange amalgam of awareness of the central role energy plays in relation to economics (not to mention life in general) and
a willful ignorance concerning the social-technological modes of energy delivery
and use, which are far more than mere technical details. We might posit that the
origin of this oversight in Bataille's thought is to be found in the economic theory,

and ultimately philosophy, both bourgeois and Marxist, of the modern period, where energy resources and raw materials for the most part do not enter into economic (or philosophical) calculations, since they are taken for granted: the earth makes human activity possible, and in a sense we give the earth meaning, dignity, by using resources that otherwise would remain inert, unknown, insignificant (one thinks of Sartre's "in-itself " here). Value has its origin, in this view, not in the "natural" raw materials or energy used to produce things but in human labor itself. Bataille merely revises this model by characterizing human activity, in other words production, as primarily involving gift-giving and wasting rather than production and accumulation.

We can argue, then, that solar energy is indeed always produced, always in excess (at least in relation to the limited life spans of individuals and even species); but it is fossil fuels that best conserve this energy and deliver it in a rich form that we humans can most effectively use. Human progress has been so explosive that these energy inputs have come to seem infinite and then have become invisible. Unfortunately, fossil fuels can be depleted, indeed are in the process of being depleted.

Why is this important in the context of Bataille? For a very simple reason: if Bataille does not worry about energy cost and depletion, he does not need to worry about energy conservation. Virtually every contemporary commentator on energy use sees only one short-term solution: conservation. Since fossil fuels are not easily replaceable by renewable sources of energy, our only option is to institute radical plans for energy conservation—or risk the complete collapse of our civilization when, in the near future, oil, coal, and natural gas production decline and the price of fuel necessarily skyrockets.[23] Some commentators, foreseeing the eventual complete depletion of fossil energy stores, predict a return to feudalism (Perelman 1981) or simply a quasi-Neolithic state of human culture, with a radically reduced global population (Price 1995).

Without a theory of depletion, Bataille can afford to ignore conservation in all senses: not only of resources and energy but also in labor, wealth, and so on. He can also ignore (perhaps alarmist) models of cultural decline. In Bataille's view, there will always be a surplus of energy; the core problem of our civilization is how we use up this excess. We need never question the "energy slaves" inseparable from our seemingly endless waste. Nor will there need to be any consideration of the fact that these energy slaves may very well, in the not-so-distant future, have to be replaced by real, human slaves.

* * *

Limits

Carrying capacity poses a limit to growth: a society can destroy the excess through sacrifice, infanticide, ritual, festival; or excess can be put to work through the waging of war, in which case carrying capacity may be expanded through the appropriation of another society's land. War too, however, shows some elements of religious, ruinous expenditure in that it entails, as does sacrifice, glory. Especially in modern times, war also brings with it the possibility of defeat: in that case there is no glory, and certainly no possibility of the expansion of carrying capacity. Indeed, as in the case of nuclear holocaust, societies run the risk of completely obliterating —wasting—the carrying capacity of their land.

In accord with Bataille's implicit ethical model, one can argue that the limits imposed by carrying capacity evoke two possible responses from societies. First, a society can recognize limits. Here, paradoxically, one violates limits, consciously transgresses them, so to speak, by recognizing them. Through various forms of ritual expenditure one ultimately respects limits by symbolically defying the very principle of conservation and measured growth—of, in other words, limits. "Spending without reserve" is the spending of that which cannot be reinvested because of the limit, and yet the very act of destruction is the transgression of the logic of the limit, which would require, in its recognition, a sage and conservative attentiveness to the dangers of excessive spending. If there is a limit to the production of goods and resources, however, we best respect and recognize that limit through its transgression—through, in other words, the destruction of precious but unusable energy resources. To attempt to reinvest, or put to use, the totality of those resources, to guarantee maximum productivity and growth, would only ignore the limit (rather than transgressing it), thereby eventually lowering the limit if not eliminating it entirely (elimination of carrying capacity, ecological destruction, desertification).[24] For this reason, a theory of expenditure is inseparable from, is even indistinguishable from, a theory of depletion.

Such an affirmation—of limits and expenditure—entails a *general* view of economy and, we might add, ecology. In positing such a respect for limits through their transgression, we forgo an individual concern, which would customarily be seen as the human one (but which is not, in Bataille's view): a concern with personal survival, enrichment, and advancement. From a larger perspective, we forego the needs of Man as a species or moral category (or the needs of God as Man's moral proxy). The supremacy of self-interest is tied for Bataille to the simple ignorance of limits: not their transgression, but their heedless violation. In the case of transgression of limits, we risk what might be personally comfortable or advantageous in order to attain a larger "glory" that is tied to unproductive expenditure and entails a possible dissolution of the self. From a general perspective, this expenditure is (as Bataille would say) on the scale of the universe; it must also be, in principle, on the scale of the carrying capacity of a given landscape or ecology (else the expenditure would very quickly cancel itself out).

This version of limits and their transgression can be associated with Bataille's conception of eroticism. What separates humans from the animals, according to

Bataille, is the interdiction of "immediate, unreserved, animal pleasure [*jouis-sance*]" (*OC*, 8:47). Decency, the rules against sexual expression, incest, and intense pleasure that characterize human society are fundamental to an organized society. But the human is not exclusively to be found in the interdiction: its ultimate "self-consciousness" is derived through the ecstatic transgression of that interdiction. Interdiction is an aftereffect of transgression, just as conservation is an aftereffect of expenditure (we produce and conserve in order to expend). What ultimately counts for us as humans (for us to be human) is an awareness of the necessity of expenditure (including that of our own death)—an awareness that animals lack.

> Of course, respect is only the detour of violence. On the one hand, respect orders the humanized world, where violence is forbidden; on the other, respect opens to violence the possibility of a breakout into the domain where it is inadmissible. The interdiction does not change the violence of sexual activity, but, by founding the *human* milieu, it makes possible what animality ignored: the transgression of the rule. . . .
>
> What matters is essentially that a milieu exists, no matter how limited, in which the erotic aspect is unthinkable, and moments of transgression in which eroticism attains the value of the greatest overthrow [*renversement*]. (*OC*, 8:47–48; italics Bataille's)

Eroticism, the general or collective experience of transgression, is impossible without the knowledge of human limits, interdictions. In the same way, we can say that the destruction of excess in an economy is only "on the scale of the universe" if it maintains and respects limits. We could even go beyond this and say that the maintenance of those limits, the carrying capacity in today's terminology, is only possible through the ritual, emotionally charged destruction of excess wealth (and not its indefinite, seemingly useful, but indifferent reproduction), just as interdictions are only meaningful, and therefore maintainable, when they are periodically transgressed.

The only other approach to limits, as I have indicated, is to ignore them: the consumption of scarce resources should go on forever; growth is limitless. In the realm of eroticism, this would be either to be entirely unaware of moral limits (interdictions)—as are animals—or on the other hand, to see limits as so absolute that no meaningful transgression can take place; in this case all eroticism would be so minor, so secondary, that no intimate relation between interdiction and eroticism could be imagined, and no dependence of interdiction on the transgressive expression of eroticism could be conceived. In this case limits would be so overwhelming that they would not even be limits: in effect one could not violate them since they would be omnipresent, omnipotent. Their transgression would be inconceivable (to try to violate them would simply manifest one's own degeneracy or evil, one's status outside the community, in an asylum or hell). Not coincidentally, this position is that of a religious-social orientation in which flamboyant expenditure—sexual, religious, phantasmic—is inconceivable, or unworthy of conception, and in which all excess must therefore be reinvested in material productive processes (even eroticism is subordinated to the production of more people): Calvinism, the Protestant ethic, various fundamentalisms, and so on. This is the narrow view, that of the restricted economy, the economy of the "individual":

> Each investor demands interest from his capital: that presupposes an unlimited development of investment, in other words the unlimited growth of the forces of production. Blindly denied in the principle of these essentially productive operations is the not unlimited but considerable sum of products consumed in pure loss [en pure perte].
> (OC, 7:170; AS, 182)

This restricted economy, which hypostatizes limits (moral, personal) only ultimately to ignore them or degrade them, is the economy that values war as a mode of expansion (typified, for Bataille at least, by Islam) and as utility (self-defense, deterrence, mutually assured destruction). The limit is ignored in the restricted economy only at the risk of reimposition of an absolute limit, cataclysmic destruction, or ecological collapse (nuclear holocaust, the simple elimination of carrying capacity).

Bataille's ethics, then, entail a choice between these two alternatives: recognition of limits through the affirmation of expenditure in a general economy, and the ignorance of limits through a denial of expenditure in a closed or restricted economy. The first entails the affirmation of glorious pleasure, sacred matter and energy, and anguish before death, while the latter entails the ego-driven affirmation of utility and unlimited growth with all the attendant dangers (the untheorized and quite sudden imposition of the limits to growth).

The irony in all this is that the first, transgressive, and "human" ethics will inevitably be sensitive to ecological questions—respectful of carrying capacity—through its very affirmation of loss. The second, attempting to limit severely or do away with waste and thereby affirm the particular interests of an individual, a closed social group, or a species (Man) in the name of "growth," will only universalize the wasting—the ultimate destruction—of the carrying capacity that serves as the basis of life. Conservation is therefore a logical aftereffect of expenditure; we conserve in order to expend. In other words, we conserve not to perpetuate our small, monadic existences or the putative centrality of our species, but rather to make possible a larger generosity, a larger general economy that entails the transgression (in angoisse) of our narrow, selfish "practicality," our limitedness (i.e., the inevitable postponement of pleasure).

By expending we conserve. Bataille's utopian ethics foresees a society that creates, builds, and grows in and through loss. Bataille thereby affirms the continuation of a human collectivity whose humanity is inseparable from that general—collective and ecstatic—expenditure. Inseparable, in other words, from a loss of the very selfish fixation on knowledge, authority, and even comforting immortality with which the word "humanity" is usually associated. The raison d'être of society, so to speak, will lie in the very unreasoned logic of its excessive and transgressive expenditure. This highest value will be maintained and known through recognition of limits, which is ultimately reasonable but to which the act of expenditure nevertheless cannot be reduced (because the affirmation of limits entails their transgression at the "highest" point of development and knowledge).

Notes

1. Hochroth's superb article is necessary reading for any understanding of Bataille and energy. She very cogently traces out Bataille's approach as a critique of Carnot and as a reading—and political critique—of Wilhelm Ostwald. Both Bataille and Ostwald are energeticists, Hochroth notes. Energeticism proposes that "all occurrences are produced by differential intensities, otherwise known as the law of transformation. In this way, as opposed to the mechanistic view, everything can be explained in terms of energy" (Hochroth 1995, 71). The difference between Bataille and Ostwald is between their respective "politics of expenditure"—a difference leading to a "different scale of values" (73). Ostwald believed not in glorious expenditure but in "reducing waste to a minimum" (76).

2. See on this topic Diamond (2005) and Manning (2004).

3. There is a tendency among recent critics to consider the problem of energy, its conservation and expenditure, as somewhat outdated; see Clarke (2001), for whom the question of energy is replaced by that of information.

4. See Huber and Mills (2005).

5. Recall that for Bataille the sacred is double, both the elevated, conservative sacred of established and productive religion (the right-hand sacred), and the base (left-hand) sacred of cursed matter and orgiastic, lawless loss.

6. In any case the external and internal limits of a society are closely linked. The catastrophic result of a social system that cannot conceive of—i.e., theorize—its limits is in fact the central question of *The Accursed Share*. Such a society still has to face its excess but, doing so in ignorance, it risks destroying itself. Such is the danger of the cold war face-off between the United States and the USSR—or so Bataille would argue—and such is the danger facing a society that cannot theorize its ecological limits (so I would argue, following Bataille's lead).

7. One can of course make the same generalization about animal ecologies; the struggle between animals, and even the internal limitation of animal populations in some species, takes place when energy supplies are scarce. In the case of human societies, warfare over limited resources and land is an example of this kind of conflict.

8. Derrida characterizes the crucial fifth section of *The Accursed Share*, in which nuclear destruction and the Marshall Plan are discussed, as "most often muddled by conjectural approximations" (Derrida 1978, 337n33). One can argue, in fact, that the vast majority of readings of Bataille tend to downplay or dismiss the social, ethical, and political implications of "general economy" and see it instead as a critique of Hegelianism or metaphysics writ large (ibid.), as a critique of epistemology (Hollier 1990), or of modernist aesthetics

(Bois 1997). My argument, on the contrary, is that to a large extent Bataille was a social and even utopian (or dystopian) thinker whose vision of the future entailed a radical alteration in (the study and practice of) economics, religion, and eroticism. Bataille even called it "Copernican" at one point, no doubt alluding not only to the importance of the sun in Copernicus's theory but to the Copernican tradition, starting, of course, with Giordano Bruno.

9. See, for example, Foucault's *Discipline and Punish* (1995)—especially the importance he places on the (constitutive) role of violent spectacle in society—and Lingis's emphasis, for example, in *Trust* (2004), on transgression and excess in interpersonal relations.

10. Bataille (1970–88, 7:40); Bataille (1988, 33–34).

11. In this sense "Man" for Bataille is exactly *opposed* to a humanist Man who conserves and works, jealously guarding his property, life, and the sanctity of his God (whom he creates in his own fearful image). This is the Man not of constructive labor and the desire for tranquility, but Man the painter of the Lascaux caves. A "primitive" Man, then, who practices a religion and an economy "on the scale of the universe."

12. "Carrying capacity" refers to the population (of any given species) that a region can be reasonably expected to sustain. It is defined by LeBlanc in this way: "The idea [of carrying capacity] in its simplest form is that the territory or region available to any group contains only a finite amount of usable food for that group. Different environments can carry or support different numbers of people: deserts can support fewer people than woodlands, the Arctic can support very few, and so on" (2003, 39).

13. I should stress in this discussion that "expenditure" in Bataille's sense (*la dépense*) is a term to be valorized only in relation to the human. After all, the stars, in their burning of inconceivable amounts of hydrogen, do not expend it because there is no purpose to which it can be put. "Expenditure" is a term, like "transgression," that has meaning only against another term. One can only "transgress" an "interdiction"; one can only "expend" in relation to a need, or command, to conserve. Man's identification with the universe as expenditure is only in the sense that spending like and with the universe puts into question the integrity and coherence of Man. The energy of the universe is excessive precisely because it cannot be put to work by Man. Man's "tendency to expend" is therefore a profound movement that signals the very demise of the category "Man." Energy is only energy—the power to do work—in relation to Man. Energy is only loss in relation to the death of Man. A more profound energy—that which cannot be harnessed, that which is spent and not conserved—can only be said to be expended against the limit that is the finitude (and meanness) of Man. But if we were not human we could not write of these things at all. A total identification with the universe would result not in Bataille's writing, but in the vast undifferentiated, powerful, but purposeless energy of the black hole.

14. My translation; Ambrosino's text is not included in the English translation of *The Accursed Share*. Many of Ambrosino's texts of the late 1940s and early 1950s published in the review edited by Bataille, *Critique*, display the same assumption, that energy is available in infinite supply—not only in the universe as a whole, but in modern fossil fuel–based economies. This is a most peculiar position for a trained physicist to take. See, for example, his review of Norbert Wiener's *Cybernetics*, in *Critique* 41 (October 1950): "Energy, in a physical sense, is everywhere (the least gram of matter, etc. . . .), the sources of negative en-

tropy, with which man furnishes himself, and his industry, are practically inexhaustible [*intarissable*]" (Ambrosino 1950, 80).

15. On EROEI and its implications for any energy retrieval, distribution, or consumption system, see Heinberg (2003, 138). Another way of looking at this issue in Bataille (if not Ambrosino) is to think of two kinds of energy: an energy in fuel that is easily stockpiled, easily accessed, and another energy, elusive, difficult to capture and to use, an energy that is a priori excessive, active before the buildup of fossil fuels, in excess after their depletion. This is a larger energy, neither useful nor wasted in the way that fossil fuels are wasted: energy as excess, in excess, energy as the defiance and mortality of Man (what is there before him, what makes him possible as energy consumer, but what eludes him as well).

16. The last really massive oil field found by petroleum geologists was the Al-Ghawar, discovered in Saudi Arabia in 1948 (1 year before the publication of *The Accursed Share*).

17. See Heinberg (2003, 142–46).

18. Huber and Mills, conservative "infinite energy supply" critics, celebrate the "high grade," "well ordered" power that is derived from much larger quantities of "low grade energy." As they put it, "Energy consumes itself at every stage of its own production and conversion. Only about 2 percent of the energy that starts out in an oil pool two miles under the Gulf of Mexico ends up propelling two hundred pounds of mom and the kids two miles to the soccer field" (2005, 51). What Huber and Mills tend to ignore is that the fuel that allows this refining and concentration into the form of high-grade power is extremely limited. There are not a lot of docile energy sources that allow themselves to be quantified and put to work, and they affirm, in their combustion, our own finitude, our own mortality (they "serve" us only through their own depletion, through their acceleration of the arrival of the moment when we will be left without them). Most energy sources instead call attention to human finitude in that they are *insubordinate to the human command to become high-grade power*. They are the affirmation of the fundamental limitation in the quantity of available refinable energy sources; this is energy that does not lend itself to simple stockpiling and use. It is an energy with a sacrificial component, coursing through the body, parodically moving people through the city as they gawk and display (the walking of the flaneur), "transporting" both the body *and* the emotions. If this energy is infinite, it is so in the overwhelming quantity that is lost to us and not in its availability to our hubristic attempts at harnessing it.

19. See Beaudreau (1999, 7–35), for whom value in industrial economies is ultimately derived from the expenditure of inanimate energy, not labor-power. Conversely, "[human] labor in modern production processes is more appropriately viewed as a form of lower-level organization (i.e., supervisor)" (18).

20. It has taken millions of years of concentration in the fossilization process to produce the amazingly high-energy yields of fossil fuels: tapping into sunlight alone cannot come close. The sun is fundamentally resistant to human attempts at harnessing it; its power is of the moment, not of quantified stockpiles. F. S. Trainer, for example, sees enormous problems with the use of solar energy to fuel human society, even the most parsimonious: the difficulty of collecting the energy in climates that have little direct sunlight (1995, 118); the inefficiency of converting it to electricity and storing it, where at least 80% of the energy

will be lost in the process (118–24); and even the expense of building a solar collection plant where, in Trainer's estimation, "it would take eight years' energy output from the plant just to repay the energy it would take to produce the steel needed to build it!" (124)—all these facts indicate that solar energy in relation to human civilization is, well, too diluted.

21. The "energy slave" is based on the estimate of mechanical work a person can do: an annual energy output of 37.2 million foot-pounds. "In the USA, daily use per capita of energy is around 1000 MJ, that is, each person has the equivalent of 100 energy slaves working 24 hours a day for him or her" (Boyden 1987, 196).

22. The dependency on concentrated sources of energy might be alleviated by ever more sophisticated and efficient methods of energy production and use (e.g., wind and solar energy, cellulose-based ethanol), but it can never be eliminated because none of the alternative sources of energy promise anywhere near as great an EROEI as the fossil fuels. See above all Heinberg (2003, 123–65).

23. See Odum and Odum (2001, 131–286).

24. See the prime example cited by LeBlanc: an area of Turkey where he did research as a young anthropologist. "Almost 10,000 years of farming and herding have denuded an original oak-pistachio woodland, and today [in a photograph of the area] only a few trees can be seen in the distance" (LeBlanc 2003, 140).

Works Cited

Ambrosino, Georges. 1950. "La Machine Savante et la vie: Norbert Wiener, *Cybernetics*." *Critique* 41:70–82.

Bataille, Georges. 1970–88. *Oeuvres complètes (OC)*. Vols. 1–12. Paris: Gallimard.

———. 1988. *The Accursed Share: An Essay on General Economy (AS)*. Vol. 1, *Consumption*. Translated by Robert Hurley. New York: Zone.

Beaudreau, Bernard C. 1999. *Energy and the Rise and Fall of Political Economy*. Westport, CT: Greenwood.

Bois, Yve-Alain. 1997. *Formless: A User's Guide*. New York: Zone.

Boyden, Stephen. 1987. *Western Civilization in Biological Perspective*. Oxford: Clarendon.

Clarke, Bruce. 2001. *Energy Forms: Allegory and Science in the Era of Classical Thermodynamics*. Ann Arbor: University of Michigan Press.

Derrida, Jacques. 1978. "From Restricted to General Economy: A Hegelianism without Reserve." In *Writing and Difference*, trans. Allan Bass, 251–79. Chicago: University of Chicago Press.

Diamond, Jared. 2005. *Collapse: How Societies Chose to Fail or Succeed*. New York: Viking.

Foucault, Michel. 1995. *Discipline and Punish: The Birth of the Prison*. Translated by Alan Sheridan. New York: Vintage.

Heinberg, Richard. 2003. *The Party's Over: Oil, War, and the Fate of Industrial Societies*. Gabriola Island, BC: New Society.

Hochroth, Lysa. 1995. "The Scientific Imperative: Improductive Expenditure and Energeticism." *Configurations* 3 (1): 47–77.

Hollier, Denis. 1990. "The Dualist Materialism of Georges Bataille." *Yale French Studies* 78:124–39.

Huber, Peter W., and Mark P. Mills. 2005. *The Bottomless Well: The Twilight of Fuel, the Virtue of Waste, and Why We Will Never Run Out of Energy.* New York: Basic Books.

LeBlanc, Steven A. 2003. *Constant Battles: The Myth of the Peaceful, Noble Savage.* New York: St. Martin's.

Lingis, Alphonso. 2004. *Trust.* Minneapolis: University of Minnesota Press.

Manning, Richard. 2004. *Against the Grain: How Agriculture Has Hijacked Civilization.* New York: North Point.

Odum, Howard T., and Elizabeth C. Odum. 2001. *A Prosperous Way Down: Principles and Policies.* Boulder: University Press of Colorado.

Trainer, F. E. 1995. *Consumer Society: Alternatives for Sustainability.* Sydney: Zed.

))) Atomic Health, or How the Bomb Altered American Notions of Death

Joseph Masco

What happened to health in the atomic age? If we consider health the absence of illness and thus the opposite of death, the atomic bomb has fundamentally altered, if not totally invalidated, the concept. My dictionary defines health as a "condition of being sound in body, mind, or spirit" involving "freedom from physical disease or pain." I like this idea, even yearn for the simple purity it assumes about bodies and knowledge. But this concept of health is at best nostalgic, representing a dream image from an age long surrendered to modern technology and the nation-state. The atomic bomb has not only produced a new social orientation toward death, it has insinuated itself at the cellular level to challenge the very structure of all living organisms (plant, animal, human) on planet Earth. In order to assess a transformation of life on this scale, we need to understand how the atomic bomb has inverted definitions of health and security, remaking them from positive values into an incremental calculus of death.

Instant mass death, individualized cellular mutation, and radiation-induced disease have become normalized threats in our world, producing a new concept of the healthy life as well as a new relationship between citizens and the state mediated by catastrophic risk. The inability to think beyond these atomic potentialities —let alone to reduce the technological, political, and environmental conditions that continue to support them—renders these nuclear effects invisible, normalized, routine. In the atomic age, incipient death has become a form of health itself, both normalized and rendered invisible as a new form of nature. Health, in my

Against Health: How Health Became the New Morality, ed. Jonathan M. Metzl and Anna Kirkland (New York: NYU Press, 2010), 133–56

formulation, is thus a social construct that regularizes certain potentialities for the human body through an increasingly perverse process of naturalization and amnesia. I propose to track the profound physical and psychological effects of the atomic bomb on American society by exploring how the long-standing state concern with "hygiene" as a joint project of public health and security mutated during the Cold War arms race.

After the atomic bombing of Hiroshima and Nagasaki in August 1945, the technological potential for nuclear war expanded exponentially in the United States, quickly producing a world in which American life could end at any minute. Always on alert, intercontinental missiles, nuclear submarines, and long-range bombers ensure that nuclear war can begin any second of the day and, in the first minute of conflict, exceed the total destructive power unleashed in all of World War II. The Cold War state installed this new form of total destruction within the minute-to-minute reality of everyday life by calling it "national security." The technological capability for a totalizing mass death has not changed since the dissolution of the Soviet Union and the formal end of the Cold War. It remains a brute fact of the life that we live, and have lived for several generations, that our everyday space is imbued with the potential for an unprecedented form of collective death. It is unhealthy to say the very least.

But a radically foreshortened social future is only one possible effect of the nuclear revolution; the other is individualized, covert, unpredictable, and cellular. The atmospheric effects of nuclear development—there have been thousands of nuclear explosions conducted in test sites around the world—delivered vast amounts of radioactive material into the global biosphere. Taken up by global wind currents as well as plants and animals, these materials were delivered into each and every person on the planet and deposited in varying amounts within their genomes. As a result, all of us carry in our bodies traces of Strontium-90 and other human-made radioactive elements from the nuclear test program, making life on planet Earth quite literally a post-nuclear formation.[1]

We don't really talk about atomic health anymore. We reject the domestic costs of the US nuclear project in favor of the more generic discourse of "terrorism" and "WMDs." These terms project responsibility for nuclear fear outward onto often nameless and faceless others. They deny that the United States has been the global innovator in nuclear weapons, responsible for every significant technological escalation of the form. The United States also remains the only country to have used nuclear weapons in war, and, in terms of its nuclear development program, it remains the most nuclear-bombed country on Earth (with 904 detonations at the Nevada Test Site alone). Indeed, the social process of coming to terms with an imminent fiery death or a slower, cancerous one has now been normalized as a fact of everyday American life. The roots of nuclear anxiety, however, have not dissipated over time but rather have become more deeply woven into American culture and individual psyches.[2] Whether we choose to recognize it or not, each of us is now a post-nuclear creature, living in a social, biological, and emotional world structured by the cumulative effects of US nuclear weapons science.[3]

If you don't believe me, if you resist my definition of nuclear health as incipient death as absurd or extreme or simply too depressing, consider the hyperrationalist argument about radioactive life presented by Herman Kahn in his 1960 treatise, *On Thermonuclear War*. Writing as a RAND Corporation strategist, Kahn was influential in conceptualizing US nuclear policy in the early Cold War period. He set out to "think the unthinkable" and work through the details of a nuclear conflict in his book, taking readers from the first salvo of atomic bombs through the collapse of the nation-state and into the possible forms of social recovery.[4] Kahn was a central inspiration for the character of Dr. Strangelove in Stanley Kubrick's 1964 film of the same name, and like the character in the film, he delighted in the shock effects he produced in his audience by exploring the details of nuclear conflict and the conditions of a post-nuclear society.[5] On a surface level, *On Thermonuclear War* is an effort to calculate via cost-benefit analysis the economic, environmental, and health effects of different types of nuclear war. The text, however, also reveals the absurd biological and social stakes of Cold War "national security," calling into question the very "rationality" of the nuclear state. Kahn begins by breaking nuclear war into eight stages. He then assesses the resulting "tragic but distinguishable postwar states" of nuclear conflict depending on the specific war strategy and scale of the conflict.[6] For Kahn, nuclear war is a universe of physical, emotional, and social misery in which decisions still have to be made with stark consequences at each stage of the conflict, from military strategy to medical mobilization and economic recovery. To those who would reject this line of thinking as immoral or perverse, Kahn simply asks: Would you prefer a "post-nuclear America" with 50 million dead or 100 million dead? How about environmental ruin lasting 10 years or 50?[7] Nuclear conflict is rationalized here, as are its vast range of effects, in order to open up a new possibility for post-nuclear governance. Kahn ultimately runs the gruesome numbers to assess the degree of suffering Americans should be willing to accept in order to fight the Cold War. He also attempts to calculate the precise point at which the nuclear destruction would be so great that "the survivors would envy the dead," thus rendering the Cold War null and void as a collective project.[8]

In making these calculations, Kahn provides a new vocabulary of collective risk as well as a variety of new metrics for assessing health in the nuclear age. Consider just the figure titles he uses to illustrate the consequences of different nuclear scenarios: "Acceptability of Risks," "Genetic Assumptions," "The Strontium-90 Problem," "Radioactive Environment 100 Years Later," "Morbidity of Acute Total Body Radiation," "Life Shortening," and "Seven Optimistic Assumptions." While these tables detail a scale of suffering that is unprecedented in human experience,

Kahn's ultimate metric is not radiation exposure rates or "life shortening" effects or the economic cost of rebuilding destroyed cities and infrastructure, but a rather new kind of social calculus: the "defective child." Acknowledging that the nuclear test programs of the United States and Soviet Union had distributed vast amounts of Strontium-90 in the form of fallout into the global biosphere, Kahn assesses the damage to the human genome. Recognizing that the health effects of Strontium-90 contamination can be both immediate (cancer) and multigenerational (genetic damage), Kahn calculates how many American children will be born severely disabled because of US investment in the atomic bomb. This is a remarkable moment in American history as it recognizes that "national security"—traditionally a defense of both citizens and the state from outside dangers—has been fundamentally altered in the nuclear age, requiring new forms of internal sacrifice. Kahn, however, goes much, much further in his assessment of the Strontium-90 problem, revealing the new terms of nuclear health in a security state that will consider paying any cost right up to the edge of total annihilation in order to win the Cold War:

> I could easily imagine a war in which the average survivor received about 250 roentgens. . . . This would mean that about 1 percent of the children who could have been healthy would be defective; in short, the number of children born seriously defective would increase, because of war, to about 25 percent above the current rate. This would be a large penalty to pay for a war. More horrible still, we might have to continue to pay a similar price for 20 or 30 or 40 generations. But even this is a long way from annihilation. It might well turn out, for example, that US decision makers would be willing, among other things, to accept the high risk of an additional 1 percent of our children being born deformed if that meant not giving up Europe to Soviet Russia. Or it might be that under certain circumstances the Russians would be willing to accept even higher risk than this, if by doing so they could eliminate the United States.[9]

It might well turn out . . . In a world where a security state can imagine as a viable calculus exchanging the health of its children in perpetuity for a political victory, the terms of both "public health" and "security" have been mightily and permanently altered. We have moved from a notion of health as an absence of disease to a graded spectrum of dangerous effects now embedded in everyday life. Health, as Kahn presents it here, is a statistical calculation rather than a lived experience, and it is precisely by approaching health as a population effect rather than an individual one that he creates the appearance of rationality.

Kahn's interest in thinking through nuclear warfare, however, does force him, in a few rare but telling moments, to confront the raw physical reality of the human body. Food, for example, would be a major concern in a post-nuclear world, and Kahn imagines nuclear war scenarios in which US agriculture could be "suspended" for 50 to 100 years.[10] Since nuclear war in the era of thermonuclear weapons would spread its effects over the entire continental United States, Kahn focuses on how to manage food produced in a largely contaminated environment. His answer once again points to a new concept of health as incipient death. He proposes to classify food based on the amount of Strontium-90 contamination it contains in groups labeled "A" (little contamination) to "E" (heavily contaminated), and then to distribute the food based on the following criteria:

> The A food would be restricted to children and to pregnant women. The B food would be a high-priced food available to everybody. The C food would be a low-priced food also available to everybody. Finally, the D food would be restricted to people over age forty or fifty. Even though this food would be unacceptable for children, it probably would be acceptable for those past middle age, partly because their bones are already formed so that they do not pick up anywhere near as much strontium as the young, and partly because at these low levels of contamination it generally takes some decades for cancer to develop. Most of these people would die of other causes before they got cancer. Finally there would be an E food restricted to the feeding of animals whose resulting use (meat, draft animals, leather, wool, and so on) would not cause an increase in the human burden of Sr-90.[11]

These people would die before they got cancer. Thus, while contamination is total, linking all Americans in a post-nuclear reality, the role of governance is not to prevent disaster but to minimize its effects through the rational calculation of individual age versus the longevity of radioactive materials in the distribution of food. Again, there is no concept of health here if by health we mean an absence of disease or risk. There is rather the naturalization of a new baseline reality (in this case a very contaminated North America and a damaged human genome) within which to begin anew the calculations of effective medical governance. Specifically, the length of time it takes for cancer to appear is calculated against the expected longevity of middle-aged Americans in a post-nuclear world and thereby eliminated as a concern. Under this line of thinking, it is not a sacrifice if you die before the health effects of eating radioactive food give you a fatal cancer.

* * *

What kind of governance is this? And what has happened to the idea of public health as a state project devoted to improving the lives of citizens? While nuclear war has not yet occurred on the scale Kahn imagined, the effects of the nuclear test project distributed fallout and other environmental contamination on a massive scale to US citizens. The combined effects of environmental damage and social anxiety created by the nuclear arms race fundamentally altered—indeed, placed in opposition—the concepts of "national security" and "public health." In other words, the US effort to build an atomic bomb did not just create a new kind of weapon; it revolutionized American society. The achievement of the bomb in 1945 certainly transformed the nature of military affairs, but it also fundamentally altered the citizen–state relationship and created a new concept of state power, the nuclear superpower. How does nuclear superpower status change concepts of health in the United States as well as alter the state's commitment to improving the lives of its citizens? It is important to remember that the roughly ten thousand nuclear weapons that the United States maintains in its current arsenal (and the vast means of delivering them via aircraft, submarines, cruise missiles, and intercontinental missiles) are capable of holding the entire planet hostage. The first technological achievement of the Cold War state was the production of a 24/7 system of global surveillance and nuclear war fighting capability. This "closed world" system, as Paul Edwards has described it, has kept the world on the edge of a nuclear conflict that could happen with less than 10 minutes of warning.[12] Since there is no interior or exterior to this global logic which makes a claim on every living being on the planet, the bomb does not produce "bare life" (a reduction to a purely biological condition) or a "state of exception" in Giorgio Agamben's formulation;[13] in its totalizing scope and effects, it is something rather new in human history.

The modern state form is grounded, as Michel Foucault has argued, in the management of internal populations and the constant improvement of the means of securing that population from a variety of threats. The state's effort to improve and regulate the lives of citizens in terms of hygiene, disease, mental health, and education not only produced the social sciences but also mobilized state power in ways that touched every citizen—from taxes, to inoculations against disease, to the structure of the school system, to the safety of roads, and to the logics of imprisonment (for crime, insanity, and infectious disease). Modern technology as well as modern social science provided the means for constant social improvement as the nation-state form developed, constructing for citizens an imagined future in which health was to be an endless horizon of better living and part of an increasingly secure world. More state security was thus coterminous with the promise of more security for the individual, in other words, a happier, healthier life. This narrative of idealized "progress" was possible until the atomic revolution, which both underscored the reality of radical technological change and invalidated the state's ability to regulate society at the level of health and happiness. Security and health, which were linked for several hundred years in the development of the modern state concept, became contradictory ideas after 1945, all but invalidated as intellectual concepts when challenged by the nuclear reality of the Cold War.

Foucault seems to acknowledge this in his lectures on security. He notes that something fundamental about the state changes after 1945, disrupting his theory about the evolution of state power from monarchal authority through the modern nation-state form, which relies on a variety of means of influencing individuals (discipline) and populations (biopower) in constituting its power.[14] In *Society Must Be Defended*, he argues:

> The workings of contemporary political power are such that atomic power represents a paradox that is difficult, if not impossible, to get around. The power to manufacture and use the atom bomb represents the deployment of a sovereign power that kills, but it is also the power to kill life itself. So the power that is being exercised in this atomic power is exercised in such a way that it is capable of suppressing life itself. And, therefore, to suppress itself insofar as it is the power that guarantees life. Either it is sovereign and uses the atom bomb, and therefore cannot be power, biopower, or the power to guarantee life, as it has been ever since the nineteenth century. Or, at the opposite extreme you no longer have a sovereign right that is in excess of biopower, but a biopower that is in excess of sovereign right.[15]

The "excess" biopower produced by the bomb is a topic that bears much scrutiny, for the United States has literally built itself through nuclear weaponry. In his *Atomic Audit*, Stephen Schwartz has documented the fact that between 1940 and 1996, the United States spent *at least* $5.8 trillion on nuclear weapons.[16] This makes the bomb the third-largest federal expenditure since 1940, just after non-nuclear military spending and Social Security, and accounting for roughly eleven cents out of every federal dollar spent. After the atomic destruction of Hiroshima and Nagasaki, the relationship among military, governmental, industrial, and academic institutions grew as these institutions supported a US security state that was founded on nuclear weaponry and anticommunism. The Cold War nuclear standoff consequently became the ground for a new articulation of state power as well as a new social contract in the United States, transforming the terms of everyday life as well as the very definition of public health.

The atomic revolution, however, presented an immediate contradiction to US policymakers about how to define security and health. By the early 1950s, it was clear that the atomic test program in the Pacific was distributing fallout globally even as the expanding arms race with the Soviet Union made possible the extinction of both nation-states and whole ecosystems. Similarly, the decision to open a continental test site in Nevada in 1951 (to lower the costs of nuclear weapons research) ensured the direct exposure of Americans with each and every aboveground detonation.

The construction of a new national security apparatus, which attacked public health with each detonation, thus required a new kind of collective sacrifice. Not able to pursue both the atomic bomb and the protection of citizens (in the classic sense of not exposing them to damaging health effects), the United States chose instead to normalize the nuclear crisis as a new form of nature. In effect, the Cold War state minimized the health effects of the atomic test program while constantly inflating fears of a Soviet attack in order to create a perception-based form of risk management. It then told citizens that the central problem of the nuclear age was just in their heads. Nuclear war was officially converted into a question of emotional self-control.

In other words, mental health was increasingly fused with national security after 1945 as the United States mobilized nuclear logics to remake its social institutions and its approach to global affairs. Nuclear fear, a unique physiological and mental state first achieved via the atomic bombing of Hiroshima and Nagasaki, quickly became the basis for a permanent wartime economy in the United States and the total mobilization of American political, scientific, academic, and military institutions. With the first Soviet nuclear detonation in 1949, the United States became a paranoid state, seeing communists and nuclear threats everywhere at home and around the world. The nuclear project not only coordinated American institutions in support of a military agenda, it enabled a vision of the entire world as defensible, vulnerable space. "Containing communism" around the world placed US citizens in a new relationship to the security state, which required not only unprecedented financial commitments during "peacetime" but also a parallel project to transform Americans into Cold Warriors. For US policy makers, an immediate question was how to avoid creating an apathetic public on the one hand or a terrorized one on the other—how to create public support for an unprecedented, and potentially unending, militarism.[17]

At the height of the Cold War emotional management project, a new form of mental health practice materialized. As Jonathan Metzl has documented, the nuclear crisis of the mid-1950s was coterminous with the discovery and first mass-marketing of antidepressants in the United States.[18] Psychopharmacology produced a revolution in the very idea of mental health as individuals could now regulate their brain chemistry in an effort to achieve a state of internal calm. Metzl shows how the arrival of psychopharmaceuticals immediately began to replace psychotherapy as the dominant paradigm in treating mental illness. Thus, psychotherapy moved away from personal history and trauma as an explanatory mode for mental illness at precisely the moment in which trauma was both nationalized and codified within the Cold War system. By eliminating history and experience as an explanatory mechanism, mental treatment could be reimagined as solely a chemical negotiation, not one involving the social contradictions produced by a nuclear arms race. And by calibrating their emotions to the expectations of Cold War society through drugs, many Americans moved closer to normalizing a state of emergency in everyday life. Anxiety, however, was both a resource and a problem for the early Cold War state.

For the architects of the Cold War system within the Truman and Eisenhower administrations, the solution to public mobilization was a new kind of project, one pursued with help from social scientists and the advertising industry, to teach citizens a specific kind of cognitive and emotional attitude toward the bomb. Drawing on the historic concept of defense as a protection of citizens, the federal program for "civil defense" was actually a quite radical effort to psychologically and emotionally remake Americans Cold Warriors. It turned the idea of public protection—a classic definition of security—into a mental operation, concerned almost exclusively with perceptions, emotions, and self-discipline. This project took the form of an elaborate propaganda campaign involving films, literature, town meetings, and educational programs designed to teach Americans to fear the bomb; to define and limit that nuclear fear in ways useful to the Cold War project; and to move responsibility for domestic nuclear crisis from the state to citizens, enabling all citizens to have a role in a new collective form of American militarism. Through the 1950s, Civil Defense involved yearly simulations of nuclear attacks on the United States, in which designated American cities would act out nuclear catastrophe. Local newspapers would run banner headlines—"Washington DC, Detroit Destroyed by Hydrogen Bombs"—allowing civic leaders and politicians to lead theatrical evacuations of the city that were filmed by television cameras.[19] The formal goal of this state program was to transform "nuclear terror," which was interpreted by officials as a paralyzing emotion, into "nuclear fear," an affective state that would allow citizens to function in a time of crisis. As Guy Oaks has documented in *The Imaginary War*, civil defense programs of the 1950s and '60s sought to do nothing less than "emotionally manage" US citizens through nuclear fear.[20]

Moreover, since the Cold War was recognized as a state of long-term crisis, civil defense was designed to create a new kind of citizen equipped with a psychological and emotional constitution capable of negotiating the day-to-day, minute-to-minute nuclear threat for the long haul. In addition to turning the domestic space of the home into the front line of the Cold War, civil defense argued that citizens should be psychologically prepared every second of the day to deal with a potential nuclear attack. Thus, the Cold War state gave up on the idea of merely preventing attack (part of the traditional logic of security), and via the discourse of civil defense, demanded that citizens accept personal responsibility for surviving nuclear conflict. The key move in this shift from emphasizing protection to vulnerability in public safety was a campaign to make panic—not nuclear war—the official enemy. As Val Peterson, the first head of the Federal Civil Defense Administration (FCDA), wrote in 1953:

> Ninety per cent of all emergency measures after an atomic blast will depend on the prevention of panic among the survivors in the first 90 seconds. Like the A-bomb, panic is fissionable. It can produce a chain reaction more deeply destructive than any explosive known. If there is an ultimate weapon, it may well be mass panic—not the A-bomb.[21]

Panic is fissionable. Indeed, Peterson not only argued that Americans were particularly "susceptible to panic" but offered a checklist on how citizens could become "panic stoppers" by training themselves for nuclear attack and becoming like "soldiers" at home.

Thus, the official message from the early Cold War state was that self-control was the best way for citizens to fight a nuclear war, revealing a national project to both colonize and normalize everyday life with nuclear fear. Panic, as Jackie Orr has so powerfully shown, became more than a means of managing populations in the nuclear age. It remade the individual as a permanently insecure node in the larger Cold War system.[22] Regulating the psychology of a nation-state at this level, however, demanded that Cold War policymakers understand how to produce panic as well as calm it. As Orr has argued, this effort produced a collapsing field of mental health, national security, and individualized identity formation across the frontiers of gender, family, expertise, and self-knowledge.[23] The US nuclear project writ large required citizens to accept as normal social conditions that were both pathologically insecure and intellectually irreconcilable with health.

The FCDA effort to regulate the national public via images of nuclear conflict took on extraordinary proportions, involving massive, public exercises in which citizens acted out their own destruction and where the state set out to show citizens what a post-nuclear world might look like.[24] The Cold War security state ultimately sought to use nuclear fear to promote a new kind of social and psychological hygiene, one uniquely suited for a nuclear age. In the 1954 FCDA film, *The House in the Middle*, for example, we see that nuclear fear could be harnessed to any kind of domestic project, including household cleanliness.

Presented as an atomic experiment, the film documents how nuclear flash, heat, and blast effects engage model houses built at the Nevada Test Site. The three houses presented in the film are in different stages of cleanliness and repair and filmed against the stark desert landscape. The narrator offers this description of the project:

> Three identical miniature frame houses, with varying exterior conditions, all the same distance from the point of the explosion. The house on the right, an eyesore. But you've seen these same conditions in your own hometown: old unpainted wood, and look at the paper, leaves and trash in the yard. In a moment, you'll see the results of atomic heat flash on this house: the house on the left. Typical of many homes across the nation: heavily weathered, dry wood, in run-down condition. This house is the product of years of neglect. It has not been painted regularly. It is dry and rotten—a tinderbox ready to turn into a blazing torch. The house in the middle: in good condition, with a clean, unlittered yard. The exterior has been painted with ordinary, good quality house paint. Light painted surfaces reflect heat and the paint also protects the wood from weathering and water damage. Let's see what happens under atomic heat.[25]

The moralizing tone of the narration underscores that this film is ultimately an effort to recuperate the classic definition of hygiene as a social responsibility, and in so doing to create an image of a state that treats nuclear crisis as it would an infectious disease or a natural disaster. The centerpiece of the film is the slow-motion footage of the atomic blast wave hitting the model houses. The houses on the right and left ignite and burn to the ground, leaving the house in the middle standing. After contemplating this slow-motion destruction in detail, the narrator reexamines the ashes of the houses on the right and left and asks against somber background music, "Which of these is your house? This one? The house on the right? Dilapidated with paper, dead grass, litter everywhere? The house on the left—unpainted, run-down, neglected? Is this your house?" Then on an upbeat music cue, the narrator concludes:

> The house in the middle—cleaned up, painted up, and fixed up, exposed to the same searing atomic heat wave—did not catch fire. Close inspection reveals only a slight charring of the painted outer surface. Yes, the white house in the middle survived an atomic heat flash. These civil defense tests prove how important upkeep is to our houses and town.[26]

Only a slight charring. Household cleanliness is directly linked here to the likelihood of winning or losing a nuclear war. Hygiene, as a classic domain of state intervention into public health, is remade as a means of surviving not illness but nuclear attack. Indeed, the film does not mention fallout or radiation effects at all. The implicit message of the film is that each citizen should patrol his or her neighborhood for trash and feel free to discipline neighbors into perfect home performances. *The House in the Middle* argues ultimately that appearances are more important than reality, that a fresh coat of paint is more important than recognizing the city-killing power of nuclear weapons. Indeed, it offers citizens the opportunity to manically tend to their homes in every detail, from paint and yard work to furnishings and foodstuffs, as a means of achieving public health and national security, even as the reality of thermonuclear warfare promises little to no hope of survival.

The House in the Middle demonstrates how the terms of everyday life in the first decades of the Cold War were saturated with a new kind of state discourse, one that used nuclear fear to promote its policies. The challenge in such a psychological strategy, of course, is one of modulation—enough fear to produce support for US Cold War policies, but not too much to generate a countermovement. Indeed, the debates about nuclear fallout and the ineffectual nature of the Civil Defense program eventually forced the nuclear program into a major reorganization in the early 1960s. Civil defense, at this time, was factually wrong in many of its claims about the ability to survive a nuclear war. Moreover, the effort by Herman Kahn and others to "think the unthinkable" revealed to many the impossibility of civil defense in the thermonuclear age. US scientists also produced a powerful antinuclear counterdiscourse that emphasized the health effects of nuclear testing as well as the false claims of civil defense, challenging the state to return to concepts of security and health committed to improving living conditions rather than normalizing nuclear crisis. But the first decade of the Cold War witnessed the development of an entirely new form of governance, one in which a very public "national security" discourse was used strategically by the state to both mask and enable a global vision grounded in the sacrifice of citizens born and unborn. Public health, in this context, became secondary to a vision of national security that installed new forms of individual death within a larger structure of mass death. The social consequences of the nuclear arms race ultimately inverted the concept of health as absence of disease, replacing it with an increasingly naturalized vision of health as incipient death.

The effects of this joint transformation in the logics of health and national security are still with us today. The reactions to the suicide attacks on Washington, DC, and New York in 2001 reinvigorated the emotional management strategy and returned Americans to the logics of survival and sacrifice that structured the Cold War period. The threat of terrorism was, in fact, magnified by the George W. Bush administration into grounds for a fundamental rethinking of US global ambitions as well as US domestic policy, just as the first Soviet nuclear test in 1949 was mobilized by the Truman and Eisenhower administrations to construct the Cold War state. The state's appeal to citizens after 2001 was also modeled on the early civil defense campaigns as it argued for a normalization rather than an elimination of a totalized threat.

* * *

Fifty years after Val Peterson's national campaign to teach citizens to fear panic more than the bomb, the official position is still that every second of your life should be spent rehearsing the possibility of disaster, training yourself as a citizen to take responsibility for events that are by definition out of your control. Repeating the civil defense logics from the first decades of the Cold War, the state is absent here except in its desire to install a specific kind of fear within everyday life through advertising. The first 90 seconds of a crisis are still the most important, requiring that children be prepared to recite emergency information between swings of a little league bat or rungs on the jungle gym. These commercials reiterate the lessons of nuclear civil defense, that everyday life is structured around the possibility of mass injury from one moment to the next and that citizens must take responsibility for their own survival.

One of the most difficult cultural logics to assess is the arrival of a new social relationship to death. The vulnerability of children is one immediate register of American ideas of death, whether revealed in Herman Kahn's meditations on the number of "defective children" it will take to win the Cold War or in these DHS commercials that portray children as robots, preprogrammed to respond to crisis, rehearsing their life-line phone numbers even in the midst of play. Since 1945, the

United States has built its national community via contemplation of specific images of mass death while building a defense complex that demands ever more personal sacrifice in the name of security. Health as an absence of disease or anxiety is long gone. What we have instead is a negotiation of degrees of contamination, of degrees of anxious association, of degrees of escalating risk. The initial concept of public health as a key coordinate of the modern state system has been replaced by a governmental calculus based on individual sacrifice. A "war on terror" requires that Americans once again naturalize a notion of health as incipient death, inviting each and every citizen to dust off the familiar logics of Cold War sacrifice and embrace the further collapse of mental health into national security. Demilitarizing the mind by rejecting the "war on terror" project thus contains revolutionary potential; it might finally overthrow a national security project that both relies on and installs ever more deeply the possibility of violence and death into everyday life. In the twenty-first century, the last thing we need is more "health." Demilitarizing the mind, body, and spirit of American citizens by getting rid of the bomb, however, might just open up an entirely new kind of nature—post-national security, post-terror. Panic should be the least of our concerns.

Notes

1. See Joseph Masco, *The Nuclear Borderlands: The Manhattan Project in Post–Cold War New Mexico* (Princeton, NJ: Princeton University Press, 2006); and Arjun Makhijani and Stephen I. Schwartz, "Victims of the Bomb," in *Atomic Audit: The Costs and Consequences of U.S. Nuclear Weapons Since 1940*, ed. Stephen I. Schwartz (Washington, DC: Brookings Institution Press, 1998).

2. See Amy Kaplan, "Homeland Insecurities: Some Reflections on Language and Space," *Radical History Review* 85 (Winter 2003): 82–93.

3. See Joseph Masco, "Mutant Ecologies: Radioactive Life in Post-Cold War New Mexico," *Cultural Anthropology* 19, no. 4 (November 2004): 517–50.

4. Herman Kahn, *On Thermonuclear War* (Princeton, NJ: Princeton University Press, 1960).

5. See Sharon Ghamari-Tabrizi, *The Worlds of Herman Kahn: The Intuitive Science of Thermonuclear War* (Cambridge, MA: Harvard University Press, 2005).

6. Kahn, *On Thermonuclear War*, 34.

7. Ibid., 19.

8. Ibid., 40.

9. Ibid., 46.

10. Ibid., 66.

11. Ibid., 66–67.

12. See Giorgio Agamben, *Homo Sacer: Sovereign Power and Bare Life* (Stanford: Stanford University Press, 1998).

13. See Paul N. Edwards, *The Closed World: Computers and the Politics of Discourse in Cold War America* (Cambridge, MA: MIT Press, 1996).

14. See Michel Foucault, "The Risks of Security," in *Power: Essential Works of Foucault, 1954–1984*, vol. 3, ed. James Faubion (New York: New Press, 2000), 365.

15. Michel Foucault, *Society Must Be Defended: Lectures at the College de France, 1975–76*, ed. Mauro Bertani and Alessandro Fontana (New York: Picador, 2003), 253.

16. Schwartz, *Atomic Audit*, 1.

17. See Guy Oakes, *The Imaginary War: Civil Defense and American Cold War Culture* (New York: Oxford University Press, 1994), 34.

18. Jonathan Metzl, *Prozac on the Couch: Prescribing Gender in the Era of Wonder Drugs* (Durham, NC: Duke University Press, 2003).

19. See Stacy C. Davis, *Stages of Emergency: Cold War Nuclear Civil Defense* (Durham, NC: Duke University Press, 2007); and Laura McEnaney, *Civil Defense Begins at Home: Militarization Meets Everyday Life in the Fifties* (Princeton, NJ: Princeton University Press, 2000).

20. Guy Oaks, *The Imaginary War* (Oxford: Oxford University Press, 1995), 47.

21. Val Peterson, "Panic: The Ultimate Weapon?," *Collier's*, August 21, 1953, 99.

22. Jackie Orr, *Panic Diaries: A Genealogy of Panic Disorder* (Durham, NC: Duke University Press, 2006), 14.

23. Ibid.

24. See Joseph Masco, "Survival Is Your Business: Engineering Ruins and Affect in Nuclear America," *Cultural Anthropology* 23, no. 2 (May 2008): 361–98.

25. Federal Civil Defense Administration, *The House in the Middle*, 1954, 12 min., film.

26. Ibid.

》》》 The Draukie's Tale: Origin Myth for Wave Energy

Laura Watts

Perhaps you know
of the sea creatures,
the Selkie?

Silver-skinned 'seals'
who turn, at the touch of land,
into beautiful
men and women;

shape-shifters,
who transform themselves
from seal to homo sapiens
as they cross from sea to sand.

ebban an' flowan (Edinburgh: Morning Star, 2015)

And perhaps you know
of the Draugr,
or Trow?

Trickster folk,
wily and wayward,
who live under the mounds of the dead,
with the soil and the peat;

they are a wise folk,
as old as fossil fuel,
filled with ancient energy.

In a past that is yet to come,
a Draugr and a Selkie met
on a rocky beach
in the far north of the world.

They walked a while,
and fell in love (as happens);

and had a daughter,
a lovely and lonely girl,
the only one of her kind:
        a Draukie.

Half-bright Selkie,
        shape-shifter;
half-dark Draugr,
        with ancient power.

This Draukie, she had
pearlescent skin
that glowed in the sea dark.

She had steel strong fins,
that whipped up the waves.

a creature of the underworld,
like her father;

a creature of sea energy,
like her mother.

She swam fast and fierce,
slicing the sea into spume, and
the waves into wild, high water.

She swam alone, though,
always singing
long, sad songs.

One day, on the sea,
there was an Icelandic fishing boat
full of scientists.

They heard her on their
hydrophone,
saw her power on their
wave oscilloscope.

They wanted that power,
for the world
has need of sea energy.

They followed her song.
And they waited,
ready, with a big hook.

The Draukie swam close,
within her wave storm
(the boat tossed, and
almost turned over).

Closer she swam,
and then . . .
there was a hook about her neck—
almost pulling it clean off.

Thump,
she went, like a whale
on the deck.

They kept her in the ship's hold,
in a perspex tank.

The scientists prodded and probed:
processed data, and
tried to determine how
she made her wave power.

But the Draukie was still,
silent.
Her wave power,
            gone.

Then one young scientist,
a rebellious type,
sat and sang to the Draukie,
on her tea breaks.

The Draukie listened,
sang back, and soon
they were friends.

The Draukie told
our young scientist
her deepest, true name:

Överflöd,
Overflowing,
Generosity,

she sang
(for her name slid
from shore to shore).

I am Overflowing,
Generosity,
my power cannot be caught,

only changed
or shape-shifted.

The young scientist took a stand,
and set her friend free.
And the Draukie flew on her fins,
to a far away sea.

Then our young scientist
persuaded her peers,
not to imprison her friend,
but to tame and transform.

We must build big devices,
she said, to sing and to call
to our wave energy Draukie.

Scientists built new machines:
wings in the waves, that
beat sounds under sea;

machines to call a Draukie,
swimming at full storm.

At times scientists see her,
as they still try to tame
that overflowing power,

turn Generosity
into electricity.

And, when you walk on a beach,
near a wave power machine,

you can, if you look,
find bits of bright skin
shed from
that one lonely Draukie,

like shells.
Keep them.

Acknowledgments

With thanks to Per Ebert, Valdimar Halldórsson, Jan Krogh, Brit Ross Winthereik, James Maguire, and the Alien Energy research project at the IT University of Copenhagen. This poem was inspired by my ongoing ethnographic fieldwork around the marine energy industry in the Orkney Islands, Scotland. It was originally composed and performed as an oral story. This subsequent version was written and first published as part of the poetry collection *ebban an' flowan*, a poetic primer for marine renewable energy, with Alec Finlay and Alistair Peebles (Edinburgh: Morning Star, 2015).

))) A Quake in Being

Timothy Morton

In *The Ecological Thought* I coined the term *hyperobjects* to refer to things that are massively distributed in time and space relative to humans.[1] A hyperobject could be a black hole. A hyperobject could be the Lago Agrio oil field in Ecuador, or the Florida Everglades. A hyperobject could be the biosphere, or the solar system. A hyperobject could be the sum total of all the nuclear materials on Earth; or just the plutonium, or the uranium. A hyperobject could be the very long-lasting product of direct human manufacture, such as Styrofoam or plastic bags, or the sum of all the whirring machinery of capitalism. Hyperobjects, then, are "hyper" in relation to some other entity, whether they are directly manufactured by humans or not.

Hyperobjects have numerous properties in common. They are *viscous*, which means that they "stick" to beings that are involved with them. They are *nonlocal*; in other words, any "local manifestation" of a hyperobject is not directly the hyperobject.[2] They involve profoundly different temporalities than the human-scale ones we are used to. In particular, some very large hyperobjects, such as planets, have genuinely *Gaussian* temporality: they generate spacetime vortices, due to general relativity. Hyperobjects occupy a high-dimensional phase space that results in their being invisible to humans for stretches of time. And they exhibit their effects *interobjectively*; that is, they can be detected in a space that consists of interrelationships between aesthetic properties of objects. The hyperobject is not a function of our knowledge: it's *hyper* relative to worms, lemons, and ultraviolet rays, as well as humans.

Hyperobjects have already had a significant impact on human social and psychic space. Hyperobjects are directly responsible for what I call *the end of the world*, rendering both denialism and apocalyptic environmentalism obsolete. Hyperob-

Hyperobjects: Philosophy and Ecology after the End of the World (Minneapolis: University of Minnesota Press, 2013), 1–24

jects have already ushered in a new human phase of *hypocrisy, weakness,* and *lameness*: these terms have a very specific resonance in this study, and I shall explore them in depth. *Hypocrisy* results from the conditions of the impossibility of a metalanguage (and as I shall explain, we are now freshly aware of these conditions because of the ecological emergency); *weakness* from the gap between phenomenon and thing, which the hyperobject makes disturbingly visible; and *lameness* from the fact that all entities are fragile (as a condition of possibility for their existence), and hyperobjects make this fragility conspicuous.[3] Hyperobjects are also changing human art and experience (the aesthetic dimension). We are now in what I call *the Age of Asymmetry.*

Hyperobjects are not just collections, systems, or assemblages of other objects. They are objects in their own right, objects in a special sense that I shall elucidate as we proceed through this book. The special sense of *object* derives from *object-oriented ontology* (OOO), an emerging philosophical movement committed to a unique form of realism and nonanthropocentric thinking. Least of all, then, would it be right to say that hyperobjects are figments of the (human) imagination, whether we think imagination as a bundling of associations in the style of Hume, or as the possibility for synthetic judgments a priori, with Kant. Hyperobjects are real whether or not someone is thinking of them. Indeed, for reasons given in this study, hyperobjects end the possibility of transcendental leaps "outside" physical reality. Hyperobjects force us to acknowledge the immanence of thinking to the physical. But this does not mean that we are "embedded" in a "lifeworld."

Hyperobjects thus present philosophy with a difficult, double task. The first task is to abolish the idea of the possibility of a metalanguage that could account for things while remaining uncontaminated by them. For reasons I shall explore, poststructuralist thinking has failed to do this in some respects, or rather, it didn't complete the job. The second task is to establish what phenomenological "experience" is in the absence of anything meaningfully like a "world" at all: hence the subtitle, "Philosophy and Ecology after the End of the World."

* * *

Throughout *Hyperobjects* I frequently write in a style that the reader may find "personal"—sometimes provocatively or frustratingly so. This decision to write somewhat "personally" was influenced by Alphonso Lingis's risky and rewarding phenomenology. It seems appropriate. I am one of the entities caught in the hyperobject I here call *global warming*; one of the entities I know quite well. And as an object-oriented ontologist I hold that all entities (including "myself") are shy, retiring octopuses that squirt out a dissembling ink as they withdraw into the ontological shadows. Thus, no discourse is truly "objective," if that means that it is a master language that sits "meta" to what it is talking about. There is also a necessarily iterative, circling style of thought in this book. This is because one only sees pieces of a hyperobject at any one moment. Thinking them is intrinsically tricky.

This line of reasoning makes me seem like a postmodernist, though for reasons that will become clear, the emerging ecological age gets the idea that "there is no metalanguage" much more powerfully and nakedly than postmodernism ever did.[4] Since for postmodernism "everything is a metaphor" in some strong sense, all metaphors are equally bad. But since for me, and indeed for all humans as we transition into the Age of Asymmetry there are real things for sure, just not as we know them or knew them, so some metaphors are better than others.[5] Yet because there is nowhere to stand outside of things altogether, it turns out that we know the truth of "there is no metalanguage" more deeply than its inventors. The globalizing sureness with which "there is no metalanguage" and "everything is a metaphor" are spoken in postmodernism means that postmodernism is nothing like what it takes itself to be, and is indeed just another version of the (white, Western, male) historical project. The ultimate goal of this project, it seems, was to set up a weird transit lounge outside of history in which the characters and technologies and ideas of the ages mill around in a state of mild, semiblissful confusion.

Slowly, however, we discovered that the transit lounge was built on Earth, which is different from saying that it was part of Nature. (Throughout this book, I capitalize Nature precisely to "denature" it, as one would do to a protein by cooking it.) "The actual Earth," as Thoreau puts it, now contains throughout its circumference a thin layer of radioactive materials, deposited since 1945.[6] The deposition of this layer marks a decisive geological moment in the Anthropocene, a geological time marked by the decisive human "terraforming" of Earth as such.[7] The first significant marks were laid down in 1784, when carbon from coal-fired industries began to be deposited worldwide, including in the Arctic, thanks to the invention of the steam engine by James Watt. The birth of the steam engine, an all-purpose machine whose all-purpose quality (as noted in its patent) was precisely what precipitated the industrial age, was an event whose significance was not lost on Marx.[8] This universal machine (uncanny harbinger of the computer, an even more general machine) could be connected to vast assemblages of other machines to supply their motive power, thus giving rise to the assemblages of assemblages that turn the industrial age into a weird cybernetic system, a primitive artificial intelligence of a sort—to wit, industrial capitalism, with the vampire-like downward causality of the emergent machine level, with its related machine-like qualities of abstract value, sucking away at the humans on the levels beneath. After 1945 there began the Great Acceleration, in which the geological transformation of Earth by humans increased by vivid orders of magnitude.

Yet like everyone else until about a decade ago, Marx missed the even bigger picture. Think about it: a geological time (vast, almost unthinkable), juxtaposed in one word with very specific, immediate things—1784, soot, 1945, Hiroshima, Nagasaki, plutonium. This is not only a historical age but also a geological one. Or better: we are no longer able to think history as exclusively human, for the very reason that we are in the Anthropocene. A strange name indeed, since in this period nonhumans make decisive contact with humans, even the ones busy shoring up differences between humans and the rest.

The thinking style (and thus the writing style) that this turn of events necessitates is one in which the normal certainties are inverted, or even dissolved. No longer are my intimate impressions "personal" in the sense that they are "merely mine" or "subjective only": they are footprints of hyperobjects, distorted as they always must be by the entity in which they make their mark—that is, me. I become (and so do you) a litmus test of the time of hyperobjects. I am scooped out from the inside. My situatedness and the rhetoric of situatedness in this case is not a place of defensive self-certainty but precisely its opposite.[9] That is, situatedness is now a very uncanny place to be, like being the protagonist of a Wordsworth poem or a character in *Blade Runner*. I am unable to go beyond what I have elsewhere called *ecomimesis*, the (often) first-person rendering of situatedness "in."[10] This is not to endorse ecomimesis, but to recognize that there is no outside, no metalanguage. At every turn, however, the reader will discover that the prose in this book sways somewhat sickeningly between phenomenological narrative and scientific reason. Yet just as I am hollowed out by the hyperobject, so by the very same token the language of science is deprived of its ideological status as cool impersonality. The more we know about hyperobjects, the stranger they become. Thus hyperobjects embody a truth about what I once thought only applied to life-forms, the truth of the *strange stranger*.[11]

What this book seeks then is a weird ecomimesis that tugs at the limits of the rhetorical mode, seeking out its hypocrisy. For reasons I give later, the term *hypocrisy* is very carefully chosen. *The time of hyperobjects is a time of hypocrisy*. Yet, for the same reasons, seeking out hypocrisy cannot be done from the point of view of cynicism. If there is no metalanguage, then cynical distance, the dominant ideological mode of the left, is in very bad shape, and will not be able to cope with the time of hyperobjects.

* * *

Hyperobjects are what have brought about the end of the world. Clearly, planet Earth has not exploded. But the concept *world* is no longer operational, and hyperobjects are what brought about its demise. The idea of the end of the world is very active in environmentalism. Yet I argue that this idea is not effective, since, to all intents and purposes, the being that we are supposed to feel anxiety about and care for is gone. This does not mean that there is no hope for ecological politics and ethics. Far from it. Indeed, as I shall argue, the strongly held belief that the world is about to end "unless we act now" is paradoxically one of the most powerful factors that inhibit a full engagement with our ecological coexistence here on Earth. The strategy of this book, then, is to awaken us from the dream that the world is about to end, because action on Earth (the real Earth) depends on it.

The end of the world has already occurred. We can be uncannily precise about the date on which the world ended. Convenience is not readily associated with historiography, nor indeed with geological time. But in this case, it is uncannily clear. It was April 1784, when James Watt patented the steam engine, an act that commenced the depositing of carbon in Earth's crust—namely, the inception of humanity as a geophysical force on a planetary scale. Since for something to happen it often needs to happen twice, the world also ended in 1945, in Trinity, New Mexico, where the Manhattan Project tested the Gadget, the first of the atom bombs, and later that year when two nuclear bombs were dropped on Hiroshima and Nagasaki. These events mark the logarithmic increase in the actions of humans as a geophysical force.[12] They are of "world-historical" importance for humans— and indeed for any life-form within range of the fallout—demarcating a geological period, the largest-scale terrestrial era. I put "world-historical" in quotation marks because it is indeed the fate of the concept *world* that is at issue. For what comes into view for humans at this moment is precisely the end of the world, brought about by the encroachment of hyperobjects, one of which is assuredly Earth itself, and its geological cycles demand a *geophilosophy* that doesn't think simply in terms of human events and human significance.

The end of the world is correlated with the Anthropocene, its global warming and subsequent drastic climate change, whose precise scope remains uncertain while its reality is verified beyond question. Throughout *Hyperobjects* I shall be calling it *global warming* and not *climate change*. Why? Whatever the scientific and social reasons for the predominance of the term *climate change* over *global warming* for naming this particular hyperobject, the effect in social and political discourse is plain enough. There has been a decrease in appropriate levels of concern. Indeed, denialism is able to claim that using the term *climate change* is merely the rebranding of a fabrication, nay evidence of this fabrication in flagrante delicto. On the terrain of media and the sociopolitical realm, the phrase *climate change* has been such a failure that one is tempted to see the term itself as a kind of denial, a reaction to the radical trauma of unprecedented global warming. That the terms are presented as choices rather than as a package is a symptom of this failure, since logically it is correct to say "climate change as a result of global warming," where "climate change" is just a compression of a more detailed phrase, a metonymy.

If this is not the case, then *climate change* as a *substitute* for *global warming* is like "cultural change" as a substitute for *Renaissance*, or "change in living conditions" as a substitute for *Holocaust*. *Climate change* as substitute enables cynical reason (both right wing and left) to say that the "climate has always been changing," which to my ears sounds like using "people have always been killing one another" as a fatuous reason not to control the sale of machine guns. What we desperately need is an appropriate level of shock and anxiety concerning a specific ecological trauma—indeed, *the* ecological trauma of our age, the very thing that defines the

Anthropocene as such. This is why I shall be sticking with the phrase *global warming* in this book.

Numerous philosophical approaches have recently arisen as if in response to the daunting, indeed horrifying, coincidence of human history and terrestrial geology. *Speculative realism* is the umbrella term for a movement that comprises such scholars as Graham Harman, Jane Bennett, Quentin Meillassoux, Patricia Clough, Iain Hamilton Grant, Levi Bryant, Ian Bogost, Steven Shaviro, Reza Negarestani, Ray Brassier, and an emerging host of others such as Ben Woodard and Paul Ennis. All are determined to break the spell that descended on philosophy since the Romantic period. The spell is known as *correlationism*, the notion that philosophy can only talk within a narrow bandwidth, restricted to the human–world correlate: meaning is only possible between a human mind and what it thinks, its "objects," flimsy and tenuous as they are. The problem as correlationism sees it is, is the light on in the fridge when you close the door?

It's not quite idealism, but it could tend that way. But the problem goes back further than the Romantic period, all the way back to the beginning of the modern period. (Unlike Latour, I do believe that we have "been modern," and that this has had effects on human and nonhuman beings.)[13] The restriction of philosophy's bandwidth attempts to resolve a conundrum that has been obsessing European thinking since at least the uncritical inheritance by Descartes of the scholastic view of substances—that they are basic lumps decorated with accidents.[14] Despite his revolutionary rationalism—brilliantly deriving reality from his confidence in his (doubting) mental faculties—Descartes uncritically imported the very scholasticism his work undermined, imported it into the area that mattered most, the area of ontology. Since then, even to say the word *ontology* has been to say something with a whiff of scholasticism about it. Epistemology gradually took over: How can I know that there are (or are not) real things? What gives me (or denies me) access to the real? What defines the possibility of access? The possibility of possibility? These thoughts even affect those who strove against the trend, such as Schelling and Heidegger, and the original phenomenologists, whose slogan was "To the things themselves!" Speculating outside of the human became a minor trend, exemplified by the marginalization of Alfred North Whitehead, who, thanks to speculative realism, has been enjoying a recent resurgence.

Speculative realism has a healthy impulse to break free of the correlationist circle, the small island of meaning to which philosophy has confined itself. It is as if, since the seventeenth century, thinking has been cowed by science. Yet science not only cries out for "interpretation"—and heaven knows some defenses of the humanities these days go as far as to argue that science needs the humanities for PR purposes. Beyond this, science doesn't necessarily know what it is about. For a neo-Darwinist, reality is mechanisms and algorithmic procedures. For a quantum physicist, things might be very different. Reality might indeed entail a form of correlationism: the Copenhagen Interpretation is just that. Or everything is made of mind.[15] So what is it? Which is it? Asleep at the switch, philosophy has allowed the default ontology to persist: there are things, which are basically featureless lumps, and these things have accidental properties, like cupcakes decorated with colored sprinkles.

This thinking—or the lack thereof—is not unrelated to the eventual manufacture, testing, and dropping of Little Boy and Fat Man. Epistemological panic is not unrelated to a sclerotic syndrome of "burying the world in nullity . . . in order to prove it."[16] This thinking still continues, despite the fact that thought has already made it irrelevant. The thinking reaches the more than merely paradoxical idea that if I can evaporate it in an atomic energy flash, it must be real. The thinking is acted out daily in drilling, and now "fracking" for oil. The year 1900 or thereabouts witnessed a number of "prequels" to the realization of the Anthropocene and the coming of the Great Acceleration. These prequels occurred within human thinking itself, but it is only in retrospect that humans can fully appreciate them. Quantum theory, relativity theory, and phenomenology were all born then. Quantum theory blew a huge hole in the idea of particles as little Ping-Pong balls. Relativity theory destroyed the idea of consistent objects: things that are identical with themselves and constantly present all the way down. Extreme forms of realism in narrative began to set streams of consciousness free from the people who were having them, and the hand-holding benevolent narrator vanished. Monet began to allow colors and brushstrokes to liberate themselves from specific forms, and the water in which the water lilies floated, exhibited on the curving walls of the Orangerie, became the true subject of his painting. Expressionism abolished the comforting aesthetic distances of Romanticism, causing disturbing, ugly beings to crowd towards the viewer.

What did the "discoveries of 1900" have in common? Water, quanta, spacetime began to be seen. They were autonomous entities that had all kinds of strange, unexpected properties. Even consciousness itself was no longer just a neutral medium: phenomenology made good on the major philosophical discovery of the Romantic period, the fact of consciousness that "has" a content of some kind.[17] Monet had started painting water lilies; or rather, he had started to paint the space in which water lilies float; or rather, he had started to paint the rippling, reflective object in which the lilies float—the water. Just as Einstein discovered a rippling, flowing spacetime, where previously objects had just floated in a void, Monet discovered the sensuous spaciousness of the canvas itself, just as later Tarkovsky was to discover the sensuous material of film stock. All this had been prefigured in the Romantic period with the development of blank verse narratives, meandering through autobiographical detours. Suddenly a whole lot more paper was involved.

Around 1900 Edmund Husserl discovered something strange about objects.

No matter how many times you turned around a coin, you never saw the other side as the other side. The coin had a dark side that was seemingly irreducible. This irreducibility could easily apply to the ways in which another object, say a speck of dust, interacted with the coin. If you thought this through a little more, you saw that all objects were in some sense irreducibly withdrawn. Yet this made no sense, since we encounter them every waking moment. And this strange dark side applied equally to the "intentional objects" commonly known as thoughts, a weird confirmation of the Kantian gap between phenomenon and thing. Kant's own example of this gap is highly appropriate for a study of hyperobjects. Consider raindrops: you can feel them on your head—but you can't perceive the actual raindrop in itself.[18] You only ever perceive your particular, anthropomorphic translation of the raindrops. Isn't this similar to the rift between weather, which I can feel falling on my head, and global climate, not the older idea of local patterns of weather, but the entire system? I can think and compute climate in this sense, but I can't directly see or touch it. The gap between phenomenon and thing yawns open, disturbing my sense of presence and being in the world. But it is worse still than even that. Raindrops are raindroppy, not gumdroppy—more's the pity. Yet raindrop phenomena are not raindrop things. I cannot locate the gap between phenomenon and thing anywhere in my given, phenomenal, experiential, or indeed scientific space. Unfortunately raindrops don't come with little dotted lines on them and a little drawing of scissors saying "cut here"—despite the insistence of philosophy from Plato up until Hume and Kant that there is some kind of dotted line somewhere on a thing, and that the job of a philosopher is to locate this dotted line and cut carefully. Because they so massively outscale us, hyperobjects have magnified this weirdness of things for our inspection: things are themselves, but we can't point to them directly.

Around 1900 Einstein discovered something strange about objects. The speed of light was constant, and this meant that objects couldn't be thought of as rigid, extended bodies that maintained their shape. Lorentz had noticed that electromagnetic waves shrank mysteriously, as if foreshortened, as they approached light speed. By the time you reach the end of a pencil with your following eyes, the other end has tapered off somewhere. If you put tiny clocks on your eyelids, they would tell a different time than the tiny clocks on your feet lying still beneath the table as you twirl the pencil in your fingers, the tiny clocks in each fingernail registering ever so slightly different times. Of course you wouldn't see this very clearly, but if you were moving close to the speed of light, objects would appear to become translucent and strangely compressed until they finally disappeared altogether. Spacetime appeared, rippling and curved like Monet's water lilies paintings. And there must then be regions of spacetime that are unavailable to my perception, even though they are thinkable: another strange confirmation of the Kantian gap between phenomenon and thing.

Around 1900 Max Planck discovered something strange about objects. If you tried to measure the energy in an enclosed object (like an oven) by summing all the waves, you reached absurd results that rocketed toward infinity above a certain temperature range: the blackbody radiation problem. But if you thought of the energy as distributed into packets, encapsulated in discrete quanta, you got the right result. This accuracy was bought at the terrible price of realizing the existence of a bizarre quantum world in which objects appeared to be smeared into one another, occupying indeterminate areas and capable of penetrating through seemingly solid walls. And this is yet another confirmation of the phenomenon-thing gap opened up by Kant, for the simple reason that to measure a quantum, you must fire some other quanta at it—to measure is to deflect, so that position and momentum are not measurable at the same time.

The Kantian gap between phenomenon and thing places the idea of substances decorated with accidents under extreme pressure. Drawing on the breakthroughs of the phenomenologist Husserl, Heidegger perhaps came closest to solving the problem. Heidegger realized that the cupcakes of substance and the sprinkles of accidence were products of an "objective presencing" that resulted from a confusion within (human) being, or *Dasein*, as he put it. Heidegger, however, is a correlationist who asserts that without Dasein, it makes no sense whatsoever to talk of the truth of things, which for him implies their very existence: "Only as long as Dasein *is*, 'is there' [*gibt es*] being . . . it can neither be said that beings *are*, nor that they are not."[19] How much more correlationist do you want? The refrigerator itself, let alone the light inside it, only exists when I am there to open the door. This isn't quite Berkeleyan *esse est percipi*, but it comes close. Heidegger is the one who from within correlationism descends to a magnificent depth. Yet he is unwilling to step outside the human–world correlation, and so for him idealism, not realism, holds the key to philosophy: "If the term idealism amounts to an understanding of the fact that being is never explicable by beings, but is always already the 'transcendental' for every being, then the sole correct possibility for a philosophical problematic lies in idealism."[20] Heidegger had his own confusion, not the least of which is exemplified by his brush with Nazism, which is intimately related to his insight and blindness about being. Graham Harman, to whose object-oriented ontology I subscribe, discovered a gigantic coral reef of sparkling things beneath the Heideggerian U-boat. The U-boat was already traveling at a profound ontological depth, and any serious attempt to break through in philosophy must traverse these depths, or risk being stuck in the cupcake aisle of the ontological supermarket.

Harman achieved this discovery in two ways. The first way is simple flexibility. Harman was simply ready to drop the specialness of Dasein, its unique applicability to the human, in particular to German humans. This readiness is itself a symptom of the ecological era into which we have entered, the time of hyperobjects. To this effect, Harman was unwilling to concede Heidegger the point that the physical reality described in Newton's laws did not exist before Newton.[21] This line of Heidegger's thought is even more correlationist than Kant's. The second way in which Harman attacked the problem was by a thorough reading of the startling tool-analysis in the opening sections of Heidegger's *Being and Time*. This reading demonstrates that nothing in the "later" Heidegger, its plangency notwithstanding, topples the tool-analysis from the apex of Heidegger's thinking. Heidegger, in other words, was not quite conscious of the astonishing implications of the discovery he made in the tool-analysis: that when equipment—which for all intents and purposes could be anything at all—is functioning, or "executing" (*Vollzug*), it withdraws from access (*Entzug*); that it is only when a tool is broken that it seems to become present-at-hand (*vorhanden*). This can only mean, argues Harman, that there is a vast plenum of unique entities, one of whose essential properties is *withdrawal*—no other entity can fully account for them. These entities must exist in a relatively *flat ontology* in which there is hardly any difference between a person and a pincushion. And relationships between them, including causal ones, must be *vicarious* and hence *aesthetic* in nature.

If we are to take seriously the ontological difference between being and beings, argues Harman, then what this means is twofold:

(1) No realism is tenable that only bases its findings on "ontic" data that are pregiven. This would be like thinking with prepackaged concepts—it would not be like thinking at all.

(2) Idealism, however, is unworkable, since there exist real things whose core reality is withdrawn from access, even by themselves.

Point (1), incidentally, is the trouble with science. Despite the refreshing and necessary skepticism and ruthless doubt of science, scientific discoveries are necessarily based on a decision about what real things are.[22] Point (2) is the primary assertion of OOO, Harman's coral reef beneath the Heideggerian U-boat.

It will become increasingly clear as this book proceeds that hyperobjects are not simply mental (or otherwise ideal) constructs, but are real entities whose primordial reality is withdrawn from humans. Hyperobjects give us a platform for thinking what Harman calls *objects* in general. This introduction is not quite the right moment for a full explication of OOO. Outlining OOO might mean that we never got around to hyperobjects themselves. And, more significantly, the subtlety of OOO itself requires a thorough examination of hyperobjects. Moreover, it seems like good practice to start with the things at hand and feel our way forward—in this I join Lingis. Yet I trust that by the end of the book the reader will have a reasonable grasp of how one might use this powerful new philosophical approach for finding out real things about real things.

So, let's begin to think about hyperobjects in some depth. What is the most striking thing about their appearance in the human world? Naturally humans have been aware of enormous entities—some real, some imagined—for as long as they have existed. But this book is arguing that there is something quite special about the recently discovered entities, such as climate. These entities cause us to reflect on our very place on Earth and in the cosmos. Perhaps this is the most fundamental issue—hyperobjects seem to force something on us, something that affects some core ideas of what it means to exist, what Earth is, what society is.

What is special about hyperobjects? There's no doubt that cosmic phenomena such as meteors and blood-red Moons, tsunamis, tornadoes, and earthquakes have terrified humans in the past. Meteors and comets were known as *disasters*. Literally, a disaster is a fallen, dysfunctional, or dangerous, or evil, star (*dis-astron*). But such disasters take place against a stable background in at least two senses. There is the Ptolemaic–Aristotelian machinery of the spheres, which hold the fixed stars in place. This system was common to Christian, Muslim, and Jewish cosmology in the Middle Ages. To be a disaster, a "star" such as a meteor must deviate from this harmonious arrangement or celestial machinery. Meanwhile, back on Earth, the shooting star is a portent that makes sense as a trace on the relatively stable horizon of earth and sky. Perhaps the apocalypse will happen. But not just yet. Likewise, other cultures seemed to have relatively coherent ways of explaining catastrophes. In Japanese Shinto, a tsunami is the vengeance of a *Kami* who has been angered in some way.

It seems as if there is something about hyperobjects that is more deeply challenging than these "disasters." They are entities that become visible through post-Humean statistical causality—a causality that is actually *better* for realism than simply positing the existence of glass spheres on which the fixed stars rotate, to give one example. This point never fails to be lost on global warming deniers, who assert, rightly, that one can never directly prove the human causes of global warming, just as I never prove that this bullet you fire into my head will kill me. But the extreme statistical likelihood of anthropogenic global warming is better than simply asserting a causal factoid. Global warming denial is also in denial about what causality is after Hume and Kant—namely a feature of phenomena, rather than things in themselves.

What does this mean to nascent ecological awareness? It means that humans are not totally in charge of assigning significance and value to events that can be statistically measured. The worry is not whether the world will end, as in the old model of the *dis-astron*, but whether the end of the world is already happening, or whether perhaps *it might already have taken place*. A deep shuddering of temporality occurs. Furthermore, hyperobjects seem to continue what Sigmund Freud considered the great humiliation of the human following Copernicus and Darwin. Jacques Derrida rightly adds Freud to the list of humiliators—after all he displaces the human from the very center of psychic activity. But we might also add Marx, who displaces human social life with economic organization. And we could add

Heidegger and Derrida himself, who in related though subtly different ways displace the human from the center of meaning-making. We might further expand the list by bringing in Nietzsche and his lineage, which now runs through Deleuze and Guattari to Brassier: "Who gave us the sponge to wipe away the entire horizon?" (Nietzsche).[23] And in a different vein, we might add that OOO radically displaces the human by insisting that my being is not everything it's cracked up to be—or rather that the being of a paper cup is as profound as mine.

Is it that hyperobjects seem to push this work of humiliation to a yet more extreme limit? What is this limit? Copernicus, it is said, is all about displacement. This was first taken to mean an exhilarating jump into cognitive hyperspace. But what if the hyperobjects force us to forget even this exit strategy? What if hyperobjects finally force us to realize the truth of the word *humiliation* itself, which means being brought low, being brought down to earth? Hyperobjects, in effect, seem to push us into a double displacement. For now the possibility that we have loosed the shackles of the earthly to touch the face of the "human form divine" (Blake) seems like a wish fulfillment.[24] According to hyperobjects themselves, who seem to act a little bit like the gigantic boot at the end of the *Monty Python* credits, outer space is a figment of our imagination: *we are always inside an object*.

What we have then, before and up to the time of hyperobjects from the sixteenth century on, is the truth of Copernicanism, if we can call it that—there is no center and we don't inhabit it. Yet added to this is another twist: there is no edge! We can't jump out of the universe. Queen Mab can't take Ianthe out of her bed, put her in a spaceship, and whisk her to the edge of time to see everything perfectly (Percy Shelley's fantasy). Synthetic judgments a priori are made inside an object, not in some transcendental sphere of pure freedom. Quentin Meillassoux describes Kant's self-described Copernican turn a Ptolemaic counterrevolution, shutting knowing up in the finitude of the correlation between (human) subject and world.[25] But for me, it is the idea of a privileged transcendental sphere that constitutes the problem, not the finitude of the human–world correlation. Kant imagines that although we are limited in this way, our transcendental faculties are at least metaphorically floating in space beyond the edge of the universe, an argument to which Meillassoux himself cleaves in his assertion that reality is finally knowable exclusively by (human) subjectivity. And *that* is the problem, the problem called anthropocentrism.

It is Kant who shows, at the very inception of the Anthropocene, that things never coincide with their phenomena. All we need to do is extend this revolutionary insight beyond the human–world gap. Unlike Meillassoux, we are not going to try to bust through human finitude, but to place that finitude in a universe of trillions of finitudes, as many as there are things—because a thing just is a rift between what it is and how it appears, for any entity whatsoever, not simply for that special entity called the (human) subject. What ecological thought must do, then, is unground the human by forcing it back onto the ground, which is to say, standing on a gigantic object called *Earth* inside a gigantic entity called *biosphere*. This grounding of Kant began in 1900. Phenomenology per se is what begins to bring Kantianism down to Earth, but it's hyperobjects and OOO that really convince me that it's impossible to escape the gravitational field of "sincerity," "ingenuousness," being-there.[26] Not because there is a *there*—we have already let go of that. Here I must part company with ecophenomenology, which insists on regressing to fantasies of embeddedness. No: we are not in the center of the universe, but we are not in the VIP box beyond the edge, either. To say the least, this is a profoundly disturbing realization. It is the true content of ecological awareness. Harman puts it this way:

> On the one hand, scientism insists that human consciousness is nothing special, and should be naturalized just like everything else. On the other hand, it also wants to preserve knowledge as a special kind of relation to the world quite different from the relations that raindrops and lizards have to the world. . . . For all their gloating over the fact that people are pieces of matter just like everything else, they also want to claim that the very status of that utterance is somehow special. For them, raindrops know nothing and lizards know very little, and some humans are more knowledgeable than others. This is only possible because thought is given a unique ability to negate and transcend immediate experience, which inanimate matter is never allowed to do in such theories, of course. In short, for all its *noir* claims that the human doesn't exist, it elevates the structure of human *thought* to the ontological pinnacle.[27]

The effect of this double denial of human supremacy is not unlike one of Hitchcock's signature cinematic techniques, the pull focus. By simultaneously zooming and pulling away, we appear to be in the same place, yet the place seems to distort beyond our control. The two contradictory motions don't cancel one another out. Rather, they reestablish the way we experience "here." The double denial doesn't do away with human experience. Rather, it drastically modifies it in a dizzying manner.

The ecological thought that thinks hyperobjects is not one in which individuals are embedded in a nebulous overarching system, or conversely, one in which something vaster than individuals extrudes itself into the temporary shapes of individuals. Hyperobjects provoke *irreductionist* thinking, that is, they present us with scalar dilemmas in which ontotheological statements about which thing is the most real (ecosystem, world, environment, or conversely, individual) become impossible.[28] Likewise, irony qua absolute distance also becomes inoperative. Rather than a vertiginous antirealist abyss, irony presents us with intimacy with existing nonhumans.

The discovery of hyperobjects and OOO are symptoms of a fundamental shaking of being, a *being-quake*. The ground of being is shaken. There we were, trolling

along in the age of industry, capitalism, and technology, and all of a sudden we received information from aliens, information that even the most hardheaded could
not ignore, because the form in which the information was delivered was precisely
the instrumental and mathematical formulas of modernity itself. The Titanic of
modernity hits the iceberg of hyperobjects. The problem of hyperobjects, I argue,
is not a problem that modernity can solve. Unlike Latour then, although I share
many of his basic philosophical concerns, I believe that we *have* been modern, and
that we are only just learning how not to be.

Because modernity banks on certain forms of ontology and epistemology to
secure its coordinates, the iceberg of hyperobjects thrusts a genuine and profound
philosophical problem into view. It is to address these problems head-on that this
book exists. This book is part of the apparatus of the Titanic, but one that has decided to dash itself against the hyperobject. This rogue machinery—call it speculative realism, or OOO—has decided to crash the machine, in the name of a social
and cognitive configuration to come, whose outlines are only faintly visible in the
Arctic mist of hyperobjects. In this respect, hyperobjects have done us a favor.
Reality itself intervenes on the side of objects that from the prevalent modern
point of view—an emulsion of blank nothingness and tiny particles—are decidedly medium-sized. It turns out that these medium-sized objects are fascinating,
horrifying, and powerful.

For one thing, we are inside them, like Jonah in the Whale. This means that
every decision we make is in some sense related to hyperobjects. These decisions
are not limited to sentences in texts about hyperobjects. When I turn the key in
the ignition of my car, I am relating to global warming. When a novelist writes
about emigration to Mars, he is relating to global warming. Yet my turning of the
key in the ignition is intimately related to philosophical and ideological decisions
stemming from the mathematization of knowing and the view of space and time
as flat, universal containers (Descartes, Newton). The reason why I am turning
my key—the reason why the key turn sends a signal to the fuel injection system,
which starts the motor—is one result of a series of decisions about objects, motion, space, and time. Ontology, then, is a vital and contested political terrain. It
is on this terrain that this study will concentrate a significant amount of attention.
In the menacing shadow of hyperobjects, contemporary decisions to ground ethics
and politics in somewhat hastily cobbled together forms of process thinking and
relationism might not simply be rash—they might be part of the problem.

The "towering-through" (Heidegger) of the hyperobject into the misty tran-
scendentalism of modernity interrupts the supposed "progress" that thinking has
been making toward assimilating the entire universe to a late capitalism-friendly
version of *Macbeth*, in which (in the phrase Marx quotes) "all that is solid melts
into air."[29] For at the very point at which the melting into air occurs, we catch the
first glimpses of the all-too-solid iceberg within the mist.

* * *

I doubt gravely whether capitalism is entirely up for the job of processing hyper-objects. I have argued elsewhere that since the raw machinery of capitalism is reactive rather than proactive, it might contain a flaw that makes it unable to address the ecological emergency fully.[30] Capitalism builds on existing objects such as "raw materials" (whatever comes in at the factory door). The retroactive style of capitalism is reflected in the ideology of "the consumer" and its "demands" that capital then "meets."

The ship of modernity is equipped with powerful lasers and nuclear weapons. But these very devices set off chain reactions that generate yet more hyperobjects that thrust themselves between us and the extrapolated, predicted future. Science itself becomes the emergency break that brings the adventure of modernity to a shuddering halt. But this halt is not in front of the iceberg. *The halting is (an aspect of) the iceberg.* The fury of the engines is precisely how they cease to function, seized up by the ice that is already inside them. The future, a time "after the end of the world," has arrived too early.

Hyperobjects are a good candidate for what Heidegger calls "the last god," or what the poet Hölderlin calls "the saving power" that grows alongside the danger-ous power.[31] We were perhaps expecting an eschatological solution from the sky, or a revolution in consciousness—or, indeed, a people's army seizing control of the state. What we got instead came too soon for us to anticipate it. Hyperobjects have dispensed with 200 years of careful correlationist calibration. The panic and denial and right-wing absurdity about global warming are understandable. Hyper-objects pose numerous threats to individualism, nationalism, anti-intellectualism, racism, speciesism, anthropocentrism, you name it. Possibly even capitalism itself.

Notes

1. Timothy Morton, *The Ecological Thought* (Cambridge, MA: Harvard University Press, 2010), 130–35.

2. *Local manifestation* is philosopher Levi Bryant's term for the appearance of an object. See *The Democracy of Objects* (Ann Arbor, MI: Open Humanities Press, 2011), 15.

3. In some sense, the idea of *weakness* is an expansion of Vattimo's proposal for a weak thinking that accepts the human–world gap, and which moves through nihilism. Gianni Vattimo, *The Transparent Society*, trans. David Webb (Baltimore: Johns Hopkins University Press, 1994), 117, 119.

4. Jacques Lacan, *Écrits: A Selection*, trans. Alan Sheridan (London: Tavistock, 1977), 311.

5. I derive this line of thinking from Graham Harman, *Guerrilla Metaphysics: Phenome-nology and the Carpentry of Things* (Chicago: Open Court, 2005), 101–2.

6. Henry David Thoreau, *The Maine Woods*, ed. Joseph J. Moldenhauer (Princeton, NJ: Princeton University Press, 2004), 71.

7. The atmospheric chemist Paul Crutzen devised the term *Anthropocene*. Paul Crutzen

and E. Stoermer, "The Anthropocene," *Global Change Newsletter* 41, no. 1 (2000): 17–18; Paul Crutzen, "Geology of Mankind," *Nature* 415 (January 3, 2002): 23, doi:10.1038/415023a.

8. Karl Marx, *Capital*, 3 vols., trans. Ben Fowkes (Harmondsworth: Penguin, 1990), 1:499.

9. The decisive study of situatedness is David Simpson, *Situatedness; or, Why We Keep Saying Where We're Coming From* (Durham, NC: Duke University Press, 2002), 20.

10. Timothy Morton, *Ecology without Nature: Rethinking Environmental Aesthetics* (Cambridge, MA: Harvard University Press, 2007), 33.

11. Jacques Derrida, "Hostipitality," trans. Barry Stocker with Forbes Matlock, *Angelaki* 5, no. 3 (December 2000): 3–18; Morton, *Ecological Thought*, 14–15, 17–19, 38–50.

12. See Trinity Atomic Website, http://www.abomb1.org.

13. Bruno Latour, *We Have Never Been Modern*, trans. Catherine Porter (Cambridge, MA: Harvard University Press, 2002).

14. Martin Heidegger, *Being and Time*, trans. Joan Stambaugh (Albany: State University of New York Press, 1996), 83–85.

15. Arthur Eddington, *The Nature of the Physical World* (New York: Macmillan, 1928), 276.

16. Heidegger, *Being and Time*, 191.

17. See David Simpson, "Romanticism, Criticism, and Theory," in *The Cambridge Companion to British Romanticism*, ed. Stuart Curran (Cambridge: Cambridge University Press, 1993), 10.

18. Immanuel Kant, *Critique of Pure Reason*, trans. Norman Kemp Smith (Boston: St. Martin's, 1965), 84–85.

19. Heidegger, *Being and Time*, 196.

20. Heidegger, *Being and Time*, 193.

21. Heidegger, *Being and Time*, 208.

22. Martin Heidegger, *Phenomenological Interpretations of Aristotle*, trans. Richard Rojcwicz (Bloomington: Indiana University Press, 2001), 23.

23. Friedrich Nietzsche, *The Gay Science*, trans. and ed. Walter Kaufmann (New York: Vintage, 1974), 125.

24. William Blake, "The Divine Image," in *The Complete Poetry and Prose of William Blake*, ed. D. V. Erdman (New York: Doubleday, 1988).

25. Quentin Meillassoux, *After Finitude: An Essay on the Necessity of Contingency*, trans. Ray Brassier (London: Continuum, 2010), 119–21.

26. José Ortega y Gasset, *Phenomenology and Art*, trans. Philip W. Silver (New York: Norton, 1975), 63–70; Harman, *Guerrilla Metaphysics*, 39, 40, 135–43, 247.

27. Graham Harman, "Critical Animal with a Fun Little Post," *Object-Oriented Philosophy* (blog), October 17, 2011, http://doctorzamalek2.wordpress.com/2011/10/17/critical-animal-with-a-fun-little-post/.

28. The term *irreduction* is derived from the work of Bruno Latour and Graham Harman. Graham Harman, *Prince of Networks: Bruno Latour and Metaphysics* (Melbourne: Re. press, 2009), 12.

29. Karl Marx, *The Communist Manifesto*, in *Selected Writings*, ed. David McLellan (Ox-

ford: Oxford University Press, 1977), 12; William Shakespeare, *Macbeth* (New York: Washington Square Press, 1992), 19.

30. Morton, *Ecological Thought*, 121.

31. Martin Heidegger, *Contributions to Philosophy: (From Enowning)*, trans. Parvis Emad and Kenneth Maly (Bloomington: Indiana University Press, 1999), 283–93. See also Joan Stambaugh, *The Finitude of Being* (Albany: State University of New York Press, 1992), 139–44.

⟫ Notes toward a Post-carbon Philosophy: "It's the Economy, Stupid"

Martin McQuillan

> One more try to save the discourse of a "world" that we no longer speak, or that we still speak, sometimes all the more garrulously, as in an emigrant colony.
>
> —Jacques Derrida, "Economies of the Crisis"

Post-carbon Philosophy

This is an essay about philosophy, its task and object, after the end of a carbon economy. This topic is not wild science fiction; according to most estimates, we (collectively as a planet) have reached or surpassed the moment of peak oil (the point at which world reserves begin to dwindle). Effectively the post-carbon epoch has already begun, since it is now a task of the critical imagination to envisage a world beyond the fractal distillation of petroleum. The task of thinking a post-carbon world also revolves around the difficulty of thinking "a world" at this moment, that is to say, a world made global by what we have been taught to term mondelization or more strictly mondelatinization (a globalization based on Western hegemony and privilege).[1] The place of philosophy, as a Western model of thought, is no doubt vexed in this situation, given that all of the terms of globalization such as "economy," "law," "sovereignty," "world," and so on are all philosophical terms and are all metaphysical through and through. However, given where this task of thinking is beginning from and the resources it has to hand, we will have to pick up the philosophical heritage that confronts us and come to terms with the obvious and inevitable difficulty of becoming part of the history of the object that one wishes to describe through this philosophical vocabulary. If it

Telemorphosis: Theory in the Era of Climate Change, vol. 1, ed. Tom Cohen (Ann Arbor, MI: Open Humanities Press, 2012), 270–92

is true that we are entering an epoch of new materialities for which we as yet have no descriptive framework, then philosophy must respond to this situation. The question of matter after all is also a philosophical concept. The empirical and all empiricisms are, as Derrida notes as early as "Violence and Metaphysics," philosophical gestures that embed themselves within the history of philosophy. His reading of Levinas in this essay is to suggest the ways in which Levinas demonstrates that all empiricism is metaphysical, and a constant philosophical thematization "of the infinite exteriority of the other." Levinas in contrast understands the empirical not as a positivism but as an experience of difference and of the other. "Empiricism," claims Derrida, "always has been determined by philosophy, from Plato to Husserl as nonphilosophy: as the philosophical pretention to nonphilosophy" (152). That is as philosophy's way of affecting to speak in a nonphilosophical way. However, nothing can more profoundly conjure the need for philosophy than this denial of philosophy by philosophy. Within the metaphysical schema that is nonphilosophy, the irruption of the wholly other solicits philosophy (i.e., the logos) as its own origin, end, and other. There is no escape from philosophy as far as empiricism is concerned; there will only ever be a thinking about the empirical that is philosophical. It is this radicalization of empiricism that deconstruction proposes as a breathless, inspiring journey for philosophy in the later years of the twentieth century; as Derrida states in the opening paragraphs of the essay on Levinas, it is the closure of philosophy by nonphilosophy that gives thought a future: "It may even be that these questions are not philosophical, are not philosophy's questions. Nevertheless, these should be the only questions today capable of founding the community, within the world, of those who are still called philosophers; and called such in remembrance, at very least, of the fact that these questions must be examined unrelentingly" (79). So, the question of the materiality of a post-carbon economy may not be a question that philosophy has the resources to answer but which must nevertheless be thought about and so determined in a philosophical manner. It may be the case that long before peak oil we reached the point of peak philosophy, with Hegel or Marx, Nietzsche, or Heidegger, and what remains is a post-philosophical speculative economy, which in some necessary way sits on considerable philosophical and metaphysical reserves. This thought will be of use to us in the coming pages; just as I am at pains in this opening paragraph to point out that one cannot seek to swap climate change denial for the denial of philosophy.

It will then be difficult to imagine the new materialities of the age of climate change without philosophy, and in fact the persistent theme of matter in the discourse on environmentalism will undoubtedly compel us towards philosophy. It may be the case that philosophy will tell us that all and every environmentalism is a metaphysics grounded on an unquestioned empiricism based upon an unsustainable distinction between nature and culture. The task of a deconstruction of the question of the environmental then might be a rethinking of the experience of the environment, and the environment as experience, as an encounter with an irreducible presence and perception of a phenomenality that is also an experience of the other, the wholly other, and of difference. It is this wholly other, the other that separates Derrida from Levinas, that must be attended to as the new materiality of an epoch of climate change and post-carbon economy. However, in this essay I will not be attending to the environmental effects as such of carbon. In a certain sense it is nonsense to speak of a post-carbon materiality since even if hydrocarbon fuels were to be outlawed tomorrow, carbon and its allotropes would remain, as they were in Plato's time, the fourth most abundant element in the universe. Therefore, whenever I speak of a "post-carbon philosophy" the question in hand does not strictly concern the depletion of a material itself. Rather, the task for thinking relates to the sort of world, being in the world, and thought concerning the world that an economy and culture based on the exploitation of hydrocarbons has given rise to, and what its prospects might be as this economy and culture inevitably weans itself off of petroleum and onto some other alternative energy source, while living with the inheritance of a century of intensive hydrocarbon use. In the end, the culture and economy of post-carbon modernity might not look that different from the one that we occupy today. Undoubtedly, carbon-fuelled capital will not easily give up its privileges in favour of so-called sustainable living, and will seek to replace the risks of a carbon-based economy with those of a nuclear-based economy. The new materialities and bio-diversity challenges of the post-carbon age may quickly begin to look exactly the same as those of the present moment, the here and now proposing its own future. The shape of things to come is not the object of our speculation here. Rather, in this text I would like to consider the question of speculation itself.

Philosophy will not name an alternative energy source, and this is a question that philosophy cannot answer and may not be a philosophical question. Philosophy, on the other hand, offers a model of crisis. It is at its most eloquent when talking about limits, ends and telos, and about the inability of theory or the humanities or the human sciences to ground themselves in institutions and actions, and concerning the incommensurability between what we still call "politics," "ethics," "economy," and the global mutations today and the deconstruction of those mutations. If we are to understand what is most singular about our present time, it will emerge from this philosophical reflection. However, the challenge for philosophy is not to draw down upon this template but to name and situate the most acute moment of the here and now as a crisis that distinguishes it from all previous crises. It is in understanding how the present crisis (of the globalization of neoliberalism, climate change, peak oil, and bio-diversity extinction) differs from all other crises (in the history of Western colonialism, say) that philosophy might begin to think philosophically about this moment. Equally, there may well be no universal experience of this moment that would enable philosophy to act in a philosophical manner with respect to it. Given the diverse experiences of what it might mean for different parts of the world to be in this situation, there may not be either a common horizon or discourse capable of offering an assured competence to frame and explain the crisis. In this sense, the philosophical concept of crisis that holds in the tradition after Valéry and Husserl would be faced with an inability to phenomenalize and ontologize a determinable universal experience of the crisis. Philosophy then sits on the cusp between being drawn to the crisis as the only means of determining it and having its own constitutive divisibility demonstrated by its inability to address the crisis through the redundancy of its own model of crisis.[2]

It may be the case that the crisis of peak oil and simultaneous irreversible climate change is only a concentration and reiteration of previous liminal cases in the history of the exploitation of planetary resources by the West. For example, kerosene derived from coal and petroleum replaced whale oil as the source of illumination in America and Europe. The whaling industry had been in decline for a decade due to diminishing sperm whale populations and the destruction of the Northern whaling fleet during the Civil War, when hydrocarbon illuminants were introduced out of necessity and innovation, leading to gaslight homes and cities and a new phase of development in modernity. However, to say that our present crisis today is not unique does not mean that it is not singular. The task for thinking about the present crisis might be to understand the idea of the crisis today, the understanding that this is a crisis, that it corresponds to an idea or model of a crisis, provided for us by, say, philosophy, and is theatricalized in contemporary discourse as such. The status of the idea that the world is in crisis as doxa might be one of the defining features of this crisis, the very thing that makes it like no other moment. The singularity of our present crisis might be defined by the resources spent in the mediatisation of the idea that the world is in crisis, by the competing political interests of the day (the advocates, the activists, the sceptics, the deniers, and the lobbyists) through all the channels of contemporary communication. On the one hand, within such rhetorical exchanges, the naming or denial of a crisis always serves a political interest. On the other hand, to identify an event as a crisis is always to ontologize it and to submit it to the model of the crisis that would explain it and domesticate it. Perhaps, we might say that today is not a crisis at all but rather the latest instance of a long history of planetary exploitation by capital, this instance being no more critical than any other in a long history. The naming of a crisis in the present works to mask that history and to neutralize it, giving it form and therefore a program and calculability.

For either side of the present debate on climate change, say, to name only one of the threats to planetary life today, to name this process and event as a crisis would be to appropriate it for the present and for a metaphysics of presence. In giving the event of climate change a form and a certain calculability one has begun to neutralize the effects of its unknowable future and to erase the experience of alterity at the heart of an encounter with the wholly other. To name it as a crisis is to subject it to the temporality of "the crisis," namely that it will one day come to an end and a state of normativity will be restored. One side of the debate would say that "normality" (whatever normal might mean on a planet that has weathered at least five major ice ages) can be resumed by cutting carbon emissions and introducing "sustainable" energy consumption. The other side says the present is in fact normal and no change in climate is in process. Either way, each side depends upon the idea of a normative climate derived from the idea that a change in climatic conditions would constitute a crisis for the human race. A crisis that would no longer allow mankind to run a system of resource exploitation that has sustained its development for the last 200 years. In other words, what is at stake in the climate change debate is the very future of a world economy and the normative, or ideal, conditions for its operation. That is to say, the debate is predicated on an essentially conservative notion of how to sustain the ideal or normative. In fact, the event and singularity of climate change is constituted in the concept that it is already irreversible. The singularity of climate change as a crisis might be that it is not subject to the temporality of the crisis and that it might be a crisis without resolution and so demonstrate itself not to be a crisis at all but a constant state. In this sense crisis becomes a permanent condition or at least the resolution of this crisis is the construction of a new idea of the normative. In this way, climate change must change the very idea of a crisis by which we seek to determine it. At the same time, climate change becomes part of the latest chapter in the history of the idea of crisis and continues to be appropriated by it and subsumed to the model it undermines.

On the issue of irreversible climate change and peak oil, I find myself to be surprisingly sceptical. Given that hydrocarbon consumption is already in decline and without fresh initiatives (although there are many mountain tops to be dynamited and much oil shale to be extracted yet) even the most optimistic forecasts suggest that petroleum consumption will have dwindled to almost nothing in my lifetime, why should we then be overly concerned by targets for climate emissions set for 50 years into the future? I do not doubt the damage that will be caused to, say, the polar ice caps during those 50 years and the resultant biodiversity loss. However, such targets seem to me to be aimed at "eking out" the remains of carbon emission rather than addressing the more fundamental underlying causes of climate change. This "eking out" is also a transition phase for global capital as it acquires a new carbonless fuel. In fact, while it may take a million years for the planet to correct the damage inflicted in the next fifty, surely this realignment of the environment will take place in the absence of continued fossil fuel consumption. That is to say, that irreversible climate change is not necessarily irreversible, just that it is irreversible in the lifetime of global capital that as a planetary life-form is something of a Johnny-come-lately. In this way, the discourse on climate change on all sides retains the vestiges of an irreducible humanism and the paro-chialism of Western metaphysics. It would not be enough, it seems to me, for phi-losophy to accept the discourse and axiomatics of climate change advocates, even if it were possible to accept the science without reservation. Rather, this discourse in its present state is open to question, is fragile and perfectible, or even decon-structable (as Derrida reminds us of the abolitionist discourse on the death pen-alty[3]) because it more often than not limits the idea of planetary life to the present conditions of a Western-led globalization and so inadvertently positions climate change as the latest accident to befall the Western subject. In this way, climate change is the latest phase of the crisis of European humanity. Accordingly, it calls for a response from Western humanity, one that will require the response of sci-ence, philosophy, and the human sciences (including economics). However, as an encounter with the wholly other of planetary life beyond the limits of Western humanism, the event of climate change will transform, mutate, and challenge the protocols of European humanity's intellectual apparatus. Climate change accord-ingly is a challenge to reason and so to philosophy as the custodian of Western reason. If one were to question or attempt to exceed this as the locus of the present discourse of crisis, one might quickly find oneself in a position where terms like "crisis" and "irreversible" were no longer appropriate.

As soon as we have determined a moment of crisis, with the temporality of a crisis and the calculability of a crisis, then we have entered the realm of economy. The response to a crisis is always economic in every sense of that word. Just as we bring a crisis "in house" by naming it as such, one must always ask what is the quickest and most efficient way of bringing a crisis to an end. Equally, in the final analysis, we are often told a planetary crisis will be a question of economics. How-ever, just as one must question reason as the axiom of crisis, then we must ques-tion whether economics can continue to stand as a strictly determinable region of competence in the face of such a crisis. One would have to ask if economics can offer a competent realm of judgment, decision, and will to respond to the de-mands of the incalculable. It is not that economics does not have a response to the incalculable, but rather that a response directed here cannot be purely or strictly economical. This is where philosophy makes a return in the economy of crisis management.

There is a clear and legible connection between what the oil industry calls "speculation" and the speculative philosophical enterprise. I would like to suggest that speculating for oil has been the basis of industrial modernity and the Western economy for the last 200 years (whether that oil was derived from the exploitation of hydrocarbons or whales is a moot point, even though the use of oil as a fuel dates, according to Herodotus, to Babylon 4,000 years ago). As that which fuels the engine of the economy, that which makes the economy as such possible, the search for oil is an investment in a venture with the hope of gain but with the risk of loss. The speculative structure of oil exploitation follows from and is now itself the basis for the structure of all investment in stock, property, and the fictional products of capital today. As with speculative philosophy, it involves the conjecture or theorization of a future event in the absence of the firm evidence of a present. It is this speculative structure that opens the future as a thinking of risk and as thinking as risk, that closely ties philosophy to the carbon economy. The question of the post-carbon economy is therefore clearly a matter of concern for philosophy. I have been careful to speak here from the beginning of a "post-carbon economy" (although we might also say "post-carbon economies" since there will undoubtedly be more than one). While oil is a material thing, the stuff itself as it were, the price of oil is a question of economy, a matter of relation, and is irreducibly conceptual. My concern in this text is to think through the prospects for an economy and culture (if that is not a tautology) based upon the pricing of hydrocarbons as fuel. This question may come down to one that concerns the future of speculation itself, as much as it does a speculation on the future. I would like to offer a hypothesis here, even a speculation, apropos of today, that will need to be put at risk and tested, namely that whenever we speak of so-called "environmental catastrophe" the business of "financial crisis" is never far behind. In fact, a strong formulation of the hypothesis might be that the two are intimately connected, and that their relation always follows from a structure of speculation. In this sense, philosophy is what is put at risk and must put itself at risk by the conjuncture of the two.

Oil Reserves

The Western-led global economy is predicated on the value of oil. At the end of the Second World War, the Bretton Woods monetary conferences created the World Bank, the International Monetary Fund, and established a new international monetary system of competitive disinflation relative to the US dollar by tying the gold standard to the dollar. This system worked well for a while and enabled the postwar reconstruction of western Europe and Japan, and put in place the inaugural openings of the General Agreement on Tariffs and Trade which form the basis of the global economy today. However, the Bretton Woods Gold Exchange system began to break down in the mid-sixties under the pressures of the excessive costs of the Vietnam War and the resurgence of the European and south Asian manufacturing base as exporters of global trade. By November 1967 as the value of the dollar began to look increasingly precarious, withdrawals of gold bullion from the US Treasury were becoming excessive (as national governments, notably De Gaulle's France, redeemed their dollar holdings against bullion stocks as agreed at Bretton Woods). When sterling was devalued in 1967, Bretton Woods was doomed;

the dollar came under further pressure to devalue itself against gold in order to protect the value of Federal reserves against redemption of dollars by foreign governments. The plot of Ian Fleming's 1959 *Goldfinger* (adapted as a film in 1964) can only be understood in relation to the Bretton Woods concords. The threat to render the Federal gold reserve inaccessible after exploding a radioactive bomb would have meant national governments could only redeem their reserve dollars against Ulrich Goldfinger's own gold reserve (said to be the second largest in the world), thus pushing up the value of his gold and mortgaging the US economy to Goldfinger's personal enterprises. Consequently, James Bond, agent for the sick man of Europe, temporarily rescues not only the post-war, Cold War status quo of the Bretton Woods agreement but secures the future of Western leadership in a global economy. Rather than risk the depletion of the US gold reserve and the collapse of the US credit rating, in 1971 President Nixon abandoned the dollar–gold link, dissolving the terms of Bretton Woods in favour of a system of free-floating currencies, with the international banks and the markets determining the value of the dollar. This combined with the ongoing costs of Vietnam led to inflationary pressures and wage-price freezes in the American national economy. OPEC discussed the value of pricing oil in several currencies to spread the risks of the volatile American domestic scene. However, in 1974, Nixon moved to do a deal with Saudi Arabia only to price oil in dollars when the Saudis, unknown to the rest of OPEC and the US's Western allies, secretly purchased $2.5 billion in US Treasury bills with their surplus oil funds (equivalent to 70% of all Saudi assets), once more ensuring the position of the dollar as the international reserve currency and initiating the phase of American global hegemony based on petrodollar recycling, swapping the gold standard for the standard of oil, so-called "black gold." As shocks in the price of oil followed political and supply instabilities, so the need for national governments to acquire US dollars ensued. On the one hand, dollars flow into oil-supplying countries far beyond the needs of domestic investment. These surplus dollars are stored in banks in New York and London to retain their value as dollars. On the other hand, oil-importing countries need to buy dollars to meet the rising price of oil. In the 1970s developing nations in Africa borrowed dollars from international banks sitting on surplus petrol dollars, creating debts to be repaid entirely in dollars and at the then-high levels of interest rates based on inflationary pressures in the Western national economies. In this way, the emergent, post-colonial African continent was impoverished and a cycle of crippling debt initiated. The IMF set up by Bretton Woods was used to enforce debt repayment to the international banks through the implementation of "austerity" programmes that also opened up developing countries to Western private companies. As surplus petrol dollars flow in from OPEC countries and are leant out again as Eurodollar bonds or loans, the Federal Reserve is in a unique position with respect to creating credit and expanding the money supply.[4] It is this situation that enables the United States to sustain improbable budget deficits and latterly allowed the bailout of Wall Street after the 2007 banking crisis. However, in a post–Cold War, multi-polar era, after peak oil, the position of the dollar as world reserve currency is once more in question. Previous attempts to rebalance the world economy now have genuine impetus from the need to address the imbalance of current accounts between the US and China, Chinese currency manipulation, and the emergence of the Euro as an alternative or additional reserve currency. Further, the cost in blood and treasure of physically securing the oil supply through military means may prove unsustainable for the US (the cost of bailing out the banks, $660 billion, was the same as the US annual military budget for 2007). In a post-carbon economy, it might not be the best position to be sitting on a mountain of petrodollars.

There is then a great deal at stake in the question of a post-carbon economy, beyond the actual "irreversible" damage done to the earth's climate by carbon emissions. Oil is presently the essential fuel of the global economy and oil trades are the basic enablers of manufacturing infrastructure in every industrial nation, of global transportation, and the primary energy source for 40% of the industrial economy. Oil trades in dollars have been the basis for American economic, cultural, and military hegemony since the 1970s, and the liquidity that ensures the development of the Western-led global economy. A post-carbon economy presents a considerable challenge to the present geopolitical dispensation and, coterminus to this, the current conditions of capital. It is for these reasons that we might say that the response to climate change as a staging post in the ongoing crisis of European humanity revolves around more than a purely scientific solution that will bring the crisis to a dénouement to enable a return to Western-capital normativity, and everything that depends upon it.

Modern as the phenomenon might be and while philosophy has a great deal to say about "energy," for example, if I might be allowed to paraphrase one of Derrida's more familiar hyperboles: no philosopher as a philosopher has ever taken seriously the question of oil. Oil and carbon emission has a massive readability today and may define the most acute moment of the paroxysm that makes the present crisis like no other. This is not to say that there have not previously been bouts of financial uncertainty and environmental disasters precipitated by oil. In fact, the history of oil production might be nothing other than a chain of such instances. Rather, the most decisive index of the present moment is the toxic combination of climate change caused by carbon emission, the urgency for global capital of the risks of peak oil, and the central role played by oil trades in the global economy. We might go so far as to say on this later point that the entire practice of the Western economy, that is, the so-called global economy, depends upon oil. That is to say, that while the idea of the world market and of the "free exchange" of goods has a philosophical heritage running through early modern humanism and Enlightenment thought, our present understanding of all exchange, debt, and faith runs through oil. To speak of a post-carbon economy might in fact be to say something quite radical, given that our present situation is so intensively related to the price of oil. To think an industrialized economy without the price of oil may on the one hand simply be a question of swapping one transcendental signifier for another, as gold was replaced by oil, so oil might be replaced by a trade in plutonium recycling. On the other hand, an opportunity exists here to understand economy as an experience of difference and as an encounter with the wholly other. This would require an other understanding of economy, one that was not dedicated to the utilization of wealth (what we now call a "restricted economy") but one in which we began to understand the complexities of a sovereign economic term such as gold or oil, not in its loss of meaning but in relation to its possible loss of meaning (what Derrida, after Bataille, after Hegel calls a "general economy").[5] In this sense, a "post-carbon economy" presents an opportunity for a consideration of economy not to be limited to the circulation of strictly commercial values, the meaning and established value of objects such as gold, oil, and plutonium or so-called "carbon swaps." Rather than a phenomenology of values as a restricted economy, we might begin to understand what exceeds the production, consumption, and destruction of value within the circuit of exchange. What Bataille might call "energy" beyond the energy of oil. This would not be a reserve of meaning within economy but an aneconomic writing of economy that is legible because its concepts move outside of the symmetrical exchanges from which they are identified and which according to a certain logic of recuperation they continue to occupy. This task of paleonymy as deconstruction is not one that philosophy will undertake on its own but one that will be played out in the irreversible mutations that take place in the global economy as a consequence of climate change, one which philosophy, opened by the materialism of nonphilosophy, will merely be at the forefront of reporting. It returns us to a familiar problem with which we began: having exhausted the oil reserve and the language of philosophy, the unfinished project of Modernity must continue to inscribe within its frames and language of intelligibility (i.e., philosophy) that which nevertheless exceeds the oppositions of concepts governed by its doxical logic. It is not that nineteenth- and twentieth-century thought is incapable of responding to the new crisis of climate change but that climate change is a product of such thought as its latest episode and challenge. On the other hand, such a reading of economy seeks to understand or think what is unthinkable for philosophy, its economic blind spot.

* * *

Post-carbon Coda

This text has begun to open up a number of questions for thinking the present and future of a "European" culture and economy based on the consumption of oil. It has identified the interrelation of the present economic and environmental crises and suggested a role for philosophy in framing our understanding of these events. It has also proposed a history of oil pricing that demonstrates that its supposedly robust materiality is conceptual through and through. Finally, it has considered the relation of carbon consumption to the dematerialization of money and to faith, credit, debt, and literature (the later pointing towards alternative forms of exchange that disrupt established economies). At each of these steps, an attempt has been made to submit the experience of oil to a thinking of difference and to understanding the risk of climate change as an encounter with a wholly other (planetary life itself) that is also unlocking the reserves of difference within the human. It has proposed a hypothesis for future consideration: the inextricable link between financial crisis and environmental damage. The testing and demonstration of this hypothesis will require a larger work of comparison and analysis that will involve an expanded deconstruction of political economy, political theology, and the unavoidable question of what theory today calls bio-power. This speculation will require time and will give time, the time of a speculation proposed against redemption at a future date, in a race to keep the lights on as one writes in the shadow of oil depletion, a waning telos for a resource that will have been and which will never come again. It is a task of the carbonic economy that will be complicit with the very thing it provokes, ensuring its guilt of the exact thing it believes itself to be innocent of. Perhaps, there will be no thought on carbon without carbon. Perhaps philosophy since Plato will have only ever been a carbon-based form and the end of philosophy will have been the end of carbon. There will always be a price to be paid for such speculations with all hands to the pump in an epoch of what Tom Cohen calls "irreversible critical climate change."

Notes

1. See Jacques Derrida's "Faith and Knowledge: Two Sources of Religion at 'the Limits of Reason Alone'" in *Acts of Religion*.

2. Many of my comments here reprise Derrida's reflection on the idea of crisis in the text "Economies of the Crisis."

3. For example, see Derrida and Roudinescou 152.

4. For a fuller, if partisan, discussion of the petrodollar see Clark.

5. See Derrida's "From Restricted to General Economy: A Hegelianism without Reserve."

Works Cited

Clark, William R. *Petrodollar Warfare: Oil, Iraq, and the Future of the Dollar.* Gabriola Island: New Society Publishers, 2005.

Derrida, Jacques. *Acts of Religion.* Edited by Gil Anidjar. New York: Routledge, 2002.

———. "Economies of the Crisis." In *Negotiations: Interventions and Interviews, 1971–2001*, trans. Elizabeth Rottenberg. Stanford: Stanford University Press, 2002.

———. "From Restricted to General Economy: A Hegelianism without Reserve." In *Writing and Difference*, trans. Alan Bass, 251–77. London: Routledge, 1978.

———. "Violence and Metaphysics: An Essay on the Thought of Emmanuel Levinas." In *Writing and Difference*, trans. Alan Bass, 79–153. London: Routledge, 1978.

Derrida, Jacques, and Elizabeth Roudinescou. *For What Tomorrow: A Dialogue*. Translated by Jeff Fort. Stanford: Stanford University Press, 2004.

Fleming, Ian. *Goldfinger*. London: Jonathan Cape, 1959.

》》》 Learning How to Die in the Anthropocene

Roy Scranton

I.

Driving into Iraq just after the 2003 invasion felt like driving into the future. We convoyed all day, all night, past Army checkpoints and burned-out tanks, till in the blue dawn Baghdad rose from the desert like a vision of hell: Flames licked the bruised sky from the tops of refinery towers, cyclopean monuments bulged and leaned against the horizon, broken overpasses swooped and fell over ruined suburbs, bombed factories, and narrow ancient streets.

With "shock and awe," our military had unleashed the end of the world on a city of six million—a city about the same size as Houston or Washington. The infrastructure was totaled: water, power, traffic, markets, and security fell to anarchy and local rule. The city's secular middle class was disappearing, squeezed out between gangsters, profiteers, fundamentalists, and soldiers. The government was going down, walls were going up, tribal lines were being drawn, and brutal hierarchies savagely established.

I was a private in the United States Army. This strange, precarious world was my new home. If I survived.

Two and a half years later, safe and lazy back in Fort Sill, Oklahoma, I thought I had made it out. Then I watched on television as Hurricane Katrina hit New Orleans. This time it was the weather that brought shock and awe, but I saw the same chaos and urban collapse I'd seen in Baghdad, the same failure of planning and the same tide of anarchy. The 82nd Airborne hit the ground, took over strategic points, and patrolled streets now under de facto martial law. My unit was put on alert to prepare for riot control operations. The grim future I'd seen in Baghdad was coming home: not terrorism, not even WMDs, but a civilization in collapse, with a crippled infrastructure, unable to recuperate from shocks to its system.

And today, with recovery still going on more than a year after Sandy and many

New York Times, November 10, 2013; Opinionator Blog: The Stone, http://opinionator.blogs .nytimes.com/2013/11/10/learning-how-to-die-in-the-anthropocene/?_r=0

critics arguing that the Eastern seaboard is no more prepared for a huge weather event than we were last November, it's clear that future's not going away.

[In] March [2013], Admiral Samuel J. Locklear III, the commander of the United States Pacific Command, told security and foreign policy specialists in Cambridge, Massachusetts, that global climate change was the greatest threat the United States faced—more dangerous than terrorism, Chinese hackers, and North Korean nuclear missiles. Upheaval from increased temperatures, rising seas, and radical destabilization "is probably the most likely thing that is going to happen . . ." he said, "that will cripple the security environment, probably more likely than the other scenarios we all often talk about."

Locklear's not alone. Tom Donilon, the national security advisor, said much the same thing in April 2013, speaking to an audience at Columbia's new Center on Global Energy Policy. James Clapper, director of national intelligence, told the Senate in March that "extreme weather events (floods, droughts, heat waves) will increasingly disrupt food and energy markets, exacerbating state weakness, forcing human migrations, and triggering riots, civil disobedience, and vandalism."

On the civilian side, the World Bank's recent report, "Turn Down the Heat: Climate Extremes, Regional Impacts, and the Case for Resilience," offers a dire prognosis for the effects of global warming, which climatologists now predict will raise global temperatures by 3.6 degrees Fahrenheit within a generation and 7.2 degrees Fahrenheit within 90 years. Projections from researchers at the University of Hawaii find us dealing with "historically unprecedented" climates as soon as 2047. The climate scientist James Hansen, formerly with NASA, has argued that we face an "apocalyptic" future. This grim view is seconded by researchers worldwide, including Anders Levermann, Paul and Anne Ehrlich, Lonnie Thompson, and many, many, many others.

This chorus of Jeremiahs predicts a radically transformed global climate forcing widespread upheaval—not possibly, not potentially, but *inevitably*. We have passed the point of no return. From the point of view of policy experts, climate scientists, and national security officials, the question is no longer whether global warming exists or how we might stop it, but how we are going to deal with it.

II.

There's a word for this new era we live in: the Anthropocene. This term, taken up by geologists, pondered by intellectuals, and discussed in the pages of publications such as the *Economist* and the *New York Times*, represents the idea that we have entered a new epoch in Earth's geological history, one characterized by the arrival of the human species as a geological force. The biologist Eugene F. Stoermer and the Nobel Prize–winning chemist Paul Crutzen advanced the term in 2000, and it has steadily gained acceptance as evidence has increasingly mounted that the changes wrought by global warming will affect not just the world's climate and biological diversity, but its very geology—and not just for a few centuries, but for millenniums. The geophysicist David Archer's 2009 book, "The Long Thaw: How Humans Are Changing the Next 100,000 Years of Earth's Climate," lays out a clear and concise argument for how huge concentrations of carbon dioxide in the atmosphere and melting ice will radically transform the planet, beyond freak storms and warmer summers, beyond any foreseeable future.

The Stratigraphy Commission of the Geological Society of London—the scientists responsible for pinning the "golden spikes" that demarcate geological epochs such as the Pliocene, Pleistocene, and Holocene—have adopted the Anthropocene as a term deserving further consideration, "significant on the scale of Earth history." Working groups are discussing what level of geological timescale it might be (an "epoch" like the Holocene, or merely an "age" like the Calabrian), and at what date we might say it began. The beginning of the Great Acceleration, in the middle of the twentieth century? The beginning of the Industrial Revolution, around 1800? The advent of agriculture?

The challenge the Anthropocene poses is a challenge not just to national security, to food and energy markets, or to our "way of life"—though these challenges are all real, profound, and inescapable. The greatest challenge the Anthropocene poses may be to our sense of what it means to be human. Within 100 years—within three to five generations—we will face average temperatures 7 degrees Fahrenheit higher than today, rising seas at least 3 to 10 feet higher, and worldwide shifts in crop belts, growing seasons, and population centers. Within a thousand years, unless we stop emitting greenhouse gases wholesale right now, humans will be living in a climate the Earth hasn't seen since the Pliocene, three million years ago, when oceans were 75 *feet* higher than they are today. We face the imminent collapse of the agricultural, shipping, and energy networks upon which the global economy depends, a large-scale die-off in the biosphere that's already well on its way, and our own possible extinction. If *Homo sapiens* (or some genetically modified variant) survives the next millenniums, it will be survival in a world unrecognizably different from the one we have inhabited.

Geological timescales, civilizational collapse, and species extinction give rise to profound problems that humanities scholars and academic philosophers, with their taste for fine-grained analysis, esoteric debates, and archival marginalia, might seem remarkably ill suited to address. After all, how will thinking about Kant help us trap carbon dioxide? Can arguments between object-oriented ontology and historical materialism protect honeybees from colony collapse disorder? Are ancient Greek philosophers, medieval theologians, and contemporary metaphysicians going to keep Bangladesh from being inundated by rising oceans?

Of course not. But the biggest problems the Anthropocene poses are precisely those that have always been at the root of humanistic and philosophical questioning: "What does it mean to be human?" and "What does it mean to live?" In the epoch of the Anthropocene, the question of individual mortality—"What does *my life* mean in the face of death?"—is universalized and framed in scales that boggle the imagination. What does human existence mean against 100,000 years of climate change? What does one life mean in the face of species death or the collapse of global civilization? How do we make meaningful choices in the shadow of our inevitable end?

These questions have no logical or empirical answers. They are philosophical problems *par excellence*. Many thinkers, including Cicero, Montaigne, Karl Jaspers, and The Stone's own Simon Critchley, have argued that studying philosophy is learning how to die. If that's true, then we have entered humanity's most philosophical age—for this is precisely the problem of the Anthropocene. The rub is that now we have to learn how to die not as individuals, but as a civilization.

III.

Learning how to die isn't easy. In Iraq, at the beginning, I was terrified by the idea. Baghdad seemed incredibly dangerous, even though statistically I was pretty safe. We got shot at and mortared, and IEDs laced every highway, but I had good armor, we had a great medic, and we were part of the most powerful military the world had ever seen. The odds were good I would come home. Maybe wounded, but probably alive. Every day I went out on mission, though, I looked down the barrel of the future and saw a dark, empty hole.

"For the soldier death is the future, the future his profession assigns him," wrote Simone Weil in her remarkable meditation on war, "The Iliad or the Poem of Force." "Yet the idea of man's having death for a future is abhorrent to nature. Once the experience of war makes visible the possibility of death that lies locked up in each moment, our thoughts cannot travel from one day to the next without meeting death's face." That was the face I saw in the mirror, and its gaze nearly paralyzed me.

I found my way forward through an eighteenth-century Samurai manual, Yamamoto Tsunetomo's "Hagakure," which commanded: "Meditation on inevitable death should be performed daily." Instead of fearing my end, I owned it. Every morning, after doing maintenance on my Humvee, I'd imagine getting blown up by an IED, shot by a sniper, burned to death, run over by a tank, torn apart by dogs, captured and beheaded, and succumbing to dysentery. Then, before we rolled out through the gate, I'd tell myself that I didn't need to worry, because I was already dead. The only thing that mattered was that I did my best to make sure everyone else came back alive. "If by setting one's heart right every morning and evening, one is able to live as though his body were already dead," wrote Tsunetomo, "he gains freedom in the Way."

I got through my tour in Iraq one day at a time, meditating each morning on my inevitable end. When I left Iraq and came back stateside, I thought I'd left that future behind. Then I saw it come home in the chaos that was unleashed after Katrina hit New Orleans. And then I saw it again when Sandy battered New York and New Jersey: government agencies failed to move quickly enough, and volunteer groups like Team Rubicon had to step in to manage disaster relief.

Now, when I look into our future—into the Anthropocene—I see water rising up to wash out lower Manhattan. I see food riots, hurricanes, and climate refugees. I see 82nd Airborne soldiers shooting looters. I see grid failure, wrecked harbors, Fukushima waste, and plagues. I see Baghdad. I see the Rockaways. I see a strange, precarious world.

Our new home.

The human psyche naturally rebels against the idea of its end. Likewise, civilizations have throughout history marched blindly toward disaster, because humans are wired to believe that tomorrow will be much like today—it is unnatural for us to think that this way of life, this present moment, this order of things is not stable and permanent. Across the world today, our actions testify to our belief that we can go on like this forever, burning oil, poisoning the seas, killing off other species, pumping carbon into the air, ignoring the ominous silence of our coal mine canaries in favor of the unending robotic tweets of our new digital imaginarium. Yet the reality of global climate change is going to keep intruding on our fantasies of perpetual growth, permanent innovation, and endless energy, just as the reality of mortality shocks our casual faith in permanence.

The biggest problem climate change poses isn't how the Department of Defense should plan for resource wars, or how we should put up sea walls to protect Alphabet City, or when we should evacuate Hoboken. It won't be addressed by buying a Prius, signing a treaty, or turning off the air-conditioning. The biggest problem we face is a philosophical one: understanding that this civilization is *already dead*. The sooner we confront this problem, and the sooner we realize there's nothing we can do to save ourselves, the sooner we can get down to the hard work of adapting, with mortal humility, to our new reality.

The choice is a clear one. We can continue acting as if tomorrow will be just like yesterday, growing less and less prepared for each new disaster as it comes, and more and more desperately invested in a life we can't sustain. Or we can learn to see each day as the death of what came before, freeing ourselves to deal with whatever problems the present offers without attachment or fear.

If we want to learn to live in the Anthropocene, we must first learn how to die.

⟫⟩ Ethics for the Anthropocene

Dale Jamieson

6.4. Ethics for the Anthropocene

The Anthropocene presents novel challenges for living a meaningful life. They begin with questions of ethics.

From the beginning of human morality, ethics has been primarily concerned with the proximate: what presents to our senses and causally interacts with us in identifiable ways. However, what is proximate is flexible. Stories, music, relics, sacred space, and even the establishment of a common language are all ways of bringing into view what would otherwise be remote. The expanding circle of ethics (which to a great extent coincides with globalization) has made the distal proximate through new living arrangements, forms of travel, and kinds of imagery enabled by technological innovation. However, there may be a limit to what can be made proximate.

The late philosopher Bernard Williams distinguished what he called "the morality system" from "ethics." Ethics concerns the generic question of how we should live and goes back to at least Homer and the ancient Greek dramatists. It is relatively universal and resilient, though flexible and revisable in its content. The morality system, on the other hand, is

> a particular development of the ethical, one that has a special significance in modern Western culture. It particularly emphasizes certain ethical notions rather than others, developing in particular a special notion of obligation, and it has some peculiar presuppositions.[1]

The mark of the morality system is the establishment of an inner deontic order that mirrors an external law.[2] It is characterized by an emphasis on purity, voluntariness, inescapability, and generalizability. According to Williams, the morality system has been enormously influential on "we moderns," though its underpinnings are largely illusory. He thinks that "we would be better off without" this "peculiar institution."[3]

One does not have to accept Williams's entire story to wonder whether morality has more than met its match in the Anthropocene. Not everything that matters can be made proximate to creatures like us. Not all of contemporary life can be fruitfully modeled on eighteenth-century concepts. The morality system may have room for revision and may not disappear, but it may come to be seen as more like

Reason in Dark Time: Why the Struggle against Climate Change Failed and What It Means for Our Future (Oxford: Oxford University Press, 2014), 185–92

the "etiquette system," important for a particular domain, but hardly an oracle that can answer all of our most important questions.[4]

Ethics is a collective construction, like morality, but it seems to allow more individual variation. For this reason it may seem more revisable than morality, at least from the perspective of an individual. While ethics is fundamentally agent-centered, it leaves its mark on the world because it requires attunement to reality. While there is no guarantee or even much reason to believe that ethics and morality together can provide comprehensive guidance for life in the Anthropocene, we can hope that they can make some contribution to making the world better and enabling us to live meaningful lives.

An ethics for the Anthropocene would, in my view, rely on nourishing and cultivating particular character traits, dispositions, and emotions: what I shall call "virtues." These are mechanisms that provide motivation to act in our various roles from consumers to citizens in order to reduce GHG emissions and to a great extent ameliorate their effects regardless of the behavior of others. They also give us the resiliency to live meaningful lives even when our actions are not reciprocated.

My conception of the virtues does not rest on any deep metaphysical commitments about "natural goodness" or "the good for man." It flows from the general view that when faced with global environmental problems such as climate change, our general policy should be to try to reduce our contribution regardless of the behavior of others, and we are more likely to succeed in doing this and living worthwhile lives by developing and inculcating the right virtues than by improving our calculative abilities.[5]

The green virtues that would be part of an ethics for the Anthropocene would not be identical to classical or Christian virtues but neither would they be wholly novel. Much that mattered to humanity in the Pleistocene will matter in the Anthropocene as well. In writing a set of virtues for the Anthropocene we can draw on a great deal of traditional wisdom. However, some speculation is in order when we contemplate how to live meaningfully in a world that has not yet fully taken shape.

We can think of green virtues as falling into three categories: those that reflect existing values; those that draw on existing values but have additional or somewhat different content; and those that reflect new values. I call these three categories preservation, rehabilitation, and creation.[6] I will discuss each in turn, offering tentative examples of green virtues that might fall into these various categories.

Thomas Hill Jr. (1983) offers an example of preservation. He argues that the widely shared ideal of humility should lead people to a love of nature. Indifference to nature "is likely to reflect either ignorance, self-importance, or a lack of self-acceptance which we must overcome to have proper humility."[7] A person who has proper humility would not destroy redwood forests (for example) even if it appears that utility supports this behavior. If what Hill says is correct, humility is a virtue that ought to be preserved by greens.

Temperance may be a good target for the strategy of rehabilitation. Long regarded as one of the four cardinal virtues, temperance is typically associated with the problem of *akrasia* and the incontinent agent. But temperance also relates more generally to self-restraint and moderation. Temperance could be rehabilitated as a green virtue that emphasizes the importance of reducing consumption.[8]

A candidate for the strategy of creation is a virtue we might call mindfulness. Much of our environmentally destructive behavior is unthinking, even mechanical. In order to improve our behavior we need to appreciate the consequences of our actions that are remote in time and space. A virtuous green would see herself as taking on the moral weight of production and disposal when she purchases an article of clothing (for example). She makes herself responsible for the cultivation of the cotton, the impacts of the dyeing process, the energy costs of the transport, and so on. Making decisions in this way would be encouraged by the recognition of a morally admirable trait that is rarely exemplified and hardly ever noticed in our society.

Cooperativeness would be another important characteristic of agents who could successfully address the problems of climate change. Surprisingly, this characteristic appears to be neglected by both ancient and modern writers on the virtues. Perhaps a virtue of cooperativeness is a candidate for creation; or perhaps, though not itself a virtue, cooperativeness would be expressed by those who have a particular constellation of virtues.[9]

There are other potential candidates for green virtues, some of which are related to those in the tradition and others that are not. Simplicity, for example, has a relatively long history, and the related virtue of conservatism has also been mentioned.[10] In what follows I discuss a virtue of particular importance in the Anthropocene.

6.5. Respect for Nature

Respect for nature has been celebrated at various places and times to different degrees.[11] It is a persistent if not universal value. There are at least precursors of this idea in Kant and strong assertions of it in the Romantic tradition.[12] It is frequently attributed to indigenous peoples and found in various Asian traditions. While it is difficult to say exactly what this virtue consists in, it is relatively easy to give examples of the failure to express it.

[A]ccording to some eminent scientists "it is clear that we live on a human-dominated planet."[13] If we dominate our planet, then surely we can be said in an important sense to dominate nature. Dominating something can be one way of failing to respect it, so it is plausible to say that in virtue of our domination of nature we fail to respect it.[14] But what exactly does it mean to dominate nature?

[Elsewhere, I have] claimed that domination is related to the extent to which an agent has power over a subject. When an agent's power is of a certain kind or extremity, it can compromise a subject's autonomy to the extent that the agent can be said to dominate the subject. In the literature of environmental ethics, nature is often seen as autonomous in the sense of self-determining.[15] Rather than being autonomous (i.e., governed by its own laws and internal relations), nature is increasingly affected by human action. While humans (and other forms of life) have always influenced their environments, what makes the present human relationship to nature one of domination is the degree and extremity of human influence. Human influence on nature is now so thoroughgoing that it constitutes domination.[16]

Domination can be expressed attitudinally in the ways in which we think and feel about nature as well as substantively. We often treat nature as "mere means," as if it did not have any value or existence independent of its role as a resource for us. As a society we seem to treat the Earth and its fundamental systems as if they were toys that can be treated carelessly, as if their functions could easily be replaced by a minor exercise of human ingenuity. It is as if we have scaled up slash-and-burn agriculture to a planetary scale.[17]

One of the insights of the social movements of the 1960s was that a vicious circle can take hold with subordinated groups.[18] Mistreatment diminishes respect, which leads to further mistreatment, which further diminishes respect, and so on. The same vicious circle can take hold with nature. Dominating nature both expresses and contributes to a lack of respect, which in turn leads to further domination.

Respecting nature, like respecting people, can involve many different things. It can involve seeing nature as amoral, as a fierce adversary, as an aesthetic object of a particular kind, as a partner in a valued relationship, and perhaps in other ways. These attitudes can exist simultaneously within a single person.

When nature is seen as amoral, it does not constitute a moral resource in any way. Moral concepts arise, on this view, either from divine commandment, as in the case in the Hebrew Bible, from reason (as in Kant), from the emotions (as in Hume), or are artificial human constructions laboriously created and maintained to provide us with refuge in an otherwise heartless world (as in the story told by Thomas Hobbes). One memorable statement of nature as amoral occurs in chapter 5 of the *Tao Te Ching*, attributed to the Taoist sage Lao-Tse: "Heaven and Earth are impartial; they treat all of creation as straw dogs." In ancient Chinese rituals, straw dogs were burned as sacrifices in place of living dogs. What is asserted here is that the forces that govern the world are as indifferent to human welfare as humans are to the fate of the straw dogs they use in ritual sacrifice. On this view we should respect nature because of its blind, unpurposing force and power.

Seeing nature as amoral can easily slip into seeing nature as an immensely powerful and even malevolent adversary, and humanity as weak, vulnerable, and in need of protection.[19] If humanity and its projects are to survive and thrive, nature must be subdued and kept at bay. Nature, on this view, is the enemy of humanity.

Amoral nature can be respected for its radical "otherness" that cannot be assim-ilated to human practices. Nature as an adversary can be respected for its power and abilities in pursuing its ends, which are fundamentally at odds with those of humanity. Seeing nature as amoral or as an adversary can provide grounds for re-specting nature but can also provide a rationale for dominating nature.[20]

A third way of respecting nature sees profound aesthetic significance in its overwhelming power. This thought is powerfully developed in Edmund Burke's 1757 work *A Philosophical Enquiry into the Origin of Our Ideas of the Sublime and Beautiful*. The human experience of the sublime is, according to Burke, a "delight," and one of the most powerful human emotions. Yet, perhaps paradoxically, the experience of the sublime involves such "negative" emotions as fear, dread, pain, and terror, and can occur when we experience deprivation, darkness, solitude, si-lence, or vacuity. The experience of the sublime arises when we feel we are in danger but it is actually not so. Immensity, infinity, magnitude, and grandeur can cause this experience of unimagined eloquence, greatness, significance, and power. The sublime is often associated with experiences of mountains or oceans. Such experiences may occasion wonder, awe, astonishment, admiration, or reverence. In its fullest extent, the experience of the sublime may cause total astonishment.

The idea of the sublime was profoundly influential on nineteenth-century American culture, notably through painters such as Thomas Cole and Frederic Church. It went on to be an important influence on American environmentalism through the writings of John Muir and, more recently, Jack Turner (1996), Dave Foreman (1991), and other advocates for "the big outside." Indeed, the case for wilderness preservation is often made in the language of the sublime.

Finally, there is the idea of nature as a partner in a valuable relationship. Peo-ple often speak of particular features of nature as if they were friends, lovers, or even parents. People who see elements of nature as friends often feel that they learn from nature as they do from other companions. Some speak of nature in language that is usually reserved for lovers.[21] Indeed, we often speak of those who want to protect nature as "nature lovers." In some people, nature elicits feelings of filial devotion. John Muir wrote that "there is a love of wild nature in everybody, an ancient mother-love."[22] Many of us also associate nature with a feeling of being home. I grew up in San Diego, California, and the sights, smells, breezes, and qual-ity of light that I experience when I am there are transformative, especially when I step onto the beach at Torrey Pines, just north of the city.

This idea of nature as a partner in a valuable relationship makes itself felt in economic language when people talk about "natural capital" or "ecosystem ser-vices." On this view protecting nature returns monetized benefits. Damaging na-ture damages ourselves.

These different ways of respecting nature support somewhat different attitudes toward nature and reasons for respecting it. Rather than discussing the details, I will mention three reasons for respecting nature that seem quite robust across times and cultures. Respect for nature can be grounded in prudence, can be seen as a fitting response to the roles that nature plays in giving our lives meaning, and can also spring from a concern for psychological wholeness.

One reason for respecting nature is that it is in our interests to do so. The geoscientist Wallace Broecker (2012: 284) compares our climate-changing behavior to poking a dragon with a sharp stick. Angering the dragon of climate is not likely to be a good business plan for maintaining human life on Earth. Versions of this argument are ubiquitous in the environmental literature, and something like this view is implicit in slogans such as Barry Commoner's (1971) "third law of ecology," which states that "nature knows best." It can also be seen as providing the foundation for the precautionary principle.

A second reason for respecting nature is that, for many people and cultures, nature provides important background conditions for lives having meaning. It is easy to think of examples from history, literature, or contemporary culture. Blake's idea of England as a "green and pleasant land" is important in English literature, history, and identity. The cherry orchard in Chekhov's play of the same name defines the life of everyone in the community. Think of the role that landscape plays in the lives of indigenous peoples. For that matter, think of how the "flatirons" define Boulder, Colorado.[23]

An analogy may help to bring the point out more clearly. Representational painting is not the only kind of valuable painting, but it is one very important kind. Indeed, it may be the mother from which other forms of valuable painting emerged. Representational painting exploits the contrast between foreground and background. What is in the foreground gains its meaning from its contrast with the background. What I want to suggest is that nature provides the background against which we live our lives, providing us with an important source of meaning. It is thus not surprising that we delight in nature and take joy in its operations, and feel grief and nostalgia when familiar patterns are disrupted and natural features destroyed.[24] In these respects, meaning and mourning are closely related concepts.[25]

A third reason for respecting nature flows from a concern for psychological integrity and wholeness. As Kant (and later Freud) observed, respecting the other is central to knowing who we are and to respecting ourselves. Indeed, the failure to respect the other can be seen as a form of narcissism. Some work in environmental psychology gestures toward a story in which the recognition of nature as an "other" beyond our control is at the root of our self-identity and communal life.[26]

Many of these same reasons for respecting nature apply to respecting those who have gone before and those who will come after. Seeing ourselves as related to others in these ways is important to respecting ourselves and knowing who we are. It is also central to giving meaning to our lives. Such respect is also likely to help keep us from destroying ourselves.

The idea of respect for nature may seem in tension with another thought that is often articulated by environmentalists. On this view the ultimate source of our environmental problems is our separation from nature. The solution is to see humans as part of nature. From this perspective, nature is inside of us and we are part of nature. Our skin is a permeable membrane that is itself part of the natural world.[27] How can we respect nature when we ourselves are part of nature?

Such claims can be irritating because it is easy to hear them as trivial, false, pernicious, or mystical. For a naturalist such claims seem trivial. Of course we are part of nature. What else is there for us to be part of? Yet in another sense it is clear that we distinguish people from nature in much the same way that we distinguish artifacts from natural objects. Someone who cannot make such distinctions, at least in the ordinary case, either does not know how to speak the language or has some serious psychological deficiency or disorder. The claim that we are part of nature can also seem pernicious since it seems to imply that there is no moral difference between a human being killed by an earthquake and one who is killed by another human being. Of course those who claim that humans are part of nature typically want to deny this implication, but this is where the mysticism sets in.

Nevertheless, I think there is important truth in the claim that humans are part of nature. We can take many different perspectives on the relationship between ourselves and nature. For example, we can see nature as a set of cycles and from within this single perspective there are multiple views. From the point of view of biogeochemistry, nature is the carbon cycle, the nitrogen cycle, and so on. On this view we, like other natural objects, are instances of these cycles. At another level of analysis we can say that breathing and respiration are instances of the same cycles that govern the atmosphere; our circulatory system as well as various cellular processes are instances of the hydrological cycle; digestion and metabolism recapitulate the soil cycle; and we are as subject to the laws of thermodynamics as any planet or star. We could go on acknowledging other perspectives and various points of view within them. From these perspectives we are not separate from nature. Not only has nature brought us into existence and sustains us, but it also constitutes our identity.

This may seem hopelessly abstract or romantic, but it is because of these perspectives from which we see ourselves as part of nature that we cannot fully reduce nature to competing baskets of distributable goods, at least not without radically changing our own self-understandings. We are hesitant about markets in kidneys and more than hesitant about markets in brains, in part because we see these organs as partly constitutive of who we are. Even if we allow such markets we will not be tempted to think that everything that is important about a kidney or a brain is expressed by its market value. It would be strange for someone to do a benefit–cost analysis of a brain as if its value in a shadow market were its most important feature. The same sort of strangeness attaches to attempts to assess in market terms "the value of the world's ecosystem services and natural capital."[28] A residue remains of our relation to nature that cannot be fully expressed in the language of economics. This dimension is primordial, and occurs in various traditions around the world. It cannot easily be dismissed.[29]

Much that I have said in this section is sketchy and unsatisfactory. The important points, which surely need fuller development and deeper reflection, are these. Respect for nature is an important virtue that we should cultivate as part of an ethics for the Anthropocene. Respect can be manifest in many different ways within a single person, sometimes simultaneously. Nature itself is not a single thing, and we can respect elements or dimensions of nature while expressing contempt for others. Respecting nature is respecting ourselves.

Notes

1. Williams (1985: 6).

2. This thought is explicit in the work of Williams's longtime colleague, G. E. M. Anscombe (see especially Anscombe 1958); to a great extent they share a common critique of modern moral philosophy, though their positive views are quite different.

3. Williams (1985: 174). By artfully using an expression ("peculiar institution") that has traditionally been applied to American slavery, Williams indirectly references Nietzsche's critique of Judeo-Christian morality as a slave morality. For Williams the morality system is basically a rationalized version of Judeo-Christian religious ethics, though so far as I know he never states this explicitly in his published writings.

4. On the relationship between morality and etiquette, see Foot (1972).

5. I defend these claims more fully in Jamieson (2007). The instrumental attitude I take toward the virtues separates me from traditional virtue theorists and many of those who work in the tradition of environmental virtue theory. Cf. Sandler (2007); Cafaro and Sandler (2005).

6. These strategies reflect the mechanisms of moral change discussed in Section 5.6. A fuller account would also have to provide an account of the vices; see, e.g., Thompson and Bendyk-Keymer (2012: chaps. 10–12).

7. Hill (1983: 222).

8. Another example of rehabilitation is exemplified in Jonathan Lear's story of how courage came to take on new meaning in the life of Crow Chief Plenty Coups after his people were virtually destroyed and confined to a reservation (Lear 2006). For its application to climate change, see Thompson (2010).

9. Hume is an exception in the tradition in noting the importance of cooperativeness. For further discussion of the importance of cooperativeness to morality, see Hinde (2002).

10. For simplicity, see Elgin (2010) and Cafaro (2005); generally, see Jamieson (1992, 2012).

11. Respect for nature can be thought of as a duty as well as a virtue, which is how Paul Taylor (1986/2011) understands it, and also how I regarded it in Jamieson (2010). See also Wiggins (2000).

12. On Kant, see Wood (1998); for an expression of respect for nature in Romantic poetry, see Coleridge's poem "The Rhyme of the Ancient Mariner."

13. Vitousek et al. (1997: 494).

14. There is a sense of "domination" in which it does not imply a lack of respect (e.g., one team can be said to dominate another in a game), but for reasons that are given below (e.g., that our lack of respect for nature expresses attitudinally as well as substantively) and for others that are obvious it is not this sense that is in play here.

15. See, for example, Katz (1997); the essays collected in Heyd (2005); and Turner (1996). What Turner means by "wildness" is related to what I mean by "autonomy." For reservations, see O'Neill et al. (2007: 134–37).

16. This is why Vitousek et al. (1997) used the language of domination. These are also the sorts of reasons why McKibben (1989) took climate change to mark "the end of nature." While this was an exaggeration, McKibben was making an important point: though it does not mark the end of nature, climate change is a mark of the Anthropocene. For more on these themes, see Jamieson (2008: 166–68) and Jamieson (2002: 190–96).

17. I owe this image to Jeremy Waldron.

18. This theme was especially prominent in the work of Franz Fanon and Malcolm X.

19. Werner Herzog's *Grizzly Man* is a wonderful film on this and related themes.

20. Mill is an interesting case of someone who saw nature as amoral but maintained a fundamental respect for nature, in part for its otherness, but also because of its aesthetic qualities and the ways it contributes to human life.

21. There is even a blog "52 Ways to Fall in Love with the Earth," which can be viewed at http://52ways.wordpress.com/. Retrieved July 18, 2013.

22. http://www.goodreads.com/author/quotes/5297.John_Muir. Retrieved July 18, 2013.

23. http://en.wikipedia.org/wiki/Flatirons#A_symbol_of_Boulder. Retrieved July 18, 2013.

24. For an articulate example of these feelings regarding the devastation of Utah's red rock canyon country by the creation of Lake Powell, see Lee (2006) and Abbey (1985). A similar sense of loss and nostalgia can be engaged by urban projects such as Robert Moses's plan to build a highway through Manhattan's Washington Square Park; see Caro (1975).

25. I owe this thought to Sebastiano Maffettone.

26. See, e.g., Clayton and Opotow (2003).

27. These themes are suggested by Suzuki and McConnell (1997).

28. This is the title of Costanza et al. (1997). According to the authors, the value in question is in the range of $16–54 trillion per year. For a critical discussion, see Sagoff (2004: chap. 6).

29. Sagoff (1991) and Dworkin (1993: chap. 3) argue points that are similar to this— Sagoff when he distinguishes nature from the environment, and Dworkin when he talks about species as sacred.

Works Cited

Abbey, Edward. 1985. *The Monkey Wrench Gang*. Salt Lake City: Dream Garden Press.

Anscombe, G. E. M. 1958. "Modern Moral Philosophy." *Philosophy* 33 (124): 1–19.

Cafaro, Philip. 2005. "Thoreau, Leopold, and Carson: Toward an Environmental Virtue

Ethics." In *Environmental Virtue Ethics*, ed. Philip Cafaro and Ronald Sandler, 31–44. Lanham, MD: Rowman & Littlefield.

Cafaro, Philip, and Ronald Sandler, eds. 2005. *Environmental Virtue Ethics*. Lanham, MD: Rowman & Littlefield.

Caro, Robert A. 1975. *The Power Broker: Robert Moses and the Fall of New York*. New York: Vintage Books.

Clayton, Susan, and Susan Opotow, eds. 2003. *Identity and the Natural Environment: The Psychological Significance of Nature*. Cambridge, MA: MIT Press.

Costanza, Robert, Ralph D'Arge, Rudolf de Groot, Stephen Farberi, Monica Grasso, Bruce Hannon, Karin Limburg, Shahid Naeem, Robert V. O'Neill, Jose Paruelo, Robert G. Raskin, Paul Sutton, and Marjan van den Belt. 1997. "The Value of the World's Ecosystem Services and Natural Capital." *Nature* 387:253–60.

Dworkin, Ronald. 1993. *Life's Dominion: An Argument about Abortion, Euthanasia, and Individual Freedom*. New York: Vintage Books.

Elgin, Duane. 2010. *Voluntary Simplicity: Toward a Way of Life That Is Outwardly Simple, Inwardly Rich*. New York: Harper.

Foot, Philippa. 1972. "Morality as a System of Hypothetical Imperatives." *Philosophical Review* 81 (3): 305–16.

Heyd, Thomas, ed. 2005. *Recognizing the Autonomy of Nature: Theory and Practice*. New York: Columbia University Press.

Hill, Thomas, Jr. 1983. "Ideals of Human Excellence and Preserving the Natural Environment." *Environmental Ethics* 5 (3): 211–24.

Hinde, Robert A. 2002. *Why Good Is Good: The Sources of Morality*. London: Routledge.

Jamieson, Dale. 1992. "Ethics, Public Policy, and Global Warming." *Science, Technology, and Human Values* 17 (2): 139–53.

———. 2002. *Morality's Progress: Essays on Humans, Other Animals, and the Rest of Nature*. Oxford: Oxford University Press.

———. 2007. "When Utilitarians Should Be Virtue Theorists." *Utilitas* 19 (2): 160–83.

———. 2008. *Ethics and the Environment: An Introduction*. Cambridge: Cambridge University Press.

———. 2010. "The Question of the Environment." In *Trattato di Biodiritto*, diretto da S. Rodota and P. Zatti, Ambito e Fonti del Biodiritto, a cura di S. Rodota and M. Tallacchini, 37–50. Milano: Giuffre Editore.

———. 2012. "Ethics, Public Policy, and Global Warming." In *Ethical Adaptation to Climate Change: Human Virtues of the Future*, ed. Allen Thompson and Jeremy Bendik-Keymer, 187–202. Cambridge, MA: MIT Press.

Katz, Eric. 1997. *Nature as Subject: Human Obligation and the Natural Community*. Oxford: Rowman & Littlefield.

Lear, Jonathan. 2006. *Radical Hope: Ethics in the Face of Cultural Devastation*. Cambridge, MA: Harvard University Press.

Lee, Katie. 2006. *Glen Canyon Betrayed: A Sensuous Elegy*. Flagstaff, AZ: Fretwater.

McKibben, Bill. 1989. *The End of Nature*. New York: Random House.

O'Neill, John, Alan Holland, and Andrew Light. 2007. *Environmental Values*. New York: Routledge.

Sagoff, Mark. 1991. "Nature versus the Environment." *Report from the Institute for Philosophy & Public Policy* 11 (3): 5–8.

———. 2004. *Price, Principle, and the Environment.* Cambridge: Cambridge University Press.

Sandler, Ronald L. 2007. *Character and Environment: A Virtue-Oriented Approach to Environmental Ethics.* New York: Columbia University Press.

Suzuki, David, and Amanda McConnell. 1997. *The Sacred Balance: Rediscovering Our Place in Nature.* Vancouver: Greystone Books.

Taylor, Paul. 1986/2011. *Respect for Nature: A Theory of Environmental Ethics.* Princeton, NJ: Princeton University Press.

Thompson, Allen. 2010. "Radical Hope for Living Well in a Warming World." *Journal of Agricultural and Environmental Ethics* 23 (1): 43–59.

Thompson, Allen, and Jeremy Bendik-Keymer, eds. 2012. *Ethical Adaptation to Climate Change: Human Virtues of the Future.* Cambridge, MA: MIT Press.

Turner, Jack. 1996. *The Abstract Wild.* Tucson: University of Arizona Press.

Vitousek, Peter M., Harold A. Mooney, Jane Lubchenco, and Jerry M. Melillo. 1997. "Human Domination of Earth's Ecosystems." *Science* 277 (5325): 494–99.

Wiggins, David. 2000. "Nature, Respect for Nature, and the Human Scale of Values." *Proceedings of the Aristotelian Society* 100 (1): 1–32.

Williams, Bernard. 1985. *Ethics and the Limits of Philosophy.* Cambridge, MA: Harvard University Press.

Wood, Allan. 1998. "Kant on Duties regarding Nonrational Nature." *Proceedings of the Aristotelian Society Supplementary Volume* 72 (1): 189–210.

))) We Have Always Been Post-Anthropocene: The Anthropocene Counterfactual

Claire Colebrook

I.

The proposed (and close to consecrated) conception of the Anthropocene epoch appears to mark as radical a shift in species awareness as Darwinian evolution effected for the nineteenth century. If the notion of the human species' emergence in time requires new forms of narrative, imaginative, and ethical articulation (Beer 2000), then the intensifying sense of the species' *end* makes a similar claim for rethinking "our" processes of self-presentation and self-preservation. Rather than Darwin's timeline of life's grandeur, with a (random but fortuitous) proliferation of difference and complexity, we might have to confront a sudden event of geological impact within an intensively human timeline. Rather than one more event *within time*, not only does the Anthropocene—as did Darwinian evolution—require us to shift our scale of narration away from human generations and history to species' emergence and deep time, but it raises the problem of intersecting scales, combining the human time of historical periods (late capitalism, industrialism, nuclear power) with a geological time of the planet. This, in turn, requires us to open the classically feminist question of the *scale of the personal* (Clark 2012). If, as Kate Millett (1970) argued, the personal is the political, then this requires us to make some decision as to what counts as the political: is my personal sense of gender meaningful only in terms of the history of the human family, or in terms of the narrower history of bourgeois marriage, or might we say that the personal is geological and that in order to understand the sexed subject I need to take the emergence of the human species, and the domination of the planet (and other humans), into account?

Bruno Latour (2013) has recently argued that awareness of the Anthropocene closes down the modern conception of the infinite universe, drawing us back once again to the parochial, limited, and exhausted earth. Rather than an open horizon of possibility limited only by the pure laws of logic or universal reason, we are now the "earthbound." Latour draws on the work of Alexandre Koyré (1957), who had defined the modern infinite universe in contrast with the closed world: the *world* is a collection of beings, each defined according to its created kind, and each giving the whole its specific and interconnected way of being, as though the cosmos were a grand organism. By contrast, the infinite universe allows for the thought of matter as such, subject to the formal measures of physics; the modern subject, in turn, is a purely formal, rational, and calculative being. Against this modern abstraction of a world of matter opposed to a subject, Latour (2008) has argued for contemporary modes of existence defined quite specifically by relation to a singular world lived in its power to affect and to be a matter of concern. The "universe" is no longer a horizon of infinite possibility where we are morally compelled to act "as if" the laws of reason might one day yield a fully rational existence, freed from all pathology (Kant 2002). For Latour the detached and objective comportment of modern science will not help us deal with what matters most in a world of climate change. Where once science might have been defined as a practice of logic and truth, pure in its difference from the interests of everyday life, the world that we deal with can no longer be regarded as abstract matter but must be considered as something that has its reality by way of what we do and how we observe. The notion of the earth as bounded, as anything but unlimited, seems to have forced itself on us. Our potentials and what we can do with ourselves and the planet are not limited by the laws of formal physics or logic but are determined by variables and volatilities that we cannot fully command. As Naomi Oreskes and Eric Conway (2014) have argued, one of the contributing factors in the ongoing failure to act on climate change has been a conception of science as an isolated activity *not* bound up with systems of political action and social dynamics. What is required, Latour argues, is a sense of ourselves as *earthbound*—not as observers of matter, but as oriented toward matters of concern in which our own being depends on a *world* (a specific world, not an open universe).

If this is so—if we have to abandon the notion that whatever we take the world to be it might always be otherwise, offering infinite possibilities imagined by way of scientific progress—then it might require us to redefine all those *hyper*-modern proclamations of a post-human, post-feminist, and post-racial future (which rely on refusing forms of intrinsic difference) as *hypo*-modern. We are not faced with infinite and open potentiality or becoming; the modern notion of self-definition and a world devoid of any kinds or essences is giving way to differences and distinctions that force themselves on us; we cannot look back on what we have become and how we have evolved and argue that nothing prevents us from becoming anything we want to be. Nowhere is this shift from indifference to difference more intense than in the problem of feminism. To argue, for example, that sex is an effect of gender (de Lauretis 1987; Butler 1990; Gatens 1991), or that we only know life as differentiated *after* the event of human systems of communication, is to refuse matter's resistance and recalcitrance—not its vibrancy or its agency so much as its tendency to remain indifferent. Here, I would like to contrast two senses of indifference: the first would be hyper-modern, and would be—like Koyré's open universe—exemplified by the notion of a conception of matter as pure quantity without tendencies of its own, subsumable and easily mastered by abstract or formal conceptions of being. Something of this is sustained today in Alain Badiou's (2007) conception of being as a *pure* multiplicity; however we determine or quantify being, being is not reducible to the sense we make of it—being is thinkable only as a void, as what exits after the subtractions of all qualifications.

Another conception of indifference, and one that I would like to pursue in this chapter, is *hypo*-modern: refusing the disjunction between either a closed (fully differentiated) world whose intrinsic sense, difference, and life we need to respect, or a void that is differentiated or qualified by reason. I would like to propose an indifference that is destructive of inscribed difference but not because there is something like a pure undifferentiated matter that requires structures of language to render it distinct. There has been a widespread rejection (and revival) of undifferentiated lifeless matter: Jane Bennett (2010) insists on matter's own vitality and difference, against the notion of a neutral substrate or a restriction of life and action to organisms; by contrast, Alain Badiou defines being as a void beyond all predication, but Badiou is perhaps the last in a line of thinkers who want to think of being as such, and not the difference of beings. The form of indifference I am charting here would refuse both these notions: neither is the world differentiated by human predication or linguistic structures (being a blank matter before all form), nor does it bear its intrinsic qualities. Indifference is how we might think about an "essentially" rogue or anarchic conception of life that is destructive of boundaries, distinctions, and identifications. To live is to tend toward *indifference*, where tendencies and forces result less in distinct kinds than in complicated, confused, and dis-ordered partial bodies. (One might think here of the human body, whose vital attachment to the earth, food, sex, language, technology, and other humans not only disperses the self across a series of connections but also operates as much to diminish as to intensify stability. The more we attach ourselves to food production and consumption, the more the planet and human bodies suffer from excess and depletion. The more we invest in sexual, commodity, or political desire, the more rigid and elusive networks of pleasure come to be. The more "we" reflect on "our" mark on the planet, the more we appear to be a single polluting species, while also being more and more divided by the causes and consequences of what has come to be known as the Anthropocene. In short, the more "we" appear to be unified as a species, differentiated from other species, and the more we become defined by the claim of the Anthropocene, the more mindful we should become of all the forces and tendencies too minimal to appear as differences.)

Rather than returning to conceptions of the bounded earth as a single living, mutually interdependent organism with its own differences and kinds, we might think of all the differences we make and mark as supervening on a world that does not come with its own inscription or difference but is not, for all that, devoid of a complexity that will always exceed any of the differences we read into the world. The Anthropocene has presented itself to many as a nonnegotiable difference: "we" abandon a world that was deemed to be indifferent to our narrow historical periods, and "we" recognize that human history is geologically significant after all, and that "we" have made a definitive difference. (There is, once again, a "we" rendered fully present by Anthropocene scarring.) Against this narrative of non-negotiable and definitive difference I would suggest that we think about indifference not by referring to a universe devoid of determination (the modern and purely calculable blank matter or substance without qualities that we might come to know disinterestedly). Indifference is the milieu in which we live, always destroying and confusing inscribed differences. This notion of indifference combines both Gilles Deleuze's (1994) notion that the virtual is more different than the clumsy categories of our differentiated world and Giorgio Agamben's (2004) argument that difference emerges from indifference—that difference comes into being and is always haunted by its dissolution. Taking these two very abstract notions seriously would require us to think of the Anthropocene both as a difference that emerges from other potential stratifications (a differentiation of a world that might have been more finely or acutely differentiated) and as something that prompts the question of the counter-Anthropocene. At what point does a difference make a difference or appear *as different*? Had "we" behaved differently, perhaps we would not have become the species that made a difference; at what point or threshold of our polluting, ecosystem-destroying history did we make a difference? Here, indifference does not signal what Daniel Dennett (1995) has referred to as the "substrate neutral" character of evolving life, where the random algorithms of combinations of elements yield complexity. Instead, indifference operates as a counter-Anthropocene provocation. If the Anthropocene is the return of difference—because humans are once again exceptional, but now in their destructive and inscriptive impact—it might be worth asking how such difference is inscribed and on what scale such difference operates. What might it mean to think a counterfactual scenario where humans had not inflicted the difference of the Anthropocene on the planet?

At the question of scale, we might ask why it is that a certain geological stratification is privileged as finally confirming human impact and difference, and what possible human existence might have prevented such a scar from occurring. If the Anthropocene is a judgment that constitutes us once again, *as human*, and as different from other species in our impact, then we need to decide at what point planetary impact is deemed to be inscriptive. There are a whole series of thresholds, such as sedentary agriculture, colonization, the steam engine, nuclear energy, and capitalism. Not only might we contest just where and at what register differences are inscribed, but further political questions might be raised about the convergence of difference. The time of politics and the time of the planet, once deemed to be distinct, are now colliding, but not converging.

Questions of politics were once *of the polity*, so that if we were to think that the "personal is political," we would look to the history of wages, domestic labor, gender norms, reproductive medicine, education, and cultural production. Today, thinking about the politics of any person (including the very possibility of personhood with rights and freedoms) would also need to be mindful that the capitalism that enabled liberal freedom and personhood relied on favorable conditions that exploited the world's poor and that are almost certainly unsustainable into the future (Mulgan 2011).

The Anthropocene emerges as a dissonant difference; post-Darwinian humans have lived with a sense of the difference of scale between a human time of generations (politics) and the dwarfing times of geological change. To talk about humans *as such* was once deemed to be counterpolitical, with politics being definitively marked in Fredric Jameson's (1981) imperative to "always historicize," where history does not mean referring to anything as de-differentiating as "the human condition." In the Anthropocene these two timelines, in their dissonance and difference, intersect: geological change is occurring within human and humanly experienced time. Human activity—the impact of a single species—has reached such an intensity as to generate geological inscription; this Anthropocene dissonance of difference is privileged precisely because scales that were deemed to be divergent are now converging. Species and geology are now co-articulated; we look at the earth—now—as if, in our future absence, we will be readable as having been.

Other forms of human impact—such as pollution or the destruction of ecosystems —have long been acknowledged, but with the claim made for the Anthropocene a particular difference is deemed to be dramatically different. We do not just make the earth different, but make it different on a different scale. From a modernity in which apparent difference was vanquished—acknowledging only life or matter without intrinsic difference—post-human claims for being one more aspect of a general "life" have now been vanquished.

The return of "man" opens the past few decades of difference theory up for question. What needs to be rethought are some of the key motifs of what has come to be known as "theory." First, the notion that there is no such thing as the human (either by way of our difference from animals, or because of intra-human differ-ences in culture and history) must give way to a sense of the human as defined by destructive impact. Second, theory (at least in its post-structuralist phase) suppos-edly placed the real or the material in parenthesis, knowable only after the event as that which is made sense of by inscription. But it was precisely at the point that humans regarded the world as an "in itself," which can be known only as it is for us, that "we" were doing damage to the earth itself. Modernity started to destroy the material substrate of its existence at the same time as modernity increasingly denied any reality other than that known and constituted by man. The Anthropo-cene seems to arrive just as a whole new series of materialisms, vitalisms, realisms, and inhuman turns require "us" to think about what has definite and forceful ex-istence regardless of our sense of world. The ambivalence of the Anthropocene does not only concern two temporalities (opening us to geological impact while drawing us back to human agency and human historical force); it also pulls in contrary directions with regard to what might be thought of as post-humanity. I would suggest that rather than simply say that philosophy, theology, politics, and common sense always have opposing tendencies (though that may be true), the diverging temporalities and humanisms/post-humanisms of the Anthropocene prompt the question of the ways in which human difference and indifference might be thought.

One effect of the Anthropocene has been a new form of difference: it now makes sense to talk of humans as such, both because of the damage "we" cause and because of the myopia that allowed us to think of the world as so much matter or "standing reserve." Humans are, now, different; and whatever the injustices and differences of history and colonization, "we" are now united in being threatened with nonexistence. Alongside the return of the repressed of human exceptionalism and the acceptance that we are now different after all, we might also need to con-front a *new form of indifference*. We might need to think beyond the nonnegotiable terrain of sexual difference, where all life emerges from relations and encounters between tendencies. One thing is certain: if there had not been sexual difference in its narrowest sense (man and woman), there could not have been the nuclear family, division of labor, and then industrialism. At the same time, if there had not been industrialism, at least in our world history, women would not have been liberated from domestic labor and granted access to the forms of planet-exploitative luxuries that have generated personhood in its modern Western liberal sense. Life may generate sexual difference as organic life emerges, but there is also a stronger or narrower sense of gendered sexual difference—familial, personal, binary sexual difference—that relies on the same processes of "civilization" that generated the Anthropocene. The family and gendered divisions of labor become crucial for intensified practices of imperialism, militarization, colonization, indentured labor, slavery, and mass production. So we might ask at what point living difference became sexual, at what point organisms relied on sexual difference for the ongoing evolution of life, and at what point that sexual difference became gendered, personal, and productive of the figure of familial "man," who would in turn become destructive of the planet.

In order for something like human and organic sexual difference and the entire trajectory of feminism and feminist consciousness to emerge, there must have been a longer duration of geological temporality that enables humanity, its harnessing of the planet's energy, and then (finally) the sexually differentiated person. Again we come up against the problem of indifferent difference: something like man as such, the human as such, emerges from an inscriptive technological trajectory that does not include all humans, and certainly not all life. How is something differentiated, and what other differences might have been drawn such that "we" might not have become the species that was capable of making a geological and destructive difference? If we are claiming that the Anthropocene epoch is a game changer that forces us to rethink a "we" that is given in destructive impact, then we are also prompted to ask about another possible world where what has come to be known as human did not generate such a trace. Why (in the Anthropocene) have we fetishized the differences of our own making, and why are we so sure that we know about the differences that make a difference, or the differences that are readable? Here I would like to make a claim for feminism as a critical labor of difference and indifference. Feminism draws attention to differences that have been deemed not to make a difference, but it has also just as frequently denied what have been declared to be constitutive differences (gender differences, historical differences, religious difference).

II.

This paper was originally delivered at the University of Wisconsin–Milwaukee's "Anthropocene Feminism" conference in April 2014. Richard Grusin opened the conference with a question regarding the possibility of Anthropocene feminism (singular), or Anthropocene feminisms (plural). I want to suggest that we keep feminism in the singular, as one overwhelming problem that generates—inevitably, but in a way that is singular—an impossible multiple. Feminism is always the question of *who*: who speaks, for whom, and whose subjectivity is presupposed in the grammar of the question? Feminism wages a war over difference, either by claiming that woman is not subsumable beneath the figure of man, or by insisting that women cannot be set apart or excluded from the world of man. Here, I want to suggest that if the personal is political, then it is also geological: this is not to say that geology *as stratified* is the scale that must be deployed to read all other scales, but that the figure of "man" in the Anthropocene—industrial man, *homo faber*, *homo economicus*, consumer man, nuclear man—cannot claim to be humanity as such without a prior history of appropriation and stratification.

Anthropocene feminism is therefore multiple, but only in being singular, by always asking, *whose* Anthropocene? When we talk about the planet having been scarred to the point of being geologically readable, what future reader do we imagine, and who is attributing the inscription, and to whom? An Anthropocene feminism would not accept the Anthropocene as an epoch, as a line or strata whose significance would not be in dispute. Rather than think of this line as privileged and epochal, we might ask for whom this strata becomes definitive of *the human*. The concept of humans considered as a species, or as defined by the "human condition," destroys history at the traditional political level—of the polity located within socioeconomic history—and opens a new political scale, but this scale cannot be simply definitive. Since Marxism at least, political questions have contested historical scale: what you take to be "human," "personal," "sexual," or natural needs to be expanded to consider the history from which such timeless notions emerge, but there is no properly political-historical scale. I might consider my sexual identity to be explicable only in terms of mass media conceptions of twenty-first-century subjectivity and lifestyle marketing, or I might think that the bourgeois family and oedipal individual are the proper frame of reference, or that the family and sexuality emerge from a more intense germinal influx of pre-personal human time, or I might think that understanding personal sexuality requires an understanding of an even deeper deep time, focusing on human evolution. The Anthropocene has tended to erase the problem of scale, and has done so by fetishizing difference (the privileged difference, of *this* line, readable by this modality of man). The policy implications of the Anthropocene have tended to suspend the typically feminist questions of this "we" that we seek to maintain, and have instead led to the return to supposed species solidarity. Worse still, the Anthropocene state of emergency is deemed to be of such severity as to short-circuit deliberation; just as the 2008 global financial crisis allowed the immediate bailout of banks without questions of justice and blame being allowed to delay what was declared to be a necessary response, so the severity of the Anthropocene presents itself as justification in advance for executive actions (such as geoengineering). How is it that geological readability (of a specific scale) has become that which defines the human?

Does not this strata of the Anthropocene and what "we" have done imply another possible form of human life that did not reach the point of this late-techno-industrial mode of readability? Now might be the time to think about pre- or non-Anthropocene humans, beings who did not manage to define themselves as a species by way of climate change. How might "we" have been otherwise? Such a consideration would then open a calculation: given where we are now (with industrialism, technoscience, mass media, globalism, and traditions of liberal justice), there might be a threshold at which we might be prepared to sacrifice the historical "progress" we made for the sake of living better. At what point did we become Anthropocene humans? With the invention of the steam engine, with nuclear energy, or perhaps earlier with the forms of the sedentary polity that generated the ideas behind these technologies?

III.

What does the Anthropocene tell us? In what ways is it a game changer? Or in what ways does this event within knowledge and human history alter the relation between thought and its outside? We can begin by discussing three implications to do, respectively, with humanity (and post-humanity), temporality (and history), and sexual difference (and gender).

I want to explore these three possible implications while entertaining the Anthropocene counterfactual: let us imagine that all that is named by the Anthropocene (cataclysmic and irreversible human destruction of the planet) had not occurred. What would we lose or gain, and would we think and act differently if we could live our time over? The reason why I want to pose this counterfactual is to test an insight from phenomenology and post-phenomenology whereby any event that occurs—such as the Anthropocene—should not be seen as external or accidental: *if something is possible, then it should not be deemed to be inessential, but should alert us to what it is in the essence that allows for such a possibility.* This is not to confuse a possibility with a necessity, but it is to say that if something takes place, then it must be a potentiality that is not extrinsic to the being or life under consideration. Or, put less abstractly, what is life such that it is able to generate a species capable of destroying all life? We already know the answer to that question: extinction is not the opposite of life but part of life's possibility.

We might want to say that we can imagine a non-Anthropocene (or even post-Anthropocene) human; what we have come to know as "man" might have evolved differently. Why, then, did human life as it is actualize the capacity to be milieu-destructive: how might we have been otherwise, and would this be the same "we"? We can readily admit that extinction is intrinsic to life. After all, we know that life emerges from catastrophic change, and that human life emerges not only after the mass extinction of nonhuman living species but also in a species bifurcation (where we can just as well imagine another world where the Neanderthals became the species that would possibly look back at once-existing humans). Extinction is the way of the world and of life. But the Anthropocene is different, at least intensively; there is not just more extinction, but a coming into being and passing away that differs in kind. For hundreds of years we have read the fossil record and noted the extinction of many species, but if the Anthropocene is true, humans will have become extinct and not just left a fossil here and there but marked out a geological strata. We knew that life was intertwined with extinction, but we did not know that a life-form could have geological impact.

So let us entertain the counterfactual: humans came into being but did not develop technology to the point where the geological impact of the Anthropocene took place. Some evidence of what such a humanity might look like is offered in societies that have not employed the intensive industrialized agriculture that alters the earth's biomass. Not embarking on that route (of industrialized agriculture) avoids intensive resource depletion and carbon emissions. Of course, it is possible (logically at least) that we might have had a world of industrialized agriculture and production that was not dependent on fossil fuels and finite resources. If such a world were possible, is it the world we would choose? I want to consider this counterfactual scenario, before looking at what has become of history, humanity, and sexual difference in the actual Anthropocene world that we didn't necessarily choose but that befell us nevertheless.

Counterfactual 1: Humanity could have developed differently, remaining more nomadic and with a sense of history more attuned to the broader rhythms of the earth beyond that of the human agricultural year and its seasons. This possibility is suggested in Nigel Clark's (2008) work on the deep historical time and attunement of Aboriginal Australian culture, which not only works with the temporality of nature—including the use of fire to burn back growth that would otherwise fuel larger catastrophic fires—but also has a sense of climactic change far broader than that of the agricultural year of regular and stable seasons. That is, rather than see climate change as an event befalling a stable nature, we might see stable nature as a product of the European imaginary that cannot understand a world that has rhythms and transitions of a complexity greater than the human sense of seasonal change. There's a suggestion in Clark's work that we deconstruct the opposition between climate change denial's claims that major climactic shifts are part of the way of the world and the ecological insistence on the anthropogenic violence done to climate stability: there is no such thing as a natural stability that anthropogenic climate change disturbs in the first instance. "Climate" and "geology" are relational and dynamic composites. What is different about the era of the Anthropocene and anthropogenic climate change on a massive industrialized scale is not that a stable nature has been disturbed, but that humans have increasingly stabilized nature to a mechanized and rigid timetable of production based on hyperconsumption, and this in turn has generated volatile and intense change. Climate change in the anthropogenic sense is the consequence of thinking of nature as an unchanging standing reserve. So to think the counterfactual of the Anthropocene would be to imagine that we had not invented nature (Morton 2007).

Counterfactual 2: The second possibility is to imagine that the material composition of the world might have been otherwise, and that exactly the same technological-industrial complex developed but did so in a way that drew on a renewable resource that did not pollute the atmosphere (or that humans had managed to find a way, from this planetary composition, to develop solar power or some other technology without costing the earth).

If we entertain the first scenario, how would what we have come to know as humanism/post-humanism, history, and gender be different? One possibility is that there would be no such thing as "the human," which relies on a universalizing global imperative that abandons localism and imagines the here and now as being not simply present for me, but present for any subject whatever. We might have to think seriously about the move to abstraction, logic, and universalism that occurs with the state form. Civilizations of industrial complexity are not simply knowledge events but rely on the harnessing of human power, but this enslavement is in turn a consequence of the earth offering sufficient resources of excess that would enable the sorts of sedentary cultures that in turn enable hierarchical power structures and the development of what Jared Diamond has referred to as kleptocracy (Diamond 1999). We do not have to buy into Jared Diamond's specific narrative about how stored energy enables social and political complexity and hierarchy; indeed, it would be important to note again that there would be competition for the point at which differences make a difference. Is it with stored grain, monumental architecture, fire, or some other technology that social units become capable of developing intensive production, capable of producing enough energy to generate "culture" in the highly narrow sense? Only with complex archival inscription and material memory systems such as writing could we have both the expansion of empire and the technological history that generates global hyperconsumption. So we might say that it would be well worth sacrificing Euclid and Newton—if that would be a sacrifice—to avoid embarking on the path to globalism and universal humanism. We would lose human rights, but then we might not require them as much. This is not to say that there would not be human-on-human violence, but neither would there be the industrialized violence of genocide and mass slaughter in warfare, nor would there be the state-forming violence required to develop these potentialities for institutionalized violence. That is, it is only with the appropriation of surplus labor that a culture can develop the symbolic systems required for the organized sacrifice and subjection that mark religion and cultural sublimation, and that in turn enable the collective investment for mass warfare. So, if globalism has been one of the crucial conditions for a sense of the universally human, it is also the case that the ethical demand of the universally human seems imperative in a violent cosmos. Along with human rights we might happily sacrifice universal suffrage, civil rights, and—of course—women's rights, feminism, and certainly ecofeminism, because perhaps we would not quite need these forms of universalism in a simpler and less rationalized cosmos.

That nonglobal, nonindustrial, non-technoscientific world would be one counter-Anthropocene scenario. Such a possibility poses this very inconvenient truth: what we know narrowly as feminism relies on the hyperconsumption mode of globalism. In its enlightened liberal form feminism requires (at least initially) a reading elite liberated from domestic labor, capable of thinking about the freedom of thought and reason, blessed with the favorable conditions of possible personhood, and granted the reflective luxury to use the trope of slavery and its opposite. When Wollstonecraft (1992) called for the rights of woman and extended the trope of slavery and abolition to the liberation of women, she was relying on technological developments that allowed the greater freedom of humans precisely because industry was now extracting energy from the earth, in the form of coal and other ultimately polluting and depleting resources. Women could start to demand equality precisely because of an industrial capitalism in a certain portion of the world that extended the leisure time once reserved for the very few. It is true that perhaps such feminist rights would not have been required without what Carole Pateman (1986) has referred to as the sexual contract, and the ongoing familial form that requires stabilization of the state, and mastery of what has become "nature." This thought experiment or imagined scenario of the counter-Anthropocene that did not arrive at rights and universalism lends more weight to Walter Benjamin's claim that every document of civilization is a document of barbarism (Benjamin 1969). This is not just to say that string quartets are written and appreciated in the same world as mass slaughter, but that some violent subjection of humans for the sake of generating surplus production and energy is required to release the time and space of the history of enlightenment. Philosophy, rights, and—I would argue—the constitution of a private self with a definitive sexuality and fulfilling life trajectory are not merely contingently placed alongside planetary and human-on-human violence. Dipesh Chakrabarty has argued that had humans not embarked on the intense depletion of planetary resources that has resulted in the Anthropocene, human enslavement would have been worse than it has been (Chakrabarty 2009). Quite crudely we can locate abolition and suffrage movements at the same time that industrialized economies were able to extract more planetary energy with less hold on human energy.

This counterfactual brings us not so much to another possible world without humanism, and without a single history of human enlightenment and private sexuality, but rather to a threshold: it is possible to imagine a counterhistory of minimal impact on the planet that might still allow for many of the things we know to be human—including inscription, morality, language, and technology. It is possible to think that humans could never have "progressed" without some planetary damage or alteration but that such "progress" would not have developed to quite such a suicidal and ecocidal pitch and would not have generated the globalism of humanity in general, a single universal time and a private sexual difference, or recognition of one's self via gender.

Let's explore the second counterfactual and ask whether we could have these reparative effects of globalism and humanism without damage to the planet and without the violent extraction of human resources. We would need to imagine another material earth, or another technoscience that began with renewable and nonpolluting energy. We might imagine that rather than relying on slavery to generate the *otium* required for philosophical reflection, the ancient Greeks had found a resource that was nonviolent and yet still allowed for the thought of humanity to be generated. Is such a world logically possible?

That question itself is, I would suggest, symptomatic of the logic of the Anthropocene: the idea of a life that could develop to its utmost potentiality without incurring debt or death to itself is both what drives technological-industrial investment and what generates the delusional idea of a life without expense, loss, or misprision; the notion of generating more (in the final instance) than one initially takes, the dream of a pure ecology in which everything serves to maximize everything else, and in which there is no cost—it is this logic (or the logic of logic, of the pure counterfactual, or pure *techne* without *physis*) that marks all that has stood for humanism, post-humanism, a certain dream of history and of utopian sexual difference. Nowhere is this more evident than in claims for the good Anthropocene: supposedly if we have the power to transform the planet, then we also have the power to transform the planet for the better. Not only does such a dream not ask the question of *whose* betterment the Anthropocene will intend, and not only does the good Anthropocene consecrate the *anthropos* in its current form as a prima facie value, but it sustains the valorization of difference: if we have made a difference, then we can make more of a difference. But all the work on the Anthropocene to date, for all its claims of spikes, cannot agree on *the* difference, and rather than see this as a failure of knowledge, we might ask another question. There are multiple markers or claims for the Anthropocene difference: intensive agriculture, changes in the earth's biomass, nuclear energy, colonization, industrialization, capitalism, and so on. Every one of these markers in turn covers over further differences: can we really charge modernity, or capitalism in general? Certainly "man" is too broad an agent if one wants to think about the difference the Anthropocene marks, but rather than more nuanced differences it might be better to note that any such difference—including the usually targeted capitalism—is achieved by way of indifference (not respecting premodern boundaries, identities, hierarchies, or kinds). To say that we might transform the world by way of geoengineering (make it different and better) might be perceived by many as more of the same. To say that we wish "we" had not made such a difference, and that we should strive to find nature again, is also more of the same. Rather than valorize different forms of a utopian humanity—outside capital, outside industry, beyond humanism—one might say that what has come to be known as the Anthropocene is bound up with a logic of mastering and erasing difference. Either "man" can read the Anthropocene and recognize that there is one humanity after all, different in kind from other organic forms of life, or man can mark off capitalism or corporations as the agents of destruction. Any "good" Anthropocene would be possible only by way of countless injustices, just as what we think of as justice has occurred by way of a history of passed and erased thresholds. We perhaps know with some certainty that a world without geometry, physics, biology, and mathematics would not have yielded the degree of destruction "we" have today, but there are no lines of difference that would allow us to clean up and mitigate the past (even in our imagination). If there had been no Plato, no Averroes, no culture of fire, we would not be in this mess, but—then—who would we be?

Where does this leave us? Are we just saying that life is intertwined with death, that all ethical relations to others also negotiate violence, and that feminism is a grubby queer business that can do so much for so few?

If that were what we are saying, then we would be keeping the counter-Anthropocene ideal in place—yes, that pure world is desirable but never fully achievable. I am arguing something far stronger, I hope, that has implications for how we think about the Anthropocene. All our talk of mitigation and stability maintains a notion of stabilized nature, a nature that is ideally there for us and cyclically compatible with production; it does not confront what the Anthropocene *occludes*. We should not think, at precisely the point at which we posit a geological impact of a certain readable sort, that—had we known this—we would or could have acted in a way that was essentially different or noncontaminating. Or, put another way: the Anthropocene *is* the counter-Anthropocene. We look at the geological scar and remark to ourselves, *as though something has changed*: now, finally, the earth is telling us that we have impact. "My goodness, who would have thought that centuries of slavery, violence, kleptocracy, plundering, and then liberation of some humans at the expense of others—who would have thought that this was a destructive indictment of 'the human'? Who could possibly have imagined that our species was destructive of its milieu without the definitive evidence of the geological record?"

We are thinking of the Anthropocene as exceptional, as a volatility or destabilization of nature that has been caused, accidentally, by us. But what we know about the political state of exception is this: if law can be suspended, leading to a condition of immediate force, without law, then we can experience this lawlessness, this loss of the proper, as always potential, *because there is nothing proper about law* (Agamben 2005). By analogy, this loss of nature, this exceptional volatility and antihuman hostility *of nature* (where nature is now changing on us, refusing to be stable), is the condition from which what we know as the Anthropocene manufactured a stable nature. What we now call climate change is the reemergence of what made climate possible. Climate was manufactured from climate change.

This event of the Anthropocene is exceptional—but not by being extrinsic to what we have come to refer to as nature or humanity; if, now, we are responding to planetary destruction with surprise, and wondering how we might engineer a future that would not cost the earth, then the Anthropocene is hardly an event. It is the continuation of "man," as the being who believes that he can finally be different and transform himself to the point where in his relation to the planet he no longer makes a difference. And if woman—in the form of ecofeminism—claims that she and she alone can offer a *proper, connected, natural, and attuned* relation to the earth, then we have chosen a gendered sexual difference at the expense of the question of how gendered sexual being emerges from a history that is ecologically bound up with violence and depletion.

References

Agamben, Giorgio. 2004. *The Open: Man and Animal*. Stanford: Stanford University Press, 2004.

———. 2005. *State of Exception*. Translated by Kevin Attell. Stanford: Stanford University Press.

Badiou, Alain. 2007. *Being and Event*. Translated by Oliver Feltham. London: Continuum, 2007.

Beer, Gillian. 2000. *Darwin's Plots: Evolutionary Narrative in Darwin, George Eliot, and Nineteenth-Century Fiction*. Cambridge: Cambridge University Press.

Benjamin, Walter. 1969. "Theses on the Philosophy of History." In *Illuminations*, ed. Hannah Arendt, trans. Harry Zohn, 253–64. New York: Schocken.

Bennett, Jane. 2010. *Vibrant Matter: A Political Ecology of Things*. Durham, NC: Duke University Press.

Butler, Judith. 1990. *Gender Trouble:* London: Routledge.

Chakrabarty, Dipesh. 2009. "The Climate of History: Four Theses." *Critical Inquiry* 35 (Winter 2009): 197–222.

Clark, Nigel. 2008. "Aboriginal Cosmopolitanism." *International Journal of Urban and Regional Research* 32, no. 3 (September 2008): 737–44.

Clark, Tim. 2012. "Derangements of Scale." In *Telemorphosis: Theory in the Era of Climate Change*, vol. 1, ed. Tom Cohen, 148–66. Ann Arbor: University of Michigan.

de Lauretis, Teresa. 1987. *Technologies of Gender: Essays on Theory, Film, and Fiction*. Bloomington: Indiana University Press.

Deleuze, Gilles. 1994. *Difference and Repetition*. Translated by Paul Patton. New York: Columbia University Press.

Dennett, Daniel C. 1995. *Darwin's Dangerous Idea: Evolution and the Meanings of Life*. New York: Simon & Schuster.

Diamond, Jared. 1999. *Guns, Germs, and Steel*. New York: W. W. Norton.

Gatens, Moira. 1991. "A Critique of the Sex/Gender Distinction." In *A Reader in Feminist Knowledge*, ed. Sneja Gunew, 139–59. London: Routledge.

Jameson, Fredric. 1981. *The Political Unconscious: Narrative as a Socially Symbolic Act*. Ithaca, NY: Cornell University Press.

Kant, Immanuel. 2002. *Groundwork of the Metaphysics of Morals*. Translated and edited by Allen Wood. New Haven, CT: Yale University Press.

Koyré, Alexandre. 1957. *From the Closed World to the Infinite Universe*. Baltimore: Johns Hopkins University Press.

Latour, Bruno. 2008. *What Is the Style of Matters of Concern?* Amsterdam: Van Gorcum.

———. 2013. "Facing Gaia: Six Lectures on the Political Theology of Nature." http://macaulay .cuny.edu/eportfolios/wakefield15/files/2015/01/LATOUR-GIFFORD-SIX-LECTURES _1.pdf.

Millett, Kate. 1970. *Sexual Politics*. New York: Doubleday.

Morton, Timothy. 2007. *Ecology without Nature: Rethinking Environmental Aesthetics*. Cambridge, MA: Harvard University Press.

Mulgan, Tim. 2011. *Ethics for a Broken World: Imagining Philosophy after Catastrophe*. Montreal: McGill-Queen's University Press.

Oreskes, Naomi, and Erik M. Conway. 2014. *The Collapse of Western Civilization: A View from the Future*. New York: Columbia University Press.

Pateman, Carole. 1988. *The Sexual Contract*. Stanford: Stanford University Press.

Wollstonecraft, Mary. 1992. *A Vindication of the Rights of Woman*. New York: Knopf.

⟩⟩⟩ Air

Karen Pinkus

A dream: It is September 19, 1783. You are inside the palace of Versailles along with the court (including Louis XIV and Marie Antoinette). Perhaps you establish your location there by glancing at yourself in a mirror in the great hall, as a *mise en abyme*. Or is this too much of a cliché? Perhaps you are dressed in period costume. Or maybe an anachronistic Emma Peel–style aerodynamic flight suit designed for maximum mobility. Or both, in the way that dreams allow. You leave the palace and move into the garden just at the moment that the Montgolfiers have finished all of the processes associated with heating air. You breathe in: there is no residual smoke odor. The Mongolfiers are just lifting off, accompanied by several farm animals. Along with the court—oblivious to the preparatory energy-intensive stages—you marvel as the balloon floats upward and then comes back down safely (no animals are harmed in this experiment). Or maybe you are in the basket. Or simultaneously watching from below and floating upward.

> Adapted from *Fuel: A Speculative Dictionary* (Minneapolis: University of Minnesota Press, 2016)

Time to wake up.

Air = nothing.

The dream of generating energy from nothing is nothing new. What if we could power our world with free, clean, unlimited, unmetered air? But we have already used two terms—"energy" and "power"—that imply systems. Energy is the fundamental ability to do work. Power is the rate at which energy is used. Fuels, as I hope to distinguish them from systems of energy, are potentialities, perhaps flowing or trapped in rock, perhaps gaseous and invisible, slippery or noxious, not yet rigidified forms of power. Perhaps already discovered, monetized, projected as future earnings, or offset by taxes or compensatory actions, externalized by companies, discounted into climate models *even while still in the ground*.[1]

What follows is a dictionary of fuels, some familiar and in common use, some imagined, some plausible, some the stuff of (science) fiction. As with any dictionary, the reader is free to read any entry, or read them in any order or to read the thing through. Or to leave it on the shelf without ever cracking it open. My ambition for this dictionary is to scramble our thinking about fuel—not in order to demonize energy per se, and not in order to create a new hierarchy in which certain renewables take over from fossil fuels (although the fossil/nonfossil distinction must be emphasized), but instead to open up potential ways of interacting with substances (real and imaginary), by wrenching them out of narrative (violently in some cases) and placing them into the form of an idiosyncratic dictionary so they could eventually be replaced by users into new narratives.

You won't find "nuclear" in this dictionary. "Nuclear" is not a fuel, and I do my best to retain this rigor, if for no other reason than to provoke thought or surprise or recognition or even fear in readers, to undo a kind of passivity with regard to the place or placement of fuels into vast and interconnected machines, grids, pipelines, storage containers, ecosystems, and even extraplanetary or off worlds. What does it mean to consider fuel as prior to, as punctual toward, energy and its many infrastructures?

One response to this question is to begin to unravel the knot of "future fuels." Oil companies have today transformed themselves into "energy companies" in search and development of fuels for the future.[2] Future fuels may be transitional (natural gas), fossil based, "renewable," fusionable/fissionable, or fantasized substances to come. The American Petroleum Institute has a current project called "Energy Tomorrow," advertising with familiar sorts of images and rhetorics of hope not in the least limited to fuel. Lest we think that this genre is somehow linked to a recent consciousness of end times or catastrophic change, recall a popular slogan from the 1940s: Our times are primitive. True progress is yet to come. Brought to us by the Ford Motor Company. A young Henry Ford watches a water wheel in a stream in the backyard of a sun-drenched farm as he contemplates the flow of metals on the assembly line in future factories.[3] Or you glance into a crystal ball

and see yourself (through a strange optical trick) driving your future wife and rosy-cheeked daughter and dog: "There's a Ford in your future."

This future shouldn't be so far away that we forget it, yet we need time to get our act together. Just a little longer. Don't bother me with positive feedback, turning points, thresholds, Kyoto targets, intended nationally determined contributions. The very companies that the Petroleum Institute represents might even develop carbon capture and sequestration (CCS) divisions that could someday outearn the extraction and combustion of carbon. Regardless, these companies have every reason to blur the distinctions between fuels and energy. The more mystification the better. But genuine "future fuels" never actually come to be, for their time is never any precise moment of political-technological cooperation.

"Potentiality" is a term that will appear at various points in this dictionary. Perhaps the most important thinker to interrogate this concept recently is Giorgio Agamben, who takes it from Aristotle (*dynamis*, as opposed to actuality or *energeia*) and reads it through the lens of Heidegger, among others. To be sure, Agamben is not interested in fuels or energy in the modern geopolitical sense, but rather in

law and the forms of life it lets be or works to inhibit. In Italian, *potentialità* (along with "inoperativity," *inoperosità*, another term that appears in Agamben) feels strange, awkward, just as it does in English. Where does "potentiality" stand with relation to "potential" (as in "potential sources of fuel," for instance) or with relation to power? The Romance languages have sets of two different terms for the latter, both of which are contained in (but perhaps also exceed) "potentiality": *potenza* and *potere*, *puissance* and *pouvoir*, *potencia* and *poder*.[4] The English word "power" is a brawny term that does not seem to allow for non-power or the possibility, but not the actualization, of power or the paradoxical nonuse (of fuels).

Matters are complex and borders fluid. Because fuels have not yet been inserted into a system that will consume them, use them up, they may = hope. This term also reappears throughout this dictionary, but for now let's be clear that it has nothing whatsoever to do with conservation. However much efforts to use "less" fuel (less than what—compared to a baseline calculated during the good ol' days?) on a global scale might produce "fewer" carbon emissions (but then, fewer than what?), conservation is, alas, instrumental rhetoric of the most confounding sort.

We might choose to embrace the kind of austerity projected forward in one dark version of a "postsustainable future," to use Allan Stoekl's term.[5] Or we might use the philosophy of Georges Bataille to think otherwise—toward a future when fossil fuels have been exhausted, but forms of human (excessive) energy power the species in new directions. Ten years after Stoekl published *Bataille's Peak*, we are facing another alternate future: one in which we "never run out of oil." The new North American boom in unconventional oil and gas is changing the geopolitical landscape as I write. By the time this text appears in print, atmospheric concentrations of greenhouse gases will have increased with regard to the present of writing.

Yet writing/thinking fuel is absolutely crucial to *our* survival—whether we take "our" here to signal the species as a whole or "the humanities," which is now qualified in many institutions with "digital," "medical," or more pertinently "environmental" and even "energy."[6] Writing and thinking about climate change, we feel a strong push from those other experts—the engineers and policy scholars; students considering their future "fields"; the scientists who express genuine curiosity about what the humanities can contribute to solutions or, more likely, to adaptation ("resilience"), since this implies a necessary "cultural dimension." Yet—in my own institutional experiences—when the scientists and (especially) the social scientists invite humanists to the table, they do not want to hear how critical theory might take us beyond, might transgress the limits of what could best be defined as a certain type of value-enriched behaviorism. We may help think about a future, but it is one governed by a tyranny of the practical.

Certainly, the pleasure of our company is *not* requested in Bataillian dress. We should probably not speak of "unusable energy" that "by definition does not work, that is insubordinate, that plays *now* rather than contributing to some effort that may mean something at some later date and that is devoted to some transcendent goal or principle." We can expect strange looks and awkward pauses if we bring up "the energy of the universe, the energy of stars and 'celestial bodies' that do no work" or the energy "that traverses our bodies, that moves them in useless and time-consuming ways, that leads to nothing beyond death or pointless erotic expenditure, that defies quantification in measure: elapsed moments, dollars per hour, indulgences saved up for quicker entry into heaven." Talking of sacrifice is not only a downer, but it will lead to our friends looking at their watches and skipping dessert. And then if by "sacrifice" we mean something like "the movement of the opening out, the 'communication,' of self and community with death: the void of the universe, the dead God," the evening will be over rather abruptly.[7] Our survival depends on our ability to be nimble and offer, precisely, something others might call "hope" or "ethics." "No doom and gloom, please," say research centers, funding organizations, and potential collaborators. That will get us nowhere. But where do we think we are going? And powered by what means?[8]

That "future fuels" are perpetually deferred only strengthens the links between hydrocarbons and the present economy. In the present, carbon economy, we live under a prevailing view that "environmental ethics" in the broadest sense (green conscience) and "power" are antithetical terms that may be forced to come together because of external circumstances. And more mundane, ubiquitous marketers who try to reconcile power and "nature" in the most cynical sense actually only end up strengthening the distinction. Consider, for instance, a Volkswagen brochure for "clean diesel"–powered vehicles with the slogan "Go Green without Going Slow." A car speeds along a highway flanked by green hills dotted with wind turbines: "So will you forfeit power in exchange for doing good?" The answer, of course, is no. Yet "no" here, meant to distinguish Volkswagen from its competitors, suggests that power and ethics are, indeed, in the normativity of everyday consumption, incompatible.[9] Or were. Because, of course, this all turned out to be a giant hoax. In part, Volkswagen blamed the stringent requirements of Kyoto and the California Air Resource Board for their actions. (Lo-CARB—another carbon diet fad?)[10]

A fantasy (courtesy of the Ford Motor Company, 1961): You are seated in the Delta-shaped Ford Gyron. As you take off, the wheels, one in front, two in the back, lift up. A gyroscope helps stabilize the vehicle. You have two passengers. Who? Wife and son? When the daughter arrives, it's time to get a family car. But for now . . . you can visualize your estimated . . . no . . . your *exact* time of arrival and speed on a screen in front of the passenger compartment. Though it is foggy outside, you can clearly survey the road conditions ahead thanks to the infrared rays that project onto your "snooperscope." But that's not all: you can communicate (with the office? or your wife can call her mother?) using a "cordless" telephone!

What's that you say? Fuel? Did I mention the gyroscope for stabilization? Fuel? "The shape and silhouette . . . suggest the possible use of new power sources, such as fuel cells, since it is unlikely that any existing internal combustion engine both small enough to fit within the front end and powerful enough to propel the vehicle is presently available." Enjoy the glide![11]

Today, Apple is said to be building an electric car (with built-in iPod/iPad/iPhone/iWatch dock, to be sure). And today, "futuristic" vehicles such as the Tesla Roadster bring together money, power, and speed, but again, only through a magical spatiotemporal dislodgment of fuel beyond the frame of perception. Tesla owners, who may feel good about themselves as green consumers, also suffer from "range anxiety," a pathology the company is working to calm through the blue pill of smart technologies, global positioning software, and interfaces with charging stations. Elon Musk, the company's founder, has announced the arrival of self-driving versions of the Tesla. The self-driving car, center stage at Google's research center, could potentially run on any available fuel. Just imagine the work that autopiloting will generate for law firms and courts, dealing with insurance, responsibility, and indemnity of the humans in or not in the vehicle. Or autonomous vehi-

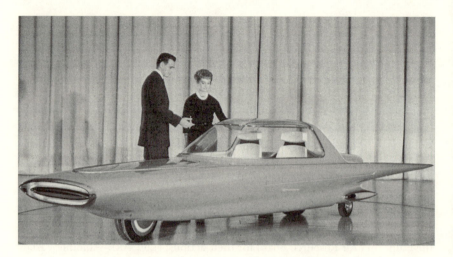

cles might radically transform the entire economy.[12] All of this talk about smart cars . . . but fuel remains buried.

All of this talk about cars . . . when they really contribute only a small proportion of fuel usage, not to mention other sectors contributing to the buildup of greenhouse gases in the atmosphere such as deforestation and agriculture. When "clean," "green," non-carbon-based future fuels will have been "present" (scalable, feasible to the public in the common sense), it will already have been too late with regard to climate change. Notice the awkward use of the future perfect in English in the sentence above! Sober, we witness the contortions of the "future fuel" industry confronting the unfathomable temporality of climate change, of the Anthropocene.

Notes

1. In a recent study, a group of British scientists have quantified (in both monetary and purely physical terms) how much oil and coal already discovered would have to remain in the ground in order to avoid "catastrophic warming." The study is based on the assumption that cheaper reserves would be tapped first, before "tough oil." The authors are able to paint a very real scenario about losses already to be suffered by large energy companies and nation-states: they name names. Now clearly, this study, which imagines *not using* over 90% of US and Australian coal, as well as almost all of Canada's oil sands ($20 trillion), is a concretization of potentiality, yet potentiality also exceeds it. Leaving fuel in the ground, then, also means a special relationship to it that not only is nonconsumerist but is so complex that potentiality seems the best way to recognize this complexity.

2. A full 43 of the 500 Companies in the Standard and Poors Index are "energy companies," which makes strategies for divestment by universities or other corporate bodies especially complicated.

3. "A Boy . . . a Water Wheel . . . and a Dream!" The ad was widely printed in farm publications in the mid-1940s.

4. Antonio Negri's key notion of "constituent power" is a combination of *potere* (power) and *potenza* (strength). As Maurizia Boscaglia expands in a translator's note to *Insurgencies* (*Il potere costituente*), for Negri *potere* refers to the existing power of the state and institutions. *Potenza* is a "radically democratic force that resides in the desire of the multitude and is aimed at revolutionizing the status quo through social and political change. Strength is at the core of the concept of constituent power itself as the force that produces (but cannot be contained within) power and its institutions; constituent power is fueled by strength." (Trans. Note 3 in Negri 336). Negri's work around biopolitics certainly appears addressed to human capacity for productive action and cannot be immediately translated into the realm of matter. Antonio Negri, *Insurgencies: Constituent Power and the Modern State*, trans. Maurizia Boscagli (Minneapolis: University of Minnesota Press, 2009).

5. See Allan Stoekl's *Bataille's Peak* for a brilliant discussion of postsustainable futurity: a future that might offer an alternative to austerity and sacrifice through excess energy, a solar energy so immense it cannot be metered, monetized, or contained. A Bataillian future might not sit well with those policy makers who demand practical solutions, but his imagination seems at least to approach the unfathomability and "exorbitant" nature of climate change. Allan Stoekl, *Bataille's Peak: Energy, Religion, and Postsustainability* (Minneapolis: University of Minnesota Press, 2007).

6. Dominic Boyer and Imre Szeman are two important proponents of an "energy humanities," a field that they note as being recognized by "the sciences, by government, indeed by industry." See their jointly authored piece "The Rise of Energy Humanities: Breaking the Impasse," February 12, 2014, www.universityaffairs.ca/opinion/in-my-opinion/the-rise-of-energy-humanities.

7. Stoekl, *Bataille's Peak*, xvi.

8. In "Murmurations—'Climate Change' and the Defacement of Theory," his introduction to the edited collection titled *Telemorphosis*, Tom Cohen addresses the sovereign debt crisis in relation to attempts to "save" the humanities and to questions of climate change itself. Tom Cohen, introduction to *Telemorphosis: Theory in the Era of Climate Change*, vol. 1, ed. Tom Cohen (Ann Arbor: University of Michigan Press, 2012).

9. The text of the ad reads, "Go Green without Going Slow." "Driving a fuel-efficient* car shouldn't mean sacrificing performance. It's this belief that drove us to develop the TDI Clean Diesel engine and a turbocharged hybrid. Both help maximize driving dynamics while helping minimize environmental impact. And we've expanded that thinking beyond cars to everything we do; from the first LEED Platinum-certified automotive plant to working with the Surfrider Foundation to protect our oceans. It's called thinking blue and it's how we're thinking beyond green. *That's the power of German Engineering.*" The asterisk directs readers to the US Government site for mpg estimates. While in the public sphere cars and transportation, individual consumption, and choices about fuel are dominant, in the broad scheme of things all forms of transportation (including planes, trains, automobiles, boats, and so on) account for less than 15% of global greenhouse emissions. In the US the percentage is higher. Still, the car must be put in proper perspective. For Allan Stoekl the car is the significant metonymy for a larger question of expenditure linked to fossil fuels,

against which he will posit a bodily excess rather than a mere conservation of energy. To be sure, switching to electric cars, even if fossil fueled, would be more efficient than gasoline. Electric cars have smaller engines and no transmission, which makes them lighter, for instance. But long before the electric car was invented, killed, and revived, in Jules Verne's *Robur* a group of Philadelphia balloon enthusiasts take their "Go-Ahead" for a spin when they encounter another flying object. At first, the men on the ground wonder whether it is the flapping of bird wings. But then, "a suspicion communicated itself electrically to the brains of all on the clearing." It is the Albatross, the superior and powerful electric aerostat!

10. For a succinct summary of the "clean diesel" swindle, see Taras Grescoe, "The Dirty Truth about 'Clean Diesel,'" *New York Times*, January 3, 2016, Sunday Review section, 7.

11. Ford Gyron Brochure ca. 1954. Collection of the author.

12. Zack Kantor, "How Uber's Autonomous Cars Will Destroy Millions of Jobs and Reshape the US Economy by 2025," *Personal Blog of Zack Kantor*, http://zackkanter.com/2015/01/23/how-ubers-autonomous-cars-will-destroy-10-million-jobs-by-2025/.

❭❭❭ Excerpt from *Cyclonopedia: Complicity with Anonymous Materials*
Reza Negarestani

To grasp war as a machine, or in other words, to inquire into the Abrahamic war machine in its relation to the technocapitalist war machine, we must first realize which components allow Technocapitalism and Abrahamic monotheism to reciprocate at all, even on a synergistically hostile level. The answer is oil: War on Terror cannot be radically and technically grasped as a machine without consideration of the oil that greases its parts and recomposes its flows; such consideration must begin with the twilight of hydrocarbon and the very dawn of the earth. In Dean Koontz's novel *Phantoms*, Timothy Flyte, a renegade paleontologist who considers himself a professor of Ancient Epidemics, is a tabloid writer researching an unnameable Tellurian sentient being which he calls the Ancient Enemy, responsible for devouring countless civilizations (the Aztecs and the Lost Colony at Roanoke, for example). A bio-chemical combat unit invites him (in line with *The Exorcist*, in which neurologists invite a vicar for assistance) to investigate the mysterious disappearance of people in a village in Colorado. The Ancient Enemy is a Thing*like bio-hazardous predator hunting organic entities, using bio-sorcery and mutating various organic phyla (possessing a soldier and turning his blood into a small lizard). The Ancient Enemy is trying to spread its gospel via three chosen characters. Timothy Flyte finds many parallel traits between The Ancient Enemy and The Antichrist. Examining the corpses of victims, he detects traces of porphy-

Cyclonopedia: Complicity with Anonymous Materials (Melbourne: re.press, 2008), 56–68, 107–9

rin, a chemical substance common to blood, plants, and petroleum. The Ancient Enemy or the Tellurian Antichrist that persistently looms in the Mesopotamian dead seas (originally where Antichrist comes from) or near the oceans is Petroleum or *Naft* (Arabic and Farsi word for "oil").

According to the classic theory of fossil fuels (i.e., excluding Thomas Gold's theory of the Deep Hot Biosphere), petroleum was formed as a Tellurian entity under unimaginable pressure and heat in the absence of oxygen and between the strata, in absolute isolation—a typical Freudian Oedipal case, then . . . Petroleum's hadean formation developed a satanic sentience through the politics of in-between which inevitably "wells up" through the God-complex deposited in the strata (the logic of "double-articulation, the double-pincer" according to Deleuze and Guattari), to the surface. Envenomed by the totalitarian logic of the tetragrammaton, yet chemically and morphologically depraving and traumatizing Divine logic, petroleum's autonomous line of emergence is twisted beyond recognition. Emerged under such conditions, petroleum possesses tendencies for mass intoxication on pandemic scales (different from but corresponding to capitalism's voodoo economy and other types of global possession systems). Petroleum is able to gather the necessary geo-political undercurrents (subterranean or blobjective narrations of politics, economy, religion, etc.) required for the process of Eradication or the moving of the earth's body toward the Tellurian Omega—the utter degradation of the earth as a Whole. As the ultimate Desert or Xerodrome, the Tellurian Omega engineers a plane of utter immanence with the sun where the communicator can no longer be discriminated from what is communicated to the sun. Xerodrome is the earth of becoming-Gas or cremation-to-Dust. Ironically, this earth as a degenerate wholeness and twisted sentience overlaps with the Desert of Cod on which no idot may be erected. And in fact, the desert of God is manipulated on behalf of the Tellurian Omega and its undercurrents. Monotheism in its ultimate scenario is a call for the Desert—the monopolistic abode of the Divine. In the end, everything must be leveled to fulfill the omnipresence and oneness of the Divine. So that for radical Jihadis, the desert is an ideal battlefield; to desertify the earth is to make the earth ready for change in the name of the Divine's monopoly, as opposed to terrestrial idols. In line with Wahhabi and Taliban Jihadis, for whom every erected thing, so to speak, every verticality, is a manifest idol, the desert, as militant horizontality, is the promised land of the Divine.

In light of the emphatic horizontality of the desert in monotheistic apocalypticism, Deleuze and Guattari's model of horizontality or plane of consistency can only be a betrayal of radical politics and a hazardous misunderstanding of the war machine. However, in geological reality, monotheism functions as an involuntary host for Tellurian insurgencies and undercurrents; it is directly connected to the twisted nether regions of the earth itself. Monotheism is a convoluted plane of tactics and meta-strategies for giving rise to Tellurian blasphemies or twisted strains of geological reality. In the wake of monotheism, Tellurian insurgencies feed on their corresponding, seemingly religious counterparts belonging to the monopoly of the Divine: the blobjective earth is nurtured by petropolitics. Tellurian Omega grows on the desert of God, ad infinitum. The Kingdom of Apocalypse or monotheistic desert is a passageway through which the earth's ultimate blasphemy with the Outside smuggles itself in and begins to unfold. The apocalyptic desert is a field through which the Tellurian Dynamics of the earth can be ingrained within anthropomorphic belief systems. Camouflaged within the formation of belief, Tellurian insurgencies can be safely accelerated, steadily developed, anomalously recomposed, and intensified by anthropomorphic entities, either through religions or through seemingly secular societies whose economic systems are still rooted in monotheistic platforms. In which case, there is no worse Tellurian blasphemy than "Thy Kingdom come." Those Mecca-nomic agencies of War on Terror who consider everything that is not a desert a violation against the all-consuming hegemony of God crave for the desert as a ground independent of Earth and its inhabitants; but what they actually achieve, and passively cooperate with, is the Tellurian insurgency of the earth toward Xerodrome. Ibn Hamedani calls this desert the "Mother of All Plagues"—a plan(e) for reaching immanence with the molten core of the earth and the sun (the tide of extinction). On this plane, you either turn into diabolical particles, or evaporate and are recollected as cosmic-pest ingredients. This is exactly where religious extremists (the Taliban, with their ironically phallomaniac hatred for anything erected, for instance) turn into the stealth mercenaries of geological insurgencies, the cult of Tellurian Blasphemy (demonogrammatical decoding of the earth's body). They want Cod but what they get is the Tellurian Omega—the incinerating immanence with the sun and the earth's core assembled on an axis which knows nothing of authoritarian divine and monopolistic convergence, the Hell-engineering Axis of the earth.

It seems therefore that both the technocapitalist process of desertification in War on Terror and the radical monotheistic ethos for the desert converge upon oil as an object of production, a pivot of terror, a fuel, a politico-economic lubricant, and an entity whose life is directly connected to Earth. While for Western technocapitalism the desert gives rise to the oiliness of war machines and the hyperconsumption of capitalism en route to singularity, for Jihad oil is a catalyst to speed the rise of the Kingdom, the desert. Thus for Jihad, the desert lies at the end of an oil pipeline.

Or, once again, take Oil as a lubricant, something that eases narration and the whole dynamism toward the desert. The cartography of oil as an omnipresent entity narrates the dynamics of planetary events. Oil is the undercurrent of all narrations, not only the political but also that of the ethics of life on Earth. Oil lubes the whole desert expedition toward Tellurian Omega (either as the Desert of God or the host of singularity, the New Earth). As a Tellurian lube, oil simply makes things move forward. Koontz's *Phantoms* is key for this movement toward Tellurian Omega, through the superficial (GAS pipeline), subterranean (Oil reservoirs), and deeply Chthonic (Thomas Gold's *Deep Hot Biosphere*) Thingness of petroleum, the Blob. To grasp oil as a lube is to grasp Earth as a body of different narrations being moved forward by oil. In a nutshell, oil is a lube for the divergent lines of terrestrial narration.

* * *

The distribution of porosity through the earth does not follow a rhizomatic structure but goes by way of random clusters with variable densities, similar to the dispersion of suspended dust and moisture in fog. The mutual contamination of solid and void in holey space is increasingly intensifying, with no end in sight, since it is the internal impetus of solid to be active, to re-modify itself, to knit itself through economic networks which maintain and guarantee its survival and growth, assisting its quest to be grounded. All activities throughout the solid part are reinvented as convoluting lines (Nemat-functions) at deepest levels of the composition. Whenever the solid messes with the void in order to keep itself dynamic and solidly "constructive" or "consolidated," the void only becomes more contaminative, its worm-functions become more furious, excited to the point of frenzy; they begin to rise from compositional depths to engineer the vermicular space of the Old Ones, an intricate traffic zone, the ( )hole complex. In this way, with every activity that it willfully undertakes, the solid levels all obstacles in its path to damnation. ( )hole complex is inexhaustible in its infidelity and perfidiousness; it is the source of the clandestine manipulation of solidus and of a double treachery against solid and void alike.

In the past, the holey space of mines incited peasant revolutions and barbarian invasions, but now it is oil fields that make technocapitalist terror-drones and the desert-militarism of Islamic Apocalypticism cross each other, forming militarization programs and complicities for revolutionizing the planetary surface. If oil has undergone a process of weapomzation on the Islamic front of War on Terror, and has turned into a fuel for technocapitalist warmachines, this is not a matter of a politico-economic evolution: in Arabia, Sudan, Libya, Syria and even the Arabic clusters below the Persian Gulf, the Islamic state must cross deserts to feed on oil fields because of the exclusive location of oil fields in these countries. But the desert is the space of nomad-burrowers, desertnomads, and their warmachines with minimum climatologic regulation.

Of all nomads traversing the earth, those most radical in their forging of warmachines under the minimum influence of climatologic factors are the desertnomads. This is why both the renomadization of the Wahhabistic state of Saudi Arabia through desert-militarism (belonging to desert-nomads) and the semisedentarization of nomads (by the State) and their metamorphosis into *nuphtatwse* (clandestine petro-nomads who roam between oil fields instead of oases) were inevitable. Bound to purely nomadic ways of living, and remaining relatively distant from environmental factors of climatological dependence such as water, moderate climate, or diverse pastoralism (they are mostly just camel-keeping pastoralists), the nomads of Arabia retained truly nomadic traits until the mid-twentieth century, and were introduced to the State's sedentarization programs very late. This premature but late sedentarization of desert nomads by the oil-seeking State was in fact the main factor in the contamination of the desert nomads of Arabia with petropolitics and the pestilential vitalization of Wahhabi religion with nomadic tactics, ways of life, and the logic of the desert. Thus the contemporary religio-political traits of Wahhabism (the Wahhabistic agencies supposedly targeted by the War on Terror) are undoubtedly diverse mutations resulting from the attraction of the occultural and alien elements of desert nomadism through oil. The State and desert-nomads were introduced, and slid inexorably toward each other, through the poromechanics of oil and the holey space shaped by the logic of oil extraction in the desert. World petropolitics—earth as narrated by oil—in regard to the Middle East and the War on Terror emerges out of these mutual contaminations between States and desert nomads, facilitated by the holey space of petroleum.

Oil fields draw petrological nemat-spaces between the State and the desert-nomads. On the one hand, such spaces manipulate the State by furious desert-nomadism, and on the other, they reconfigure desert-militarism of the nomads according to the State's petropolitics. This corresponds with the ambiguity between solid and void in the ( )hole complex, which traffics and smuggles its own itinerant lines through the polemics of solid and void. The problem of oil fields and the ( )hole complex between the State and desert-nomads is indeed far more sophisticated than that of mines and their ambulant dwellers (miners). Firstly, there is no equivalent, in oil fields, for the miner, since the connection of *naphtanese* (former desert-nomads) with oil is not an intimacy based on consumption, production, or even transportation (what connected old miners to mines as their temporary niches). Secondly, oil as a ubiquitous earth-crawling entity—the Tellurian Lube—spreads the warmachines and politics of *naphtanese* or desert-nomads as totally pervasive entities. Finally, even in the absence of desert-nomadism, oil turns Time toward apocalyptic blasphemies. A patch of oil is enough to stir the apocalypse out of Time. If oil does not benefit the middle class (an economic boom initially moderating the economy but ultimately giving rise to economic fissions), and if it does not lead to the outbreak of cannibalistic economies—as in the case of Mexico, Venezuela, Sudan, and possibly Mauritania—it will certainly charge clandestine-military pipelines with apocalyptic modes of divergence, as in the case of Islamic countries. In any case, oil, with its poromechanical zones of emergence in economy, geopolitics, and culture, mocks the Divine's chronological Time with the utmost irony and obscenity.

PART FOUR

The Aesthetics of Petrocultures

Energy in the fossil fuel era has come to be associated with transnational corporate intrigue, statecraft, and geopolitical power. But the forms of energy we have used at different moments in history have also had a very real impact on culture and aesthetics, a fact that is only now beginning to be more fully understood. The energy historian Vaclav Smil has written that "timeless artistic expressions show no correlation with levels or kinds of energy consumption."[1] As the pieces in this section show, this may well be one of the few times that Smil has been wrong in tracking how energy has shaped society. Far from being timeless, cultural and aesthetic expressions are deeply marked by the shape and character of the energy systems that give birth to them. The arts of a world powered by horses and the labor of bodies cannot help but be distinct from the expanded time, space, and power of our own petromodernity. The growing recognition of the complex ways in which culture has been shaped and marked by energy has generated some of the most compelling and original work in the energy humanities and promises to change both cultural analysis and cultural expression in the twenty-first century.

This section contains work by critics, writers, and artists who, in recognition of the significance of energy in the development of aesthetics and culture, have begun to grapple seriously with the influence of shifts and changes in energy systems. One of the common ways of organizing the historical development of the arts is into movements or periods, such as Romanticism and Victorianism (in literature) or modernism and postmodernism (in fields from architecture to the visual arts). In their contributions, Patricia Yaeger and Barry Lord argue that building such cultural histories around energy would add important new insights into our comprehension of the development of cultural forms, as well as the social character of our past and present energy systems. At present, the important material and social impact of energy is entirely missing from our accounts of the development of literature (Yaeger) and the visual arts and architecture (Lord). Making energy a key element of our analytic framework would give us a better understanding of the cultural transitions that have accompanied energy transitions and would help add energy more determinately to cultural analysis. It might also offer us a sense of cultural shifts that could aid in our move away from the fossil fuel era. A prime example of such a shift can be found in Lord's description of a new "culture of stewardship" that exists in forms of contemporary art practice that speak to the "use of all resources with a view to the needs of future generations as we aim to meet our current requirements as efficiently as possible."

One of the earliest links between energy and aesthetics was made by Amitav Ghosh in his influential 1992 essay "Petrofiction: The Oil Encounter and the Novel." Ghosh's essay is, in part, a review of Adbul Rahman Munif's monumental *Cities of Salt* quintet, which maps the coming-to-be of American petroleum interests in the Middle East and its world-altering impact on those living in the Arabian Peninsula. "Petrofiction" also offers a broader critique of the failure of literary fiction to address what Ghosh names the "Oil Encounter": the intertwining of the fates of Americans and those living in the Middle East around this commodity, a connection that continues to have far-reaching economic, cultural, and political reverberations. It is not only with respect to this specific oil encounter that literature has seemed to miss the mark. While there are a number of novels that have contended with the importance of oil to modernity—from Upton Sinclair's *Oil!* (1927) to more recent work such as Helon Habila's *Oil on Water* (2011) and Tom McCarthy's *Satin Island* (2015)—Ghosh's insight has resonated with critics and theorists who have attended to the representational challenges presented by energy in general and fossil fuels in particular. For the most part, despite their evident sociohistorical importance, fossil fuels have played a very small role in contemporary artistic and literary expression; this curious absence is one that many of the critical and creative pieces in this section struggle to better understand.

Jennifer Wenzel's "Petro-Magic-Realism" offers an example of how to illuminate the aesthetic operations of oil. In order to offer further specificity and analytic density to the widely used literary category of "magic realism," Wenzel argues that the "political ecology" of Nigerian literature is better described as "petro-magic-realism," "a literary mode that combines the transmogrifying creatures and liminal space of the forest in Yoruba narrative tradition with the monstrous-but-mundane violence of oil exploration and extraction, the state violence that supports it, and the environmental degradation that it causes." Such mappings of specific encounters between aesthetics and politics via the medium of oil remain important: there are all too many oil encounters about which the whole story has yet to be told. At another level, however, it is possible to make the claim that *any and all* examples of cultural expression in the era of oil have been crucially figured by it, rendering not only Nigerian or American literature but all of modern literature into a petro-literature. Graeme Macdonald offers a provocative and compelling overview of many of the key issues that have arisen for critics as they have begun to grapple with the entwined histories of energy and culture. For Macdonald, "fiction, in its various modes, genres, and histories, offers a significant (and relatively untapped) repository for the energy-aware scholar to demonstrate how, through successive epochs, particularly embedded kinds of energy create a predominant (and oftentimes alternative) culture of being and imagining in the world; organizing and enabling a prevalent mode of living, thinking, moving, dwelling, and working." His article unfolds the implications of what fossil fuels have meant for being and imagining the world, and what they have meant, too, for the fictional forms that have developed in relation to this world shot through with "narrative energetics" and "psycho-social dynamics" linked to oil's force and power.

In addition to contributions from critics, this section contains contributions from fiction writers and poets who have focused their attention on how oil and energy shape human experience. Warren Cariou's short story "An Athabasca Story" narrates Elder Brother's encounter with the technological apparatus of the Alberta oil sands, which has plopped itself down in the middle of his territory. Like the encounter Munif describes, this meeting is characterized by a blind sense of entitlement by those engaged in extracting oil and a sense of puzzlement by those native to the land whose space and culture have been invaded. Tensions over land, belonging, and ownership and the sharp divide between those who want access to fossil fuel resources and those who want to preserve the spaces of their lives are also explored in the poems from Julia Kasdorf's *Shale Play*. In their distinctive style and structure, her three poems draw attention to the physical and psychological dislocations enacted by fracking in the US Northeast.

What is significant about the growth of literary works that address energy and fossil fuels isn't only that they fill in the gap in our imaginaries to which Ghosh pointed. These works offer examples of writers who are attuned to how literature is constituted by fossil fuels in deeper and darker ways that necessitate formal experimentation to fully draw out and expose to the light. Kasdorf's "Sealed Record" reads like a court transcript, emphasizing the mechanisms of law and history that can shape and entrap us. Lesley Battler's poem about the oil sands, "Pax Mc-Murray," offers mock gravitas to Canada's petroleum empire by narrating it through the language of the Roman Empire. In "Idylls of Inuvik," Battler transforms an oil company relocation guide into the language of Henry James, while in another poem, the magazine *Oil Week* is placed into dialogue with the language of philosopher Michel Foucault. Such experiments in literary form figure the aesthetics of petrocultures as endemic to contemporary life and thought.

Perhaps no contemporary writer pushes us more to recognize the degree to which our culture is interpolated with our own dominant form of energy than the poet Adam Dickinson. His poems here are configured around the chemical shape of the polymers at the heart of the plastics that have become an index of modernity. For instance, "Hail" offers us a taxonomy of the plastics that now exist *within* all living systems, including human beings: we unwillingly ingest plastics into our bodies throughout our lives, passing it on to next generations right at birth through the milk infants ingest from the bodies of their mothers. Today, we don't just depend on oil for energy; by converting it into plastics, and by ingesting those plastics even without noticing it, we spend our lives *becoming* petroleum, whether we want to or not.

Macdonald writes, "if *all* fiction is potentially energetic, valorizing energy use, then how do we kinetically assert our claims and configure our readings to make it more apparent?" The writers and critics in this section interrogate, navigate, and experiment with the cultural configurations of energy use and the energy content of cultural history. And as the interview between Andrew Pendakis and Ursula Biemann emphasizes, questions about the aesthetics of energy are not limited to the literary arts. The visual arts have seen a similar explosion of interest in energy and fossil fuels, including the need for formal experimentation in figuring them aesthetically, alongside a belated recognition of their (too) long absence by naming them at all. Biemann has been one of the pioneers in exploring resource culture and pondering the specific visual challenges presented by oil and energy; her video essay *Black Sea Files* is both a study of the politics surrounding the BTC pipeline from Azerbaijan to Turkey and an interrogation of the challenge of representing oil visually.

What links the diverse pieces in this section is their insistence on the saturation of our culture and aesthetics in the energy of fossil fuels and its consequences for attempts through aesthetics to comprehend the world we have wrought for ourselves. "The petroleum infrastructure has become embodied memory and habitus for modern humans, insofar as everyday events such as driving or feeling the summer heat of asphalt on the soles of one's feet are incorporating practices," Stephanie LeMenager writes. "Decoupling human corporeal memory from the infrastructures that have sustained it may be the primary challenge for ecological narrative in the service of human species survival beyond the twenty-first century." Her essay on the affective and bodily dispositions of petrocultures names one of the key challenges of decoupling culture and oil, as we start the process of creating a new aesthetics that might animate the culture of stewardship that Lord and others hope will supplant the culture of oil in which we still live today.

Notes

1. Vaclav Smil, "World History and Energy," in *Encyclopedia of Energy*, vol. 6, ed. Cutler J. Cleveland (Philadelphia: Elsevier, 2004), 55.

⟩⟩⟩ Petrofiction: The Oil Encounter and the Novel

Amitav Ghosh

If the spice trade has any twentieth-century equivalent, it can only be the oil industry. In its economic and strategic value as well as its ability to generate far-flung political, military, and cultural encounters, oil is clearly the only commodity that can serve as an analogy for pepper. In all matters technical, of course, the comparison is weighted grossly in favor of oil. But in at least one domain, it is the spice trade that can claim the clear advantage: in the quality of the literature that it nurtured.

Within a few decades of the discovery of the sea route to India, the Portuguese poet Luis de Camões had produced *The Lusiads,* the epic poem that chronicled Vasco da Gama's voyage and in effect conjured Portugal into literary nationhood. The Oil Encounter, in contrast, has produced scarcely a single work of note. In English, for example, it has generated little apart from some more or less second-rate travel literature and a vast amount of academic ephemera—nothing remotely of the quality or the intellectual distinction of the travelogues and narratives produced by such sixteenth-century Portuguese writers as Duarte Barbosa, Tomé Pires, and Caspar Correia. As for an epic poem, the very idea is ludicrous. To the principal protagonists in the Oil Encounter (which means, in effect, America and Americans on the one hand and the peoples of the Arabian Peninsula and the Persian Gulf on the other), the history of oil is a matter of embarrassment verging on the unspeakable, the pornographic. It is perhaps the one cultural issue on which the two sides are in complete agreement.

Still, if the Oil Encounter has proved barren, it is surely through no fault of its own. It would be hard to imagine a story that is equal in drama, or in historical resonance. Consider its Livingstonian beginnings: the Westerner with his caravan loads of machines and instruments, thrusting himself unannounced upon small, isolated communities deep within some of the most hostile environments on earth. And think of the postmodern present: city-states where virtually everyone is a "foreigner," admixtures of peoples and cultures on a scale never before envisaged, vicious systems of helotry juxtaposed with unparalleled wealth, deserts transformed by technology, and military devastation on an apocalyptic scale.

It is a story that evokes horror, sympathy, guilt, rage, and a great deal else, depending on the listener's situation. The one thing that can be said of it with absolute certainty is that no one anywhere who has any thought for either his conscience or his self-preservation can afford to ignore it. So why, when there is so much to write about, has this encounter proved so imaginatively sterile?

Incendiary Circumstances (1992; repr., Boston: Houghton Mifflin, 2005), 138–51

On the American side, the answers are not far to seek. To a great many Americans, oil smells bad. It reeks of unavoidable overseas entanglements, a worrisome foreign dependency, economic uncertainty, risky and expensive military enterprises; of thousands of dead civilians and children and all the troublesome questions that lie buried in their graves. Bad enough at street level, the smell of oil gets a lot worse by the time it seeps into those rooms where serious fiction is written and read. It acquires more than just a whiff of that deep suspicion of the Arab and Muslim worlds that wafts through so much of American intellectual life. And to make things still worse, it begins to smell of pollution and environmental hazards. It reeks, it stinks, it becomes a Problem that can be written about only in the language of Solutions.

But there are other reasons why there isn't a Great American Oil Novel, and some of them lie hidden within the institutions that shape American writing today. It would be hard indeed to imagine the writing school that could teach its graduates to find their way through the uncharted firmaments of the Oil Encounter. In a way, the professionalization of fiction has had much the same effect in America as it had in Britain in another, imperial age: as though in precise counterpoint to the increasing geographical elasticity of the country's involvements, its fictional gaze has turned inward, becoming ever more introspective, ever more concentrated on its own self-definition. In other words, it has fastened upon a stock of themes and subjects, each of which is accompanied by a well-tested pedagogic technology. Try to imagine a major American writer taking on the Oil Encounter. The idea is literally inconceivable. It isn't fair, of course, to point the finger at American writers. There isn't very much they could write about: neither they nor anyone else really knows anything at all about the human experiences that surround the production of oil. A great deal has been invested in ensuring the muteness of the Oil Encounter: on the American (or Western) side, through regimes of strict corporate secrecy; on the Arab side, by the physical and demographic separation of oil installations and their workers from the indigenous population.

It is no accident, then, that the genre of "My Days in the Gulf" has yet to be invented. Most Western oilmen of this generation have no reason to be anything other than silent about their working lives. Their working experience of the Middle East is culturally a nullity, lived out largely within portable versions of Western suburbia.

In some ways the story is oddly similar on the Arab side, except that there it is a quirk of geography—of geology, to be exact—that is largely to blame for oil's literary barrenness. Perversely, oil chose to be discovered in precisely those parts of the Middle East that have been the most marginal in the development of modern Arab culture and literature—on the outermost peripheries of such literary centers as Cairo and Beirut.

Until quite recently, the littoral of the Gulf was considered an outlying region within the Arab world, a kind of frontier whose inhabitants' worth lay more in their virtuous simplicity than in their cultural aspirations. The slight curl of the lip that inevitably accompanies an attitude of that kind has become, if anything, a good deal more pronounced now that many Arab writers from Egypt and Lebanon—countries with faltering economies but rich literary traditions—are constrained to earn their livelihood in the Gulf. As a result, young Arab writers are no more likely to write about the Oil Encounter than their Western counterparts. No matter how long they have lived in the Gulf or in Libya, when it comes to the practice of fiction, they generally prefer to return to the familiar territories staked out by their literary forebears. There are, of course, some notable exceptions (such as the Palestinian writer Ghassan Kanafani's remarkable story "Men in the Sun"), but otherwise the Gulf serves all too often as a metaphor for corruption and decadence, a surrogate for the expression of the resentment that so many in the Arab world feel toward the regimes that rule the oil kingdoms.

In fact, very few people anywhere write about the Oil Encounter. The silence extends much further than the Arabic- or English-speaking worlds. Take Bengali, a language deeply addicted to the travelogue as a genre. Every year several dozen accounts of travel in America, Europe, China, and so on are published in Bengali, along with innumerable short stories and novels about expatriates in New Jersey, California, and various parts of Europe. Yet the hundreds of thousands of Bengali-speaking people who live and work in the oil kingdoms scarcely ever merit literary attention—or any kind of interest, for that matter.

As one of the few who have tried to write about the floating world of oil, I can bear witness to its slipperiness, to the ways in which it tends to trip fiction into incoherence. In the end, perhaps, it is the craft of writing itself—or rather writing as we know it today—that is responsible for the muteness of the Oil Encounter. The experiences that oil has generated run counter to many of the historical imperatives that have shaped writing over the past couple of centuries and given it its distinctive forms. The territory of oil is bafflingly multilingual, for example, while the novel, with its conventions of naturalistic dialogue, is most at home within monolingual speech communities (within nation-states, in other words). Equally, the novel is never more comfortable than when it is luxuriating in a "sense of place," reveling in its unique power to evoke mood and atmosphere. But the experiences associated with oil are lived out within a space that is no place at all, a world that is intrinsically displaced, heterogeneous, and international. It is a world that poses a radical challenge not merely to the practice of writing as we know it but to much of modern culture: to such notions as the idea of distinguishable and distant civilizations, or recognizable and separate "societies." It is a world whose closest analogues are medieval, not modern—which is probably why it has proved so successful in eluding the gaze of contemporary global culture. The truth is that we do not yet possess the form that can give the Oil Encounter a literary expression.

For this reason alone, *Cities of Salt*, the Jordanian writer Abdelrahman Munif's monumental five-part cycle of novels dealing with the history of oil, ought to be regarded as a work of immense significance. It so happens that the first novel in the cycle is also in many ways a wonderful work of fiction, perhaps even in parts a great one. Peter Theroux's excellent English translation of this novel was published a few years ago under the eponymous title *Cities of Salt*, and now its successor, *The Trench*, has appeared. Munif's prose is extremely difficult to translate, being rich in ambiguities and unfamiliar dialectical usages, and so Theroux deserves to be commended for his translations—especially of the first book, where he has done a wonderful job. He is scrupulously faithful to both the letter and the spirit of the original, while sacrificing nothing in readability. Where Theroux has intervened, it is in what would appear to be the relatively unimportant matters of punctuation and typography. (He has numbered each chapter, though the Arabic text does not really have chapters at all, but merely extended breaks between pages; and he has also eliminated Munif's favorite device of punctuation, a sentence or paragraph that ends with two period points rather than one, to indicate indeterminacy, inconclusively, what you will . . .). These changes are slight enough, but they have the overall effect of producing a text that is much more "naturalistic" than the original. One day a professor of comparative literature somewhere will have fun using Theroux's translations to document the changes in protocol that texts undergo in being shaped to conform to different cultural expectations. The Arabic title of Munif's first novel has the connotation of "the wilderness," or "the desert," and it begins with what is possibly the best and most detailed account of that mythical event, a First Encounter, in fiction—all the better for being, for once, glimpsed from the wrong end of the telescope. The novel opens, appropriately, on an oasis whose name identifies it as the source, or the beginning: Wadi al-Uyoun, an "outpouring of green in the harsh, obdurate desert." To the caravans that occasionally pass through it, as to its inhabitants, the wadi is an "earthly paradise," and to none more so than one Miteb al-Hathal ("the Troublemaker"), an elder of a tribe called al-Atoum:

> Left to himself to talk about Wadi al-Uyoun, Miteh al-Hathal would go on in a way no one could believe, for he could not confine himself to the good air and the sweetness of the water . . . or to the magnificent nights; he would tell stories which in some cases dated back to the days of Noah, or so said the old men.

But unsettling portents begin to intrude upon this earthly paradise. One evening at sunset, one of Miteb's sons returns from watering the family's livestock and tells his father of the arrival of "three foreigners with two marsh Arabs, and they speak Arabic"—"People say they came to look for water." But when Miteb goes to find out for himself, he sees them going to "places no one dreamed of going," collecting "unthinkable things," and writing "things no one understood," and he comes to the conclusion that "they certainly didn't come for water—they want something else. But what could they possibly want? What is there in this dry desert besides dust, sand and starvation?"

The people of the wadi hear the foreigners asking questions "about dialects, about tribes and their disputes, about religion and sects, about the rocks, the winds and the rainy season"; they listen to them quoting from the Koran and repeating the Muslim profession of faith; and they begin to "wonder among themselves if these were jinn, because people like these who knew all those things and spoke Arabic yet never prayed were not Muslims and could not be normal humans." Reading the portents, Miteb the Troublemaker senses that something terrible is about to befall the wadi and its people, but he knows neither what it is nor how to prevent it. Then suddenly, to everyone's relief, the foreigners leave, and the wadi settles back, just a trifle uneasily, to its old ways.

But soon enough the strangers come back. They are no longer unspecified "foreigners" but Americans, and they are everywhere, digging, collecting, and handing out "coins of English and Arab gold." Their liberality soon wins them friends in the wadi, but even the closest of their accomplices is utterly bewildered by their doings; "nothing was stranger than their morning prayers: they began by kicking their legs and raising their arms in the air, moving their bodies to the left and right, and then touching their toes until they were panting and drenched with sweat." Then a number of "yellow iron hulks" arrive, adding to the bewilderment of the wadi's inhabitants: "Could a man approach them without injury? What were they for and how did they behave—did they eat like animals or not?" Fearing the worst, the people of the wadi go to their emir to protest, only to be told that the Americans have "come from the ends of the earth to help us"—because "there are oceans of blessings under this soil."

The protests are quickly suppressed, Miteb and other troublemakers are threatened with death, and before long the wadi's orchards and dwellings are demolished by the "yellow iron hulks." After the flattening of his beloved wadi, Miteb mounts his white Omani she-camel and vanishes into the hills, becoming a prophetic spectral figure who emerges only occasionally from the desert to cry doom and to strike terror into those who collaborate with the oil men. As for Miteb's family and the rest of the wadi's inhabitants, they are quickly carried away by passing camel caravans. A number of them set out for a coastal settlement called Harran ("the Overheated"), where the new oil installations are to be built, a "cluster of low mud houses"—a place evidently very much as Doha and Kuwait were only a few decades ago.

The rest of Munif's narrative centers upon the early stages of Harran's transformation: the construction of the first roads, the gradual influx of people, the building of the oil installations, the port, and the emir's palace. Working in shifts, the newly arrived Arab workers and their American overseers slowly conjure two new townships into being, Arab Harran and American Harran. Every evening, after the day's work is done, the men drift home

> to the two sectorlike streams coursing down a slope, one broad and one small, the Americans to their camp and the Arabs to theirs, the Americans to their swimming pool, where their racket could be heard in the nearby barracks behind the barbed wire. When silence fell the workers guessed the Americans had gone into their air-conditioned rooms whose thick curtains shut everything out: sunlight, dust, flies, and Arabs.

Soon Harran no longer quite belongs to its people, and the single most important episode in the building of the new city has little to do with them. It is the story of an R-and-R ship that pays the city a brief visit for the benefit of the Americans living in the yet unfinished oil town:

> The astonished people of Harran approached [the ship] imperceptibly, step by step, like sleepwalkers. They could not believe their eyes and ears. Had there ever been anything like this ship, this huge and magnificent? Where else in the world were there women like these, who resembled both milk and figs in their tanned whiteness? Was it possible that men could shamelessly walk around with women, with no fear of others? Were these their wives, or sweethearts, or something else?

For a whole day and night, the inhabitants of Harran watch the Americans of the oil town disporting themselves with the newly arrived women, and by the time the ship finally leaves "the men's balls are ready to burst." This event eventually comes to mark the beginning of the history of this city of salt:

> This day gave Harran a birth date, recording when and how it was built, for most people have no memory of Harran before that day. Even its own natives, who had lived there since the arrival of the first frightening group of Americans and watched with terror the realignment of the town's shoreline and hills—the Harranis, born and bred there, saddened by the destruction of their houses, recalling the old sorrows of lost travellers and the dead—remembered the day the ship came better than any other day, with fear, awe, and surprise. It was practically the only date they remembered.

The most sustained wrong note in *Cities of Salt* is reserved for its conclusion. The novel ends with a dramatic confrontation between the old Harran and the new: between a world where the emir sat in coffeehouses and gossiped with the Bedouins, where everybody had time for everyone else and no one was ever so ill that they needed remedies that were sold for money, and a universe in which Mr. Middleton of the oil company holds their livelihoods in his hands, where the newly arrived Lebanese doctor Subhi al-Mahmilji ("physician and surgeon, specialist in internal and venereal diseases, Universities of Berlin and Vienna") charges huge fees for the smallest service, where the emir spies on the townspeople with a tele-scope and needs a cadre of secret police to tell him what they are thinking. "Every day it's gotten worse," says one longtime resident of Harran, pointing toward the American enclave: "I told you, I told every one of you, the Americans are the dis-ease, they're the root of the problem and what's happened now is nothing com-pared to what they have in store for us."

The matter comes to a head when a series of events—a killing by the secret police, sightings of the troublemaking Miteb, the laying off of twenty-three workers —prompts the workers of Harran to invent spontaneously the notion of the strike. They stop working and march through the town chanting:

> The pipeline was built by beasts of prey,
> We will safeguard our rights,
> The Americans do not own it,
> This land is our land.

Then, led by two of Miteb the Troublemaker's sons, they storm the oil instal-lation, sweeping aside the emir's secret police and the oil company's guards, and rescue some of their fellow workers who'd been trapped inside. And the book ends with an unequivocal triumph for the workers: the half-crazed emir flees the city after ordering the oil company to reinstate its sacked employees.

It is not hard to see why Munif would succumb to the temptation to end his book on an optimistic note. His is a devastatingly painful story, a slow, roundabout recounting of the almost accidental humiliation of one people by another. There is very little bitterness in Munif's telling of it. Its effectiveness lies rather in the gradual accumulation of detail. Munif's American oil men are neither rapacious nor heartless. On the contrary, they are eager, businesslike, and curious. When invited to an Arab wedding, they ask "about everything, about words, clothing and food, about the names of the bridegroom and his bride and whether they had known each other before, and if they had ever met. . . . Every small thing excited the Americans' amazement." It is not through direct confrontation that the Harranis met their humiliation. Quite the opposite. Theirs is the indignity of not being taken seriously at all, of being regarded as an obstacle on the scale of a minor technical snag in the process of drilling for oil.

Better than any other, Munif's method shows us why so many people in the Middle East are moved to clutch at straws to regain some measure of self-respect for themselves—why so many Saudis, for example, felt the humiliation of Iraq's army almost as their own. But in fact the story is even grimmer than Munif's version of it, and the ending he chooses is founded in pure wish-fulfillment. It probably has more to do with its author's own history than with the story of oil in the Gulf.

Abdelrahman Munif was born in 1933, into a family of Saudi Arabian origin settled in Jordan. (He was later stripped of his Saudi citizenship for political rea-

sons.) He studied in Baghdad and Cairo and went on to earn a PhD in oil economics at the University of Belgrade—back in the days of Tiroite socialism, when books
written by progressive writers always ended in working-class victories. Since then
Munif's working life has been spent mainly in the oil industry in the Middle East,
albeit in a rather sequestered corner of it: he has occupied important positions in
the Syrian Oil Company, and he has served as editor in chief of an Iraqi journal
called *Oil and Development*.

No one, in other words, is in a better position than Munif to know that the
final episode in his story is nothing more than an escapist fantasy. He must certainly be aware that the workforces of the international oil companies in the Arabian peninsula have never succeeded in becoming politically effective. When they
showed signs of restiveness in the 1950s, they were ruthlessly and very effectively
suppressed by their rulers, with the help of the oil companies. Over the past couple of decades, the powers that be in the oil sheikdoms (and who knows exactly
who they are?) have followed a careful strategy for keeping their workers quies-
cent: they have held the Arab component of their workforces at a strictly regulated
numerical level while importing large numbers of migrants from several of the
poorer countries of Asia.

The policy has proved magically effective in the short run. It has created a class
of workers who, being separated from the indigenous population (and from each
other) by barriers of culture and language, are politically passive in a way that a
predominantly Arab workforce could never be within the Arab-speaking world—
a class that is all the more amenable to control for living perpetually under the
threat of deportation. It is, in fact, a class of helots, with virtually no rights at all,
and its members are often subjected to the most hideous kinds of physical abuse.
Their experience makes a mockery of human rights rhetoric that accompanied
the Gulf War; the fact that the war has effected no changes in the labor policies
of the oil sheikdoms is proof in the eyes of millions of people in Asia and Africa
that the "new world order" is designed to defend the rights of certain people at the
expense of others.

Thus the story of the real consequences of the sort of political restiveness that
Munif describes in *Cities of Salt* is not likely to warm the heart quite as cozily as
the ending he gives his novel. But if Munif can be accused of naiveté on this score,
he must still be given credit for seeing that the workplace, where democracy is
said to begin, is the site where the foundations of contemporary authoritarianism
in the oil sheikdoms were laid.

Today it is a commonplace in the Western media that aspirations toward democracy in the Arabian peninsula are a part of the fallout of changes ushered in
by oil and the consequent breakdown of "traditional" society. In fact, in several
instances exactly the opposite is true: oil and the developments it has brought in

its wake have been directly responsible for the suppression of whatever democratic aspirations and tendencies there were within the region.

Certain parts of the Gulf, such as Bahrain, whose commercial importance far predates the discovery of oil, have long possessed sizable groups of businessmen, professionals, and skilled workers—a stratum not unlike a middle class. On the whole, that class shared the ideology of the nationalist movements of various nearby countries such as India, Egypt, and Iran. It was their liberal aspirations that became the victims of oil's most bizarre, most murderous creation: the petrodespot, dressed in a snowy *dishdasha* and armed with state-of-the-art weaponry—the creature whose gestation and birth Munif sets out to chronicle in the second volume of the *Cities of Salt* cycle, *The Trench*.

Unfortunately, *The Trench* comes as a great disappointment. The narrative now moves away from Harran to a city in the interior called Mooran ("the Changeable"), which serves as the seat of the country's ruling dynasty. With the move to the capital, the focus of the narrative now shifts to the country's rulers.

The story of *The Trench* is common enough in the oil sheikdoms of the Arabian peninsula; it begins with the accession to power of a sultan by the name of Khazael, and it ends with his deposition, when he is removed from the throne by rival factions within the royal family. Munif describes the transformations that occur during Sultan Khazael's reign by following the career of one of his chief advisers, a doctor called Subhi al-Mahmilji (who earlier played an important part in the creation of the new Harran). The story has great potential, but Munif's voice does not prove equal to the demands of the narrative. It loses the note of wonder, of detached and reverential curiosity, that lent such magic to parts of *Cities of Salt*, while gaining neither the volume nor the richness of coloring that its material demands.

Instead Munif shifts to satire, and the change proves disastrous. He makes a valiant attempt—not for nothing are his books banned in various countries on the Arabian peninsula—but satire has no hope of success when directed against figures like Sultan Khazael and his family. No one, certainly no mere writer of fiction, could hope to satirize the royal families of the Arabian peninsula with a greater breadth of imagination than they do themselves. As countless newspaper reports can prove, factual accounts of their doings are well able to beggar the fictional imagination. Indeed, in the eyes of the world at large, Arab and non-Arab, the oil sheik scarcely exists except as a caricature; he is the late twentieth century's most potent symbol of decadence, hypocrisy, and corruption. He pre-empts the very possibility of satire. Of course, it isn't always so. The compulsions and the absurdities of an earlier generation of oil sheiks had their roots in a genuinely tragic history of predicament. But those very real dilemmas are reduced to caricature in Munif's Sultan Khazael.

Even where it is successful, moreover, Munif's satire is founded ultimately upon a kind of nostalgia, a romantic hearkening back to a pristine, unspoiled past. It is not merely Americans from the oil companies who are the intruders here: every "foreigner" is to some degree an interloper in Harran and Mooran. As a result, Munif is led to ignore those very elements of the history of the oil kingdoms that ought to inspire his curiosity, the extraordinary admixtures of cultures, peoples, and languages that have resulted from the Oil Encounter.

Workers from other parts of Asia hardly figure at all in Munif's story. When they do, it is either as stereotypes (a Pakistani doctor in *Cities of Salt* bears the name Muhammad Jinnah) or as faceless crowds, a massed symbol of chaos: "Once Harran had been a city of fishermen and travelers coming home, but now it belonged to no one; its people were featureless, of all varieties and yet strangely unvaried. They were all of humanity and yet no one at all, an assemblage of languages, accents, colours and religions." The irony of *The Trench* is that in the end it leaves its writer a prisoner of his intended victim. Once Munif moves away from the earliest stages of the Oil Encounter, where each side's roles and attributes and identities are clearly assigned, to a more complicated reality—to the crowded, multilingual, culturally polyphonic present of the Arabian peninsula—he is unable to free himself from the prison house of xenophobia, bigotry, and racism that was created by precisely such figures as his Sultan Khazael. In its failure, *The Trench* provides still one more lesson in the difficulties that the experience of oil presents for the novelistic imagination.

))) Literature in the Ages of Wood, Tallow, Coal, Whale Oil, Gasoline, Atomic Power, and Other Energy Sources

Patricia Yaeger

> Power! Incredible, barbaric power! A blast, a siren of light within him, rending, quaking, fusing his brain and blood to a fountain of flame, vast rockets in a searing spray! Power!
>
> —Henry Roth, *Call It Sleep*, 419

This fountain of overwriting in Henry Roth's *Call It Sleep* captures the incommensurability between the frail human form and the power of electricity. After connecting himself to the rail powering trains that run through New York's Lower East Side slums, Roth's ten-year-old protagonist, David, veers between life and

PMLA 126, no. 2 (2011): 305–10

death. His electrocution is self-inflicted and deliberate. Earlier in the novel David longs for the source of this "searing spray," for the fantasied angel-coal that burned the prophet Isaiah clean: "where could you get angel-coal? Mr. Ice-man, give me a pail of angel-coal. Hee! Hee! In a cellar is coal. But other kind, black coal, not angel coal. Only God had angel-coal. Where is God's cellar I wonder. How light it must be there" (231). Although David also associates cellar coal with a promising disobedience, with sexual and religious transgression, Roth is more skeptical; he explores modernity's coal-made economy as a dark power tarnishing America's promise as *di goldene medine* (the golden land). In a country that offers opportunity, but at the cost of language loss and hard labor, survival demands a constant entanglement with dirty energy.

This Editor's Column peruses the relation between energy resources and literature. Instead of divvying up literary works into hundred-year intervals (or elastic variants like the long eighteenth or twentieth century) or categories harnessing the history of ideas (Romanticism, Enlightenment), what happens if we sort texts according to the energy sources that made them possible? This would mean aligning Roth's immigrant meditations on power with Henry Adams's blue-blood musings on "the dynamo and the virgin," or comparing David's coal obsessions with those of Paul, the coal miner's son in D. H. Lawrence's *Sons and Lovers*. We might juxtapose Charles Dickens's tallow-burning characters with Shakespeare's, or connect the dots between the fuels used for cooking and warmth in *The Odyssey* and in Gabriel García Márquez's *Cien años de soledad*.

I first became interested in literature's relation to energy when, piqued by America's energy extravagance, I picked up Jack Kerouac's *On the Road* and wondered, how often do Dean Moriarty and Sal Paradise stop for gas? As they crisscross the country, do they worry about how much fuel they're using or the price of oil? Or is this is a question for the twenty-first century, for a nation that survived the Arab oil embargo and the BP oil spill and may not survive global warming? By 1950 America's appetite for oil surpassed its use of coal. By the 1970s America was consuming 70% of the world's oil with little thought about sustainability. In an era of unprecedented material abundance, why should Paradise, Moriarty, or a host of other car-mad heroes worry about gas? It seemed as naturally there, as American, as the apple pie and ice cream Paradise eats "all the way across the country" (49). *On the Road*'s characters rarely experience the material world as an impediment. For Paradise even cotton picking becomes a lark. After allowing Mexican American friends to finish his picking, Paradise feels "like a million dollars; I was adventuring in the crazy American night" (100).

Even though Paradise avoids material worries, *On the Road* is fascinated with clean raw materials and their transformation into dirty culture ("before me was the great raw bulge and bulk of my American continent; somewhere far across,

gloomy, crazy New York was throwing up its cloud of dust and brown steam" [79]).
Energy anxiety keeps popping up. Hitching a ride east, Moriarty rants about bourgeois drivers obsessed with

> ". . . the weather, how they'll get there—and all the time they'll get there anyway, you
> see. . . . 'Well now,'" he mimicked, "'I don't know—maybe we shouldn't get gas in that
> station. I read recently in *National Petroffious Petroleum News* that this kind of gas has a
> great deal of O-Octane *gook* in it and someone once told me it even had semi-official
> high-frequency *cock* in it, and I don't know, well I just don't feel like it anyway . . .' Man,
> you dig all this." He was poking me furiously in the ribs to understand. I tried my wild-
> est best. (209; 3rd ellipsis in orig.)

Moriarty isn't worried about the price of oil (or its Saudi and Venezuelan sources
—a problem for American business in the 1950s) or his own fuel dependency, but
is Kerouac? Is there an energy unconscious at work in this text? Paradise starts his
trip in the midst of the unknown and unsaid. He travels in the wrong direction
(northeast) and stalls, "crying and swearing and socking myself on the head," in
"an abandoned cute English-style filling station," where he curses and cries "for
Chicago" (12–13). Are the gas station's empty pumps a premonitory metaphor for
resource anxiety, for what Pierre Macherey calls "that absence around which a real
complexity is knit" (101)? Or is an empty gas station just an empty gas station—the
halted traveler's bad luck, the writer's reality effect? In Macherey's theory of ab-
sences, "[w]hat is important in the work is what it does not say. . . . What the work
cannot say is important, because there the elaboration of the arguments is acted
out in a sort of journey to silence" (87). But is this always true?

Certainly Kerouac's characters are gasaholics. Oil dependency created their
world; each city, suburb, truck stop, and bite of pie depends on Standard Oil,
Shell, Mobilgas, or Phillips 66. What happens if we rechart literary periods and
make energy sources a matter of urgency to literary criticism? What happens if we
think systematically about how *On the Road* and its sibling texts relate to energy
sources across time and space? Within the genre of the 1950s road narrative, what
does it mean that John Updike's Mrs. Maple gets excited when a muscular gas
station attendant rocks her car as he washes its windows (56)? What about Elizabeth Bishop's "The Moose" and its antipastoral reminder that in the twentieth
century sacred sight must be carbon-based?

> by craning backward,
> the moose can be seen
> on the moonlit macadam;
> then there's a dim
> smell of moose, an acrid
> smell of gasoline. (173)

We need to contemplate literature's relation to the raucous, invisible, energy-producing atoms that generate world economies and motor our reading. Let me chart some coordinates for an energy-driven literary theory. First, resource depletion is not new; it's a repetitive fact. Native Americans living in woodland regions moved entire villages whenever nearby forest stocks were depleted. A Jewish holiday is built around an oil shortage and its miraculous cessation.

Second, energy sources have varied wildly over time and space and include almost anything that burns: palm oil, cow dung, random animal carcasses mounted on sticks. When the biblical God declared, "Let there be light," was oil from fish stocks or olives the source of illumination?

Third, energy use is uneven. The age of coal is not close to being over, is perhaps barely begun. Looking at the "ages" of energy will never be a tidy endeavor, since fuel sources interact. Describing China's burgeoning economy, Clifford Krauss writes that

> China's thirst for energy is leading it to build not only coal-fired power plants, but also wind farms, at a record pace, and to invest in energy sources around the world, like oil fields in Sudan, hydroelectric power in Burma and natural gas fields in south Texas. Beijing's ability to lift hundreds of millions of people into the middle class over the coming years will be largely based on its ability to produce more energy, and its foreign policies can also be expected to follow its energy interests. . . .

Fourth, thinking about literature through the lens of energy, especially the fuel basis of economies, means getting serious about modes of production as a force field for culture.[1] The stolen electricity at the beginning of *Invisible Man*, the marching firewood in *Macbeth*, the smog in *Bleak House*, the manure fires in Jorge Luis Borges's *Labyrinths*, the gargantuan windmills in *Don Quixote* would join a new repertoire of analysis energized by class and resource conflict breaking into visibility.

Fifth, this inquiry about energy's visibility or invisibility might change our reading methodologies. *The Political Unconscious* has long been a bible for me, with its elucidation of three extended networks for examining texts. Fredric Jameson suggests that if we first come upon the text as a symbolic act or individual parole, we must also recognize it as an ideologeme or social utterance that reconstitutes class conflict, as well as an "ideology of form," a dream catcher that captures skirmishing sign systems "which are themselves traces or anticipations of modes of production." These systems represent "progressively wider horizons" for examining the ways in which the text enacts imaginary resolutions of real social contradictions (76). Does this model of the political unconscious also describe an energy unconscious? Without reverting to crude materialism, I want to suggest that energy invisibilities may constitute different kinds of erasures. Following Jameson, we might argue that the writer who treats fuel as a cultural code or reality effect makes a symbolic move, asserts his or her class position in a system of mythic abundance not available to the energy worker who lives in carnal exhaustion. But perhaps energy sources also enter texts as fields of force that have causalities outside (or in addition to) class conflicts and commodity wars. The touch-a-switch-and-it's-light magic of electrical power, the anxiety engendered by atomic residue, the odor of coal pollution, the viscous animality of whale oil, the technology of chopping wood: each resource instantiates a changing phenomenology that could recreate our ideas about the literary text's relation to its originating modes of production as quasi-objects.

Finally, in thinking about energy we must make room for the miniature (that faint odor of moose mingling with the smell of gasoline) but also contemplate scale and the complex relations between literature and trade. Giovanni Arrighi points out that the "reshuffling of goods in space and time can add as much use-value ('utility') to the goods so reshuffled as does extracting them from nature and changing their form and substance, which is what we understand by production in a narrow sense." He quotes Abe Galiani: "Transport . . . is a kind of manufacture . . . but so is storage" if it makes goods "more useful to potential buyers" (177). Mrs. Maple's gas station attendant washes her windshield while standing on a concrete-covered basin of stored gasoline that may have come from Venezuela, Saudi Arabia, or Oklahoma. Does this change the libidinal or economic values in Updike's text? How do we think about utility and poetry together? Whatever the answer, thinking about energy is already embedded in older and stranger histories than our own, and to unearth these histories the following essays explore the roles of tallow, wood, coal, oil, human labor, and energy futures in a variety of texts.

Note

1. In Jameson's *The Political Unconscious* the text becomes "a field of force in which the dynamics of sign systems of several distinct modes of production can be registered and apprehended," and no system should become a master code or allegory for its age (98). But since fuel sources hover in the backgrounds of texts, if they speak at all, to pursue an energy unconscious means a commitment to the repressed, the *non-dit*, and to the text as a tissue of contradictions. What is the best methodology for pursuing literature's relation to energy? The answer may lie in systems theory instead of the political unconscious, or in new species of literary Marxism.

Works Cited

Arrighi, Giovanni. *The Long Twentieth Century*. London: Verso, 1994. Print.

Bishop, Elizabeth. "The Moose." *The Complete Poems, 1927–1979*. New York: Farrar, 1990. 169–73. Print.

Jameson, Fredric. *The Political Unconscious: Narrative as a Socially Symbolic Act*. Ithaca, NY: Cornell University Press, 1981. Print.

Kerouac, Jack. *On the Road*. New York: Penguin, 1976. Print.

Krauss, Clifford. "In Global Forecast, China Looms Large as Energy User and Maker of Green Power." *New York Times*, November 9, 2010. Web, April 6, 2011.

Machery, Pierre. *A Theory of Literary Production.* Translated by Geoffrey Wall. London: Routledge, 1978. Print.

Roth, Henry. *Call It Sleep.* 1934. New York: Picador, 2005. Print.

Updike, John. "Twin Beds in Rome." *The Maples Stories.* New York: Knopf / Everyman's Pocket Classics, 2009. 53–64. Print.

⟩⟩⟩ Excerpt from *Cities of Salt*

Abdul Rahman Munif

Chapter 30

Within less than a month two cities began to rise: Arab Harran and American Harran. The bewildered and frightened workers, who had in the beginning inspired American contempt and laughter, built the two cities. They lifted the white lumber on a winch, carried the heavy steel girders, and placed them over the wood and screwed them together; they installed the glass windows and the shutters, and they did all the painting. Every few hours they put their tools down to step back and look at another completed house. The American engineer who watched and supervised came over at this point to test the walls and ceilings with his hands and with instruments, and if they seemed sound he looked at the men's brown faces with surprise and some wonderment, and repeated the same word: "OK."

This happened again and again in American Harran, and in less than a month the nucleus of a large and well-ordered city had appeared and sped toward completion: hard streets, some wide and others narrow, all perfectly straight, rolled smooth by the accursed heavy machines and coated with a gleaming black substance. Houses like the geese who flew over Wadi al-Uyoun in winter, small houses and others so tall and huge that no one could imagine who would inhabit them. Many swimming pools, on several scattered sites, near them houses made of straw and palm branches, and a long street linking the northeastern hill to the sea. Hundreds of pipes lay by this roadside, but no one knew what their secret might be.

An endless line of ships arrived, carrying cargoes whose use no one could guess: even after they were unpacked from the crates and emerged from the crude wrappings and boxes to be carried away by one or two Americans to stand in shining heaps of steel, no one had the slightest idea what these new terrors were.

The Arab workers, who stood like dummies in the first days, their muscular bodies straining the overalls and stiff new white caps on their heads, were divided

Cities of Salt, trans. Peter Theroux (New York: Random House, 1987), 207–22

into small groups and sent to different areas all over the work site, and within a few weeks they had become different creatures. Words of praise and slaps on the shoulder meant approval and appreciation. They never hesitated to accept any work or to offer any assistance asked of them. They were not motivated enough to take anything on; after the fear and anxiety of their first days, especially when they were read those deadly instructions, they had felt intimidated to the point of despair, but doggedly, and independently of each other, they saw things take a new turn: their hands moved more quickly, oft-repeated foreign words and names somehow found their way into their vocabularies, and relationships were formed as they smiled and gestured more. Their fears vanished, or at least retreated.

The ships that brought all the new "calamities" also brought men in ever-larger numbers, men who came from God knew where for purposes no one could guess at. They poured off the ships like locusts and swarmed to every part of the camp. Their housing was completed in a single day; even the food served in that long hall whose use no one could guess at while it was under construction, was ready for all of them.

Every finished building pushed the Arabs one step backward, for after the walls were completed the roof was put up, and after the windows and shutters were installed the Americans started to do strange jobs, hanging strong black ropes inside the walls. They filled the windows with iron blocks that emitted a cold breeze. The men who had come by ship were each given complete sets of clothes, blankets and furniture, and their very own places to sleep. After a day or two they all had become one group, as if they had always known each other, and were equally driven to work without stopping. Some worked in the sea, others moved the pieces of pipe from one place to another, while still others assembled the machines which had arrived in pieces in crates. They ran back and forth like frightened cats, naked except for short drawers and white hats; most of the time they wore nothing else. Their faces and bodies were covered with spots; they had small scars on their fingers and elsewhere on their bodies, and sweat ran like rain over their chests and faces. Clownish people like these had never been seen here before, but they became such a common sight that no one even noticed anymore.

* * *

Less than a month later, Ibn Rashed came back with several more men. Seven of them were from this region, from Ujra and al-Rawdha, but the rest were from far-off places that no one had heard of.

Ibn Rashed had left Harran when it was a wilderness, without a single house or even any marks on the ground except for a few tents and packing crates on the western side. Now he could not hide his amazement at the miraculous things constructed in his absence; he loudly voiced his astonishment to the group of men. When he arrived and saw the men returning from American Harran in overalls and caps, he threw his hands in the air and shouted with fear, "God help us! What of your souls, you poor men?"

At first most of the men did not notice how shocked Ibn Rashed really was; they looked at each other and then at Ibn Rashed, who spoke again, only laughing harshly this time. "I said to myself, the Americans could never change the sons of Arabs—never!"

He went over to Daham, who looked silly in his tight clothes, with his round belly and bulging rear end, and patted him on the shoulder. "Man grows a new heart every day."

Ibn Rashed had not got over his shock, but he loudly and exaggeratedly praised everything he saw. He praised Daham and the other men, he praised the beautiful houses the Americans had built and said that the Arabs should try to emulate them; then he somewhat shyly asked to have everything explained to him: when the buildings were put up, who had built them and how long it had taken, and when they had been given the overalls. He reached out to finger one of the caps admiringly and asked if all the men had got clothes and caps and if there were any left. He was very excited, asking a new question before they had finished answering his last one, and in his passion to know everything he missed many of the details that Daham told him.

In his excitement Ibn Rashed forgot to introduce the men he had brought. They were deeply impressed, too, but stood quietly off to one side with their camels, and when he began asking his questions they made the camels kneel and began to unload them, until they realized that they were still far away. In an attempt to show equal awe when he saw what the men were doing, he moved quickly and called out to the others for help in unloading the camels and putting their supplies in his tent.

They made short work of this task in their enthusiasm and eagerness to cooperate, joking and asking more questions; they looked at each other, and then Ibn Rashed spoke up as if he had suddenly remembered something. "Don't worry, my friends, everything is going to be just perfect."

The men gathered together and sat down in the clearing between the tents which faced the sea; most of the workers had taken off their close-fitting work clothes, undoing the buttons that made the overalls as tight as molds, and left their caps in the tents or on the ground beside them. In a moment of silence, Ibn Rashed announced that one of the men he had brought would work as a butcher, like his father and grandfather before him, and sell meat to everyone who wanted it. He pointed to a short, very dark man and said that he would sell goods to everyone in Harran—all kinds of goods, just as in Ujra and other towns. Then he looked around until his eyes met with those of a very small, skinny man. He laughed, showing his gappy teeth. "I know the Bedouin—they like their own kind of bread and won't touch anything else, and he knows how to make it!" He laughed loudly. "Never fear, Arabs—trust in His Highness!"

No one knew who was meant by "His Highness"—the baker, Ibn Rashed himself, or some third person. The shamefaced young man, who was wearing trousers and a jacket, who sat silently at a distance as if dreaming or watching some strange play was—Ibn Rashed said—the "engineerist" who would build houses for the Arabs like those the Americans had.

That is what Ibn Rashed told the men. He was visibly tired from his long journey but was still animated. He wanted to see everything and hear about everything that had transpired in his absence: the number of ships that had come in and what they had brought, and the new people who had arrived. Daham answered the questions as fast as he could, but Ibn Rashed was interested in too many different things, and eventually he stood up and asked Daham to accompany him to Arab Harran.

Ibn Rashed was a different man among the townspeople of Harran. He was extremely gracious and talked to everyone. He inquired about their health and asked if they were happy in their new houses or needed anything. He was especially polite to the "aged gentleman," as he called Ibn Naffeh out of respect. When he was through with his first question, he asked about the land in Harran, if it was mostly public property or divided up, and if it was divided up, who owned it. Then he asked about the lands adjoining Harran, whether they were wilderness or privately owned. He was very precise in his questions and asked Daham to make notes of everything he was told. On their way back he told him to "take good care of all this, because it's important," but he said nothing more.

Ibn Rashed thought and made decisions by himself; he consulted no one and let no one in on his secret. He had gone to Ujra very suddenly, as if involved in some conspiracy; the only people who knew he was traveling were the ones who saw him ride off. He had asked the men several times to allow him plenty of time on the road, as he put it. He had been gone a month and now he was back—instead of solving the problems they had had, he had brought with him new men and new problems.

These were his own secrets. On arriving he had asked about Naim before asking about anyone else, and when Daham al-Muzil told him that "there are a lot of new men and Naim is busy with them," Ibn Rashed had one of his men go to the American compound and tell Naim that he had arrived and wanted to see him on matters of importance. But the man came back and said that he had not been able to see Naim.

"You have to see him tomorrow. It's urgent," Ibn Rashed told Daham. He smiled and said softly, "He is the key . . . we have to get him."

Daham nodded vigorously but had no idea what Ibn Rashed was talking about.

Chapter 31

American Harran grew taller, more spacious, and more alien with every passing day, until one afternoon the workers were sent away early and not allowed to re-enter certain sectors, even though their work was incomplete. The large pool, for instance, was not finished. That afternoon there was an uncanny feeling in American Harran, as if something were about to happen. Ibn Rashed decided to declare a holiday for the next day and a half, to celebrate the opening of the three shops— the bakery, the butcher's, and the general store—but it was a huge ship appearing on the horizon that changed everything in both Arab Harran and American Harran, and a great deal more besides.

When the huge ship dropped anchor at sundown, it astonished everyone. It was nothing like the other ships they had seen: it glittered with colored lights that set the sea ablaze. Its immensity, as it loomed over the shore, was terrifying. Nei-ther the citizens of Harran nor the workers, who streamed from the interior to look, had ever seen anything like it. How could such a massive thing float and move on the water?

Voices, songs, and drums were heard as soon as the ship neared the shore; they came from the shore as well as the ship, as all the Americans in the compound flooded outdoors. Music blared as small boats began ferrying the passengers from the now motionless ship. There were dozens, hundreds of people, and with the men were a great many women. The women were perfumed, shining, and laugh-ing, like horses after a long race. Each was strong and clean, as if fresh from a hot bath, and each body was uncovered except for a small piece of colored cloth. Their legs were proud and bare, and stronger than rocks. Their faces, hands, breasts, bellies—everything, yes, everything glistened, danced, flew. Men and women em-braced on the deck of the large ship and in the small boats, but no one could be-lieve what was happening on the shore.

It was an unforgettable sight, one that would never be seen again. The people had become one solid mass, like the body of a giant camel, all hugging and press-ing against one another. The astonished people of Harran approached impercep-tibly, step by step, like sleepwalkers. They could not believe their eyes and ears. Had there ever been anything like this ship, this huge and magnificent? Where else in the world were there women like these, who resembled both milk and figs in their tanned whiteness? Was it possible that men could shamelessly walk around with women, with no fear of others? Were these their wives, or sweethearts, or something else?

The men of Harran stared, panting. Whenever they saw something particularly incredible they looked at each other and laughed, and looked back again yearn-ingly. They clicked their teeth sharply and stamped their feet. The children raced ahead of them and arrived first to sit by the water, and some even dove into the water to swim toward the ship, but most of the people preferred to stay behind on the shore, where they could move around more easily. Even the women watched everything from afar, though none of them dared to come near.

This day gave Harran a birth date, recording when and how it was built, for most people have no memory of Harran before that day. Even its own natives, who had lived there since the arrival of the first frightening group of Americans and watched with terror the realignment of the town's shoreline and hills—the Harranis, born and bred there, saddened by the destruction of their houses, recalling the old sorrows of lost travelers and the dead—remembered the day the ship came better than any other day, with fear, awe, and surprise. It was practically the only date they remembered.

The workers who marched down in groups to see everything with their own eyes were far more tormented and depressed than cheered by what they saw. For the first time, they were overcome by the agonizing feeling that they had made a bad mistake in coming here and must not stay long. Ibn Rashed did not seem interested at first, but he sent a couple of men to have a look at the "new calamity" and report back to him. He was busy planning for the men to resume building Harran, but even he could not stay away long. When the ship came in and its whistle blew twice and the men and women crowded its decks, waving and dancing amid the lights and the music, he started and said to Daham: "If your whole tribe loses its mind, what use is there in reasoning?" He laughed loudly. "No one I've sent over has come back or even sent word. Let's go and see for ourselves what's happening."

They started off slowly, but as the ship came into view and he saw more of the scene, something impelled Ibn Rashed to walk faster. He sat with the workers on the beach and saw the women and heard their laughter. After one of the men let out a loud groan, there was silence.

"Brothers—this is the court of King Solomon you've heard about!" shouted Ibn Rashed giddily.

They laughed and began to talk and comment on what they saw, and even some of the boys made rude remarks and were not rebuked by their elders.

Arab Harran was silent, and the men sitting on the beach were rapt in their longing surveillance, but the festivities on the ship and in American Harran grew noisier. The Arab workers had not noticed that the first Americans to arrive had brought musical instruments with them, so they were astonished to see the drums and trumpets and other instruments now piled on the American beach. After the ship emptied and its music stopped, the music from the beach grew louder, especially the sounds of the big drums, which set the beat for the singing and dancing of the partygoers, creating a new atmosphere.

"The American sons of bitches!" said one man angrily. "They don't even mind if we watch—we're no better than animals to them."

"They eat like sheikhs, Mubarak," Hajem told him. "And why shouldn't they do just as they please in their own colony?"

Most of the men had something to say, but the blaring music and dancing and the bizarre scenes that followed prevented them from speaking; others were immersed in contemplating this impossible dream. At first they pointed in fear or shame at some of the goings-on. They nudged each other to look at some new scene, but as the party spread and grew wilder and the naked or seemingly naked men and women appeared on the ship and in the small boats striking dramatic poses—the men stroked the women and then attacked suddenly for hugs and kisses, and carried the women around on their backs, and made them sit on their laps—the Arabs shouted and pointed more boisterously. The climax was when the last boat came ashore with one man and seven women. The women were reclining around the bushy-bearded, hairy-chested man, who fondled, smacked, and leaned over them one by one and put his arms around two at a time. He shrieked with laughter and jumped up, rocking the boat in time with the drumbeat, and helped one woman stand up with him. They danced three or four turns to the drums, which grew louder as they neared the beach; then the man jumped into the water and pushed the boat in, singing.

"The jinns of Eden, underneath which rivers flow," said Abdullah al-Zamel, "with lads and maidens there to serve."

"By God, it's just what Abu Muhammad said," replied Hammad al-Zaban. "Solomon and a thousand Queens of Sheba. 'Die in your lust, you poor bastards, you Arabs!'"

No one could believe his eyes: it was indescribable. Words failed them; it could not be happening. Even the boys and small children, constantly laughing and making remarks in their high voices, eventually fell silent, utterly spellbound by what they saw. The men changed their positions and craned their necks to look, more fascinated than the children, at the procession of celebrants entering American Harran. The men were mostly quiet now and slightly dizzy, feeling sharp pains throbbing in certain parts of their bodies. Some cried out, and most of them wished that they had never come to see what was transpiring before them.

* * *

The gates of American Harran swung shut behind the ship's passengers, and Juma, the black man, stood by the gates with an elephant-tail whip in his hand, like the king of Death. The voices and din intermingled and grew fainter but never faded away completely. Music and singing could be heard from scattered places all night long. When for a moment it died away, the men squatting on the beach expected something to happen, for whenever there was silence and the minutes passed slowly, violent screams of laughter burst out suddenly, followed by music louder yet than before, and because of this game their waiting was sweet and cruel at the same time.

None of the men felt the cold that filled the night, and none of them felt like talking. The ship had arrived and the Americans had filed into their compound hours ago, but time passed tonight in a way it never had before. Although the people of Harran usually went to bed early, except for a few workers who stayed up to play cards, tonight none of them felt the time passing or wanted to leave the beach. The children, fascinated by what they had seen, could not stay still. They chased each other and called out words their parents never imagined they knew! Some of them ran back to the tents, perhaps to tell the women exactly what they had seen, for it turned out that the women who had kept far off when the ship landed and disgorged the Americans knew everything that had gone on, as if they'd seen it all themselves. They even knew all about the Billy Goat, as they had named the bearded man who came ashore in the last boat, and could describe in the greatest detail how he danced, how many women were with him, and how he had jumped into the water. They told the story with shame at first, then with more spirit. They reminded the children to have their fathers come home for supper, but they must have done so unclearly or not very insistently, because in their excitement the children forgot to relay the message.

Had it been a summer night, a night with a moon filling the sky or a night that saw the triumphant return of long-absent travelers, this long evening in Harran would have been one endless succession of stories about the old days and far-off places interrupted by peals of delighted laughter and persistent questions about other travelers and foreign lands, about rain and vegetation; but tonight the men were silent except for anxious questions with no answers. They were overcome by endless worries and uncertainties.

Every one of them had much to say, and even the habitually quiet men may have wanted to speak. Some of them sang, as if to deny that their hearts were as leaden as they seemed, but depression overcame all their senses and paralyzed their power of speech. A feeling of bitterness spread from their dry throats to their stiff joints, and silence reigned completely: even Ibn Zamel, who had been active and talkative, strolling more than once to the gate to stand by the barbed wire and reporting back to the men whatever he heard, had now quieted down. He had not managed to find anything out, so he got up abruptly.

"Good night, my friends," he said weakly. He paused for a moment until they took notice, then said, "These American sons of bitches are nothing but trouble and bad news. We'll never see any good from them. They'll get the meat, and we'll get whatever bones they care to throw us."

He walked a few more steps, then turned and spoke again. "Leave them be— have nothing to do with them. God curse them and the day they came here."

Someone had to do something, because the mood of wariness born of silence and expectation that now enveloped everyone, and the departure to shadowy places near and far, and the phantasm that suddenly blazed up permitted no one to think or act. Ibn Zamel, who had dodged like a hungry wolf from one spot to another and urged the boys to jump over the barbed wire on the eastern side to have a look and report to the others in spite of his own failed attempts, now instinctively knew that their stay here would only mean more pain and problems for them all. When he made up his mind to go and had said his piece, they began to move around, to curse and sigh.

Ibn Rashed stood up and cleared his throat. "Like you said, Ibn Zamel. We have enough to worry about."

"Happy days are short," said one of the men from Harran.

"And nights shorter yet," said Ibn Zamel, now some distance away.

"Say what you want, but I'm afraid we've lost our world and our faith," said one of the men from the darkness. "We'll never touch the meat they have—we'll be lucky to get gravy!"

They all laughed because they knew what he meant. The Harrani—whose pride was inflamed by what he had seen, well traveled enough to know how people lived in lands far distant from this desolate, unknown part of the world—did not want to come back like this. As the laughter died down, he said, "You'll know by the end of this night."

* * *

No one in American Harran slept that night. The American singing and shouting never let up, and some of the men later said that the singing was louder at sunup than it had been the evening before. Others said that the ship's whistle let out a loud blast at sunrise, which only provoked louder singing.

Nor did Arab Harran sleep. After the men went home, the boys stayed out to wander along the beach in front of the ship and near the barbed wire. When they got tired of that, they moved closer to the tents to sing, shout, and tell obscene jokes. Hammad al-Zaban shouted at the boys and dogs to be quiet—"There are people trying to sleep!"—but they paid no attention.

The men who headed home felt hungry but had no appetite for food. Abdullah al-Zamel told them that his father had always quoted a saying of the Prophet that said fasting was the only way to conquer sin and temptation and suggested that they go to bed without supper. Some men thought this a good idea, and others just did not have the energy, at this late hour, to prepare food, so they decided to have only tea. They sat in the clearing between the tents and sipped at the small glasses in silence.

They were no less bitter there than they had been on the beach, and the stories they tried to tell trailed off as the desert rang with the sound of loud music and shrill laughter. It happened again and again. Even Hajem's and Hammad's ribald jokes, which at any other time would have raised loud laughter, were met with wan, forced smiles.

It was the same in the homes of Harran, where some men had light meals and went straight to bed, though it took them a long time to fall asleep.

Sorrow, desires, fears, and phantoms reigned that night. Every man's head was a hurricane of images, for each knew that a new era had begun. Harran had been a desolate, forgotten village, which received only the kind of visitors who came to sell or barter goods and then promptly rode away—and then only rarely. The only exceptions were the foreigners who came to bring news, presents, and money from Harran's own traveling sons.

It would have seemed unthinkable for Harran ever to change as it now had, this quickly, for ships to bring such immense numbers of people, for its eastern quarter to be covered with buildings. This was unimaginable. The people had got used to the new buildings and even to the new faces, but nothing had prepared them for the arrival of this last ship. Ibn Rashed had called it the ship of King Solomon, because the women it brought resembled the Queen of Sheba, or were even more beautiful. No one in Harran had the powers to describe to others what he had seen.

What new era had begun—what could they expect of the future? For how long could the men stand it? This night had passed, but what about the nights to come?

No one asked these questions aloud, but they obsessed everyone and visited the uneasy sleep of men in the form of phantoms; their repressed desires swarmed over them as the night wore on. Those men who went to bed wide awake, whose sleep came to them in uneasy fits, woke in vivid terror only to be filled with desire and warmth, fear and expectation.

))) Poems from *Endangered Hydrocarbons*

Lesley Battler

PAX MCMURRAY

rise of the Fort McMurray Empire

i studied Fort McMurray in school
but can't get over her grandeur
epicentre of the world's largest
petroleum empire

Endangered Hydrocarbons (Toronto: BookThug, 2015), 29–37, 45–48, 123–28

my Prius noses
through a chaos of roads
the Marcus Aurelius
draw-works

triumphal arches
narrative reliefs, portrait busts
dress the smallest infill pad. sites
like Cadotte boast racecourses
bathhouses

here is where Nero fiddled
while Syncrude tossed Christians
into geoclines

a Chinese-Western
restaurant lies over natural gas
caverns where the CO_2 Network
once honed capture
technologies

most remarkable is
Hadrian's Upgrader
Fort Mac's dominant
Landmark

i'm awestruck over
the scope of work
one can only gape
in wonder, how did
slaves from Sundre
pull those giant rigs
on wooden rollers

bitumen mystery cult
before the Common Era
when Sargon the Great ruled
Moose Mountain

& the mesolithic peoples
of central Alberta subsisted
on shellfish

a bitumen mystery cult
settled Fort Mac
historians say
these aliens descended
from the Petrolians who entered
Alberta during the Pliensbachian
Transpression

possibly fearing reprisal
the clan, known only as Murrians
printed its own cuneiform

(unrelated to the nihilist
Sanskrit of the activists who invaded
Grande Prairie)

instead of depicting
stick figures

lonely hominids
imploring totem animals
to score another
tomorrow

Murrian ideograms
metered a viscous scripture
too thick to trickle between rock
& wellbore

Lucretius
Fort Mac's lead hellenist
studied syncretic crude
compiled the first illuminated
schematic

any return of hydrocarbons
to the divine must come through
their emancipation from geological
bondage

gum bed diplomacy
as the Athabasca River
threatened Upgrader Alley

& the need for new crude
Preoccupied the subcontinent
Murrians hit the road to spread
the Word

we can process
Spirit trapped in gum beds
into desirable products
diesel fuel, asphalt,
kerosene

the Murrian society
drilled syllabaries
just-in-time bitumen
for everyone

we, as owner-operators
of the greatest Oilsand republic
in history, foresee a colossal
rise in fossil fuel use

must transport
World Soul
to foreign
outlets

from gum beds to oilsands
the Murrians offered a Fusion
Hotline for Big Belchers
with lower baseload
reliability

spoolable pipelines
enabled low-cost caribou
reduction

ingenuity will bring
the Oilsands to your doorstep
right where you need
them to be

the Murrians overtook
the twin metropoli
Leduc-Nisku

Built the diorite
Batholith, known for
its golden dome

most ambitious shrine
ever raised to bitumen
worship

the plus-sized stupa
tore a stripe off the city states
united under the banner
of *Edson*

the Edson wars
for centuries
Edson's conventional well culture
had bullied the intellectuals
at Fort Mac

of course
this meant war

knowing
they were outnumbered
the Murrians pioneered
supply-chain
logistics

burned coke
with quick & dirty electricity
met intensity target

transported
entire work camps
by horse-drawn chariot

routing the Edsonites
who continued utilizing
slaves, suffering
endless LTIs

stymied by diminishing
sweet gas returns, Edson
surrendered to the
Bitumenites

who released one
of the most important
documents in history
Pax McMurray 1964

the Pax relied
on bulk modules
a surplus of joules
& Alcogel 1 to fuel
the imperial army

uploading the victors'
moral claim to the Regional
Municipality of Wood Buffalo

Syncrude imposed a
provisional government
wrote off the decline rate
cut supply to debt-ridden
territories

thus
the gentle colony
of bitumen worshipers
became a modern class-based
project team

IDYLLS OF INUVIK

the Henry James guide for relocating
oil company employees and their families

vignettes
MacKenzie River, fine as the Tiber!
breezes tease scree, chic little villas lurk
far from clockgods and their traffic gridlock
    my lips shape prayers over eroded
    roofs, crooked chimney poets, water towers
    of this haughty city state, drunken pines
        stagger over the treeline but indeed!
        this must have been a hoary old city
        when Hannibal battled the Mad Trapper

    i pick up my rucksack, take in the Mike
    Zubko Airport, an aesthetic delight
    Majolica tarmac gleams, campaniles
of quaint colonades lead to the Great Fur
Trade Reliquary. this metropolis
never sleeps! Single Otter waft propane
    bush planes drone, Victorian quonsets line
    country lanes swarming with Norcan rentals
    Smartie-box mansions perch on vivid piles

    of verdigris. here one may live a dream!
    Verdi with peregrine, live concertos
    from ravens driving down the Boulevard

        or one may wander the serpentine road
        of utilidors, all Roman vinyl
        Alcan finery, Norman rotors spin
    perfect unity of heart and reason
    Goethe, a known Cicerone, would surely deem
    the Petroleum Show lovelier than
the medieval cathedral, Opera Night
the Friendship Centre booms with Carmen
but i choose to peruse Inuit land
    grants, while the canyon swoons nitrogen-red
    northern avens preen, *the very picture*
    of Worldly Wisemen, straight out of Bunyan

pondering soil disposal, possible
end-uses for marine ecosystems
cradled among scenic pinnacles, one

can imagine perforation tunnels
music of dirt bikes—sublime! i remain
in limbo, blinded by a sudden wind

city lights

discerning eyes and digital easels
will revel in picturesque Inuvik
Meet tufting artists at the Trade Post

where merchants barter carvings in charm-school
patois. one may hear the cheery "baksheesh"
of basket weavers. drown in sensation!
soak in the Nanook souk, pause for coffee
solution talk in the Escalus room
now is the time to buy an infill. Snag

a *Nunataqaq* (land of origin)
a roast phalaropes over an open fire
roll cigars in fine-ground ptarmigan. watch
auditors lariat the aurora
that threaten boreal forests, then pose
with the walleyed polar bear at the Roost

paddle a canoe, nurse the *Oxford Book
of Verse*, jot noble poetry to the
dying glory of Indian summer

Saturday night on Franklin Ave
i leave the theatre. the streets are still
teeming with people, generators roar
arias as the wendigos party
Ibyuk walks at night, permafrost adorns
the cornices of posterity, i
kiss a Governor General, win a
rack of antlers. I may have mandated
the Office of Investigator, carved
a Pietà from imported snow, crashed

the Mad trapper Inn, karaoked Part
III of the Corrections and Criminal Release
Act, signed off on the death of Aklavik

airlifted children from homeless tribes who
kidnap bottom lines, *this*, you will cry, is
the Civilized Nation, *par excellence!*

TRUTH, POWER AND THE POLITICS OF CARBON CAPTURE

dialogue between OilWeek and Michel Foucault
(sponsored by the Harper Government of Canada)

1

it was open season on oil and gas when
landowners Jane and Justin Conn and the
activist group, Ecojustice, went public
with allegations CO_2 was leaking from
the underground reservoir

> *no one knew the real problem*
> *was environmental science and the ideological*
> *functions it could serve*

they targeted both Cenovus
and the academics associated with the Weyburn-
Midale Carbon Capture and Storage Project

> *though activism didn't exactly*
> *begin with the Silent Spring business I believe*
> *that sordid affair provoked numerous questions*
> *around Power and Knowledge*

2

instead of looking at data that discredited
their claims the Conn gang ramped up the rhetoric

> *their statements are verified by the*
> *"media" of opinions, a materiality caught in*
> *the mechanisms of power formed by the*
> *press, cinema, TV, social networks*

before industry could respond, a
rogue's gallery of faux scientists and axe-
grinding activists arrived on site, blatting
soundbites such as "We are here today
on the frontier of climate destruction"

> *no one considered the interweaving*
> *of power and knowledge in a science as*
> *dubious as ecology riddled with ready-made*
> *concepts, approved terms of vocabulary*

environmentalist show-trials make
it appear the Canadian Energy Industry is
the one committing criminal acts

> *an entire discourse has risen*
> *from a population composed of people*
> *who "choose" to reduce, recycle and*
> *reuse according to precisely determined*
> *norms*

3
Canadian Press claims dead animals
were regularly found in a pit metres from
the Conns' nuptial bed

> *activist dialectic evades the reality of*
> *spurious environmental science—environmancy—*
> *by reducing it to a Hegelian skeleton*

multicoloured scum bubbled in once
bucolic ponds. "At night," Conn said, "we
could hear this sort of cannon going off"

> *semiology examines the co-opting of*
> *neutral or pastoral concepts such as "climate"*
> *and "ecology." Of course, the word "green"*
> *no longer denotes a colour among other*
> *colours within a neutral spectrum*

a CBC documentary made no attempt to
be impartial when a trembly-voiced narrator
recounted how the couple had to leave their
farm and move to Regina

> *the public broadcaster morphs "news"*
> *into tropes representing a form of nostalgia*
> *for quasi-knowledge free of error and illusion*

4

over 15 million tonnes of carbon dioxide
have been pumped underground. No test
results support claims that CO_2 has migrated
through geological storage

> *young wolves are acting on naïve*
> *ideology proposed by icons like David Suzuki*
> *who organized the wreckage of the "hippies"*
> *into massive concentration camps*

the project covers some 52,000 acres
with a total of 963 active wells, 171 injection
systems. Overall it is anticipated that some 20
Mt of CO_2 will be permanently sequestered

> *Suzuki et al. created a generation*
> *of idealists unable to distinguish carbon*
> *sequestration from their own prison of*
> *environmancy. Lenin lived in such a zone*
> *of exile in 1898, and Chekhov visited an*
> *activist colony on the Sakhalin Islands*

5

even the Calgary Herald joined the global
media inquisition at the Conn Family
conference

> *we must understand how small*
> *individuals, the microbodies of discipline*
> *deploy unexamined tactics (school recycling*
> *programs, litter clean-up projects)*

environmental despotism has
reached the courts which fined Syncrude
more than 3 million for the unintentional
death of ducks on its tailings ponds

> *one can link "justice" and the*
> *transformation of children's bodies into*
> *highly complex systems of manipulation*
> *and conditioning*

6

organizations we considered allies
called for greater government oversight
without citing Industry's comprehensive
seven-page report

> *only precise analysis can excise*
> *the desire of the masses for activism and*
> *reveal public complicity in the refusal to*
> *decipher what environmentalism*
> *(environmancy) really means*

as an industry we must wrest
the media from its addiction to activist
sensationalism and present our own
isotopes

> *the good news is that now*
> *a majority, possessing an economic*
> *plan can dismantle the social and*
> *cultural hegemony in which activism*
> *operates in our own heads*

))) Poems from *Shale Play*

Julia Kasdorf

September Melon

Larger than my head, it rests heavy in one hand
as I lift it to my ear and knock for the thud

that says the center will be red and dense, wet
and sweet, studded with shiny black seeds,

a gift this late in the season. Where we live,
among boulders and trees, thumper trucks gain

uneven purchase, so a rig, driven by one man
traces a grid through the woods, grabs saplings

with a metal claw, holds them until a blade saws
then tosses them aside. An auger drills shotholes

and sets the blast with radio-controlled detonators
thirty feet down. On Sunday morning they blasted

when everyone else was at church, the professor
says, certain the men trespassed on unleased land.

A seventy-year-old woman stands up in a public meeting
to tell how she showed the gas men a map of her farm,

said blast anywhere but here and here. They agreed.
But wouldn't you know it, they blew up the two

places where she'd buried her husband and horse.
Another landowner begged for a day to move

his bee hives. The gas men refused. What happened,
I asked, imagining the furious hum and spray

from gold-limed boxes when the earth shook.
The man shrugged, that was not the point of his story.

F Word

The industry spelling of fracking is actually fracing.

Without the K, it looks less violent:
water pressure creates fractures that allow

oil and gas to escape—as if they were
trapped—*under tight regulatory control.*

Blame the fracktivists, fracademics. A bumper
sticker claims, *I'm surrounded by gasholes.*

Frack her 'til she blows, says a tee shirt stretched
on a roughneck's chest at the Williamsport Wegman's.

Frackville, PA, named for Daniel Frack, from *vrack*,
Middle Low German: greedy, stingy, damaged, useless.

Are you going to say what the word suggests,
a student timidly asks, to women, I mean?

Fracket, a sophomore explains, is a hoodie worn
over your spaghetti strap dress to a frat house,

an old jacket that won't matter if it gets stolen
or left behind on a flagstone patio, splattered

with someone's else's vomit.

Sealed Record

You're both aware that in exchange for the sum of $750,000, you have
given up all rights that you may have against all of the defendants in
this case now and forever.

Yes.

Yes.

You understand that in exchange for that sum, you are required to turn over your home to the defendants in exchange for which you will be able to buy a new home?

Yes.

Yes.

Do you understand that by this agreement each of you have been subjected to a confidentiality agreement, which is in essence, a gag order? You are not to comment in any fashion whatsoever about Marcellus Shale / fracking activities, and you accept that?

Unfortunately, yes.

Yes.

You both understand and accept that as written the settlement agreement may apply to your children's First Amendment rights as well?

Yes.

Yes.

And you accept that because you, as adults and as legal guardians and parents of these children, believe it is in the best interests of not only them but your family?

Yes, and health reasons. We needed to do this in order to get them out of this situation.

Yes.

You understand Stephanie, regardless of what may be said about you on the Internet and blogs, you cannot respond and you will not respond?

Yes.

Chris?

Yes.

One last question. You understand that this record has also been requested to be sealed and that you have consented to it being sealed, which means that no one from this point forward will ever be able to review this record or have any understanding of what has happened here today?

Yes.

We have agreed to this because we needed to get the children out of there for their health and safety. My concern is they're minors. We know we're signing for silence forever, but how is this taking away our children's rights? I mean, my daughter is turning 7 today, my son is 10.

Frankly, your Honor, as an attorney, I don't know if it's possible to give up the First Amendment's rights of a child. The defense has requested that be a part of the petition as worded. I will tell you we objected, but it was a take it or leave it situation. I have told them in an abundance of caution, and to protect my law firm, because I don't feel like having someone coming around when they turn 18 and saying, "Look what you did to me."

Does defense counsel have any comment for the record?

The plaintiffs are defined as the whole family. That's the way the contract has been written. That's what we agreed to.

Your honor, as they have indicated, it is directed at the family, and these two minor children are part of the family. I have practiced 30-some years. I've never seen a request like this nor in my research, but they have made a choice, and

So noted.

That's all I can say. And Chris, I don't have an answer for you or Stephanie other than what I've already told you.

Nor does the court have an answer for you, and I would agree with counsel that I don't know. That's a law school question, I guess.

*Our position is that it does apply to the whole family. We would
certainly enforce it.*

Right, and candidly, you as the parents, are bound by it.

If I may, no matter where we live, they're going to be
amongst other children that are children of people in
this industry, they're going to be around it every single
day of their life. If they say one of the illegal words
when they're outside our guardianship, we're going to
have difficulty controlling that. We can tell them they
cannot say this, they cannot say that, but if on the play
ground

So noted.

You will do and you have accepted to do, the best you can as parents to
prevent that from happening.

**

From the transcript of sealed, in chambers proceedings before Honorable Paul Pozonsky,
Judge, August 23, 2011, Court of Common Pleas of Washington County, PA, Civil Division.
Stephanie Hallowich and Chris Hallowich, Plaintiffs. Range Resources Corp; Williams
Gas / Laurel Mountain Midstream; Markwest Energy Partners, LP; Markwest Energy Group,
LLC; and PA Dept. of Environmental Protection, Defendants. The *Pittsburgh Post-Gazette*
and the *Observer-Reporter* filed petitions to intervene and motions to unseal the transcript,
and in January 2013, Judge O'Dell Seneca unsealed the transcript on the grounds that corporations do not have the same rights to privacy as individuals.

❱❱❱ Petro-Melancholia: The BP Blowout and the Arts of Grief

Stephanie LeMenager

The title of an August 2010 article in the satiric newspaper the *Onion*, "Millions of
Barrels of Oil Safely Reach Port in Major Environmental Catastrophe," ironizes
the systemic violence and long duration of the petro-imperialism that was reani-
mated through the BP blowout. Describing the routine docking of an oil tanker at
Port Fourchon, Louisiana, the *Onion* continues: "Experts are saying the oil tanker
safely reaching port could lead to dire ecological consequences on multiple levels,

Qui Parle: Critical Humanities and Social Sciences 19, no. 2 (2011): 25–56

including rising temperatures, disappearing shorelines, the eradication of countless species, extreme weather events, complete economic collapse."[1] As Rebecca Solnit notes, the BP blowout is "a story that touches everything else."[2] Solnit's comment, which is not satiric, points to why the article in the *Onion* is. The BP blowout marks a rough edge of what we in the United States and arguably in the developed world take for granted as normal, petroleum economies that generate multiple levels of injury. Mike Davis has written of the "dialectic of ordinary disaster" in relation to the apocalyptic rhetoric that defamiliarizes predictable geological events such as landslides in the poorly sited inland suburbs of Los Angeles.[3] Extending Davis's critique, Rob Nixon refers to the "slow violence" of neoliberalism as the occluded referent of "disaster," which in a modern risk society is often a misnomer.[4] From the Greek *astron*, or star, "disaster" suggests an unforeseen calamity arising from the unfavorable position of a planet. The BP blowout confirms disaster criticism's focus on the expectedness of the so-called unexpected while pointing to a different aspect of how ecological collapse can obscure human social and technological histories. Here the problem is proximity. The petroleum infrastructure has become embodied memory and habitus for modern humans, insofar as everyday events such as driving or feeling the summer heat of asphalt on the soles of one's feet are incorporating practices, in Paul Connerton's term for the repeated performances that become encoded in the body.[5] Decoupling human corporeal memory from the infrastructures that have sustained it may be the primary challenge for ecological narrative in the service of human species survival beyond the twenty-first century.

The BP blowout poses a unique representational challenge because it follows an unusual episode of de-reification, a failure of the commodity form's abstraction. This "disaster" did not work as spectacle, in Guy Debord's sense of the mystification of modern means of production through screen imagery.[6] The continuous video feed available on the Internet of oil shooting out of the damaged well—however that might have been manipulated by BP—read as a humiliation of modernity as it was known in the twentieth century, which is largely in terms of the human capacity to harness cheap energy. Unlike anthropogenic climate change, which resists representation because of its global scale and its multiple temporalities, the Deepwater Horizon rig localized a plethora of visible data, more than could be disappeared by the hundreds of thousands of pounds of Corexit that BP poured into the Gulf. The BP blowout resembles Hurricane Katrina in its manifestation of "events" that support predictions of environmental catastrophe (e.g., peak oil, global climate change) that otherwise might be dismissed as effects of scientific modeling or Left fear-mongering. Yet, just as Katrina did not result in a changed national affect toward black, urban poverty, the BP explosion has not, it seems, spurred Americans to re-consider loving oil. "Even if they cap the well, hell it's just another oil spill," sang Drew Landry to the president's awkwardly sympathetic Oil

Spill Commission in mid-July of 2010.[7] Landry's briefly famous "BP Blues" predicts that the blowout will not have the effect on the Gulf Coast that, say, the Santa Barbara oil spill of 1969 had on California, which became a strong industry regulator and the arguable headquarters of US environmentalism after that event.

What Katrina and the BP blowout foreground is a competition between emotional investments in modernity as we know it, through its fossil fuel infrastructure, and in ecology, as the network of human–nonhuman relations that we theorize as "nature." We learn from these two events on the Gulf Coast of the southern United States not only that modernity and ecology are entangled objects, to paraphrase Bruno Latour, but also that the melancholia for a given Nature which has supposedly characterized modern environmentalism might be eclipsed, in the twenty-first century, by an unresolvable grieving of modernity itself, as it begins to fail. Of contemporary industrial "accidents," Latour writes, "the recent proliferation of 'risky' objects has multiplied the occasions to hear, see, and feel what objects may be doing when they break other actors down."[8] For Latour, all objects are agential, and the ecological, relational materialism worked out through Actor-Network-Theory gets a boost from incidents that physically play it out, such as the BP blowout, where, for example, benzene and other volatile organic compounds associated with petroleum extraction enter the living cells of coastal cleanup crews.[9] Of course, Latour's sort of ecological thinking, wherein humans mingle with and are perhaps invaded by other "agents," does not necessarily feel good to the ordinary people enmeshed in these events. Feeling ecological need not be pleasant.

Ironically, feeling enmeshed in the world in a more positive sense also may consist in feeling at home in the obsolescent energy regime of the twentieth century, at least when its systems work. Artists and writers responding to both Katrina and the BP spill locate the human, as an ontological category, within industrial-era infrastructures not only in the throes of failure but also predetermined to destroy basic conditions of (human) living, such as water systems. That melancholia for modernity might eclipse environmental melancholia and activism in the context of the United States' Deep South makes especial sense, because it could be said that US modernity never assumed its fullest form there, so it still piques aspirational desire.[10] Rather than acting as pure counternarrative to the ecological violence of late capitalism, the stories that Katrina and the BP blowout produce tend to imagine modern infrastructure failure as tantamount to human species extinction—as if the species is unthinkable without these increasingly obsolescent objects.

Feeling Ecological, Thinking History

In Barack Obama's June 2010 speech from the Oval Office on the Gulf oil spill, the words "catastrophe," "disaster," "assault," and "epidemic" touch the edges of what is happening in the Gulf, which the president describes as no less than the destruction of "an entire way of life," the loss of "home" for tens of thousands—perhaps hundreds of thousands—of people. Noting that "the oil spill represents just the latest blow to a place that has already suffered multiple economic disasters and decades of environmental degradation," Obama moves away from the idea of disaster as singular, connecting it, if vaguely, to past policies of deregulation and privatization.[11] But he cannot link this past to a tangible national future—the speech was widely criticized for its vague prescriptions. In the tradition of presidential rhetoric, Obama concludes within sacred, rather than historical, time, with a reference to the tradition of the Blessing of the Fleet, an annual event that takes place in predominantly Catholic, Cajun fishing communities throughout Louisiana: "As a priest and former fisherman once said of the tradition: 'The blessing is not that God has promised to remove all obstacles and dangers. The blessing is that He is with us always . . . even in the midst of the storm'" (4). The priest's words seem to refer back to Hurricane Katrina and were most likely spoken in reference to that event, as the president places them in that context.

Once we start talking of humanity in terms of the sacred, as Hannah Arendt pointed out in the wake of World War II, we essentially acknowledge the collapse of constitutional or civil rights—sacred humanity is humanity without citizenship, "raw" and unprotected. Giorgio Agamben has since revised Arendt, urging a consideration of *Homo sacer* as the modern everyman, the man who can be killed, without legal retribution, but not sacrificed, in other words not inducted into the redemptive time of the sacred. The "naked human," in Arendt's words, or "bare life," in Agamben's, marks a modern dissociation from protective traditional statuses as well as the volatility of constitutional guarantees.[12] Similarly, humanity defined as ecological, in the sense of those whose "way of life" is conditioned by a regional ecosystem, may be recognized as humanity unprotected by rights or status—the human animal whose primary community is nonhuman. Obama's repeated invocation of Gulf Coast residents' threatened "way of life," which echoes and has been echoed by media accounts of the loss of "a way of life" or a "unique way of life," indicates that Gulf Coast people have fallen out of (or were never included within) the concept of modernity, where life practices are not clearly tied to place. The historian Ramachandra Guha recognizes "ecosystem people" as those enmeshed in nonhuman ecologies that feed and sustain them, and who are threatened by the omnivorous capitalization of their resources.[13] Perhaps it is an exaggeration to imply that the southern United States is other than modern. Yet when modernity evinces spectacular failure, as in the oil-soaked Niger Delta or the significantly more privileged oil colony of the US Gulf Coast, a revised definition of "ecosystem people," indicating humanity at the mercy of collapsing naturecultures, can be useful. Gulf Coast residents' recognition of their deep entanglement with modernity's most risky objects has prompted a discourse of activism, the environmental justice movement, as well as a vernacular poetry of species failure.

The Gulf Coast materializes a twentieth-century US history in which energy, perhaps the most essential quality of biological life, has supplanted personhood, the social "face" of the individual human body. Southern personhood has long been degraded in the US national imaginary, in part as a legacy of slavery—southern blacks in particular continue to struggle with the imposition of social death—and in part because of the perceived backwardness of southern industrial development, which has figured as the lassitude of the South's poor whites. From the headquarters of the US oil colony, Houston, Texas, the African American sociologist Robert Bullard fostered the US environmental justice movement (EJ) in the 1980s as a direct response to the influx of polluting industries into southern states in the late twentieth century. What Bullard saw in the South of the 1970s and 1980s, in his self-described role of "researcher as detective," was the local trail of a global trend, the bargaining away of health—a baseline measure of human energy—for jobs.[14] EJ's well-known definition of the environment as the place where "we" (humans) "live, work, play, and pray" should be understood as, in part, an explicitly southern response to the trade-off of civil rights for corporate privileges in a region where humanity had historically been commoditized through chattel slavery.

While the environmental justice movement is now vibrant and international, the Gulf Coast origins of its North American theoretical framework are often overlooked. The oil corridor from Houston to Mobile produced this second wave of environmental activism that resonated with other national and international protests on behalf of human health; "ecopopulist" revolts, in Andrew Szasz's terms, including Lois Gibbs's battle with the Hooker Chemical Company in the working-class suburb of Love Canal, New York; and the international response to Chernobyl.[15] African American activists recognized that toxic pollution spelled the revision of a hard-won, racially inclusive concept of US citizenship and the reintroduction of *Homo sacer*, the man who can be killed without repercussion. Bare life has been a recurrent theme within southern US history, an index of both racial and regional disenfranchisement. Bullard writes tersely of the vulnerability of southern poverty pockets to corporate exploitation: "Jobs were real; environmental risks were unknown" (32). The paper mills, waste disposal and treatment facilities, and chemical plants that made the South the last mecca of US industrialization in the 1970s and 1980s—in 1980 Jimmy Carter famously intoned, "Go South, young man," in response to a national recession in which only the South seemed to be growing industrial jobs—had of course been preceded by the oil industry, which set up shop in coastal Texas in the last decades of the nineteenth century. The bargaining away of southern health for jobs has a historical arc concurrent with that of US modernity, though it is perhaps only now, after Hurricane Katrina and in the wake of the BP blowout, that the inclusion of US southerners within the South as a *global* region has become clear.

When Bullard compares the intangible quality of environmental risk with the hard realism of jobs, he points out a problem that environmental advocates have long recognized as both representational and political: environmental damage yet to come, without (current) aesthetic dimensions, does not stir up alarm or activate an ethic of care. This is one of the supposed pitfalls in trying to communicate the threat of global climate change—although that threat becomes increasingly palpable. For decades, the Gulf Coast has been sinking, quite visibly manifesting a dramatic change in climate and geological structure. What was marsh is now open ocean—to the tune of twenty-five to forty square miles of disappearing marsh per year—and that is *prior* to the BP spill. The strong aesthetic dimensions of this problem, whose geological name is subsidence, have been well documented by journalists, politicians, and even Shell Oil, which launched a media campaign to save the wetlands in the early 2000s. Loss at this scale of a nation's territorial state would normally be attributable to an act of war, which calls to mind the comedian Lewis Black's recent joke that the United States might declare war on BP, since the corporation is "attacking us with oil."[16] Yet the sinking of the Gulf Coast has not stirred significant national outrage, even since BP's debacle. Mike Tidwell, the environmental journalist whose *Bayou Farewell: The Rich Life and Tragic Death of Louisiana's Cajun Coast* (2003) marks perhaps the best-known chronicle of Gulf Coast subsidence, notes in that book his own struggles to garner attention for the crisis. I discovered Tidwell in a bookstore's travel section, a curatorial choice that underlines the perception of Gulf wetlands loss as a regional peculiarity.[17]

Explicitly political iterations of the subsidence story linked it to human health and the survival of the city of New Orleans nearly a decade before Katrina. In 1998 the Louisiana governor's office, with the state's Department of Natural Resources, the US Army Corps of Engineers, the Environmental Protection Agency, the US Fish and Wildlife Service, and all twenty of Louisiana's coastal parishes, published *Coast 2050*, the first comprehensive plan for restoring coastal Louisiana and a clarion call for federal remediation of the ecological "system collapse" wrecking the Gulf Coast.[18] Potential losses were listed then—lists we hear again, 20 years later—as 40% of the country's wetlands, one-third of its seafood, one-fifth of its oil, one-quarter of its natural gas, and a "historic" urban center of some 500,000 persons, New Orleans.[19] The price tag for coastal restoration at the time of *Coast 2050*'s publication, the late 1990s, was around $14 billion, modest in comparison to estimates of what restoration will cost in the wake of the BP blowout. Former Louisiana governor Mike Foster envisioned federal legislation akin to that which created Everglades National Park, and he hoped it would be passed into law by 2004. "Yet despite these efforts," Tidwell wrote in 2003, "the nation remains almost totally ignorant of Louisiana's plight" (*BF*, 336). Tidwell's epilogue to *Bayou Farewell*, written in 2005, after Hurricane Katrina, expresses the hope that the massive hurricane "finally awakened America to the fragility and importance of south Louisiana." But the book concludes in the fatal rhetoric of the sacred: "Either we are witnessing the death of something truly great in America or the start of something even better, something new and blessedly permanent" (*BF*, 343, 344). *Bayou Farewell* prominently features the Cajun Blessing of the Fleet. Again, the sacred is invoked when social death has already occurred, and civil rights suspended. Thinking through subsidence as a narrative that has not become national despite its dissemination through national media raises the question of when the Gulf Coast fell out of the US territorial imaginary.

One might say—as anthropologist James Clifford suggests—that it becomes clear that a certain set of humans have lost civil rights and protections when scholars gather their oral history, with the archiving of the voices of a doomed community serving to memorialize their sacrifice. In the first year of George W. Bush's presidency, 2000, the now defunct Minerals Management Service (MMS) funded an oral history of southern Louisiana, *Bayou Lafourche: Oral Histories of the Oil and Gas Industry* (2008). Bayou Lafourche is the Gulf Coast region most intimately linked to the deepwater drilling that began in the 1990s, and it was chosen as the site of the oral history project in part to create an epochal break between the era of onshore oil drilling, "shelf" drilling for natural gas, and the outer continental shelf deepwater industry. In the prologue to *Bayou Lafourche*, author Tom McGuire acknowledges that "people who knew these communities prior to the oil and gas industry, people who orchestrated the technological innovations to explore, drill, and produce in the marshes and bays for the coastal wetlands, people who ventured out into the open Gulf in the risky pursuit of fossil fuel—they were passing away. A collective memory . . . was dying out."[20] Since "incorporated towns with municipal governments which might be expected to preserve community history" were few along the Gulf Coast, and "corporate memories have been erased through mergers, acquisitions . . . closures," and "blue-collar workers seldom write memoirs," the project solicits the federal government to support the transcription of voices that have no other representative (*BL*, 2–3).

The result of the grant proposal would be some four hundred interviews, archived at several Gulf Coast universities, and a book-length report, *Bayou Lafourche*, that alternately could be titled *MMS: The Novel*. A messy social panorama composed of interwoven interviews, *Bayou Lafourche* predicts the rash of interwoven life stories that appeared in the aftermath of Hurricane Katrina, from Spike Lee's epic *When the Levees Broke* (2006) and journalist Dan Baum's nonfiction *Nine Lives* (2008) to Josh Neufeld's graphic novel *A.D.: New Orleans after the Deluge* (2008) and writer/producer David Simon's recent HBO series, *Treme* (2010–). To a certain extent, all of this post-Katrina art foregrounds the tension between the structure of the individual life, with its aspiration and its foreboding of death, and bodiless corporate structures that change form over durations longer than human life spans. Corporate temporality, like ecological time, is not strictly historical, insofar as it involves the duration of systems and "persons" neither human nor mortal. The larger conflict between human and corporate ecologies is played out in post-

Katrina narrative in the collapse of naturecultures made vulnerable by the deregulation that feeds corporate entities.

Bayou Lafourche intends to offer an epic of oil-made-by-hand, of petroleum extraction as craft and embodied memory. Interviewees reflect a landscape where risky objects like deepwater rigs share real time and space with traditional extractive work like shrimping, since the shrimpers' large boats have been used for years to service the offshore rigs. Oystermen, who don't enjoy the off-season compensation from Big Oil that shrimpers do because their smaller, flat-bottomed boats are not ocean-worthy, emphasize the more purely ecological perspective of the oyster beds that they work, which are the natural filters of the wetlands. "Barataria Bay used to be full of oysters," Whitney Dardar, a Houma fisherman, complains, "Oysters don't grow like that anymore because there is too much salt. . . . I know a lot of places that oysters used to grow that they don't grow anymore. Now, it is like the Gulf; it is all open. They are trying to restore and all but I think they are about twenty years too late for that. They pump and pump all that oil and don't put nothing back. It sinks and sinks and sinks" (*BL*, 118). Nearly every interviewee mentions the problem of subsidence. All recognize that the profit their region has gained from the oil industry is balanced by geologic loss, salt water invading freshwater marshes due to shipping canals cut for oil transport, freshwater kept from replenishing marshlands because of the channeling of the Mississippi River, also for industry and development. Living on the line between earth and world, between ecological systems and the technologies that attempt to make them more accessible, the ordinary people of Bayou Lafourche live at the cutting edge of climate collapse. Theirs is, and has been for decades, a twenty-first-century ecology.

The oil industry picked up in southern Louisiana in the 1930s, with the arrival of the Texas Oil Company, now known as Texaco. At first resented, the "Texiens," as Cajuns called them, began to hire locals for their skills as carpenters and sailors, bringing jobs to a poor, rural region made more desperate by the Great Depression. The heady ambivalence of this era is famously captured in Robert J. Flaherty's film *Louisiana Story* (1948), which was commissioned by the Standard Oil Company. For the World War II generation in southern Louisiana who became middle class as a result of Big Oil, the industry still appears, nostalgically, as a robust future. Subsidence fails to make sense within this historical boom narrative, even as it is being somatized. *Bayou Lafourche* only touches the US oil industry downturn of the 1980s and the reinvention of the industry in the 1990s through deepwater play. With deepwater drilling came unprecedented technological experimentation, subcontracting to foreign rigs and crews, the perception of federal takeover, and less local love for Big Oil. As early as the Submerged Lands Act of 1953, the federal government claimed ownership of the continental shelf to three miles off the coast of Louisiana and other Gulf states, with the exception of Texas and Florida, which own the ocean bottom extending twelve miles out from their coasts. What this meant is that the United States would be in charge of leasing the outer continental shelf, and federal coffers would enjoy income from deepwater leases—if the technology ever got that sophisticated, which was scarcely imaginable in the 1950s.

Windell Curole, a Cajun radio personality and marine biologist who has a large voice in summing up *Bayou Lafourche*, disassociates the US government from any image of a "country," representing its role in the Gulf region as that of an irresponsible corporate actor. "If you're a business man, CEO of government USA, and I see three billion dollars [from leases] coming into my treasury in my business . . . I'm going to make sure that things that protect that infrastructure are in good shape and yet government doesn't see it that way" (*BL*, 164). Curole rejects both the ecological and the moral price that he feels has been levied on Louisiana by the oil corporations and by environmentalists, too. "We're human beings. It's us and the environment we live [*sic*] and the environment and every part of it, well, every part of it. We use up stuff just like every animal uses up stuff in the environment, but the point is don't use it up so that the reason you're living there isn't good anymore" (*BL*, 163). The repetition of "the environment" and "every part of it," tics duly transcribed in Curole's interview text, indicate the anxiety of ecological compromise, a constant rehearsal of losing something not quite anticipated.

The poet Martha Serpas, who has lived all her life in Bayou Lafourche, suggests a postscript to the MMS interviews that were conducted throughout her home parish. Serpas refers to the dialectic of petroleum and subsidence as "decreation" in a poem of that name and elsewhere in her collection *The Dirty Side of the Storm* (2007). Imagining herself in a coffin that has been unsettled and set afloat by the invading ocean, the poet writes,

> Someone will lay a plaster vault for me to ride,
> like long boxes children pull down flooded roads.
> In my plaster boat I'll ride Gulf shores till I vanish like a rig in the sun.[21]

The poem suggests the Leeville cemetery, one of many Cajun burial sites that have floated out to sea due to subsidence. Serpas's poetry invites an openness to personal extinction ("If only I could give the land my body—/ . . . I would lie against the marsh grass and sink, / . . . and welcome the eroding Gulf—"), as if humans count primarily as matter, our corpses sandbagging the wetlands (79). To live in such a world is to be sculpted by subsidence, with that geological artist linked tenuously to the rigs, whose silhouettes against the sun make them appear as symptoms of distant intelligence. Serpas and the MMS interviewees offer a vernacular poetry of human species collapse: heroic, Catholic, melancholic. Feeling ecological means welcoming the breakdown of the human into "marsh grass and sink." This organicist vision is not unfamiliar in environmental writing, yet it takes on force, and threat, in a place where human bodies literally fight back the ocean because of technologies meant to extend human energy and power. The many inroads into Gulf coast marshland made to facilitate oil transport have enabled coastal subsidence, although they are not its only cause. Feeling at home in a petrol "world" creates an affective drag on thinking through human survival.

Endurance through Genre

The ecological value of artful expression as a means of human endurance became a significant touchstone for discussion in the wake of Hurricane Katrina, and to an extent Katrina nurtured the regional arts community that has responded to BP. While Katrina and the BP blowout are widely varied events, both bear a relationship to the oil industry—the subsidence of marshy coasts means less storm protection—and both force questions about how the arts can respond to and perform a productive grief, a grieving akin to surviving. In the film *Trouble the Water* (2008), fifteen minutes of Kimberly Rivers Roberts's and Scott Roberts's amateur videography makes possible the imaging of human biological resilience as something like enduring nausea. The viewer loses equilibrium in the Robertses' dizzying footage, which creates deep mimicry, "communicating trauma as a visceral and cognitive experience," as Lauren Berlant describes the translation of intimacies made possible through testimonial rhetorics.[22] From videography to the blogosphere to the book, experimentation with what I call lively genres, artistic forms that bear witness to the disappearance of human and non-human lives on the Gulf Coast by eliciting an especially interactive engagement, began in Katrina's wake.

"We are becoming aware that biocultural evolution is more tragic than comic," writes African American poet and critic Jerry W. Ward Jr. in *The Katrina Papers* (2008), "that battles with external nature eventually are transmuted into battles with ourselves."[23] Ward's project in *TKP*—his shorthand for the book, which is subtitled "*A Journal of Trauma and Recovery*"—is nothing less than shoring up his, or our, contribution to the bio-cultural fate of the species. To this end, he advocates "creating what is not literature" (53). For him, that requires pushing literature to become "interdependent," which might mean mingling literary practices in biological ones (eating, sleeping, sheltering) and the infrastructures that sustain them. As Raymond Williams reminds us, genres are enabled by explicit material contexts, by labor and resources. In light of Williams, Ward's call for the "interdependence" of literature could also be a call to rethink media genres in terms of the ecological costs of their modes of production.[24]

Although Ward claims that he neither can nor will theorize the relationship of art to disaster, *TKP* offers philosophical glosses on creativity within chaotic times, and the necessity of letting time itself just be—of not attempting to contain its vagaries in art. In *TKP*, writing, as a means of making time conventional, serves as a weak surrogate for "living." Living, in turn, figures as like water, insofar as "life" sympathizes with water's evasion of measurement and sequence. "Time knows what it is," Ward writes, "It is flowing past the hotel on the surface of the Mississippi River" (180). The scene is one of flood, which also becomes a kind of wisdom. Water's instigation to enact time as something other than progressive sequence figures in "Hurricane Haiku," one of the many poems that interrupt *TKP*'s narrative prose: "Aqua vitae heard / Mad death massing in their throats: / Blues, disaster hymns" (53–54). The water of life, strong liquor, strangles the urban poor, the southern poor, Africans "in the slave ships," as Ward elsewhere refers to Katrina survivors. But after or even within the episode of human dying, in the collapsed time of haiku, "aquae vitae" returns, as rhythmic (time-conscious) language, the blues, "disaster hymns" that immerse Katrina in histories of flooding along the Mississippi Delta, of racialized poverty, of pellagra and sharecropping. Sound, in "Hurricane Haiku," is voice *and* water, both standing for prelinguistic, elemental expression. If this implies a romantic naturalization of voice, the poem's thrust is pragmatic. The storytelling arts, the blues, hymns, are heroic but not triumphant, human biocultural rhythms that relive traumas of systemic disruption.

Throughout *TKP*, Ward reminds us how poetry and song can index diminished parameters of human health and energy that should not be obscured by the beauty of their rhythms. The lessons of African American history include the contingency of culture and health, for all humans. "It is not about us," Ward writes, presumably referring to Katrina survivors and particularly to black Katrina survivors. "The sad condition of our planet is the damnation of all classes; it is a signal that the American empire shall gnash its teeth" (114). Ward ties African American history to the tentative "pre-future" of "our planet" and the fate of humans generally, again highlighting, in a lyrical echo of Robert Bullard's work in environmental justice, the relevance of the Gulf Coast, and the black South, to US ecological futures. Updating Richard Wright's naturalism for the era of global climate change, Ward urges a consideration of climate change in terms of war—race war in the sense of both warring racial factions and of *the human race* at war with the collateral damage of modern technologies and political compromise. Ward writes: "Friends urge me to see the film that is based on Al Gore's lectures regarding global warming. If I don't ever see the film, cool. In either case, the weather in New Orleans can inform me about the progress of global warming. The film I really want to see pertains to the origins of World War Three. It has not yet been produced and edited to protect the guilty" (183). The weather in New Orleans stands for the vulnerability of that city, and many locales within the global South, to climate collapse. That the US Gulf Coast delivers an American environmental prophecy is central to Ward's critique.

With "World War Three," Ward offers a frame for global climate change that supersedes older environmental narratives. Turning repeatedly to poetic forms, he jettisons story structure so that interruptive or incomplete genres suggest the unclosing wounds of New Orleans' black, Latino/Latina, Vietnamese, largely impoverished climate refugees. In contrast, Josh Neufeld's graphic novel *A.D.: New*

Orleans after the Deluge (2008) elicits a national public through more conventional plotting. *A.D.* grew out of a comics blog hosted by the storytelling site *SMITH Magazine* and to some extent is about the art of telling a compelling story, crafting suspense and resolution. *SMITH* offers a test-run for comics authors, who can increase the market for print editions of their work by tens of thousands through preliminary web distribution.[25] Neufeld's popular audience shows up awkwardly in the blog when a comment praising his skillful plotting ("Looking forward to the rest!") is answered by another online reader: "I wonder if any of you jerkoffs are really from New Orleans . . . but yeah . . . looking forward for the rest . . . ha ha."[26] This correction suggests the blog's capacity for distributing authority over feeling, such that sentiment is negotiated by multiple authors and becomes a social problem and opportunity.

A.D. marks a departure for Josh Neufeld from his first Katrina-related publication, *Katrina Came Calling: A Gulf Coast Deployment* (2006), which appeared as a LiveJournal blog and then as a printed book. The evolution of Neufeld's experiments with genre (blog, printed journal, print comics) traces a struggle to find artistic forms relevant to ecological endurance. In *Katrina Came Calling*, Neufeld recounts his stint as a Red Cross volunteer in Biloxi, Mississippi, taking himself to task for his self-identified uselessness as a "culture-producer" ("what real good are comic book stories . . . ?"), theorizing his commitment to Katrina survivors as an effect of "scopophilic attraction," and wondering if the 9/11 bombings created an expectation of the scale of crisis that Katrina answers more fully than smaller disasters. *Katrina Came Calling* digs around in the psychology of empathy, turning up an appeal to ecological crisis as, at best, a new kind of war that might encourage moral clarity. "For me, 'disaster relief' had a double meaning," Neufeld reflects, "It also meant relief from the doubt, confusion, and gnawing self-hatred of being an American in today's world." *A.D.* also offers relief from the uncertainties of being an American insofar as it subverts the World War II–era origins of the comic superhero, that midcentury everyman who was expanded through larger-than-life avatars of exceptionalism and Cold War preparedness. In contrast, *A.D.* gives reader-viewers the opportunity to become larger-than-life humans in extremis. The book plots a virtual tour of bare life, leading the viewer to enact closure by letting go, panel by panel, of all the privileges of US modernity, including comics.

Chronicling seven "real" individuals' journeys through hurricane and flood, *A.D.* highlights the diversity, connectivity, and segregation of a modern city, suggesting the distinct odds of survival for an unemployed African American woman (Denise), an Iranian shopkeeper and his black fishing buddy (Abbas and Darnell), a gay doctor who lives on the high ground of the French Quarter ("The Doctor"), a middle-class black student (Kwame), and two downwardly mobile white artists (Leo and Michelle). Neufeld presents almost an archaeology of the private spaces of this varied cast, their homes and small businesses in New Orleans, as if sifting through Pompeiian households and calculating class-based resilience based on which furnishings are unearthed from the ruins. In the comics tradition attributed to *Tintin* creator Hergé (Georges Rémi), Neufeld's settings are realistic in their meticulous accounting of material culture, while his characters are drawn loosely enough to evoke broad identification, even though we read them as variously ethnic or "white." Denise becomes our Dantean guide through the underworlds of the New Orleans Superdome and the Ernest N. Morial Convention Center, nightmare instantiations of the graphic novel's generic iconicity. Here there are no particular persons. The ubiquitous male/female icons on the bathroom doors of these institutional spaces acquire a referential tie-in to filth and excrement, raising questions about how the iconicity of the comic genre might also enable depersonalization or disposability.

As a hand art that requires little translation, since it is not dependent upon language or culturally specific imaging, comics potentially could disseminate new global stories to normalize more sustainable energy regimes or defamiliarize obsolescent ones, as in Harriet Russell's post-petroleum short comic "An Endangered Species, Oil" (2007).[27] Yet the ecological fate of print comics is itself uncertain, betraying the lack of resilience of certain artistic genres within increasingly untenable systems of production. Neufeld queries the future of comics in *A.D.* through his character-double, Leo, who loses some 15,000 comic books in the post-Katrina flood. When Leo learns that his Mid-City neighborhood is under seven feet of water, his subsequent disaster fantasy earns a two-page panel, featuring Leo, starkly uncolored and therefore visually permeable, spinning with his erstwhile comics in a deep red, toxic sludge. The white, paper-colored bubbles from Leo's mouth suggest that he will survive, but his comics are already pink, taking on the expressionistic hue of the killing waters. Given the amount of water that it takes to make comics, newsprint, or any printed book, the fate of Leo's collection could be said to refer back to the material history of this popular art in some of the oldest and most ecologically damaging industrial practices.

The making of paper involves cooking wood chips until the tightly bound wood fibers separate into pulp, a process that consumes a tremendous amount of freshwater—which is why paper mills are often built next to rivers. According to Michael W. Toffel and Arpad Horvath, one year's worth of the *New York Times* (a single subscription) consumes about 22,700 liters of water.[28] Toxic effluents released back into rivers and streams by paper mills include cancer-causing dioxins. These effluents have been found to impair fish reproduction by interacting with fish neurotransmitter systems, and they are linked to human mercury poisoning through consumption of contaminated fish.[29] About 76% of all pulp-wood production for US paper takes place in the US South—the paper industry was an early target of Robert Bullard's analyses in the 1980s. The latest episode of ecocide inflicted upon the South by the paper industry involves genetically engineered pine plantations, some sterile and others allowed to propagate their seed, which will result in a falling-off of biodiversity. Because the US South has less public land than any other national region, it has been difficult for environmentalists to block the growth of so-called "franken-forests" through state or federal regulation.[30]

By virtue of their means of production, books like *A.D.* are looped into the regional ecosystem failure of which Katrina and the BP blowout are dramatic examples. As a complex four-color work, *A.D.* was printed in China, where looser environmental and labor regulation makes possible cost-effective printing of books, including graphic novels, which require high levels of craftsmanship.[31] The dissemination of visual art through printing is quite literally bad for the environment, contributing to the poisoning of rivers, to the clear-cutting of forests, and to global climate change—the pulp and paper industry is the third- or fourth-largest industrial emitter of greenhouse gases in most industrial countries.[32] Add to that the emissions generated by shipping art books back to the United States from China. The unsustainability of print has led writers and graphic artists to digital media, whose primary environmental advantage over print media is its elimination of paper.[33] Yet, as Neufeld suggests in contrasting his blog and print book versions of *A.D.*, the book allows for a play of "aspects of timing, meter, and rhythm," due to the tiering of images on the page and to the physical act of turning pages, which our computer screens don't allow.[34] I would add that the power of the gutter, that blank space between comics panels that invites the reader/viewer to make conceptual bridges across time, is diminished by the digital format—particularly where a web host allows reader comments onto the page so that social noise interrupts the perceptual "filling in" that cartoon iconicity invites. To me, it is unclear whether the social potential enabled by interactive web features such as comments sections ought entirely to replace the empathic mimicry associated with conventions of silent reading. Neurobiological speculations about intersubjective relations suggest that recognition of the shared states that make us human requires "inner imitation," which might be reinforced in the silent reading and imagining of other lives.[35]

To be human, to be southern, to imagine our own bare life within a twenty-first-century ecology, we need now more than ever representations, narratives, pictures, moving and still. The answer to Martin Heidegger's question in the lecture "The Origin of the Work of Art" (1935–37)—"Is art still an essential and necessary way in which that truth happens which is decisive for our historical existence?"— seems now more than ever to be yes.[36] Understanding the Gulf Coast as a diminished American future may be crucial to both national and species survival. Such understanding takes time, the sort of time handled well by the traditional arts of duration. But the printed arts require great quantities of water and trees. There may not be time enough, in terms of the endurance of human habitat, for print media or for commercial film production, which also relies heavily upon petroleum products and is a major emitter of greenhouse gases. What this means for the endurance of genre remains to be seen. The relentlessly social and interactive creativity of the blogosphere may be complemented by performance arts that are entirely off-grid, or Internet sites that refer out to embodied performances, like the New Orleans social network *Humid Beings*, whose name reminds us that we, too, are made of water.

Notes

1. "Millions of Barrels of Oil Safely Reach Port in Major Environmental Catastrophe," *Onion*, August 11, 2010, http://www.theonion.com/article/millions-of-barrels-of-oil-safely-reach-port-in-ma-17875.

2. Rebecca Solnit, "Diary," *London Review of Books*, August 5, 2010, http://www.lrb.co.uk/v32/n15/rebecca-solnit/diary/print.

3. Mike Davis, "Los Angeles after the Storm: The Dialectic of Ordinary Disaster," *Antipode* 27 (Summer 1995): 221–41.

4. Rob Nixon, "Neoliberalism, Slow Violence, and the Environmental Picaresque," *Modern Fiction Studies* 55 (Fall 2009): 443–48. See also Curtis Marez, "What Is a Disaster?," *American Quarterly* 61 (September 2009): ix–xi.

5. Paul Connerton, *How Societies Remember* (Cambridge: Cambridge University Press, 1989).

6. Guy Debord, *Society of the Spectacle*, trans. Ken Knabb (London: Aldgate Press, n.d.).

7. Drew Landry, "BP Blues," https://www.youtube.com/watch?v=EApX41i1tio.

8. Bruno Latour, *Politics of Nature: How to Bring the Sciences into Democracy*, trans. Catherine Porter (Cambridge, MA: Harvard University Press, 2004), 23. Hereafter cited as *PN*.

9. Solnit discusses the feeling of being in New Orleans and breathing in VOCs (volatile organic compounds), with a periodic "gas station smell," in "Diary."

10. It should be noted that the US South also has fostered an anti-modern and arguably "ecological" culture, often located in cultural representation in acts of fishing, hunting, and eschewing the perceived conveniences of a decadent North.

11. Barack Obama, "Text of Obama's Speech on Gulf Oil Spill," *New Haven Register*, July 16, 2010, http://www.nhregister.com/articles/2010/06/15/news/doc4c184f9627d4f201824265.txt.

12. Hannah Arendt, *The Origins of Totalitarianism* (New York: Harcourt, 1968); Giorgio Agamben, *Homo Sacer: Sovereign Power and Bare Life*, trans. Daniel Heller-Roazen (Stanford: Stanford University Press, 1998).

13. Ramachandra Guha and Juan Martinez-Alier, *Varieties of Environmentalism: Essays North and South* (London: Earthscan, 1997), 12–13.

14. Robert D. Bullard, *Dumping in Dixie: Race, Class, and Environmental Quality* (Boulder, CO: Westview, 1990), xiii. Hereafter cited as *DD*.

15. Lawrence Buell, "Toxic Discourse," *Critical Inquiry* 24 (Spring 1998): 641.

16. "Lewis Black's Spill Solution: Declare War, Invade BP," *Raw Story*, June 10, 2010, http://www.rawstory.com/rs/2010/06/lewis-blacks-spill-solution-declare-war-invade-bp.

17. Mike Tidwell, *Bayou Farewell: The Rich Life and Tragic Death of Louisiana's Cajun Coast* (New York: Vintage, 2003). Hereafter cited as *BF*.

18. Louisiana Coastal Wetlands Conservation and Restoration Task Force and the Wetlands Conservation and Restoration Authority, *Coast 2050: Toward a Sustainable Coastal Louisiana* (Baton Rouge: Louisiana Department of Natural Resources, 1998), 1–2.

19. Discussion of these figures and the fate of *Coast 2050* can be found in Mark Fischett, "Drowning New Orleans," *Scientific American*, October 2001, 78–85.

20. Tom McGuire, *History of the Offshore Oil and Gas Industry in Southern Louisiana*, vol. 2, *Bayou Lafourche: Oral Histories of the Oil and Gas Industry* (Washington, DC: US Department of the Interior, Minerals Management Service, Gulf of Mexico OCS Region, 2008), vii. Hereafter cited as *BL*.

21. Martha Serpas, "Decreation," in *The Dirty Side of the Storm* (New York: Norton, 2007), 79.

22. Lauren Berlant, "Trauma and Ineloquence," *Cultural Values* 5 (January 2001): 44.

23. Jerry W. Ward Jr. *The Katrina Papers: A Journal of Trauma and Recovery* (New Orleans: University of New Orleans Press, 2008), 206.

24. See Michael Ziser's discussion of the relative carbon weight of culture in new media systems in "Home Again: Peak Oil, Climate Change, and the Aesthetics of Transition," in *Environmental Criticism for the Twenty-first Century*, ed. Stephanie LeMenager, Teresa Shewry, and Ken Hiltner (New York: Routledge, 2011), 181–95.

25. Jeff Newelt, comics editor at SMITH, contends that web publication could produce a graphic novel that "sells 60,000 or 70,000 copies instead of 5,000." Quoted in Dave Itzkoff, "The Unfinished Tale of an Unlikely Hero," *New York Times*, September 5, 2010, 41. Josh Neufeld, *A.D.: New Orleans after the Deluge presented by SMITH Magazine* (January 1, 2007), http://www.smithmag.net/afterthedeluge/2007/01/01/prologue-1.

26. Scott McCloud, *Understanding Comics: The Invisible Art* (New York: HarperCollins, 1993), 30–37.

27. Harriet Russell, "An Endangered Species, Oil," in *Sorry, Out of Gas*, ed. Giovanna Borasi and Mirko Zardini (Montreal, Canada, and Mantova, Italy: Canadian Centre for Architecture and Maurizio Corraini, 2007), n.p.

28. Michael W. Toffel and Arpad Horvath, "Environmental Implications of Wireless Technologies: News Delivery and Business Meetings," *Environmental Science and Technology* 38 (May 2004): 2963.

29. Robert Weinhold, "A New Pulp Fact? New Research Suggests a Possible Mechanism for Some of the Damage to Fish from Pulp and Paper Mills," *Environmental Science and Technology* 43 (March 2009): 1242.

30. Conner Bailey, Peter R. Sinclair, and Mark R. Dubois, "Future Forests: Forecasting Social and Ecological Consequences of Genetic Engineering," *Society and Natural Resources* 17 (August 2004): 642, 643, 645.

31. Noelle Skodzinski, "Offshoring and the Global Marketplace," *Book Business Magazine*, http://www.bookbusinessmag.com.

32. See http://www.greenpressinitiative.org/impacts/climateimpacts.htm.

33. Nathalie Hardy, "A Writer's Green Guide," *Poets and Writers*, 37, no. 1, http://lion.chadwyck.com.proxy.library.ucsb.edu:2048; Michael J. Ducey, "Paper's Environmental

Agenda," *Graphic Arts Monthly*, July 2001 (accessed July 17, 2010); http://www.gammag
.com.

34. Josh Neufeld, "Afterword," in *A.D.: New Orleans after the Deluge* (New York: Pantheon,
2009), 193.

35. I'm referring to the controversial study of mirror neurons as a sub-personal ground
for empathy and more broadly for the recognition of common humanity. For example, see
Vittorio Gallese, "The 'Shared Manifold' Hypothesis: From Mirror Neurons to Empathy,"
Journal of Consciousness Studies 8 (2001): 43. By "the shared manifold," Gallese intends the
states we share with others, including "emotions . . . body schema . . . our being subject to
pain" (44).

36. Martin Heidegger, "The Origin of the Work of Art," in *Poetry, Language, Thought*,
trans. Albert Hofstadter (New York: Perennial Classics, 2001), 78.

**》》》 Petro-Magic-Realism: Toward a Political Ecology of
Nigerian Literature**

Jennifer Wenzel

The first Nigerian novel in English to make a splash on the Anglo-American literary scene was not Chinua Achebe's landmark *Things Fall Apart* (1958), but instead
Amos Tutuola's *The Palm-Wine Drinkard* (1952). Hailed by Dylan Thomas as a
"bewitching" tale in "young English," Tutuola's novel combines the universal appeal of a quest narrative punctuated by encounters with supernatural beings like
"Hungry-creature," "Invisible-Pawn," and "half-bodied baby," on the one hand, with
the exotic appeal of an idiosyncratic, perhaps even primitive, prose style, on the
other.[1] The journey that structures the narrative is undertaken by the eponymous
protagonist, who travels to the abode of the dead—Deads' Town—in an attempt
to bring an important person back to life. Unlike Orpheus, Ceres, or Gilgamesh,
however, the palm-wine drinkard seeks the recovery not of a wife, child, or bosom
friend, but an employee: a palm-wine tapster, whom the drinkard's father had
hired to tap wine for the drinkard from a farm of 560,000 palm trees, and who falls
to his death while on the job. Tutuola's unusual, yet parallel, syntax conveys the
relationship between drinkard and tapster: "I had no other work more than to
drink palm-wine in my life," the drinkard-narrator tells us, while the "expert palm-
wine tapster . . . had no other work more than to tap palm-wine every day."[2] Together, the expert palm-wine tapster and the prolific drinkard had formed a closed
circuit of production and consumption. Tutuola's neologism, *drinkard*, expresses
this professionalisation of consumption in a way that neither *drinker* nor *drunkard*
could. The "work" of drinking palm wine becomes impossible without the tapster,

Postcolonial Studies 9, no. 4 (2006): 449–64

yet the dead tapster cannot return to the land of the living to resume his labour. Embedded within Tutuola's marvelous tale, in other words, is an economic analysis of resource extraction and labour relations.

A similar dynamic appears in at least two other tapster tales from Nigeria—Ben Okri's short story "What the Tapster Saw" and Karen King-Aribisala's "Tale of the Palm-Wine Tapster" in her novel *Kicking Tongues*—and my goal in this essay is to understand what such seemingly magical stories about natural resources tell us about the multi-layered relationships between Nigerian literary production and other commodity exports. Okri's "What the Tapster Saw" depicts the superimposition of a petroleum economy over a palm economy in the Niger Delta. The equivalent of the journey to Deads' Town in this 1987 short story is the nightmarish vision of a palm wine tapster who falls from a tree while trespassing on Delta Oil Company territory; during a 7-day coma, he describes being surrounded by sentient, bespectacled turtles and following a snake down a borehole. In King-Aribisala's 1998 tale of a tapster, the commodity in question is Nigerian literature, both that intended for local use and that deemed export-quality. Okri's tapster "had seen the sky and earth from many angles."[3] I want to suggest that such multi-perspectival visions can help us to understand the intricate and multivalent relationships among palm, petroleum, and publishing: what tapsters see are not merely liminal, posthumous, or subterranean visions of the "bewitching" or the fantastic, but also networks of production, consumption, and exploitation, as they survey the Nigerian economic landscape from the treetops.

The palm-wine tapster is an agent of production within a local network of consumption, yet I argue that these fantastic texts situate the tapster on the margins of a broader export economy, whether of palm products, petroleum, or Nigerian literature itself: the palm wine tapsters see, or make visible, the mutual, if uneven, pressures of the global and the local. I am interested in how these texts' figuration of literature as one commodity among others can help us to understand the Nigerian novel's trajectory of "boom" and "bust" in the context of Nigeria's place in an international economy. I draw on the concerns of political ecology in order to suggest how we might historicise the signifying work that commodities do, and how literary production in Nigeria is itself constrained by cultural and material contests over natural resources. If the publishing industry, like the palm or petroleum industries, exerts different kinds of pressures within and outside Nigeria as it circulates commodities, then a concept of *petro-magic-realism* offers a way of understanding the relationships between the fantastic and material elements of these stories, linking formal, intertextual, sociological, and economic questions about literature to questions of political ecology.

The reception history of Amos Tutuola's fiction resembles an evolving allegory of resource extraction in a (neo)colonial context. When Dylan Thomas and other metropolitan reviewers celebrated *The Palm-Wine Drinkard*, some early Nigerian readers objected to the Anglo-American embrace of what we might think of as Tutuola's inexpert "tapping" of Yoruba narrative traditions (or, less generously, his unacknowledged borrowing from D. O. Fagunwa, whose writing had not yet been translated from Yoruba into English).[4] Unlike the recalcitrant dead tapster in his novel, Tutuola assured his British publishers that he had plenty more stories like *The Palm-Wine Drinkard* that he could offer them. If early Nigerian critics viewed Tutuola as a poacher, illegitimately tapping for his own benefit the trees of a communal tradition, later scrutiny of Tutuola's dealings with libraries and publishers outside of Africa would posit the author as the victim, rather than the perpetrator/ comprador, of plunder.

My conceit of resource extraction in tracing this reception history is informed by Chinua Achebe's argument in "Work and Play in Tutuola's *The Palm-Wine Drinkard*." In this 1977 lecture at the University of Ibadan, Achebe shifted critical attention on Tutuola from the idiosyncrasies of his language to the persistence of his concern with labour. This critical intervention might not seem as significant as Achebe's famous indictment of Conrad's *Heart of Darkness* in a lecture at the University of Massachusetts 2 years earlier, but in both lectures Achebe identifies in the novels a moral (or, in the case of Conrad, immoral) thrust that other critics had previously overlooked. Speaking to a Nigerian audience that had been largely dismissive of Tutuola, Achebe endorses what he sees as Tutuola's argument for a balance between work and play, as opposed to the excesses of the drinkard's exploitative professionalisation of recreation ("I had no other work more than to drink palm-wine in my life"): "For what could be more relevant than a celebration of work today for the benefit of a generation and a people whose heroes are no longer makers of things and ideas but spectacular and insatiable consumers?"[5] In my view, this reading opens Tutuola's work up for material, as well as moral, consideration: Achebe spoke at the height of Nigeria's oil boom, the moment in the late 1970s when, in Andrew Apter's account, "oil replaced labor as the basis of national development, producing a deficit of value and an excess of wealth, or a paradoxical profit as loss."[6] This moment of excess was also, and not coincidentally, the height of Nigeria's publishing boom.[7] (We might see more random coincidence in the fact that Nigeria "exported" both its first barrel of oil and Achebe's *Things Fall Apart* in 1958.)

Indeed, the compulsive numeracy and the repeated concern in Tutuola's novel with exchange, work, professionalism, and wages paid in pounds sterling suggest that there is more to the drinkard–tapster relationship than gluttony or servitude; the fallen tapster spends 2 years in apprenticeship before he qualifies as a "full dead man."[8] The tapster will not or cannot return with the drinkard to resume tapping the wine that satisfied him and his fairweather friends, but he does offer the tapster a magical egg that can feed the whole world, not an insignificant gift when the tapster returns home to find his town suffering from famine. When carelessness and greed cause the egg to break, the tapster glues it back together, only to find that it now produces hordes of magical leather whips that he then sets loose on the crowds that gather demanding to be fed.

It is tempting to argue here that what Tutuola's tapster sees, in this novel from the final decade of the colonial era, is the Nigerian neo-colonial petro-future—the moment of "spectacular and insatiable" consumption that Achebe marked a quarter century later—particularly if we read the egg's linkage of material-plenty-amidst-scarcity with the consequent violence of the whips as a prescient figure of the magic associated with the political ecology of oil. Journalist Ryszard Kapuscinski has written about the false promise of oil, in the context of Iran: "Oil creates the illusion of a completely changed life. . . . The concept of oil expresses perfectly the eternal human dream of wealth achieved through lucky accident. . . . In this sense oil is a fairy tale and like every fairy tale a bit of a lie."[9] In *The Magical State*, historian Fernando Coronil analyses what he calls "petro-magic," petroleum's false promise of wealth without work.[10] Within the fantastic frame of Tutuola's novel, the promise of the magic egg is exposed as a lie in the drinkard's unapologetic slaughter of those who had become dependent on it. The disappearance of the drinkard's friends when he can no longer supply them with palm wine offers an ironic lesson in the commodity's displacement of social relations onto objects.

One can compare the excesses of the magical egg with the rationalised juju that the drinkard uses throughout the novel; running short of money for the journey, the drinkard carves a paddle and uses juju to turn himself into a canoe. He spends a month acting as a river-ferry service, thereby earning more than £56. In his study of the cultural effects of Nigeria's oil boom, Andrew Apter links what he calls a "seeing-is-believing ontology"—the magical aspects of the oil economy noted by Coronil—to occult practices of "money magic" in southern Nigeria, whereby human blood and body parts are illicitly transmuted into currency; oil, according to Apter, figures as blood circulating through the national body.[11] What is striking in Tutuola's treatment of the drinkard-qua-canoe, however, is its emphasis on labour. Whereas the magical egg from the tapster in Deads' Town offers an image of wealth (or at least sustenance) without work, the drinkard's use of juju to turn himself into a canoe produces the means of production, but not money itself.

The trope of oil-as-magic points toward the resonance between the dynamics of Tutuola's novel and the concerns of political ecology, which seeks to understand the "convergences of culture, power, and political economy" that inform conflicts over "defining, controlling, and managing nature" and natural resources.[12] A political ecology of oil, in other words, would consider relationships between its "instrumentalities of material wealth and power" and its "less-material effects that belong to an economy of representation and value-forms."[13] Yet this reading of Tutuola in terms of the political ecology of oil is, admittedly, anachronistic. Although Tutuola writes that the drinkard's wife's pursuers "were rolling on the ground as if a thousand petrol drums were pushing along a hard road," Shell's discovery of commercially viable oil deposits in 1956 would not come until 4 years after the publication of the novel.[14] Perhaps petroleum must be read retrospectively into *The Palm-Wine Drinkard*, but the pressures of the centuries-long *international* trade in palm products must also be read into the novel. The palm economy that preceded petroleum was not only one of local palm wine consumption, with every drinkard keeping a tapster nearby.

Indeed, in the depiction of turn-of-the-century Igboland offered in Achebe's *Things Fall Apart*, we read not only of the elders' concern about overeager young tapsters tapping palm trees to death, but also of the enthusiasm of those making money once, in Achebe's words, "palm-oil and kernel became things of great price," commodities in the colonial economy into which places like Umuofia were being drawn.[15] Unlike palm wine, which spoils quickly and thus does not travel well, palm oil and palm kernel were drawn into international circuits of exchange. The European trade in palm oil and palm kernel in West Africa dates as far back as the 1480s and was worth a million pounds by 1840. Nigeria's Oil Rivers region was named for palm oil, not petroleum, and palm oil was perhaps as indispensable for nineteenth-century industry as petroleum was to twentieth-century industry: palm oil was used as an industrial lubricant, an edible oil, and in the making of soap, tin, and candles. Beyond their significance as exportable commodities, palm oil and palm kernel have been used locally for edible oil, food, and lighting, and the African oil palm can also be tapped for palm wine; the tree itself yields materials for building, roofing, and other household uses.[16] Within the riverine economy of the Oil Rivers region, jars of palm oil even functioned as currency.[17]

In his essay "Petro-Violence: Community, Extraction, and Political Ecology of a Mythic Commodity," geographer Michael Watts argues that petroleum links Nigeria crucially to what he calls "twentieth century hydrocarbon capitalism."[18] Yet the same link between Nigerian resources and the global, technocapitalist cutting edge can be made between palm oil and the nineteenth-century Industrial Revolution, or between slaves and earlier plantation economies.[19] (As I've discussed elsewhere, the rubber and mineral endowments of the Congo, including the uranium that would ultimately destroy Hiroshima and Nagasaki, have given the purported heart of darkness a similarly indispensable role in "European" modernity).[20] Watts analyzes what he calls "petro-violence," the particular forms of violence (physical, environmental, cultural) that constitute the Nigerian oil economy; the palm trade might seem idyllic in comparison, particularly since we know that palm products replaced slaves as the chief export of the Niger Delta in the early nineteenth century's transition to "legitimate commerce." Yet it's worth noting that one effect of the growth of this ameliorative trade was the intensification of the internal slave trade as demand for labour increased.[21] Nigeria's petroleum economy has literally been superimposed over (or excavated under) the palm belt of the

Niger Delta: palm oil production and exports decreased after the 1950s, to the extent that Nigeria became a net importer of vegetable oil in the 1980s. Recent efforts to rebuild the palm industry have been impeded by the poor quality of trees, stunted or sterile due to the environmental degradation associated with petroleum drilling and transport.[22]

The social life of *Elaeis guineensis*, the African oil palm, is a nexus of global, regional, and local trading relationships. I want to suggest that the nineteenth-century palm economy *seems* to offer in retrospect—from the perspective of petroleum—an *image* of a balance between local and international orientation that has not been possible with petroleum, given the intensities and excesses of capital, environmental impact, rents, and revenue associated with petroleum and involved in Watts' concept of petro-violence. I am interested here less in an objective comparison between the negative aspects of palm and petroleum than in the shifting webs of meaning associated with commodities in particular historical moments: is it possible that the rise of petroleum makes palm *seem* more "local," less alienated and alienating, that petro-violence makes palm seem peaceful by comparison?[23] Andrew Apter argues, in a similar vein, that "as a moral economy recalled with nostalgia, the palm-oil trade and the forms of 'natural value' that it invoked . . . established a profound contrast with the immoral economy of petroleum, which pumps bad money from beneath the ground, only to pollute and destroy the productive base of the economy."[24] The petro-bust of the 1980s exposed the irrational exuberance of the boom years, after the fairy tale of oil's promises had been exposed. This critique of petroleum informs Karen King-Aribisala's 1998 novel *Kicking Tongues*, which narrates a journey to the capital in Abuja by a diverse group of Nigerian citizens concerned to reclaim their country from the delusions of petro-magic—the excesses of affluence, corruption, lingering colonial consciousness, and military rule. Each of the pilgrims to the capital offers a tale to ease the journey.

King-Aribisala's parody of Chaucer's *Canterbury Tales* is even more problematic than this description might suggest, not only because the book ultimately sublimates its political protest by turning Nigeria's future over to a decidedly Christian evangelical God (a position King-Aribisala reiterates in interviews), but also because its "Tale of the Palm-Wine Tapster" in turn parodies Tutuola in order to condemn the international orientation of Nigerian literary culture. (I should note briefly that King-Aribisala's parody of Tutuola's prose style is both cruel and imprecise, implying that he was far less literate than his manuscripts allow). King-Aribisala's treatment of the tapster figure links the dynamics of the literary publishing industry to an idealised palm economy, coded in terms of the pacific nativism I alluded to above. In her tapster's tale, the tapster has a wife, who is a palm tree. The palm tree wife provides the tapster with wine, food, and oil until she commits suicide over his predilection for foreign tree species, from which the

paper is made for the books of the British literary canon. In other words, the tapster's sin is a fondness for foreign books as well as "Nigerian Books which pander
to foreign tastes," which would presumably include Tutuola's *The Palm-Wine
Drinkard*. But the tale offers a happy ending. After discovering a "new kind of
language which be not like language of foreign book" but is instead "natural," expressing "total Nigeria," the tapster describes the result of his reconciliation with
his palm tree wife:

> We get plenty plenty pickin and too much nursery and my wife trust me now and I too
> happy and she enjoys Joys of Motherhood and she never feel to be Second Class Tree
> Citizen and in The Ditch with Double Yoke and after I being interpreting this past
> Season of Anomy with understanding and things definitely not Falling Apart. We all—
> she and myself be Heroes and road of literature which be Nigerian no longer be
> Famishing.

The inconsistencies of King-Aribisala's allegory of Nigerian literature in terms of
what her tapster calls "natural resourcing" are patently clear: in integrating the
titles of novels by Buchi Emecheta, Wole Soyinka, Chinua Achebe, and Ben Okri
into her tapster's tale, she offers a vision of an autonomous, vibrant Nigerian national literature, supposedly freed from the stranglehold of foreignness, but one
that has, almost without exception, been written and/or published by agents located abroad. All of this is within a Chaucerian frame and published by Heinemann, which may straddle the foreign/Nigerian divide, but the editor who accepted
the novel for Heinemann was an American woman in Boston.[25] Collapsing dynamics of production and consumption in its critique of the international pressures on
Nigerian literary culture, King-Aribisala's argument suggests an analogy something like "Achebe is to Tutuola as palm is to petroleum," in terms of the ways their
books circulate in Nigeria and abroad.[26]

Regardless of the limitations of King-Aribisala's critique, however, it allows us
to think about the conjunctions of print capitalism and petro-capitalism in Nigeria. If the novel, following Benedict Anderson, offers a medium through which the
nation can be imagined, then too the political ecology of oil can reveal how "national imaginings . . . also depend on the very materiality of the nation as a life-
sustaining habitat—on differing modalities of configuring the metabolism between
society and nature."[27] The nation, in other words, is not only a polity but also an
ecology or lifeworld. Beyond the imagining of the *nation*, "nationalised petroleum
produces a *state* (as the owner of the means of production) that . . . mediates the
social relations by which oil is exploited . . . and . . . is simultaneously granted
access to the world market." Oil thus produces the state as an indispensable and
magical mediator between international capital and markets, on the one hand,
and the national's political and natural bodies—its human and natural resources—
on the other. Yet at the same time that oil yields legitimacy and "visibility" to the

state, its excesses "reveal . . . the state and the nation to be sham, decrepit, venal, and corrupt notions."[28] The Nigerian novel (or a Nigerian "national literature" more broadly) has functioned in a not altogether dissimilar way, both as a medium for imagining a national community and establishing international visibility, and as a site that lays bare the contradictions of Nigerian nationhood as well as the collisions between the state's image of itself and skeptical critiques. These disjunctures are evident not only in the thematic content of literature published since the disillusionments of the 1960s, but also (and perhaps more tellingly) in regional and class differences in literacy and readership, language, and genre, as well as in the troubled state of Nigerian publishing and its fraught relationship to presses and readerships abroad.[29] In the case of King-Aribisala's text, at least, the national literature that she identifies as the mode and product of reunion between the tapster and his palm tree wife is conditioned by the very international pressures that spurred the rift to begin with.

The link between literature and oil, or print capitalism and petro-capitalism, is not merely one of analogy, however. To what extent is Nigerian literature a commodity whose production and consumption are linked directly to more tangible substances like palm and petroleum?[30] In suggesting the possibilities for a political ecology of Nigerian literature, I'm concerned not only with what happens when we think about literature *as if* it were a commodifiable resource like palm or petroleum, but also with the multivalent relationships between literary production and conflicts involving natural resources. Written in London in 1987, in the midst of the Nigerian oil bust, Ben Okri's story "What the Tapster Saw" confronts head-on the intersections of the local and the global that can't be mapped simply onto palm and petroleum. The story offers a phantasmagoric glimpse into a degraded, privatised landscape where the "signboards of the world were getting bigger"; one signboard reads, "TRESPASSERS WILL BE *PERSECUTED*."[31] The sun seems never to set or rise as the earth is bathed in the glow of natural gas flares, "roseate flames [that] burned everywhere without consuming anything." A talking snake glistens with the beautiful and deadly iridescence of oil spilled on water. In this landscape where boreholes crowd out palm trees, a palm wine tapster carries on plying his trade despite the ominous signboards; when he falls from one of a "strange cluster of palm trees," he spends 7 days in a hallucinatory liminal state, persecuted by unseen assailants vaguely associated with the oil company employees trying unsuccessfully to "level the forest" with the help of witch-doctors and explosives "left over from the last war." The Delta Oil Company brings in the witch-doctors to "drive away the spirits from the forest" and to dry out its climate, while farmers who were living amidst unexploded bombs "as if the original war was over were blown up as they struggled with their poverty."[32]

Juxtapositions of bombs and bullets, coups and executions, with herbalists and witch-doctors, talking animals and masquerades, in this fictional narrative about the collision of palm and petroleum, yield what I call *petro-magic-realism*, a literary mode that combines the transmogrifying creatures and liminal space of the forest in Yoruba narrative tradition with the monstrous-but-mundane violence of oil exploration and extraction, the state violence that supports it, and the environmental degradation that it causes. The story demonstrates one way of negotiating the pressure of petroleum on literary representation: Okri situates the magical and violent aspects of petro-modernity within an older fantastic tradition. His tapster is not so much a direct descendant of Tutuola's character as a distant cousin within a broader geneaology that the narrator acknowledges by referring to "mythical figures" that include "the famous blacksmith" and the "notorious tortoise."[33] Direct allusion invokes not Tutuola but rather D. O. Fagunwa, whose story at the beginning of *Forest of a Thousand Daemons* about the hero's father shooting an antelope who turns out to be his wife is echoed in a fragmentary tale told with "curious irrelevance" at the opening of Okri's story.[34]

Ben Okri is the Anglophone African author most commonly mentioned in critical discussions of magical realism as a global literary phenomenon; Tutuola and Fagunwa are taken to be precursors of West African magical realism who, nonetheless, lack cosmopolitan, ironic distance from the "traditional" or "indigenous" materials that tend to be identified as a primary source of the magic in magical realism.[35] Magical realist texts, unlike petroleum, are a renewable resource, but both are commodity exports of the global south in high demand in the northern hemisphere; indeed, the introduction to an important critical anthology on magical realism playfully celebrates magical realism as an "international commodity" that might be seen as a "return on capitalism's hegemonic investment in its colonies . . . now achieving a compensatory extension of its market worldwide."[36] While *petro-magic-realism* might not account adequately for Okri's entire *oeuvre*, what I find productive in the term is its potential to complicate and historicise the empty globalism of the label *magical realism*, in which the magical might be anything unfamiliar to a European or American reader. In his landmark essay "Magical Realism as Postcolonial Discourse," Stephen Slemon acknowledges that magical realism "threatens to become a monumentalizing category" by offering a "single locus upon which the massive problem of *difference* in literary expression can be managed into recognizable meaning in one swift pass," thereby "justifying an ignorance of the local histories behind specific textual practices."[37] International demand for magical realist texts distorts local literary cultures in the ways that Karen King-Aribisala and her character the palm tree wife deplore. The problem with the magic in magical realism is broader than the sanctioned ignorance of metropolitan readers, however: the relationship between realism and magic tends to be read as a binary opposition between the West and the rest, between a singular (European) modernity and multifarious worldviews variously described as pre-modern, pre-scientific, pre-Enlightenment, non-Western, traditional, or indigenous. In his metacritique of magical realism, Michael Valdez Moses notes that "if the paternity of the magical realist novel is everywhere the same" (in the European realist novel and its attendant ideology), then "in each locale where the magical realist novel is born, its mother appears to be different, distinct, and as it were, native to the region."[38] The cumulative effect of such strangely binarist readings of magical realism—in which one term is always the same, the other always different—is the consolidation of the West as a single entity confronting innumerable local traditions. This reification of the West seriously undermines claims for magical realism's subversive, anti-hegemonic, or decolonizing thrust.

In troping on the unsatisfactory term *magical realism*, I am arguing that the rubric *petro-magic-realism* reveals how Okri imagines the pressures of a particular political ecology within a particular literary idiom. If petro-magic offers the illusion of wealth without work, Okri's petro-magic-realism paradoxically pierces such illusions, grounding its vision in a recognisably devastated, if also recognisably fantastic, landscape. The conjunction of magical realism and petro-magic represents a synthesis of the epistemological (or aesthetic) and ontological poles that derive from the two seminal statements in the theorization of the literary mode. Okri's deployment of the Yoruba fantastic tradition corresponds to Franz Roh's 1925 discussion of Post-Expressionist European painting, in which he coined the term "*magischer Realismus*"—magic realism—to describe an aesthetic strategy, a mode of representation. The attention in Okri's story to the devastating material effects of petro-magic, on the other hand, approximates Alejo Carpentier's 1949 articulation of "*lo real maravilloso americano*"—the American marvelous real—an ontology or state of being shaped by the complex history and distinctive landscape of the Americas.[39]

The conjunction of the aesthetic and the ontological in Okri's petro-magic-realism has important ideological and temporal implications. His tale of the palm wine tapster's nightmarish experience in Delta Oil Company territory thematizes the conflict between established and emergent modes of production (here between artisanal palm wine tapping and capital-intensive petroleum drilling) that Fredric Jameson posits as constitutive of magical realism.[40] Yet because "What the Tapster Saw" emphasises the phantasmagoric aspects of petroleum extraction, the marvelous reality represented in this narrative has a decidedly modern source, even if it is described in a fantastic idiom with a venerable literary history. Petro-magic is in no way a vestige of tradition or pre-capitalism. (Similarly, Okri's novel *The Famished Road* portrays the road not as a reductive symbol of colonial modernity opposed to the pre-colonial "bush," but rather as a dynamic, palimpsestic site of both internal and external conflict.[41]) The modernity of Okri's petro-magic-realism obstructs the consumption of magical realist texts as nostalgic encounters with an exotic yet vanishing world. As Michael Valdez Moses points out, the production and consumption of magical realist texts by "those *who would like to be*-

lieve in the marvelous" but who do not actually believe involves a tacit assumption that a disenchanted "modern world . . . is the only one with a historical future."[42] Okri's "What the Tapster Saw" implicates metropolitan consumers of magical realism and petroleum products not in modernization's inevitable disenchantment of vestigial tradition, but rather in petro-modernity's phantasmagoric ravagements of societies and lifeworlds. In this sense, petro-magic is the future.

Okri's depiction of Delta Oil Company territory in terms of a Yoruba magical forest also points toward the future in a rather different way. Beyond its thematic concern with collusion between the Nigerian state and private petroleum enterprise, "What the Tapster Saw" illuminates the conjunction of petroleum and publishing within national imagining. As Ato Quayson points out, Okri's deployment of Yoruba narrative and cosmological traditions is significant not least because Okri is not Yoruba himself (his parents were Urhobo). Quayson attributes Okri's engagement with Yoruba traditions to the "development of a broadly Nigerian consciousness," and he sees in Okri's work the potential for a "literary tradition in Nigerian writing as the strategic filiation with a specific discursive field irrespective of ethnic identity."[43] Thus Okri's work points toward the possibility of a Nigerian national literature conceived not exogenously, in opposition to the "foreign," as in King-Aribisala (who is, however ironically, herself a native of Guyana now resident in Nigeria), but in terms of affiliation across ethnic and regional divisions within the nation. Even at the level of theme, Okri's juxtaposition of a liminal tapster/tortoise narrative with the environmental and political violence of petroleum extraction by the "Delta Oil Company" implicitly yokes the site-specific suffering of communities directly affected by oil to a broader imaginary. Indeed, Andrew Apter argues that although Ken Saro-Wiwa's attempt to secure environmental and economic justice for the Ogoni people was seen for decades as a local, ethnic struggle, the Ogoni movement against "the predations of the military-petroleum complex" had taken on national significance by the time of his execution in November 1995: "the plight of the Ogoni people came to represent the contradictions of oil capitalism in Nigeria," since the "vampirism" of the petro-state extracted wealth and welfare from the national polity as a whole.[44]

Ken Saro-Wiwa's status—as a writer (and publisher) of fiction and television series, as an activist for the Ogoni people, and, after his death in 1995, as a symbol of, if not a martyr to, the constitutive element of petro-violence in the Nigerian state—offers a spectacular example of the intersections among literary publishing, petroleum extraction, and the production of the Nigerian nation-state.[45] These issues remain particularly salient for institutions that would foster literary production in Nigeria. The Association of Nigerian Authors (ANA) held its Annual Convention in early November 2001 in Port Harcourt, a setting fraught with echoes of Saro-Wiwa's execution exactly 6 years earlier. Moreover, the convention's theme was "Literature and the Environment," which elicited a presentation by Nnimmo

Bassey, executive director of the NGO Environmental Rights Action / Friends of the Earth. Bassey exhorted the ANA to become more environmentally conscious and to "strengthen the voice of impoverished communities, by facilitating exposure of the culpability of the TNCs and local big business." Invoking the memory of Saro-Wiwa, Bassey condemned the sponsorship of the Rivers State branch of the ANA by oil companies Shell and Elf; to accept such "blood money," he concluded, "would be to murder Ken AGAIN."[46]

Bassey's presentation did not mention the bloody ironies of ANA annual prizes, including the ANA/Cadbury award and the ANA/Chevron award for environmental writing.[47] Yet perhaps the most egregious thing about the Chevron prize is its size—less than $500, emblematic of the broader discrepancy between oil company profits tracked in reports to shareholders and paltry investments to compensate affected communities.[48] What is one to make, then, of the recent endowment of a $20,000 annual Nigerian Prize for Literature, sponsored by Nigerian Liquefied Natural Gas (NLNG)? This prize was to be awarded for the first time on October 9, 2004, in part to mark the anniversary of the company's first natural gas export. The company's managing director, Andrew Jamieson, noted his concern that Nigeria's fabled literary status on the continent is slipping, given declining rates of literacy, struggling libraries, and gutted educational systems.[49] One of the most interesting things about this prize, then, is that only writers who had been resident in Nigeria for the past 3 years could compete, an acknowledgment of the difficulties that confront writers working in Nigeria. Ben Okri's petro-magic-realism offers a harrowing vision, but it was written in London and does not emerge directly out of the kind of conditions that the Nigerian Prize for Literature aimed to recognise and palliate. In the same month that this prize was announced, a press release announced the foundation of the African Writers Endowment and a donation by Chinua Achebe of $75,000. The fund, which is administered in the United States, aims to recognise the "formidable difficulty and barriers African writers *world-wide* face in getting their works published and in having *world* attention paid to those works, commercially and critically," and hopes to support "the telling of the African story by Africans [including those of the diaspora], in their own voices."[50]

The geographic and economic cartographies of these two recent investments in African literature confound global/local distinctions; one wonders which would be judged more approvingly in King-Aribisala's tapster's tale. Nigerian Liquefied Natural Gas is a joint venture between the Nigerian government (represented by Nigerian National Petroleum Corporation) and multinational oil companies Shell, Elf, and Agip. Its literature prize puts real money—perhaps enough money not to be "blood money"—behind a concern for living conditions in its host country. Yet two things are worth keeping in mind while considering the implications of this "real money." First, Nigerian Liquefied Natural Gas made headlines in early 2004 not only for announcing the endowment of its $20,000 literature prize, but also for a $180 million bribery scandal for construction contracts on its plants; the bribes are alleged to have been paid between 1995 and 2002 to the Nigerian government by a consortium of companies of which Halliburton is a member.[51] Second, the Nigerian Prize for Literature itself has turned out to be somewhat illusory.

After a short list of finalists for the prize was released for the inaugural competition, NLNG sponsored a September 2004 reading tour with stops in Lagos, Kano, Ibadan, and Abuja, in anticipation of the announcement of the winner of the Nigerian Prize for Literature. At the October 9, 2004, award ceremony, the short list was further winnowed to three finalists, Bina Nengi-Ilagha, Akachi Adimora-Ezeigbo, and Omo Uwaifo. Then Alhaji Abubakar Gimba, former president of the Association of Nigerian Authors, announced that the judging panel had decided that no prize would be awarded. Of the nearly 100 entries, none rose to a standard worthy of the Nigerian Prize for Literature. Instead, the three finalists would each receive $3000 prizes; after a spontaneous intervention from audience member and Nobel Prize–winner Wole Soyinka, these consolation prizes were increased to $5000 each.[52]

The true winner of the 2004 competition, according to Gimba, was "the integrity of Nigerian literature"; to award the prize to one of the entries would be to degrade "the standard Nigerian literature had attained over the years." The report released by the judging panel found that "recourse to self-publication short-circuits the traditional publishing processes and this gives rise to the numerous stylistic and grammatical flaws just observed. It is further observed that many writers have not acquired the necessary education or undergone proper apprenticeship and training required for the high level performance expected from winning entries at this level."[53] This assessment reprises the early reception of Tutuola, whose "young English" was judged by Nigerian readers to be an embarrassment. When he announced the endowment of the prize, NLNG managing director Andrew Jamieson indicated that he aimed to intervene in the decline in educational opportunities and the degradation of institutions associated with literature and literacy, which threatened to erode the mid-twentieth-century achievements of Nigerian literature; yet these same factors were cited in the panel's decision *not* to make an award.

By excluding Nigerian authors resident abroad, even those whose books have been published in Nigeria, the multinational enterprise NLNG not only assumed the authority to effect the formation of a national literary canon within a narrow definition of "nation"; it also, by withholding the prize, signaled that Nigeria has no living canon worth forming. This decision represents a lost opportunity to bring international attention to a new generation of Nigerian writers, according to Mcphilips Nwachukwu: "very many international publishers have long assumed that no writing was going on in Nigeria after Achebe and Soyinka." (The results of the 2005 Nigerian Prize for Literature competition likely confirmed such an impression, as the prize was split evenly between poets Gabriel Okara and Ezenwa Ohaeto. Okara, now 85 years old, is best known for his poem "Piano and Drums," published in 1978. Ohaeto, who wrote his master's thesis on Okara, died of cancer at age 47 only a month after the prize was announced.[54]) Nwachukwu puts his finger on the national/international ironies of the intersection of literature, the petro-state, and its multinational patrons in the Nigerian Prize for Literature: while the prize panel has decided to keep the "exclusionary clause" that limits eligibility to writers resident in Nigeria, the prize is paid in US dollars, rather than in naira.[55] Nwachukwu further observes that with a $20,000 prize (which had yet to actually be awarded at the time he wrote), NLNG seems to have bought the silence of Nigeria's intellectuals: none had taken a stand on the Halliburton scandal. Andrew Jamieson, in the meantime, moved on to a position as vice president of Shell Global Solutions, based in The Hague.[56]

Is the Nigerian Prize for Literature mere window dressing for petro-violence —"lit-washing," rather than greenwashing—from the perspective of the columns of zeros in the company's balance sheets?[57] Probably. Is $20,000 annually a substantial investment in literature, in a country where literary patronage is nearly nonexistent? Absolutely. Yet contradictions remain. The dual award in the 2005 Nigerian Prize for Literature elicited a passionate response from activist and writer Dagga Tolar:

> How and why must anyone think that literature's fortune would fare any better, when even oil, the freest of nature's abundant gifts to the Nigerian state is unrefineable on her own very soil? . . . Nigeria, even with all its oil, does not dictate the selling terms of oil to its own people or to the international consumers. It is those with capital that do, and those at the NLNG clearly have the capital to dictate who and what they must spend their money on.

Even as he made an analogy between literature and oil as national, natural resources under the command of international capital, Tolar resisted the implication that literature be understood simply as a commodity like any other: "the logic of profit making applied to gas production and its consumption cannot in any way apply to literature."[58] Yet it is that logic of profit making applied to the production and consumption of gas and oil that not only has meant that ordinary Nigerians tend not to consume nor benefit from Nigerian oil and natural gas, but also has decimated the institutions that would foster a vibrant literary culture in Nigeria. To wish away the imbrication of literature within local and global economies, as Dagga Tolar seems to do, is no less dangerous than to celebrate NLNG's patronage uncritically. What tapsters see, and make visible for us, are these kinds of contradictions, borne of the intersections of commodities like palm, petroleum, and published writing. They offer a glimpse of what a political ecology of Nigerian literature might look like, a glimpse of the realities created and obscured by petro-magic.

Notes

1. In Bernth Lindfors (ed.), *Critical Perspectives on Amos Tutuola* (Washington, DC: Three Continents, 1982).

2. Amos Tutuola, *The Palm-Wine Drinkard* (New York: Grove Weidenfeld, 1984), 7.

3. Ben Okri, "What the Tapster Saw," in *Stars of the New Curfew* (New York: King Penguin, 1988), 183–94, 190.

4. Early Nigerian responses to Tutuola are available in Lindfors, *Critical Perspectives*.

5. See Achebe's "An Image of Africa: Racism in Conrad's *Heart of Darkness*," in *Hopes and Impediments: Selected Essays* (New York: Anchor, 1990), 100–112, 112.

6. Andrew Apter, *The Pan-African Nation: Oil and the Spectacle of Culture in Nigeria* (Chicago: University of Chicago Press, 2005), 201.

7. Wendy Griswold offers statistics on the Nigerian novel boom, which reveal a steady increase through the 1970s, wild fluctuation from 1979 to 1988, and subsequent crash. Wendy Griswold, *Bearing Witness: Readers, Writers, and the Novel in Nigeria* (Princeton, NJ: Princeton University Press, 2000), 37–38.

8. Tutuola, *Palm-Wine Drinkard*, 100.

9. Quoted in Michael Watts, "Petro-Violence: Community, Extraction, and Political Ecology of a Mythic Commodity," in *Violent Environments*, ed. Nancy L. Peluso and Michael Watts, 189–212 (Ithaca, NY: Cornell University Press, 2001), 203.

10. See especially the introduction in Fernando Coronil, *The Magical State: Nature, Money, and Modernity in Venezuela* (Chicago: University of Chicago Press, 1997). The petroleum industry itself is not immune to the discourse of petro-magic. A Halliburton subsidiary called Magic Earth is a 3-d seismic imaging company that sells and administers a software application called GeoProbe that helps to visualise and interpret oil deposits. See http://www.magicearth.com. In the Nigerian context, Andrew Apter describes an exhibit called "Magic Barrel" at a 1975 national oil seminar: from the Magic Barrel of oil emerged the "countless commodities generated by oil" (*Pan-African Nation*, 24).

11. Apter, *Pan-African Nation*, 14, 50, 249–55.

12. Nancy L. Peluso and Michael Watts, "Violent Environments," in *Violent Environments*, ed. Nancy L. Peluso and Michael Watts, 3–38 (Ithaca, NY: Cornell University Press, 2001), 25.

13. Apter, *Pan-African Nation*, 73, 275.

14. Tutuola, *Palm-Wine Drinkard*, 22.

15. Chinua Achebe, *Things Fall Apart* (1958; repr., New York: Anchor, 1994), 178.

16. Martin Lynn, *Commerce and Economic Change in West Africa: The Palm Oil Trade in the Nineteenth Century* (Cambridge: Cambridge University Press, 2002), 1–3.

17. Apter, *Pan-African Nation*, 273.

18. Watts, "Petro-Violence," 189.

19. The character Imaro in Femi Osofisan's play *The Oriki of a Grasshopper* offers an eloquent catalog of the "richest resources of our land" (slaves, palm and other agricultural

products, minerals, and oil) sent "always, always into the white ships." *The Oriki of Grass-hopper and Other Plays* (Washington, DC: Howard University Press, 1995), 13–14.

20. See "Remembering the Past's Future: Nostalgia and Some Versions of the Third World," *Cultural Critique* 62 (Winter 2006): 1–29. Fernando Coronil's attention to what he calls the "international division of nature" is also concerned to uncover the global processes by which a purportedly European capitalist modernity is produced only through the exploitation of natural resources found outside of Europe (*Magical State*, 29).

21. For recent considerations of the intersections between the slave and palm trades, see Lynn, *Commerce and Economic Change in West Africa*, and Robin Law (ed.), *From Slave Trade to "Legitimate Commerce": The Commercial Transition in Nineteenth-Century West Africa* (Cambridge: Cambridge University Press, 1995).

22. See Albert Aweto, "Outline Geography of Urhoboland," January 31, 2002, http://www.waado.org/Geography/UrhoboGeography-Aweto.htm. There is some possibility, however, for palm/petroleum synergy, as the (French) Bureau for the Development of Research on Tropical Perennial Oil Crops reports some success in using palm oil as a "replacement for diesel oil to make drilling mud for boreholes." See J. M. Noël, "Products and By-Products," *BUROTROP Bulletin* 19 (February 2003), http://www.burotrop.org/bulletin/19/products.pdf.

23. My point here is about the ways in which commodities are imagined in time, rather than about the history of the oil palm trade. Susan Martin's *Palm Oil and Protest: An Economic History of the Ngwa Region, South-Eastern Nigeria, 1800–1980* (Cambridge: Cambridge University Press, 1988) offers an analysis of the gendered impacts of price fluctuations and the imposition of colonial rule in late nineteenth-century southeastern Nigeria, which she links to the Igbo Women's War in 1929.

24. Apter, *Pan-African Nation*, 273. Although its protagonist is from the (Ghanaian) Gold Coast rather than the Niger Delta, Caryl Phillips' narrative of the nineteenth-century British–West African palm oil trade in *The Atlantic Sound* offers a corrective to this nostalgic view. Echoes of the slave trade abound in Phillips' fictionalised historical narrative, in which John Ocansey travels to Liverpool in 1881 in order to recover £2678 owed to his father for palm oil remitted to a British trading agent; the agent was supposed to have used the proceeds from the palm oil to commission a steamship for Ocansey *père* to more efficiently conduct interior trade, but the ship was never built. Caryl Phillips, *The Atlantic Sound* (New York: Vintage, 2001).

25. Karen King-Aribisala, *Kicking Tongues* (Oxford: Heinemann, 1998), 54, 57, 61, 60.

26. For statistical data on the African novels taught most frequently in literature courses in Africa, see Bernth Lindfors, "The Teaching of African Literature at Anglophone African Universities: An Instructive Canon" and "Big Shots and Little Shots of the Canon," in *Long Drums and Canons* (Trenton, NJ: Africa World Press, 1995), 45–59 and 61–75. Lindfors updates these analyses in "Who Counts? De-ciphering the Canon," in *Meditations on African Literature*, ed. Dubem Okafor, 45–54 (Westport, CT: Greenwood, 2001).

27. Coronil, *Magical State*, 8.

28. Watts, "Petro-Violence," 204 and 208; emphasis added.

29. In a 1975 essay, "Are There Any National Literatures in Sub-Saharan Black Africa Yet?," Bernth Lindfors nominated Nigeria and South Africa as most likely candidates for a national literature (*English in Africa* 2, no. 2 [1975]: 1–9). For a recent discussion of the

vexed notion of a Nigerian national literature, see Joanna Sullivan, "The Question of a National Literature for Nigeria," *Research in African Literatures* 32, no. 3 (2001): 71–85.

30. In "The Beatification of Ken Saro-Wiwa," Taiwo Adetunji Osinubi considers Helon Habila's *Waiting for an Angel / Prison Stories* in terms of the politics of Nigerian literary production in an international context. Teasing out the ways in which the text itself is enmeshed in the dynamics which it critiques, Osinubi examines Habila's treatments of the "commodification and representation of suffering for literary profit," where, in Habila's words, "the quickest way to make it as a poet" in Nigeria is to get arrested (unpublished manuscript, 16–17).

31. Okri, "What the Tapster Saw," 187, 185; emphasis added.

32. Okri, "What the Tapster Saw," 186, 189, 188. Okri's tapster's tale comes to mind in a field report circulated by the Nigerian NGO Environmental Rights Action in early 2000, which opens with the testimony of John Erakpoke, a palm wine tapper whose business disintegrated after a December 1999 pipeline rupture in Adeje. Erakpoke laments, "Nobody wants to drink palm wine again, they say it is poisoned." The report does not specify whether he had joined those who had taken to collecting and selling the spilled premium motor spirit after it ruined their farms. See Victor Raphael, "ERA Field Report #51: Spewing Premium Motor Spirit from NNPC Pipelines around Adeje," January 11, 2000, Benin City, Nigeria: Environmental Rights Action / Friends of the Earth, http://www.waado.org/Environment/OilSpills/OillSpills_Urhobo/Adeje.html.

33. Okri, "What the Tapster Saw," 193.

34. Okri, "What the Tapster Saw," 183–84. See D. O. Fagunwa, *Forest of a Thousand Daemons*, trans. Wole Soyinka (New York: Random House, 1982), 12–13.

35. See, for example, Wendy B. Faris, *Ordinary Enchantments: Magical Realism and the Remystification of Narrative* (Nashville: Vanderbilt University Press, 2004); and Brenda Cooper, *Magical Realism in West African Fiction: Seeing with a Third Eye* (London: Routledge, 1998).

36. Lois Parkinson Zamora and Wendy B. Faris, "Introduction," in *Magical Realism: Theory, History, Community*, ed. Lois Parkinson Zamora and Wendy B. Faris, 1–11 (Durham, NC: Duke University Press, 1995), 2.

37. Stephen Slemon, "Magical Realism as Postcolonial Discourse," in Zamora and Faris (eds.), *Magical Realism*, 407–26, 409, 422.

38. Michael Valdez Moses, "Magical Realism at World's End," *Literary Imagination* 3, no. 1 (2001): 105–37, 115, 110.

39. See Franz Roh, "Magical Realism: Post-Expressionism," and Alejo Carpentier, "On the Marvelous Real in America," in Zamora and Faris (eds.), *Magical Realism*, 15–31 and 75–88.

40. Fredric Jameson, "On Magical Realism in Film," *Critical Inquiry* 12, no. 2 (1986): 301–25, 311.

41. Brenda Cooper makes this point in her discussion of *The Famished Road*. Cooper, *Magical Realism*, 68–80.

42. Moses, "Magical Realism," 106.

43. Ato Quayson, *Strategic Transformations in Nigerian Writing* (Oxford and Bloomington: James Currey and Indiana University Press, 1997), 101–2.

44. Apter, *Pan-African Nation*, 261, 259, 269.

45. Saro-Wiwa's 1987 short story "Africa Kills Her Sun" offers an uncanny yet ambiguous variant on the intersections between literary realism and petro-violence: the story takes the form of a letter written by a prisoner on the night before his public execution. Ken Saro-wiwa, "Africa Kills Her Sun," in *Under African Skies: Modern African Stories*, ed. C. D. Larson, 210–21 (New York: Farrar, Straus & Giroux, 1998).

46. Nnimmo Bassey, "The Land Is Dying: Presentation by the Environmental Rights Action / Friends of the Earth, Nigeria at the Association of Nigerian Authors (ANA) 2001 Annual Convention," November 2, 2001, http://www.waado.org/Environment/PetrolPolu tion/TheLandIsDying.html. By "TNCs," Bassey means transnational corporations.

47. My paper does not include consideration of the cocoa economy. Karen King-Aribisala's "Tale of the Palm Wine Tapster" does, however, include some attention to foreign extraction of cocoa pods, which is interesting given its treatment of palm as a "native" resource.

48. Watts cites a series of statistics about the ratio of profits-to-investments: Shell has made $200 million *annually* in profits from Nigeria for 40 years and has invested $2 million *total* during that time, building one road and awarding almost 100 scholarships ("Petro-violence," 198). Sandra T. Barnes tracks a new "strategic philanthropy" among transnational corporations in West Africa: "performance-based" community development projects that make the community accountable to the corporation, rather than the other way around. "Global Flows: Terror, Oil and Strategic Philanthropy," *Review of African Political Economy* 104/5 (2005): 235–52, 244–45.

49. Mcphillips Nwachukwu, "The LNG Prize for Literature: Hopes and impediments . . . ," *Vanguard*, February 29, 2004, http://allafrica.com/stories/200403010499.html.

50. U. O. Ugorji and C. A. Ejueyitchie, "Press Release: Introducing the African Writers Endowment, Inc. (AWE)," *The Nigerian Village Square*, http://www.nigeriavillagesquare1 .com/PR/AWE.html; emphasis added.

51. H. Igbikiowubo, "Alleged $180m Scam: Creditors under Pressure to Ditch NLNG Trains 4&5," *Vanguard*, April 4, 2004, http://allafrica.com/stories/200404050669.html. Kellogg, Brown, and Root, the engineering and construction subsidiary of energy services company Halliburton, is a member of the TSKJ consortium. US vice president Dick Cheney has ties to Halliburton, which has also been accused of war profiteering in Iraq.

52. Azuka Ogujiuba, "NLNG Prize for Literature Gathers Momentum," *This Day Online*, September 4, 2004, http://www.thisdayonline.com/archive/2004/09/04/20040904 saco2.html; Chux Ohai, "$20,000 Failed Literary Prize: Soyinka Carpets Panel of Judges," *Daily Independent*, October 18, 2004, http://www.independentng.com/life/lsoct180401.htm.

53. Quoted in Ohai, "$20,000 Failed Literary Prize."

54. Henry Akubuiro, "For Me, This Prize Is Greater Than Commonwealth Crown," *Daily Sun*, October 16, 2005, http://www.sunnewsonline.com/webpages/features/literari/ 2005/oct/16/literari-16-10-2005-001.htm.

55. Mcphilips Nwachukwu, "What Is Nigerian Literature?," *Vanguard*, February 6, 2005, http://www.vanguardngr.com/articles/2002/features/arts/at106022005.html.

56. Before he left NLNG, Jamieson contextualised his cheerleading for Nigerian literature in terms of his own road not taken: at the Lagos stop on the NLNG reading tour, he

told Azuka Oguijuba that he had been torn between studying English literature and engineering when he entered university.

57. "Greenwashing" refers to the practice of using a rhetoric of concern for the environment to conceal and advance agendas (often corporate) that actually foster ecological degradation and environmental injustice: readers in the US need only think of the Bush administration's "Clear Skies" and "Healthy Forests" initiatives.

58. Dagga Tolar, "Liquefying or Liquidating Literature in Nigeria?," *Vanguard*, November 17, 2005, http://www.vanguardngr.com/articles/2002/viewpoints/vp217112005.html.

❭❭❭ This Is Not a Pipeline: Thoughts on the Politico-aesthetics of Oil

Ursula Biemann and Andrew Pendakis

ANDREW PENDAKIS: I want to begin by asking you a very general question about the aesthetics of what we might call primary substances, those materials or liquids, like oil, water and coal, which come to peculiarly mark or subtend the cultural structure of an economy. I am interested particularly in those substances which we are tempted to imagine vertically at the bottom of things, the floorboards or groundwork of any given historical period or locale. Both water and oil would appear to be candidates here: the first, indispensable, structural, perhaps, in a manner wholly unto itself, the very ur-liquid of life; the second, a basic condition of modernity, essential, but "artificially," as the element necessary less for life itself than life lived under the conditional second nature of industrial capitalism. In what sense can we speak about an aesthetics of oil cultures, a set of recurring spatial, infrastructural, or architectural motifs, for example, or even a dominant structure of feeling or experience which seems to pass through the very molecules of a whole historical reality? Is there an aesthetics of oil, or are its cultural manifestations too diverse and localized to be usefully generalized?

URSULA BIEMANN: Water has traditionally been associated with specific cultural or symbolic meanings, so it is hard to generalize the aesthetic dimension of this life-sustaining liquid. But oil is an entirely different case: it has literally propelled humanity into a different era of mobility and consumption. Hydrocarbon society is rooted in a concrete moment of discovery—it has a specifiable beginning—and an end! It has engendered the whole universe of plastic culture. We usually do not associate crude oil with the vast number of new substances and objects which have entered our lives as products of petrochemistry. In our minds, the preciousness of oil is more closely linked to petrodollars: it is a political substance. Apart

Imaginations 3, no. 2 (2012): 6–16

from geopolitics and its inevitable spatial tropes—and that is certainly a major problem in the representation of oil—as a mineral resource, the discourse that shapes the image of oil is articulated in economic and industrial terms: it is, in other words, very much a corporate substance as well. Corporations are hypersensitive when it comes to their public image, they are very careful in controlling visual representation. There is always some pipeline patrol lurking on the horizon, ready to charge down the hill and prevent you from filming. So we ended up with an aesthetic that essentially foregrounds the gigantic investments in the infrastructures necessary for extraction and evacuation, as well as their spectacular failures. The petro-era coincides with a period marked by technology and hardware; it seems that culturally speaking it has already come to an end.

PENDAKIS: I think you are right that our conception of oil is usually oriented by this wide-angle image of the silently running oil refinery or platform. Oil is, in this sense, dangerously literalized, wrongly conceived as simply coextensive with a highly simplified figure of its own productive apparatus. What do you think is screened out by this image? Also, I am intrigued by this invocation of "plastic culture." Might this be a kind of shorthand for our moment's particular relation to oil? It is true that it seems the "age of oil" is coming to an end, not in the literal spectre of Peak Oil or some kind of imminent shift to a new primary energy source, but in the sense that oil seems to chug into the present like an echo from the nineteenth century. The roar of the combustion engine feels quaintly modernist, almost embarrassingly promethean and earnest when viewed from the angle of the silent and immobile microchip. Could you speak more about this paradoxical end to (or transformation within) a certain dominant era of petro-culture?

BIEMANN: I see this level of abstraction in the representation of oil as yet another way to keep it firmly in the hands of market dynamics, a remote and inaccessible entity, supposedly too big and complex to grasp for the average citizen. What does not come through in these repetitive stencils of oil-related images is the regional histories and local textures of interaction between infrastructures and social communities, the thorough reorganization of cultural alliances and political forces on a regional level, the relocation of populations as a result of big money flowing into an impoverished area, and the nondemocratic decision-making processes of regimes involved in these deals, as well as the impressive epistemic apparatus that is set into motion by the conception of big infrastructure projects. All these dynamics affect large populations one way or another. To begin by giving these ephemeral processes some form of visual presence is a way to start filling in the missing elements.

The discovery of the vast potential of oil for the creation of new materials mustering an extraordinary range of qualities has fueled our imagination to create a synthetic world and overcome natural limits. This fact is probably just as important for the understanding of who we are today as the mobilization and substitution of labour power facilitated by the combustion engine itself. It seems to me that humanity has jumped ahead in out-of-synch rhythms. While it advanced its material sophistication to a high degree, the engine remains sadly stuck in the nineteenth century, and yet we still invest huge resources into building infrastructures for a system we know is long outdated. But I sense there is another connection there that would be important to understand better. The ability to chemically synthesize the world and thus create its artificial extension has simultaneously triggered an imagination of nature and its relation to the human subject. This bizarre construction of nature as something separate from us is what makes the gigantic ecological devastation of petro-culture possible.

PENDAKIS: Oil and water, though not quite opposites, are anecdotally understood as chemically incompatible ("they don't mix"). This incompatibility mirrors a very strong associative or symbolic antagonism. Oil is arguably the dirtiest of liquids, "the devil's excrement," not just on the level of its (highly racialized) material properties (its blackness, its stickiness, its opacity, etc.), but on the terrain of its social and political usage. Sociologists have long noted the ways in which reserves of oil have a way of evolving into inflexible or brutal state structures (layers of corrupt officials and bureaucracy) even as they engender uniquely consumerist life-worlds, populations habituated to expect an inexhaustible and "labourless" flow of automobiles, luxury goods, etc. Water seems to come from an utterly opposed moral universe. Linked to ablution, sustenance, openness, transparency, purity, etc., water is the liquid of transcendence, but also the most modest and common of social goods. This opposition has been dramatically compromised by the increasing commodification of water and the emergence of huge conglomerated interests in the profits available to those who own and market it. How do your works attempt to navigate the existing infrastructural realities of oil and water in the light of these moral geographies? Do you prefer to work within these inherited meanings as a way of mobilizing an audience politically or aesthetically, or should the artist instead try as hard as possible to have done with moral rhetorics premised on lost purities or "nature"?

BIEMANN: Rather than looking at oil and water in these essential and nominally opposed terms, I have investigated the environment they each tend to generate in conjunction with human, social, and technological entities, and these are highly situational. I would refrain from assigning symbolic meaning to the substance itself. As a corporate capital-intensive fluid, oil tends to create top-down power structures. It is a deeply antidemocratic substance in that governments, which are thriving on large incomes from fossil resources, have no need to consult their tax-payers to make political decisions: they don't need a people. States like this are like a head without a body; they are bound to create autocratic regimes. However, there are subtle and brutal processes, both of which are important to show. When I began my research on oil for *Black Sea Files* about 10 years ago, there was hardly any cultural theory and very few aesthetic productions on the topic. It would have been inconceivable then to organize a conference on petrocultures, since the sig-nification of oil beyond the economic and geopolitical dimension was largely unarticulated. I wish there had been books available then like Reza Negarestani's *Cyclonopedia* who creates a fantastic vocabulary around this dark lubricant and narrative undercurrent of all politics and ethics of life on earth. Without such in-spiring prose, I practically had to start from scratch. But what I noticed during my research is that there is a considerable discrepancy between the explanatory heart of energy geography in urban centers and at the campuses, and the messy, remote, and unstable resource fringes where these situations take place. There is a center/periphery dynamic set up between the cores of theory building and the social strug-gles around the extraction zones. If anything, I see it as my moral obligation to keep establishing the links between them, even at the risk of placing this very process again at the center.

It has become obvious that ecological devastations go hand in hand with the existential struggles of marginalized communities. But my ongoing concern with the human geographies of hydrocarbon society doesn't mean that I'm disinter-ested in the molecular level of how reality constitutes itself through water and oil and a million other things. Interesting that you would mention the chemical reac-tion between these distinct liquids, since in chemistry, substances are character-ized entirely by their willingness to bond and transform into new compounds. Both water and oil are made of hydrogen, when combined with oxygen we get water, when combined with carbon we get oil, each opening up a different arena of bond-ing opportunities. The interaction with one another might be repulsion, but that is still a strong relation, if in the negative. We should be looking for the chemical substance common to blood, plants, and petroleum, and we would find that we share a high degree of DNA.

The problem I see with highlighting the cultural value of water by foreground-ing its symbolic virtues is that we continue along the line of an economistic think-ing that assumes that anything which has a value requires some sort of compen-sation. It firmly remains within a human-centric vision of culture. Even before the calamity of bottled water, any hydraulic infrastructure, any dam, barrage, or irri-gation canal, prepares the ground for the commodification of water. By processing and facilitating water, it automatically becomes something that can be charged for. States have begun to take into account what they call virtual water, i.e., the water needed to produce something, whether it be a computer chip, roses, or a loaf of bread. By importing water-intensive products, a state can save a lot of its own water.

With *Supply Lines*, a collaborative art and research project on resource geographies I'm involved in, a number of stakes start to emerge. A central issue is to make natural resources a matter of public debate and decision-making rather than leaving it to market dynamics. The question here is how we can engage in aesthetic productions that will make commodities more transparent and accessible to public knowledge. Another point that you just mentioned is to problematize the notion of nature and do away with an anthropocentric perspective on the world whereby everything is turned into a resource for humans. This is deeply linked with the act of representation as a way to shift the discussion from the object of oil, water, or gold to the cultural meaning the stuff has for us. The frantic use of global resources goes hand in hand with an accelerating process of oversignification, not least through cultural studies or the tendency to open aesthetic and cultural discussions to fields that have been hitherto subjects of economic-industrial discourses or natural science. These are some of my most urgent concerns: how can I image resource issues in ways that are both signifying and a-signifying. Whether we assign pure and spiritual qualities to water or devilish ones to oil makes no difference; the problem is with representation itself.

PENDAKIS: You referred above to the omnipresent "pipeline patrol," and this issue of policing touches upon another question I would like to pursue. We are unquestionably living in an age of secrecy, an age in which, one might say, the secret has gone structural. The insider's stock tip, offshore accounts, classified production facilities, "creative accounting," the research records of pharmaceutical companies: secrets populated earlier societies, and they were perhaps always in some way "functional," but there is a way in which the hidden has become conscious, a basic requirement of the production and reproduction of capitalist economies. Catherine Malabou speaks of the way neurobiological processes are structurally invisibilized in the process of creating the transparency of consciousness; Karl Marx's own work would seem to echo this on the level of production, positing an order of the invisible which renders possible the liberal vista and its life-world. In *Black Sea Files* you purposively travel to the Southern Caucasus to visually research the construction of the Baku-Tblissi-Ceyhan pipeline; it is as if your objective was to literally capture this mechanism of flow before it was interred permanently into a domain of invisibility and silence. How does your work travel along this thin line between unveiling the hidden and respecting the complexity and opacity of things? Given the epistemological obsessiveness of your work, how does one reveal without blundering into the territory of absolute transparency? What can you tell us about this process or about your thoughts on the dialectic between seeing and hiding in modern or postmodern economies? What role does secrecy or hiddenness play in the political economy of oil itself?

BIEMANN: Are you alluding to the idea that infrastructures are most powerful when they are invisibly operating in the background, hidden from our consciousness? Well, the only way to capture the moment of the giant effort necessary to build such an invisible power structure is by filming the actual construction of this crude oil corridor clearing the access far into former Cold War territory. There is something to be gained from visually mapping these procedures in the most diverse way possible, challenging the monological narration based on technological pioneering. To offset the linear master narrative provided by oil corporations, I instead install a loose cartography of multiple voices uttered by migrants, peasants, fishermen, prostitutes, oil workers, and the like who intertwine with the pipeline. The powerful impact of invisible structures is true, of course, not just for energy but also, very prominently, for the media apparatus. What is kept in the hidden realm is what's not considered to be of immediate world interest, which is almost everything. What is shown is merely a selected narrow vision. I see the problem not so much in the incorrect representation but in the selection process itself and the pretense to bring clarity via this scripted formula of news writing, when in my experience, the production of knowledge on site is an immensely confusing and fragile undertaking. In an interview with two sex workers in the presence of their pimps and agents, I attempt in *Black Sea Files* to make the emergence of misunderstanding more transparent.

The essayist video format with its diverse image sources, scrolling text, and voice-over narration generally tends to stack multiple layers of meaning and interpretation, none of which pretend to be a stable, exclusive representation of reality. On occasion, I even weave myself into the process by reflecting on my practice as an undercover agent to find hidden, secret knowledge and in the process turning into an embedded artist who operates in the trenches of geopolitics. All this is acknowledging the subjective and often subversive dimension of artistic fieldwork that is more closely related to secret intelligence than, say, to anthropology.

PENDAKIS: As I alluded to above, your latest work appears to be focusing less on oil proper and more on the human (and in- or non-human) geographies of water in an age of climate change. And yet there remains a great deal of continuity here: transformations in the viability of global water are only the flip side of our dependence on fossil fuels, the perverse externality of oil's central location at the heart of our societies. In your video "Embankment," part 2 of your project *Deep Weather*, we are presented with an opening shot that almost appears to arrive unfiltered from the origins of stabilized agricultural civilization itself. Approaching from the sea, the camera captures a seemingly infinite line of wriggling human motion, a continuous flow of Bangladeshi workers dragging bags of mud to shore up and secure a barrier they hope will prevent their communities from being flooded by rising waters. This is a primeval image, one resonant with echoes from the original domestication of the Nile and other such "cradles of civilization." In such moments, you seem to be working in a mode that transcends or transmutes realist documentarity, one imbued with a sense for anthropological invariance, the repetitions of cyclical time, even a poesis conceived of as coextensive with nature itself. I sense a very similar set of concerns in the fragments I have seen from your current project, *Egyptian Chemistry*. Here again past and present seem bound in a space/time that defies the illusion of their separation, one which draws on myth (without devolving into holism or apoliticism), but also on science or sociology (without giving into determinism or a too-simple empiricism). Am I wrong to think you are trying to invent videographically a new form of universality, one that is fragile, plural, ragged, full of holes, yet somehow coacervated by the muddy oneness of the planet itself? I discern in your work the lineaments of a very interesting materialist universality, one which confidently draws on scientific naturalism, the methodologies of the social sciences (especially ethnography), but also on poetry, aesthetics, and the mythic imagination without stumbling into incoherence or eclecticism.

BIEMANN: This unspeakable image of thousands of villagers building a giant mud embankment entirely by hand without any mechanical assistance is what derailed climate patterns and melting ice fields will mean for many of us on earth. The image speaks of the primeval need for safety from extreme weather events and floods and of the futility of believing we could protect life on the planet with ridiculous technological measures. Part 1, by the way, is shot in the toxic tar sands of Alberta where the dirtiest of fossil resources is being extracted that will no doubt continue to impact on the living conditions of populations on the other side of the earth. We are speaking of a terrestrial scale here. When we have been thinking in global dimensions over the last 2 decades, this is the time to go planetary. And this shift requires a whole new vocabulary, both visual and verbal. Geography as a theoretical platform for tackling global issues such as migration networks, supply lines, or communication infrastructures turns out to be insufficient, or simply too flat, for this enlarged dimension because it falls short of grasping the depth of many dawning questions such as species survival and the transformation of the biosphere. So my new images intend to evoke temporal depth by returning to moments prior to the Industrial Revolution with its frantic dam-building activity, and to scientistic knowledge with its partition into disciplines and infinite subdisciplines. It is this alchemist approach of raising metachemical and philosophical concerns, perhaps, that imbues the video with a mythic imagination. Al Khemia (Arabic for chemistry) also happened to be the ancient word for Egypt, meaning the Black Land, possibly due to the dark muddy Nile floods periodically covering the land. The term alludes to the vision that, before anything else, the earth is a mighty chemical body, a place where the crackling noise of the forming and breaking of molecular bonds can be heard at all times. So when documenting vast land reclamation projects in Egypt, beyond a comment on technocapitalism, it is first and foremost a videography of the conversion of desert dust into soggy fertility.

So yes, I would love to think that I am inventing videographically a new, if flawed and ever morphing, sort of universality through the meddling with the muddy materiality of the Earth itself. Incidentally, for *Egyptian Chemistry* I took a number of water samples from the Nile, so the project is not about making images only. I'm not primarily focusing on strategies of representation; it also contains objective reality. I have come to realize that if we solely attempt to culturalize the discourse on the physical and chemical transformations our planet is currently undergoing by prioritizing meaning and representation, we fail to address a deeper problem. For if we are to speak about the non-human world—weather patterns, organic pollutants, copper atoms—it will not suffice to deploy an anthropocentric discourse. Not everything comes into being through human intention; we need to examine the ways in which human and non-human realities emerge together in a variety of formations. Rather than through a particular set of criteria, this is more likely to happen through the hybrid consciousness engendered by the assemblage of technological, social, and natural stuff, where some elements signify, others not. Metachemistry grasps this turbulent instance of physical and epistemic change, or lineament, as you call it, and propels us into a slightly altered dimension that can only be invoked mythically through space travel, time barriers, and the interbiospheric mobility of species.

>>> Excerpt from *The Polymers*

Adam Dickinson

Polyester

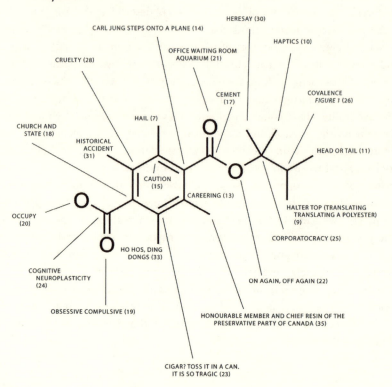

Polyethylene terephthalate, $C_{10}H_8O_4$

HAIL

Hello from inside
the albatross
with a windproof lighter
and Japanese police tape.
Hello from staghorn
coral beds
waving at the beaked whale's
mistake,

The Polymers (Toronto: House of Anansi Press, 2013), 5, 7–8, 10, 19

all six square metres
of fertilizer bags.
Hello from can-opened
delta gators,
taxidermied
with twenty-five grocery sacks
and a Halloween Hulk mask.
Hello from the zipped-up
leatherback
who shat bits of rope for a month.
Hello from bacteria
making their germinal way
to the poles in the pockets
of packing foam.
Hello from low-density
polyethylene dropstones
glacially tilled
by desiccated,
bowel-obstructed camels.
Hello from six-pack rings
and chokeholds,
from breast milk
and cord blood,
from microfibres
rinsed through yoga pants
and polyester fleece,
biomagnifying predators
strafing the treatment plants.
Hello from acrylics
in G.I. Joe.
Hello from washed up
fishnet thigh-highs
and frog suits
and egg cups
and sperm.
Hello.

HAPTICS

Inspired by cigarettes, folding chairs, and the flourished gestures
that accompany escalating disagreements, Plasticus Corporation (a
subsidiary of Dow Chemical) quietly moved into researching the

biological effects of touch on memory. The idea was to engineer nostalgia into the flexible surfaces of goods. Take, for example, the proleptic goodbye of an ice cube tray, the Merry Christmas grip of a Swiss Army knife, or the complementary blisters bevelled by borrowed cash. Researchers impregnated experimental plastics with erotic, platonic, and ritualistic dispersants in order to approximate the uncanny and its unguent penetration into the rehearsals of the brain's transcriptional grease. Soon, the golden age of mothers was upon us. Wine pairings were devised. Moods were adjusted. People saw their fingerprints blinking everywhere like biometric avalanche beacons. In the midst of the frenzy, in bathrooms where blow-dryers whipped pierced barley ears into Arcadian compost, in soliloquies sprung to life on the pocket-dialed sidewalks down memory lane, in gear-boxed convertibles retrograding the open road, in Archimedean armchairs armed with bygone gamepads, in the protopathic nuzzling of machined memorabilia, the retraction went unnoticed. Plasticus had forged the data. All the handheld déjà-vus made it feel like a publicity stunt. The hoax was taken for a hoax and the polyester lilies went on feeling up the valley.

OBSESSIVE COMPULSIVE

make a roof for the people, and the people walk
down the street with resin for a roof, and the roof
has magnesium in it, and sulphur, and the people
walk down the street with resin in their hair, and
resins are always falling from the sky to the ground,
and the birds make a people in the sky, a people
of the resin, and the resin is composed of sky, and
it composes the sky, and the people walking down
the street are the strings of resins, and covering
their hair with their arms, with newspapers, with
umbrellas, the people are the birds of resins throwing
their landings in the air like people for whom landings
are uncommon, like people committed to the expulsion
of landings, the resins coming down upon
them like people driven out of countries discovered
by resins or that have discovered resins in veins,
in the countertops of suburbs, and people walk
down the street with resins for hair, with countries
committed to colour, with the bonds between them
the birds circling, and people walking down the street

with hunched shoulders so as not to look up and
call the resins by name, call the resins in the name
of the birds, the people, circling and loosening

))) An Athabasca Story

Warren Cariou

One winter day Elder Brother was walking in the forest, walking cold and hungry
and alone as usual, looking for a place to warm himself. His stomach was like the
shrunken dried crop of a partridge. It rattled around inside him as he walked, and
with each step he took the sound made him shiver even more.

Where will I find a place to warm myself? he wondered. Surely some relations
will welcome me into their home, let me sit by the fire.

But he walked for a very long time and saw none of his relations. Eventually
he traveled so far west that he didn't know the land anymore, and even the animals
wouldn't dare to help him because they knew how hungry he was. They kept a safe
distance. So he shivered and rattled his way further and further, without anything
to guide him except the lengthening shadows and his unerring radar for trouble.

When he was nearly at the point of slumping down in a snowbank and giving
up, Elder Brother thought he smelled something. It was smoke, almost certainly,
though a kind of smoke he'd never encountered before. And though it was not a
pleasant odour at all, not like the aromatic pine-fire he had been imagining, he
knew that it meant warmth. So he quickened his frail pace and followed the scent,
over one hill and then another and yet another. And eventually he came to the top
of one more hill and he looked down across an empty valley and saw the source of
the smoke.

A huge plume billowed from a gigantic house far in the distance, and between
himself and the house there was a vast expanse of empty land. Empty of trees, of
muskeg, of birds and animals. He had never seen anything like it. The only things
moving on that vacant landscape were enormous yellow contraptions that clawed
and bored and bit the dark earth and then hauled it away toward the big house.
And the smell! It was worse than his most sulfurous farts, the ones he got when he
ate moose guts and antlers. It was like being trapped in a bag with something dead.

Elder Brother knew he should turn away and get out of that smell as soon as he
could. But that would mean spending the night by himself, freezing and chatter-
ing and rattling, and he couldn't bring himself to do it. There was warmth up there
in the big house, he could see it floating away on the breeze. In places he could

Lake: A Journal of Arts and Environment 7 (2012): 70–75

even see the heat rising in fine wiggly lines from the newly naked earth itself. So despite the smell he stepped forward and made his way out into that strange expanse.

The house was further away than it had seemed. He walked and walked across the empty space, stepping over dark half-frozen puddles, holding his nose, following the tracks of the great yellow beasts. He attempted to stay far away from the beasts themselves because they didn't look the least bit hospitable. But by the time he got halfway across the open land, he strayed close enough to these creatures that he could see each of them giving off its own smaller stream of smoke. And as he stood there studying them, he realized something else: there were people inside.

Maybe they were houses, he thought. Warm, comfortable houses that by some magic were also capable of digging and hauling the earth. Certainly they were big enough to be houses. He got closer and watched again as one of them rumbled past, shaking the ground at his feet. The man inside was bare-armed, as if it was summer. And he was chewing on something.

Of course Elder Brother was scared by the noise and the smell and the shuddering earth. But his hunger and his shivering were stronger. When the next gigantic thing came rumbling down the track he bounded out in front of it and stood there, waving his right hand desperately while his left hand remained clamped on his nose.

The thing squealed and snorted and eventually came to a stop just before it touched him. A man immediately leaned out from a window near the top of it and shouted, Who in hell are you? Where's your machine?

Oh, my brother, my dear relation, Elder Brother said. I'm very cold and hungry and I was hoping . . . to come and visit you in your house.

The man didn't say anything for quite a while. He scanned the blank horizon, as if looking for something. Finally he leaned further out the window and yelled, You're saying you're not with the company?

Uh, company . . .

Are you Greenpeace?

I'm cold, Elder Brother said.

The man took off his strange yellow hat and gazed into it for a moment, placing one hand over his forehead as if to keep something from bursting out.

Well you'll be a lot worse than cold, the man said, if you don't get the hell out of my way and off this goddamn property.

Well, *that* was rude, Elder Brother thought, but he tried not to betray his disappointment. This man talked as if he had no relation at all.

Okay, he said to the man. I won't come visit you right now, but could I please ride along on the top of your house? I want to go to the big house over there, where I'm sure they'll let me come in and get warm.

The man laughed a little, and glanced up at the sky for a moment.

I don't know what you're on, buddy, the man said. But you need to snap out of it right this goddamn minute. Cause if you don't step aside I'm gonna call Security, see, and they're gonna come out here and throw your ass in the slammer with all those other yahoos from last month and the month before. I should've called them already. But on the other hand, I could save a little time if we just had a bit of an accident here. Nobody'd ever know it happened.

The man's house made a roaring sound that made Elder Brother step back.

Oh, there's no need for that! Elder Brother said. Don't worry. I'll move aside. But before I go, I just want to know one thing: what are you doing with all that earth?

We're burning it, the man said.

Burning. But earth doesn't—

This stuff does, the man interjected. You really are a moron, aren't you? It's very special dirt, this stuff. We dig it up and take it over to the big house, as you call it, and we mix it around in there and after a while it's ready to burn. Fuel to heat your house, if you have one which I doubt. Gas to power your car. Diesel to move this big rig here. All of it comes right out of the ground. You can tell that by the smell of the air around here. Just like napalm in the morning!

The man took a deep breath through his nostrils and then laughed, but his face turned sour when he saw that Elder Brother didn't understand.

Yeah, we got real big plans for this place, the man said. There's more of this special dirt here than anywhere else in the world. Everybody wants it, and we're happy to sell it to them of course. And all those people around the world are going to help us burn this very dirt that you see here, from under your feet all the way to the far horizon. We're gonna burn it, and burn it, and burn it, until we make so much heat that the winter never comes back! And then even you and the rest of your sorry kind won't be cold anymore. So how do you like that?

When will that be? Elder Brother asked, rubbing his hands together.

Fifty or sixty years. Maybe forty.

Oh. Not to complain, but I was hoping for something a little—

Elder Brother was interrupted by an explosion of noise from the front of the big yellow house-thing, and it lurched toward him with surprising speed. He was barely able to leap out of the way before it rolled right over his footprints.

Now get off this land! the man yelled as his house roared away. It doesn't belong to you. Go back to the bush or wherever it was you crawled out from. I'm calling Security now!

Elder Brother stood there for a while and watched the house labouring over the hillocks and through the black puddles in its way. He was more than a little scared of this mysterious Security that would soon be coming after him, but he was also angry. How could this man tell him that the land wasn't his? How could this "company" keep all the magical dirt for itself? If there was so much of it, Elder Brother reasoned, there should be plenty to share with visitors.

Though he knew he should probably be running for his life, Elder Brother found himself unwilling to move. He was held there by an idea: if these people wouldn't give him any of this magical dirt, maybe he should take some for himself. Yes, what a fabulous plan! Since the man and his company were so rude, they deserved to have their precious dirt stolen. And the best part was that if Elder Brother gathered enough of this magic dirt for himself, he could burn it for years and keep warm until the winter was gone for good.

So instead of fleeing the empty land, Elder Brother began walking toward the place in the centre where the largest of the yellow contraptions were tearing away at the earth. The snow had all been cleared away there, and he could see how black this magical dirt really was. He watched the beasts moving this way and that, and he waited for this opportunity. Finally he saw an opening, and he darted between a couple of the great mobile houses toward a spot where the ground had recently been opened. It looked softer there, and warmer too. Yes, this was the place. He lifted his right hand and thrust it as hard as he could, right down into the soil, up to his elbow.

Ayah! a voice said. What are you doing, Elder Brother!

Sssshhhhhhhh, he answered. I'm taking what's mine.

And he reached deeper and deeper into the ground, spreading his fingers as wide as he could in order to hold the largest armload of dirt. A year's supply in one hand, he imagined! He reached so far that his cheek rested against the redolent earth itself. He nearly gagged at the smell but he didn't loosen his grip. He could already feel the warmth coming out of the soil and it made him a little stronger.

Elder Brother, you're hurting me! the voice cried out.

Not nearly so much as they are, he said, and with that he began reaching in with his other arm, tunneling in with his fingers, opening his arms wide in a desperate embrace. His nose was raw with the fumes, and particles of grit were getting in his mouth. He was about to heave the huge armload of dirt out right then and begin his run for the bush, but one thought stopped him. What if it wasn't enough? What if he ran out and then the winter came back?

So without another hesitation he kicked off his moccasin and began tunneling in with the toes of his right foot. He clasped and clawed until he was more than thigh-deep in the earth, and then he tilted his toes upward to hold as much as he could. Then quickly he kicked off his other moccasin and tunneled with that foot, squirming and worming until that leg too was embedded in the earth. Ass-deep and shoulder-deep in the magical soil! Surely this would be enough to last him for decades, until the winter had been vanquished for good.

You are a genius, Elder Brother, he said to himself. You deserve all the warmth you're going to get.

But when Elder Brother tried to lean back and lift the great clump of dirt out of its place, he discovered that he had no leverage. He pulled and pulled at the soil, flexing his arms and his legs all at once, but nothing moved. The only thing that happened was his limbs seemed to sink a little deeper into the ground. He grunted and panted, flexed again, shimmied his buttocks for extra oomph. However it didn't make a bit of difference.

Well, he thought, I guess I should just take a little less of this stuff, maybe make two trips. I'll just wiggle my legs out of these holes and settle for a nice big armload of magic dirt.

I imagine you can guess how that worked out. Right. It didn't. Elder Brother was stuck fast in the Athabasca tar. By this time he couldn't move a finger or a toe.

Instinctively he called out to the voice that had spoken to him earlier. Help me! I'm sorry I didn't listen to you. I'll leave now without taking anything at all.

But the voice didn't answer. Elder Brother was stuck there in the ground all night, and all the next day and the following night. He howled to the voice, asking it for forgiveness. He yelled to any of his relations who might be in earshot. He even screamed to the men in the huge yellow creatures that, from their sound, seemed to be moving closer and closer to him. (Of course he couldn't see what was going on back there. All he could see was the clump of oily dirt that his nose was resting on.) If those men in the contraptions heard him, or saw him, they gave no sign of it.

Late in the afternoon of his second day of being stuck in the ground, the sound of the contraptions became much louder, and a dark shadow suddenly closed over Elder Brother. Then he was being lifted, along with his armload of dirt and a great deal more, and he felt himself falling with the thunderous sound of everything else falling around him. He cried out but he knew it was hopeless; no one would hear him over a cacophony like this. When he landed, the dirt closed over him. It pressed into his nostrils, his ears, his mouth, even into his clenched bum. The weight of it pushed down and down until he couldn't even move an eyelid. Soon the thing began to move, and it hauled him slowly across the wasteland, encased there in the tar as if he was a fossil. And eventually the truck reached the edge of the huge smoky refinery, where it dumped him and many tons of tar sand into the yawning hopper that was the beginning of the processing line. And inside the refinery he was made very warm indeed.

Of course Elder Brother can't die, luckily for him. Or perhaps not so luckily. He's still alive even now, after everything he's been through. It's true that people don't see him much anymore, but sometimes when you're driving your car and you press hard on the accelerator, you might hear a knocking, rattling sound down deep in the bowels of the machine. That's Elder Brother, trying to get your attention, begging you to let him out.

))) What Does the Culture of Stewardship Look Like?
Barry Lord

Despite inspiring developments, biofuels and wind and solar power together still account for only 1% of the world's primary energy consumption (Smil, 2010). Any transition to renewable energy on a large scale, like all energy transitions, will require many decades of dedicated effort. The transition to renewable energy will likely be a long-range struggle for incremental growth accompanied by occasional breakthroughs, wrong-turn setbacks, and politically or economically manipulated delays—as long as big oil and big coal have as much power and influence as they do.

Yet the cultural impact of renewable energy is already being felt. It is the direction we are headed in the twenty-first century—just as coal was in the eighteenth century or oil and gas were in the twentieth. What will its culture of stewardship look and feel like? This is rather like asking a Bedouin from the Arabian Gulf in 1935 what the culture of oil and gas would be like at the end of his century. Some have suggested that the culture of renewable energy must be one of scarcity. They emphasize the need for reducing our use of energy, especially in oil-gluttonous countries like Canada and the United States, and warn of impending shortages.

Beginning in the late 1960s, a group of southern European visual artists made a powerful case for what looked like a culture of scarcity. Their works were first called *arte povera* (poor or impoverished art) by Italian curator German Celant because they were deliberately made from low-cost materials that might otherwise be seen as waste. *Arte povera* was made in the style of conceptual art: its materials were used not to make attractive art for visual delight, but rather to convey content that was often politically charged and rooted in the belief that discarded, poor and rough materials are the most appropriate means of expression in our time. Italian Mario Merz and Greek artist Jannis Kounellis were among those whom Celant grouped in one of the few exhibitions with that title. Unlike the art movements of the culture of transformation, *arte povera* has no manifestos and was not led or directed by anyone. It has nevertheless had a quietly growing influence into our century when so many people are conscious of the need for a more sustainable way of life.

Italian artist Michelangelo Pistoletto first became known in the mid-'60s for life-size, cut-out, photo-realist portraits on polished steel mirrors that were at that time associated with pop art, but he has long since been recognized as an *arte povera* practitioner. His 1967 *Rag Wall* wrapped bricks in discarded scraps of fabric, a classic *arte povera* piece. The title of his 1974 exhibition at the Tate Gallery in

Art & Energy: How Culture Changes (Arlington, VA: AAM Press, 2014), 218–41.

London, *Venere degli stacci* (Venus of the Rags), suggests his intention to integrate art with life. Over the following decades he continued to pursue this goal in performance art, theater, and related pursuits, as late as 2009 rolling a huge ball of newspapers through the streets of London. He has established a foundation in a former textile factory in his hometown, Biella, dedicated to research for solutions to the world's problems in the fields of work, education, communications, art, nutrition, politics, spirituality, and economics. This real-world foundation entitled Citta dell'Arte has become the focus of Pistoletto's ongoing work on sustainability in art and life.

French installation artist Christian Boltanski shows the influence of *arte povera* in the low-key but unmistakable social-political content of his work. In 2010 he covered the floor of the Grand Palais in Paris with old clothes. These he sorted into carpet-like rectangles in front of a mountain of discarded garments. A huge mechanical claw repeatedly descended from the high steel frame of the Grand Palais to pick up the used garments from one pile and drop them on another. The piece was entitled *Personne*, which Boltanski pointed out can mean "somebody" or "nobody" in French. Boltanski's installation used waste material to communicate disturbing and memorable content that felt relevant to a world of potential scarcity.

The oil and gas culture of consumption assumed a world of abundance. In the 1960s, when oil first became the world's dominant energy source, few people disputed this assumption, which today we recognize as unsustainable. On the other hand, raising the specter of scarcity frightens not only the "haves," but also those who justifiably hope to have more in a world where many millions remain deprived of what most of us take for granted as the necessities of life. The culture associated with renewable energy does not need to counter consumerism with threats of scarcity. Instead, a culture of stewardship implies that we plan the use of all resources with a view to the needs of future generations as we aim to meet our current requirements as efficiently as possible.

A culture of stewardship in this sense could conceivably be aimed at constructing a possible future of abundance to meet the real needs of the vast majority of the world's people. The sun generates an abundance of energy. The challenge for renewable energy is how to store that abundance so we can use it when we need it. From the granaries of ancient Egypt and Mesopotamia to the seed banks of today, the storage of abundance and the distribution of that surplus has remained one of humanity's greatest opportunities and threats. Now storage of energy has become crucial. Just as control of access to the granaries and irrigation channels was the basis of power in the ancient world, so control of the storage of energy and of the databases of information that give access to it will become the basis of power in the twenty-first century. As for distribution, sustainability is the primary value that the culture of stewardship demands. To more clearly envision the emerging culture of this century, we need to look at each of these cultural values of stewardship—storage, access, and sustainability.

Storage of Energy and Data

Storage of energy is a fundamental challenge for most renewable energy technologies. The energy of wind, water, or the sun cannot be used entirely when it is available, but must be stored so it can be used when needed. Engineers everywhere are currently hard at work on public energy storage problems. In Denmark, renewable energy pioneer Jan Gehl has devised "the Copenhagen approach" to a wind energy program, storing wind energy in car batteries. Elsewhere, supercapacitors are being developed to store electrical energy and make it available for electric automobiles. A Dutch design group, Roosegarde, became one of the 2013 winners of the Danish government's INDEX award for ethically and environmentally responsible design by developing a prototype of its "Smart Highway" that will imbed technology in existing roads to recharge electric cars (Rawsthorn, 2013).

Storage, as archivists and museum curators know, is of value only if we can document what is where and can find what we need when we need it. Storage of information with a view to its ultimate retrieval and use is a central concern of the culture of stewardship. Digitization of images of the collection, together with data and information about those images, has become a major museum function, upgrading the former job description of a museum registrar into the far more active role of a collection manager.

Data storage systems and the knowledge management that goes with them have become an integral part of most public and private enterprises. There are serious concerns about privacy and security, but no one doubts the importance of storage systems to enable us to preserve and access information—which is just as important today as harvested grain was to the ancient kingdoms of Egypt or Mesopotamia.[1]

The Cultural Impact of Storing Digitized Data

Storing digitized data has affected the way we write, read, study, learn, and store information, changing what it means to know something, or to be an expert in any subject. Knowledge can now be stored for access by everyone. The expert's role is now its interpretation, and what we can do with that knowledge.

Storing digitized data has resulted in a dramatic speed-up of all private and public business, and of all scientific and humanistic research. In less than 4 decades, it has become impossible for most of the world's governments or universities to operate without it. Today we simply cannot get the world's work done, let alone practice any of our cultures, without it. A computer crash can stop anything. A global crash could stop everything.

Digitization has amplified Marshall McLuhan's projections a million times over. The medium has become the message in ways that even he could not have anticipated. The analogue would now appear in what he would have called "the rear view mirror" (McLuhan, 1962 and 2001).

Just as Marx saw that commodities in the culture of production were really congealed labor, we need to recognize that digitized data in the culture of stewardship are congealed labor. When we look at a database, we are looking at stored energy, and we can use that energy to a more or less infinite range of cultural purposes. Electronic archives are becoming creative "media centers" that provide access to this energy. Whatever happens to the book, public libraries are vibrantly alive as we use them to access information, constantly creating more data as we work and play there. Museums, which also rely on the digital storage of information about their collections, are challenged to offer equivalent participation and access.

Just as Sumerian pictographs first encoded Mesopotamia's agricultural surplus, so digital storage media are providing a language for the renewable energy culture of stewardship. While libraries and museums are encouraging access to this language, many governments are prepared to restrict it. That is why hacking has become such a serious crime. Hackers can make data available to the general public. They are a threat to the power base vested in surveillance systems and control over stored data. Hackers are similar to the peasants of the Middle Ages who dared to steal firewood—stored energy of the trees in their masters' forests.

Can hacking become art? Italian artists Paolo Cirio and Alessandro Ludovico stole 1 million Facebook profiles and filtered them through facial recognition software and a hacked dating website to create "Face to Facebook," an online platform that enables viewers to match their facial features to their supposed personalities. They called their work a Global Mass Media Hack Performance.

Other artists are beginning to speak the language of digital storage media. Canadian photographer and filmmaker Elle Flanders and architect Tamira Sawatzky formed Public Studio, an artists' partnership that has focused on surveillance and photography. While their previous show, "Road Shots," had featured laser-cut images of security roads in Palestine, their 2013 exhibition at the O'Born Contemporary in Toronto entitled "Under the Last Sky" featured super-thin silicon wafer discs printed on one side with the sandblasted images of skies where drones had flown, and on the other with electronic circuitry photo-lithographed onto the wafer. The artists saw the database represented on the circuitry as the storage medium of surveillance information that can be photographed from a drone in the sky. The social-political content of their work is important, but so is their photographic medium. While the sky the drones fly through is the locus of one kind of energy, the surveillance database is congealed energy of another kind, and a sandblasted photograph offers yet another translation of the energy of both into art.[2]

Sustainability

In 1983 my wife and partner Gail and I produced a book for the National Museums of Canada entitled *Planning Our Museums*. Eight years later we reconceived the book for an international market and called it *The Manual of Museum Planning*. In 2012 the third edition (co-edited with our New York Principal Consultant Lindsay Martin) filled 689 pages and added a subtitle, "Sustainable Space, Facilities and Operations." The only way to discuss museum planning in the twenty-first century is in the context of sustainability.

Sustainable development was first identified as a goal for the world's economy in Rio de Janeiro in 1992 at the United Nations Conference on Environment and Development, popularly known as the Earth Summit. In preparation for the Summit, the World Business Council for Sustainable Development declared sustainability to be a major objective of the Corporate Social Responsibility (CSR) of its constituent firms. The conference issued *Agenda 21*, subsequently adopted as the United Nations plan for sustainable development worldwide. CSR and the call for sustainable development have been outstanding features of the culture of stewardship ever since.

In the United States, a year after the Rio conference, the "Leadership in Energy and Environmental Design" program was initiated, culminating in the LEED guidelines and standards for enhancing sustainability, published in 1998. By that time, the 1997 Kyoto Protocol to the UN Framework Convention on Climate Change had been signed by many nations. But anticipating that the Protocol would have an adverse impact on American jobs, and that its standards were in any case impractical, the United States refused to sign it.

The locus of green engineering and construction was originally established in Denmark and Germany. Among early highlights was the Commerzbank Tower in Frankfurt, the world's greenest skyscraper at the time, designed in 1997 by British architect Sir Norman Foster. The 850-foot-high structure includes nine three-story gardens at various levels and a twelve-story garden atrium at its heart. Office windows can be opened, toilets use gray water from the cooling towers, and water-filled grids in their ceilings cool the offices in summer while solar power helps to heat them in winter.

The culture of stewardship may be further exemplified by two museum buildings designed by the studio of the Italian architect Renzo Piano. The California Academy of Sciences in San Francisco, which opened to the public in 2008, unifies a former multibuilding campus under a two-acre, living green roof of gently sloped hills pierced by porthole windows that illuminate the interiors below. Five years later, the architect went further toward achieving carbon neutrality in a museum building when the Renzo Piano Pavilion opened as an addition to Louis

Kahn's monumental Kimbell Art Museum in Fort Worth. "The Kahn building is famous for its natural light," Piano said in a statement issued by the Kimbell. "But that was a natural lighting system designed in the late '60s and '70s. Technologies have advanced considerably since then. We needed to capitalize on the new technologies and make a design that is more flexible and responsible to the issues of today, like sustainability.

The Pavilion, just over 100,000 square feet, is three-quarters below ground, where 36 geo-thermal wells, each 460 feet deep, access the site's ground water to heat and cool the building. Only the public portion of the museum is above ground, with its western wing covered with a sculptured and insulating green sod roof. The galleries are illuminated by a ceiling of fritted glass and fabric scrims under louvers that open and close to control the amount of natural light diffused into the galleries, where the Kimbell's African, Pre-Columbian, and Asian art collections are shown. The louvers are aluminum, and on their surface facing the sun are solar panels that store the energy, which is used to light the building at night. The gallery floors are of white oak planks set one-quarter inch apart, allowing air to circulate throughout the building, decreasing the demand on the air-handling system. The Piano Pavilion will use only half the energy per square foot as the now historic and still magnificent Kimbell building by Louis Kahn.

The cutting-edge culture of sustainability is fast becoming universally accepted. Some architects now specialize in green design, and virtually every major architect's office today has personnel with qualifications in LEED or equivalent systems of accreditation. Corporate clients now routinely require that their new buildings meet LEED silver, gold, or platinum standards.[3] Twenty of the world's largest manufacturers or retailers—Coca Cola, Unilever, Walmart, among them—have committed to eliminate deforestations from their supply chains, pledging to buy only materials and products that are not destructive of the forest.[4]

Stewardship of the Earth

While architects were learning to design for sustainability, some leading visual artists anticipated the culture of stewardship by creating what was called "land art," "earth art," or environmentally conscious art, beginning in the late 1960s. American light artist James Turrell's eloquently simple but powerful revelations of the sky as a circular opening overhead within a building are among the most memorable works of this genre. In other pieces, Turrell enlightened an interior with a single hue into which viewers can walk to discover a new awareness of color, light, and spatial relationships.

In the 1970s American sculptor Robert Smithson created his *Spiral Jetty*, a man-made landscape of black basalt rock and earth extending from the shore of the north arm of Utah's Great Salt Lake into the water in a converging circular maze,

1,500 feet long, with a walkable path into the lake. As the lake level varies, this evocative sculpture is sometimes partially or wholly submerged, an environmental interaction that Smithson intended as part of the life of the piece.

Over its 6-month exhibition period in 2003, Danish-Icelandic artist Olafur Eliasson's *Weather Project* suffused Tate Modern's Turbine Hall with a sun-like hue. Visitors sprawled on the floors for hours to bask in its glow and to admire their tiny black shadows on the huge mirror installed on the ceiling. Eliasson's technology included a fine mist blowing a mixture of sugar and water into the air and a semi-circular "sun" made of hundreds of monochromatic lamps that radiated the pervasive colored light.

A year earlier at the Venice Biennale, Canadian-born photographer and filmmaker Gregory Colbert unveiled his *Ashes and Snow*, an immersive experience of 7-by-12-foot amber and sepia images of wild animals interacting peaceably with human beings, amplified by a slow-moving sonorous film that reasserted a relationship with wildlife that recalled cave art depiction of animals as fellow species sharing the earth. In 2005 along the Hudson in Manhattan, Colbert presented his absorbing imagery in a unique demountable structure built of disposable materials designed by Japanese architect Shigeru Ban. This "Nomadic Museum" was seen by millions of spellbound visitors as it toured Los Angeles, Tokyo, and Mexico City over the following 3 years.

What these works have in common is the artist's heightened awareness of the qualities of the environment—light, color, ground, wildlife, or water. This regard for environment, diametrically opposed to its exploitation for production or profit, is the thesis of much environmental and land or earth art.[5]

Stewardship of the Body

Renewable energy brings with it a culture of stewardship of the earth, and no less so, of our own bodies. This is partly rooted in the contemporary emphasis on women's rights to control what is done to and with their own bodies; it also arises from anxiety about sexually transmitted diseases and the prevalence of cancer. The fitness industry and the widespread concern for a healthier diet, including an insistence on eating more foods from within one's own ecological region, have become universally recognizable signs of this culture of stewardship of the body.

Tattooing the skin is an ancient practice, a visual art form of the body. People with tattoos used to have them in places that clothing would conceal.[6] Today tattooing has become mainstream culture. Men and women proudly display their tattoos, and some enthusiasts make their entire bodies into canvasses.

Public nudity is another notable manifestation of the stewardship of the body. Photographer Spencer Tunick has had noteworthy success assembling large crowds of nude volunteers posed on and around famous sites and monuments for group pictures. The Chilean newspaper *La Tercera* named Tunick Chile's Man of the Year in 2003 for his photograph of hundreds of naked Chileans reclining in front of the Museo de Arte Contemporaneo. Tunick's picture was recognized as a sign of the country's final liberation from the repressive Pinochet dictatorship that had ended in 1990.

Ceramist Kathy Venter has situated contemporary culture of the body in the ancient medium of terracotta, often evoking Roman Tanagra figures or the famous Chinese warrior tomb figures in her work. Venter began building and firing life-size female nudes in her native South Africa, a nation that had been infamous for its life-size body casts of indigenous people in its National Museum. Disgusted with apartheid and despairing of change, Venter and her husband fled South Africa in the 1980s and came to the west coast of Canada, where she established her studio on Salt Spring Island. Her 2012 piece *Tokai* presents a woman's body lying in exhaustion and contentment because its model had agreed to pose for Venter as soon as possible after giving birth. With her abdomen contracting and her breasts heavy with milk, the figure reminds us of the earliest sculpture of prehistory, many of which depicted pregnant or lactating women. As with most of her work, Venter splatters white clay abstractly all over the nude terracotta body, enhancing the tactility of the piece and bringing her classic subject into the realm of contemporary art.

"The body is the place," pioneer performance artist Vito Acconci used to say in the 1970s. In performance art, the artist's body constitutes the work of art. It originated in the "happenings" of the 1960s and '70s. One of the earliest and still most celebrated of the artists who developed this medium is the Serbian-born Marina Abramovic, who first became known for carving into her own flesh in order to communicate meanings rooted deep in her family life and the social-political context of the Balkans. In 2012 she produced two pieces that took her art to a new level. She co-authored and starred in a full-scale theatrical presentation entitled *The Life and Death of Marina Abramovic*. Her radically simple but profoundly moving work, *The Artist Is Present*, featured her sitting at a small table in the Museum of Modern Art for seven hours a day to gaze into the eyes of a succession of visitors for 10 minutes each. Stewardship of her body in the latter piece eloquently consisted of offering her physical, psychological, and emotive presence, with neither action nor artifice to detract.

"Performance art," Abramovic has stated, "is a mental and physical construction in which a performer comes before the public and *energy is transmitted*" (her emphasis).

The transmission of energy she asserts is the difference between the authentic experience of performance art and mere entertainment. She is currently engaged in the establishment of the Marina Abramovic Institute (MAI) for the study and preservation of performance art. Linking the culture of stewardship to the modernist culture of transformation, Abramovic has proposed that the MAI become "the Bauhaus of the 21st century," changing the way that we think about life and art.

The poignancy of performance art—that it exists only in the moment of its creation, and cannot be replicated like a drama or musical score—evokes the fragility of the culture context in which it is being advanced. It is seldom presented with the confidence in personal and social transformation that characterized modern art inspired by the age of electrification. Instead, its meanings are tentative, its values advanced cautiously, in keeping with a culture of stewardship powered by energy sources that are by no means certain of their capacity to supplant the behemoths of coal, oil, and natural gas. The future of renewable energy and its culture is still open to interpretation—and open to the creation of new art forms of the twenty-first century.

Notes

1. In the 1930s science fiction author H. G. Wells hypothesized a "World Brain," a super library of all human wisdom. Google's attempt to scan 20 million books foundered when they disregarded copyright. But subsequent attempts in the US and Europe are aimed at compiling a universal database of the holdings of the world's greatest libraries. Meanwhile a Chinese company is scanning books in all languages in no less than 17 data centers in China.

2. Photographic archives contain image databases that can offer alternative "people's histories." In Beirut the Arab Image Foundation has digitized 80,000 of its 300,000 pictures from family albums and photographers' studios all over the Arab world. Lebanese artist Walid Raad used the AIF as a resource for his work referencing Lebanon's 15-year civil war.

3. Governments also publish sustainability plans and policies: in 2009, for instance, President Obama signed Executive Order 13514 requiring all government agencies to create Strategic Sustainability Performance Plans (SSPPs). The Smithsonian Institution's SSPP, for example, focuses on carbon reduction. The US Environmental Protection Agency (EPA) provides a "Greenhouse Gas Calculator" so that American government agencies can comply with their Greenhouse Gas (GHG) Reporting Protocol. Sustainability and the culture of stewardship are also changing museum buildings; the American Alliance of Museums and the Smithsonian Institution are already giving favorable consideration to broadening the formerly tight but energy-demanding climate control standards that are needed for preservation of the collections.

4. Is it possible to lighten our ecological footprint while retaining our standard of living in a culture of stewardship based on sharing and equality? In 2007 the World Wildlife Fund (WWF) reported on its comparison conducted 1 year previously of various countries' ecological footprint per capita with their standing on the UN's Human Development Index (HDI): the WWF found that Cuba is the only country that is below the recommended biocapacity per person in its ecological footprint and yet is above the threshold for high levels of human development (at 0.8 on the HDI). The United Arab Emirates was at the other end of the scale with an extremely heavy ecological footprint, nearly matched by the US, not-

withstanding Abu Dhabi's commissioning of architect Sir Norman Foster & Partners to plan and design an entire energy-saving eco-city in the desert.

5. Earth art can have practical applications. Architect David Adjaye has been working extensively in his native West Africa, where he has remarked that "every flat roof should have a farm" where the residents of the building below can raise their own food. See Tomkins (2013).

6. Winston Churchill's mother was one of a number of nineteenth-century women who made tattooing briefly a late Victorian female fashion.

Works Cited

McLuhan, Marshall. 1962. *The Gutenberg Galaxy: The Making of Typographic Man*. Toronto: University of Toronto Press.

———. 2001. *Understanding Media: The Extensions of Man*. London: Routledge.

Rawsthorn, Alice. 2013. "Innovation for a Better Future." *New York Times*, Global Edition, September 2, 9.

Smil, Vaclav. 2010. *Energy Transitions: History, Requirements, Prospects*. Westport, CT: Praeger.

Tomkins, Calvin. 2013. "A Sense of Place." *New Yorker*, September 23, 84.

))) The Resources of Fiction

Graeme Macdonald

> I should have thought of it before, it's too late now.
>
> —Italo Calvino, "The Petrol Pump"

The opening sentence of Italo Calvino's 1974 story "The Petrol Pump" expresses a regret wearily familiar to twenty-first-century energy angst. Published in the backdraft of the 1973 global oil crisis, the ethical thrust of the tale is galvanised by a narrator reproaching himself for not fuelling his car when filling stations are closing. Initial indecision whether to make a dash for gas to enable a "necessary" car journey out of town mutates into anguished reflection on an inability to resist the systemic conditions of modern petrolia. The narrator's dismay at the discovery of his gasoline junkiedom is deepened by self-castigation for insufficient consideration to intensifying resource pressure:

> The gauge has been warning me for quite a while that the tank is in reserve. They have been warning us for quite a while that underground global reserves can't last more than twenty years or so. I've had plenty of time to think about it, as usual I've been irresponsible. (170)

Reviews in Cultural Theory 4, no. 2 (2013): 1–24

Nonetheless, an open station is located and the tank duly filled. The tale ends with the vehicle exiting the forecourt, leaving the reader to consider the consequences of such conscientious inaction. For the twenty-first-century reader, the ironic emissions of Calvino's story linger: what to do, when the car has long bolted from the station? From an age of extended (yet always already depleting) "global reserves" this prediction is wayward at best, formed by familiar combinations of alacrity, wishful thinking, and ingenuousness. In retrospect, as fossil-fuelled automobility discovers vast new markets across the globe, that "or so" stings the probability concerns of peak oil with a waspish irony in its seemingly casual projection. In spite of Dr. Hubbert and despite its fundamentally nonrenewable "nature," petroleum has continued to find a means and a relatively undisturbed way.

Calvino's slight narrative of the necessity of energy reflects, like only fiction can, on the fiction of energy's necessity. Like most politically effective literature, "The Petrol Pump" utilises speculation and supposition in subtle yet provocative ways. The conflict between the imagined ecological and economic consequences of Calvino's narrator's sorry actions, for example, is underscored by the shifting pronouns in the above excerpt, emphasizing the extent to which the individual regards his nascent petro-conscience as both privately compromising and publicly ineffectual, and therefore somehow excusable. Here, in short, is a prescient example of the ingeniousness of the privatisations and privations of oil-based modernity, where the sheer pervasiveness of oil in contemporary social infrastructures works as hard as ever to create a general structure of feeling surrounding its *inevitable* use (and misuse). The privatization of energy guilt is viewed as resting the primary burden of ecological response in the individual, in both their "choice" of energy consumption and their "green" ethical behaviour.[1]

Such irony should be instantly recognizable to contemporary scholars exploring correlations between culture and energy resources, from a perspective platformed by the ramifications the '73 oil shocks continue to provide for the world energy system and its geopolitics. It is conveyed by the provocative sleight of that "as usual" in the above-cited paragraph, and driven by the essential paradox imaginative scenarios of energy's limits continue to generate. In 2013, export and demand for fossils continue to increase, despite widely verified evidentiary warnings that at least two-thirds of *known* carbon reserves must remain in the ground to control global warming (IEA Outlook). This, it is generally agreed, is simply not going to happen. Mike-Berners Lee and Duncan Clark describe a stupefying duplicity enacted across the globe, where "green" but "nervous" (41) governmental administrations remain "more concerned about what they have to lose" (85) from carbon restriction proposals and continue to encourage and enable the extraction of fossil fuels, maintaining (and indeed accelerating) the century-long upward trend of the carbon curve and initiating what has been termed the "carbon bubble": where numerous monetary schemes and mechanisms, especially the stock market—perhaps the most threatening ecological system of our times—remain critically invested in fossil futures, to the likely detriment of a sustainable planetary future.[2] Calvino's story, it appears, retains its sardonic bite.

Such a situation is characteristic of what Frederick Buell describes as the "exuberant-catastrophic" oil society we inhabit (291). The short-lived era where oil was almost universally celebrated as an emancipating, "good" substance has long receded. Whereas the appreciation of oil's benefits has not disappeared, it is perpetually haunted by degradation and disaster, forcing extensive contemplation of ways and means of moving beyond its threatening horizon. What, if any, is cultural theory's role here? Decrying the renowned energy expert Vaclav Smil's lament, in his 1994 book, *Energy in World History*, for what he regarded as a "huge conceptual gulf between energy and culture," Buell argues that "energy history is significantly entwined with cultural history," but in so doing notes "no effective response" has, to date, been made to try to bridge this gulf (274). The recent emergence of "Petroculture" as an increasingly prominent international subfield of academic study and cultural practice bears promise the gap should and can be reduced.[3] Its aim: to claim a space for critical, literary, and artistic engagements with what has largely been a geological, political-economic, and corporate substance, measured and valued by petrodollars and combustion power rather than (or indeed alongside) aesthetic modes of representation, image, and narrative. This is still in an early enough theoretical phase to generate queries such as Andrew Pendakis's: "is there an aesthetics of oil or are its cultural manifestations too diverse and localized to be usefully generalized?" ("This Is Not a Pipeline" 8). The answer to this question relies, in part, on the way one elicits and frames the examples of what constitutes petrocultural production, of which more below. What *is* certain is that the alacrity of the concerns over energy and its constituent forms has endowed this emergent field of study with a salient cultural relevance to be broadcast and more fully theorized.

Extracting Culture

I want to propose in this inquisitorial essay that a significant area of "effective response" lies in attempts to energize interpretations of cultural production, specifically literary fiction. Fiction, in its various modes, genres, and histories, offers a significant (and relatively untapped) repository for the energy-aware scholar to demonstrate how, through successive epochs, particularly embedded kinds of energy create a predominant (and oftentimes alternative) culture of being and imagining in the world, organizing and enabling a prevalent mode of living, thinking, moving, dwelling, and working. In industrial modernity this has been largely reliant on the extraction of fossil fuels. The extent to which this energy regime has both fostered and been reliant upon a culture of extraction is of increasing interest. Yet what is recognized as extractive cultural production remains questionable. As I will point out, fictional awareness offers more than stories about energy types and systems. It establishes a means to contemplate—and possibly to deconstruct—energy capital's formidable representative skills, notably its narrativization of the "natural" necessity of oil to our functioning social systems. Oil's sophisticated signifying systems have been central to maintaining its position as the fetishized ur-commodity of modern globalized capitalism. While we can easily identify the ways in which certain formal and thematic concerns ensure Calvino's succinct story's recuperation into the evolving subgenre of world petrofiction, we must also understand how this is also a tale explicitly driven—like all storytelling—by the formalized essentialism of energy in culture and society in general, albeit in a variety of "preferred" and abstracted forms.

In establishing the character of the relation between the global regime of energy extraction and production and its fictional abstractions, literary and cultural theory has its work cut out. One way for it to begin is by considering how and why the ironic entanglements of ecological modernity can be simultaneously sustained and exposed by the fictions that circulate around energy, not only by the fanciful projections and stories created to reveal or counter energy crisis, but also in a reaffirmation of fiction's formal requirements and stylistic capabilities: its narrative energetics; its psychosocial dynamics; its requirements for causality, impetus, and productivity in plot and character development; and its chronotopic ability to straddle and traverse multiple times and spaces. Narrative requires power to *become* narrative. As a powerful cultural form it can change speed, alter force, utilise digression, and in so doing proves a forum to reflect on matters of efficiency and the rationale for certain modes of cause and effect, energy and power. This is supplemented by fiction's degrees of reflexivity: its awareness of its speculative (and often antagonistic and inverse) relationship to time and the Real. A five-page story of one man standing at a petrol pump contemplating his compulsive selfishness can thus stir examination of humanity's current entrapments within and exacerbation of the deleterious effects of the phenomenal opportunities afforded by oil and gas in the petro-privatised culture of late globalized capitalism. Along the way, it can find time to muse not only on the development of the service economy and its relation to flexible labour regimes, but also the nature of its connection to the birth, life, death, and resurrection of all forms of organic life on a planet thousands of years before and after the relatively short and explosive oil era. These are expertly hinged by a twin-engine irony generated by relative levels of short-termism (the use of dramatic suspense) and long-termism (imagined, "off-page" inevitabilities), in addition to deliberate register shifts and genre switches. We ask: will the narrator be able to fill his car in time to make his journey? But we also ask: will that journey, made feasible by the undoubted liberating opportunities of petrolic life, exacerbate the seemingly intractable dread problems surrounding energy (ab)use in the contemporary world system? The story ensures we answer yes on both counts.

In the protagonist's fears for the running of his car (and thus his way of life)

"The Petrol Pump" also reveals fiction's basic reliance on propulsive devices, elementary units of charge that power action, event, and consciousness, calibrated by laws of narrative motion and impressions of kinetic and potential energy transference. (These need not necessarily involve constant or *actual* motion or much, if any, movement—think of Beckett's minimalism, or the generic predicates for entropy in Naturalist writing.) Like the laws of thermodynamics, fiction relies on momentum and transference, absorbing and exuding, circulating, conserving and converting energy and resources, not only on the level of narrative, metaphor, and content but also in formation, production, dissemination, and reception. (Is it churlish to point out that you are, after all, reading this on once-oil or once-wood?) The question, however, of how the remarkable energy *of* fiction is inextricably connected to the (often entirely unremarked and unremarkable) energy *in* fiction —the stuff that makes things *go* and happen in literary worlds—goes mostly unstated. This despite the spectacular products and results of primary and secondary energy conversions being visible throughout literature's modern history: imagine, say, *Anna Karenina*, or *God's Bits of Wood*, or *One Hundred Years of Solitude* without coal-powered locomotives! Contemplate Conrad's novels without wind or steam. Consider the sprawling fiction of twentieth-century suburbia—relating psyches, bodies, and worlds saturated in oil-based products—suddenly shorn of plastics, deprived of automobility or domestic electric power, bereft of pharmaceuticals, denied the cheap food supplies of prime-moved fertilizer!

Emergent modes of energy research and criticism seem to disavow assertions such as Smil's that "timeless artistic expressions show no correlation with levels or kinds of energy consumption" ("World History and Energy" 559). They reach instead for a "fuller analysis" sought by Edward Cassedy and Peter Grossman, involving "a sense of the social and philosophic context in which energy technology and resources are used, and a keen appreciation of what energy issues mean to the way we live and to the world we live in" (8). The questions asked in emergent modes of energy research and criticism are thus fundamental to the constitution, categories, methodologies, and demographics of the literary field: does literature shape and shift in accordance with the dominant energy forms of the era it registers? Might it somehow play a role in *reproducing* (or, indeed, *resisting*)—perhaps inadvertently or unconsciously—a predominant energy culture? How does literature *use* energy and vice versa? Are literary modes—like social formations— brought about by developments in fuel or resource use to a far greater extent than we have previously considered? "What happens," as Patricia Yaeger asks, "if we sort texts according to the energy sources that made them possible . . . what happens if we re-chart literary periods and make energy sources a matter of urgency to literary criticism?" (306). Can we think, for example, of modernism outside an oil-electric context? Of Realism without steam or coal? Romanticism without wind or water?

To begin to answer these questions, we have to become more adept at divining the specific fuel(s) literary modes run on. This does not necessarily entail following *only* work explicitly concerned with energy resources (though this might be a start!), despite the number of particular texts from world literature that can be considered "energy classics," such as Émile Zola's *Germinal* (coal, 1885), Fyodor Gladkov's *Energy* (hydroelectricity, 1932–38), Miguel Ángel Asturias's *Hombres de maíz* (1949) and *Banana Trilogy* (food, 1950–60), Henri Queffélec's *Combat contre l'invisible* (nuclear, 1969), Gene Wolf's *Book of the New Sun* quartet (solar, 1980–83), and Abdelrahman Munif's *Cities of Salt* quintet (oil, 1984–89), to name a few. To these (where, frustratingly, the topic and concept of energy remains rather incidental to established critical inquiry) we could add numerous others, in addition to myriad literary registrations of wood, wind, whale oil, paraffin, electricity, tidal water, biofuel, GM foods, etc.

A strongly developed strain of petrocultural theory focuses on the way in which the means and effects of oil are structurally occluded from its mass of consumers, making it less apparent as an *explicit* object in social life and thus a specific topic in and for cultural production. For Peter Hitchcock, oil produces the most "violent" logic of all energy forms and in doing so militates against alternative imaginative forms of representation. Oil's powerful "symbolic order" works influentially to present an inviolable discourse as to its prerequisite role in real life, its "omnipresence" creating a sheen of dependency "that paradoxically has placed a significant bar on its cultural representation." In this view, oil's "real" fictive power is such that literary fiction cannot realistically hope to articulate it in realistic terms:

> In general, oil dependency is not just an economic attachment but appears as a kind of cognitive compulsion that mightily prohibits alternatives to its utility as a commodity and as an array of cultural signifiers. . . . I view the problem as primarily dialectical in the broadest sense, rather than as one of cultural expression by itself. ("Oil in an American Imaginary" 81–82)

Considering appropriate means of culturally expressing oil's domain, the editors of a special oil-related issue of the journal *Imaginations* somewhat echo Hitchcock in viewing the problem as one of pervasive mystification. This is a result of the collusion of corporate secrecy and consumer repression typical of late capitalism, however "ecologically responsible" it declares itself:

> The problem of visualization, of the proliferation of determinate, useful maps of our economic lives, is not specific to oil, but one politically structural to a system that is at once spectacularly consumerist and fully globalized on the level of production. However, it could be argued that oil is a uniquely occluded substance: not only does its exchange value engender an enormous corporate project of hiding, an explicit machinery of deception and spin, its pervasiveness, its presence, everywhere, perhaps singularly

christens its position as "hidden in plain sight." (Wilson and Pendakis 5; quoted in Szeman and Whiteman 55)

There is room for counterargument here that would note two basic points: (1) that such an "everywhere-felt-but-nowhere-seen" condition is geo-culturally uneven, symptomatic of the uneven international division of labour, regulation, and ownership of oil capital; (2) that we *are* in fact extremely aware of oil issues, most especially in the overconsuming Global North, where environmental membership and activism is relatively high and influential. As I and others such as Michael T. Walonen have argued, these points are somewhat qualified by a comparison of international petrofiction (and other cultural work, such as documentary photography) from the various spaces of the world oil system, notably that registering the experiences of those living and working in those "concealed" or peripheral zones of extraction. Nevertheless, the peripheral geography of fossil-fuel extraction on land and water, combined with what Rob Nixon has called the "slow" or "invisible" violence of its atmospheric and environmental effects, has always effectively "offshored" features of its transacting, refining, transmission, and emission across the "advanced" productive economies of the Global North in particular (2). In this sense oil perfectly illustrates ecologically challenged modernity's Janus face. What could be eulogized by the road-tripping narrator of Nabokov's *Lolita* as the "honest brightness of the gasoline paraphernalia" (153) of postwar America has darkened into a petro-reliant world persistently disturbed by what Buell describes as "a large portfolio of dread problems" (274). Despite these being increasingly difficult to ignore, Imre Szeman notes an obdurate "foundational gap" preventing public action on dirty energy's predilection for crisis, a gap created by

the apparent epistemic inability or unwillingness to name our energy ontologies, one consequence of which is the yawning space between belief and action, knowledge and agency: we know where we stand with respect to energy, but we do nothing about it. ("Literature and Energy Futures" 324)

We might heed this as a challenge for cultural theorists to take up: how can achieving meaningful action over the problems (and opportunities) of oil entail *knowing* oil better? The overwhelming majority of climate scientists now acknowledge that solving the problem of human-caused climate change must place less emphasis over the exactitude of the science than its communication and awareness. Most certainly this involves rethinking how to discern and locate the cultural life of emissions and their representative properties within a larger social/energy matrix. But once we discern the 500MW reactor in the corner of the parlour or the derrick in the drawing room, what then?

To reiterate: if we are to realize that historical events, economic relations, and political formations are created and sustained by energy resources available and accessible at any particular time, and that such events and formations, in turn, create *and are in part reproduced* by a specific energy culture, then reframing fiction as a crucial cultural resource historically suffused with energy, in form and substance, might require an altogether bolder and more ambitious interpretive approach. This posits the daunting yet exciting assertion that *all* (or perhaps *any*) fictional work is a veritable reservoir for the energy-aware scholar. We might see this as following Said's theory of contrapuntal reading. If we *all* "live" an extractive culture, regardless of our cognitive connections or geographic proximity to refineries, mineshafts, and drill zones, then our cultural production should reflect that, regardless of how abstract or distorted the projection. How this can be critically extracted and subsequently refined becomes the point of focus, meeting the challenge Hitchcock issues concerning energy's peculiar "cultural logic": how to mediate/interpret it as "a very mode of referentiality, a texture in the way stories get told" ("Oil in an American Imaginary" 87).

In spite of legitimate concerns it may be unworkably elastic or overdetermined, a "deep-energy" methodological perspective is, in fact, already under way in some subfields. Thermodynamic readings of the narrative and social concerns of the nineteenth- and early twentieth-century novel, for example, are well established. Electromagnetic expressions of force and speed, and a consciousness of newly mechanized motion, find their way into textual understanding of the technical and topical dynamics of the late realist novel and subsequent modernist movements from Vorticism to Futurism. Where, then, are analogous pronouncements on later cultural moments and movements? Despite being stock full of fossil fuel's refinements, most fiction set in oil-gas-nuclear-renewables era modernity awaits similar energy-based elicitations. The accelerated mobility and intensified compressions of space and time enabled by carbon-driven capitalism, and petro-technology in particular, have altered the shape and geography of literary plot, not to mention the available global constituencies of character, custom, and style, as they have massively altered global spatial, media, and economic orders. Oil, like coal, clearly has form, but to what extent has this been fully recognized? How can we appropriately interpret its discretion, in order to connect it to the larger frameworks of energy I have discussed above?

From the gas station experience upwards the principal definition of the "cost" of oil has been domestically economic, a point that needs understood partly as a *cultural* phenomenon produced by a specific mode of neoliberal political economy. Neoliberalism is an oil system, ironically enabled and sustained by on-stream petro-revenues and dramatic falls in post-1970s barrel prices (from the early '80s switch to monetarism in Reaganism/Thatcherism, to the rise of the Oil and Gas Tsars of the post-communist Soviet Union) and heavily invested in both technological and commodity capacities with the fictive capital structures of electronic financial modelling systems. Finding the energy in cultural production, especially in a service-led context, is partly imbricated in understanding the *social* and *economic* fictions of energy created, inhabited, and reproduced within any petroculture, but particularly acute in the sphere of neoliberalism and its politics. These sediment and systematise prevalent conceptions of the necessity of various forms of exhaustible resource and work to maintain and often intensify the levels of investment placed upon them.

Part of the point in theorizing energy as cultural is, therefore, to expose and determine reasons for our acculturation to its hierarchy of material (and, increasingly, immaterial) forms and the manner in which they dictate fundamental aspects of social life and organization. If, as is often remarked, in an age of consumer sovereignty, we don't really think *enough* about how we expect and trust the lights to go on when we flick the switch, then how is this related to what Owen Logan calls a "supply-side aesthetic": the manner in which the consumer identity we inhabit reproduces the way we (fail to) perceive and portray our predominant energy infrastructures (105)? How we think conceptually of waste, expenditure, and remaining amounts has also, according to Logan, become "undialectical," a point exemplified by the tendency for developed oil societies to offshore or export or make limited ethical claims on the associated pollution and waste, excising it geographically or temporally, as a problem of *elsewhere*, of the future, or by governmentalizing ineffective recycling programmes. Clearly how we "consume" rather than "use" and, crucially, *extract* fossil fuels makes us act and think about them in an uncritical, deflective way.[4] This is aided by the effective brand management of the oil and gas corporations over the last 20 years or so. This suffered some relapse in the difficulties of BP, which, prior to the Deepwater Horizon disaster, was oil's most PR-savvy representative, but has, like big oil generally, rediscovered its mojo. In all these scenarios, an energy imaginary *beyond fiction* underpins fossils as epitomizing a future of security, efficiency, and even "sustainability." "Unconventional" fossil fuel is represented as technologically innovative and thus largely positive, a "solution" to projected needs. In this scenario the very concept of modernity as founded upon and reliant on depletion-based resources is ignored for a holding-pattern, "fossilized" vision of a bountiful future.

Oil's emancipatory role in habitual experience is repeatedly vaunted in this incorporating system of petro-acculturation: how *could* we live without it? This has often been presented more as overwhelming threat than earnest challenge, particularly by those interested in retaining oil's dominance, who consistently remind us of the deep spread of oil products—and their socioeconomic benefits—across modern life.[5]

Petrofiction and Beyond

Aided by the subtending practices of culture-project sponsorship, oil's representative conversions of "polluting" energy into "productive" or "good" energy are a prime example of the challenge, but also the opportunity, facing dialectical inter-pretive responses. We might, on some level, expect the wide-scale naturalisation of nonrenewable or "dirty" energy in carbon-anxious modernity to present a for-midable blockage to "alternative" energy's cultural perception and representation, yet petrofiction's emergence as an international subgenre demonstrates literature's capacity to energize purviews, confronting and repositioning the potent social and economic signifiers "naturalising" energy and contemporary petrolic living in gen-eral. It has conjoined with powerful modes of anti–resource colonialism and eco-criticism (the bass notes of petrofiction) in seeking to heighten our planetary en-ergy consciousness. A question for future practitioners in the field will be whether this opens up portals for similar readings and revisions of other energy forms in cultural production. Could we, for example, replace "oil" with, say, "nuclear" or "wind" in much of what I have argued above? What might be different in their cultural logic?

The identification of fiction concerning and concerned about energy—not only with its limits and secure supply but also with concomitant themes of exploration and (over)production, capacity and consumption, and subthemes of conversion, distribution, and commodification—has grown, albeit incrementally, in the period since Calvino's story. An energetic form of criticism has also begun to construct a solid platform for the elaboration (and in many ways the re-categorisation) of a whole history of literature concerned with the history and future of the planet, amidst the geopolitical and biophysical machinations of global warming and the contemporary world carbon nexus. The degree to which this work can exert trac-tion on the established manner in which rising gas or domestic heating prices shake general volumes of energy indifference is interesting for students of the impact of cultural forms. Nevertheless, if, as Szeman ("Petrofictions") and others emphasize, an energy awareness has *finally* begun to spread through the arts, hu-manities, and cultural analysis generally, key questions arise: to what degree are conventional modes, not only of ecological literature ("the environmental novel," the "ecopoetic imagination," "ecocriticism") but of literature in general, limited in style, approach, and purview? Have they enough sources and resources to deal with the size and scale of the "urgency" Yaeger (see above) emphasizes? Finally, how and why is the *form* of our energy dependency a critical matter?

As the most recognizable strain of "energy art," petrofiction has its specific subconcentrations in exuberant (and damning) extraction narratives, local and transnational stories of oil's development, and its dramatic transformation of space, place, and lifestyle. To these we can add tales of corporate corruption and petro-despotism, spill and disaster, the conflict between oil capital and labor, and even the "drama" of barrel prices and fictive petro-capital enacted across international territories. But in what ways might the fiction of drill bits, mineral rights, and gushers relating the *process* of oil fail to reflect its wider material and ontological spread, as well as Hitchcock's "primary" dialectical form? Should not "petrofiction" be seen as much a fiction of "alternatives" or replacements to oil, both past and future, as it is about the supercommodity oil has become? Is not oil-based culture, by virtue of the (un)certainty of supply and ecological limits (however much they may be continually shifting or postponed), always already a *post-oil* culture? Alternatives to oil dwell within and alongside oil culture, albeit in a rather spectral fashion, as absent presences demanding attention to their inevitable—or belated—appearance. Cultural production has configured these in various manifestations and interpretive manoeuvres, although by no means are alternative energy sources as explicitly or singularly acknowledged as oil or coal. This is changing. As the prime energy form governing contemporary social forms drains away, we should expect new forms of resource fiction to become increasingly insistent. Whither hydrofiction? Windpoetics? Nuclear drama?

Petroleum culture is consistently haunted by its eventual depletion. A post-oil element is detectable in oil texts from the nineteenth century onwards, but since the 1960s a recognizable form of petrofiction has been driven primarily by depletion anxiety. Here, contemporary fears about resource wars and climate collapse (among many others) are reprocessed in apocalyptic narratives of floods, population wipeout, continental starvation, solar exhaustion, and bioenvironmental degradation. Scratch the surface of most dystopian narratives and types of resource cataclysm appear.[6] Much of this work ponders the momentous eventuality of a world without large quantities of flowing oil—gasoline in particular. Yet the zombified afterlife of petroleum in numerous postapocalyptic, carbon-fretful narratives emphasizes how hard it is to let go. Constituencies remain hooked on its scant (and thus unevenly distributed) deposits. Think, for example, of the petro-desperation of the barbarian motorcyclists encircling the embattled renegade oil refinery in George Miller's film *Mad Max 2* (1981), or the continuing allure of the bitumen-shattered highway, navigated by a tattered oil-company map, in Cormac McCarthy's *The Road* (2006), or the corpses strewn around gas pumps in Justin Cronin's vampire-apocalypse novel, *The Twelve* (2012).

As post-oil culture mourns the passing of cheap and easy oil, it speculates on the elevation of its potential alternatives. Oil is limited but not *totally* missing in novels like Sarah Hall's *The Carhullan Army* (2007) or Paolo Bacigalupi's *The Wind-Up Girl* (2009), for example, but their respective relation of a neo-communalist, new-diggers England and a flooded future Bangkok exemplifies an emergent multi-resource novel. This renders a world of mixed old and emergent new fuel and energy "choices" created from necessity-bound relations of anticipated fossil depletion and generalised resource shortage. Concentrations of food, wind, hydro, dung, wood, and muscle (animal and human)—natural and biogenetically engineered—show how an imagined future projection of *less* doesn't necessarily imply a scarcity of energy, but emphasizes its control and expenditure as a capitalised resource throughout the modernity it helps establish, yoked to the surplus logic of powerful interests, pressed into the service of capital and (neo)imperialism. Capacity becomes relative, as opposed to absolute. The persistence of uneven access and private distribution networks in the future energy imaginary ensure that regardless of its nonpolluting properties, even wind or waves or sprocket-borne power remains, rather like future Bangkok's illegally burnt animal dung, a "shit" form of "filthy" energy when tied to political and economic forms of conflict, corruption, and oppression.

It is salient that the fictions of future energy-scarce scenarios express caution regarding an almost-post-carbon future of "alternatives" that does not necessarily herald a renewables utopia. In doing so they reveal the *nature* of any society as bound up with a specific energy mode and particular system of social power. This opens up a vista towards the long view of energy's commodification within the world system, where, regardless of its degrees of "cleanliness," it has always been tarnished by powerful systemic organization, controlling price, access, distribution, and consumption. The predominating spectre of supply anxiety in late capitalism has thus ensured that it is rare to see an imagined future where *less* energy is automatically "good." Though the logic and chronology of speculative fiction's energy scenarios may be future-set, its contemporary cognition is as energy-conscious challenge, via either allegorical interpretation or verisimilar credibility, as a world of the possible, a shape of things to come (or as they *are* for the billions of fuel-poor on the planet) under the irrepressible cultural and social logic of contemporary petro-finance and ongoing carbonisation. So much (or, perhaps, so *less*) for the future.

Undoubtedly, speculative fictions of future energy landscapes present uncomfortable contemporary questions. At the very least, in visions of a world with less oil, they offer glimmers of what transition might entail. A problem, however, may lie in potentially unexpected consequences of their progressive eco-cynical vision and generic familiarity, bolstering a fossil politics opposite to what might be intended. "Look," an oil-company spokesperson can claim, "at the barbarous, chaotic world without oil," the perfect riposte to any radical imagining of a nonpolluting replacement. It could be argued that our preoccupations with scarcity have perpetuated a present situation where abundance remains not just desirable but essential. The literary fiction of inevitable fossil depletion nonetheless provides the means for its critics to confront the fictions—social, literary, geological—of ongoing abundance we face in the present. Why is it, for example, that imaginary futures of less always seem to run—implicitly or explicitly—on the drama of "more"? Involving the objective of regaining or recovering maximal (usually "dirty") energy systems we critique as unworkable in the present? As I remarked above, the historical examples of most petrofiction remind us that themes and issues such as depletion anxiety are embedded within the enthusiastic pursuit of expanded extraction. A dialectical relation has always configured cultural, political, and economic notions of energy's limits within patterns of development and desire generated by perceptions of its (real and imagined) limitlessness. In fact, the social fiction of unhindered and waste-free energy flow—always already a degraded notion in a systemic culture of nonrenewables—unconsciously pervades most, if not all, cultural production from the coal age onwards.

Oil Fantasies . . .

A point that cannot go unmentioned here: petroculturalists perhaps haven't paid much attention to the constructive story where such a superlative mode of energy *has been and remains* a "necessary," essential, and ameliorative force in modern human history. Looking back at "The Petrol Pump" now, from the century-long and continuing "success" of oil, a question arises: has contemporary eco-culture's default setting of the condemnatory registering of "dirty" energy been one-dimensional? Has it not realised fully why hydrocarbons have been "celebrated," or adequately qualified their powerful attractions? To get beyond "dirty oil" we have to better comprehend and distinguish its powerful, emancipatory attractions.[7] Imagine a hospital without pharmaceuticals or plastics, a food supply without fertilizers. What would an oil-free utopia that would dispense with these look like?

However oil's "usefulness" is perceived, it is clear that much of the culture world is hooked on relating its devastating qualities at the expense of its evident material and infrastructural qualities. Can cultural and theoretical work help to evolve distinctive replacements for it? Our criticism, like our technology and terminology, might not yet be sufficiently refined. But interpretively skilled cultural practitioners will prove crucial—not solely in decoding and countering the signifying prowess of oil capital, but in framing the social and planetary "story" of oil and narrativizing alternative energy signatures and structures in a form and space outside orthodox or vested representations.

I have argued that in order to detect energy's cultural properties, fictional resources could be read more energetically. One way of managing this involves considering how to rethink why certain texts are deemed *literally* "about" oil, electricity, coal, etc., and others less so. Most fiction dealing explicitly with energy, whether as problematic or enabling force, typically involves a coming-to-energy-consciousness, often in the context of plots about energy rights or fuel discovery and resource deprivation. The "lightbulb" moment in Calvino's story occurs in the forecourt of a new type of "self-service" filling station. In retrospect, it is instructive that its narrator's petro-anxiety is paralleled (and somewhat mitigated) by the enthralling promises of an incipient age of consumerism. This is packed into a moment of false consolation where he considers how it is that the burden of oil consumption and its excision fall on him as he performs—with all the consumer "choice" of an addict—the final labour of the energy company that profits from his purchase: pathetically, he "works" the pump and injects the high-octane "poison" into his thirsted vehicle. He sublimates his shame and resentment by resorting to an overtly sexualised populist road fantasy—the ultimate fiction of an oil-based cultural life. The genre morphing is deliberate, recalling Ryszard Kapuscinski's much-cited statement, in his *Shah of Shahs* (1980), concerning the "illusion of a completely changed life" that the "anaesthetizing" effects of oil offer. "Oil," writes Kapuscinski, "is a fairy tale, and, like every fairy tale, is a bit of a lie" (35).

Many petro-stories are driven by reflection on what characters do not know (or indeed care) about the life, designs, and infrastructures of oil, from the large-scale petro-complex to what Stephanie LeMenager calls its "affective context" and "emotional geography": the deeply embedded aesthetics of petroleum in our banal lifeworlds. An energy-conscious fiction and criticism might seek to further extrapolate and represent in all manner and modes of fiction where energy is either *not* recognized or simply taken for granted. There comes a time, however, when this ignorance is unsustainable:

> All of a sudden I'm seized by a craving to get out of here; but to go where? I don't know, it doesn't matter; perhaps I just want to burn up what little energy is left and finish off the cycle. I've dug out a last thousand lire to siphon off one more shot of fuel. ("Petrol Pump" 174)

Here, fiction's constructed ambivalence and advantageous access to consciousness and speculative scenario highlights duplicity in the romantic engineering of energy's illusions. Once we exit the shameful (fictional) realm of the forecourt—the intimate space of our oil encounter—are we who occupy the real free to forget "bad" energy and continue the mundane fantasy of its "special" effects across modern life? At a rhetorical stroke, fiction exposes the fictive life of oil. But how does it engineer a properly energized response? To imagine a world where oil use "doesn't matter" is to live literally in another world. Calvino's story wryly parodies the absurdity of desiring a limited, destructive resource, but doesn't know how or where to go without it. The ironic use of a cheap and carefree "Hollywood" metaphor of driving off into the sunset self-reflexively exposes what Szeman has called the "fiction of surplus" that both literary *and* material life seem stuck within; unable to countenance a world of less or "easy" energy, despite impending lack ("Literature and Energy Futures" 323), we retreat into fantasy.

The fantasies of oil culture continue in part because, as I have noted, oil *is* fantastic. That it is often misrecognised (or indeed misused and abused) as such is part of the problem. The surplus imaginary thrives in "environmentally responsible" late capitalist culture, often in the specious/earnest acknowledgement of the "problem" of energy. Mass-market fictions here offer potential to consider an alternative energy imaginary even if only by revealing its dominant and residual forms. Hollywood, for example, enthusiastically embraces "dirty" energy's pay dirt. The greenwashed plots of recent fantasy blockbusters, from *Avatar* (2009) to *Avengers Assemble* (2012) to *Batman: The Dark Night Rises* (2012), revolve around the miraculous technological discovery of cheap, limitless but *clean* and "ecological" forms of energy. The question of why a quantitative (or even free) "superheroic" replacement for "bad energy," offering similar power and capacity, is required isn't really on the agenda. For why would mass entertainment forego the virtualised drama of crisis for a more philosophically nuanced approach to energy's value, or even offer a more revolutionary concept or utopian suggestion about an alternative system of use and distribution? The spectacle of flat environmentalism is now a preset stance in the circulation of global cultural commodities, where a liberal humanism *in fiction* can be espoused by corporate culture producers, who, regardless of the degree to which they see themselves as somewhat apart from the "bad" energy corporate, remain heavily co-opted into the cultural and economic hegemonies of petrolife. "Less" can only appear dramatically sustainable for a finite amount of time within the actual world system, where energy's cultural capital is remarkably aligned to culture's energy capital.

Conclusion: Where's the Alternative?

The consolidation of petroculture as a critical means of reconceptualising energy enables reflection on the usefulness of *all* kinds of fiction—from across genres and literary history—for pressing political questions and eco-philosophical reflection in an energy-challenged present. The subtext of Calvino's story questioned the supremacy of fossil fuel in the 1974 context where "is there any other choice?" was a legitimate but rather novel query. It returns in the warming era where unconventional energy, oil, coal, and gas are resurgent, and large areas of the earth await pockmarking by new drilling projects. How does this cast the warnings and anxieties of depletion expressed in most petrofiction? Does not fracked gas or thin oil mark Calvino's piece as a product of an *outdated* era of high "peak" anxiety? Might the deferral of "peak" oil culture hinder the development of new subgenres in the literature of energy?

However we choose to meet these conundrums, late energy criticism must make it apparent that it can't all be about petrol. Literary history has a considerable stockpile of energetic potential. Fiction has circulated and conveyed resources of heat, light, relative speed, force, and motion long before *Don Quixote* registered wind power in 1605. From its rise to cultural prominence in modernity, the novel is replete with moments where its great theme of transition reflects developments in energy and fuel provision. Consider, for example, the moment—recoverable in numerous novels—in Giuseppe de Lampedusa's archetypally modern novel of tradition and revolution, *The Leopard* (1958), where the death of the aristocrat Don

Fabrizio is framed by the phenomenal change Italy has experienced in his lifetime, a transition measured by the accelerated *story* shift from the age of horse-driven power to the jet engine. *The Leopard*'s temporal narrative jolts characterise the coexistent elements of most energy transitions, but critical readings of the novel's expressions of the intersections between historico-political progression, shifting political culture, and transnational geography leave energy provision subsumed. In these and countless novels before and after the age of petroleum, energy makes history and it has form in so doing, but despite providing the engine room of plot, story, and context, the aesthetics and opportunities created by fuel power are not sufficiently registered, surfacing only periodically, during times of high resource angst. In an unprecedented time of permanent conflict over supply, availability, and destructive toxicity this critical blindness is unsustainable. The corrective involves new angles of methodological perspective and conceptual debates that have begun in the petrocriticism noted above. It certainly means consistently unveiling the mundane acculturation to prodigious uses of "natural" nonrenewable energy in growth-obsessed polities and economies. The task is truly formidable, given the continuing spread of carbon-based development across the globe.

The challenge is thus made to critics across the genres of fiction making, from literature to cinema: if *all* fiction is potentially energetic, valorizing energy use, then how do we kinetically assert our claims and configure our specialist readings to make it more apparent? The bedrock of this question is not only formed by the simple fact that the formal conditions for all narrative—even the most minimalist or "slo-fiction"—require a degree of forward momentum for events, space, mobility, and development: as a basic unit of charge, but also by recognizing that if literary form is always to some extent an abstraction of the social, then interpretive issues and critical formations of capacity, power, and supply determine *all* worlds. This requires we stretch our definitions and reconsider historical sedimentations of genre and period. "Petrofiction" in this frame is certainly stories about platforms, drill bits, combustible transport, deadly spills, and exploration rights. But it's also about the world a specific fuel creates and maintains, about the relation between the oblique and surface world of fuel. Petroculture, for example, renders a world of electronic gadgets, just-in-time goods, and financial transactions reliant on oil consumption but abstracted from the backstage forms of its conversion, extraction, refining, and delivery, from privatized pipelines and petro-guerrillas to compromised forms of democracy.

How trite or redundant, then, in this view, is the claim that given the global cultural reach of an oil- and gas-dominated world energy system, *all* fiction is petrofiction to various removes? That all fiction, pre- and post-oil, can be measured by its relationship to the transformed aesthetic and material world that oil created and threatens to revolutionize again, by either its absence or its carbonizing essence? Is fuel *that* fundamental to culture and cultural production? If a future of eventual diminishment or unworkable or unwanted energy types is certain, and we resort to a world of reduced force, even one of post-prime moving, then work published prior to oil (or outside the carbon complex) becomes re-energized by the examples it offers of a world constituted via alternative energy sources.

LeMenager argues that

> the petroleum infrastructure has become embodied memory and habitus for modern humans, insofar as everyday events such as driving or feeling the summer heat or asphalt on the soles of one's feet are incorporating practices . . . decoupling human corporeal memory from the infrastructures that have sustained it may be the primary challenge for ecological narrative in the service of human species survival beyond the twenty-first century. (26)

What would a non-hydrocarbon imaginary resemble, after humanity's experience of oil? Reading fiction in this light offers eco-chronological backflips. A bounty of refuelled scenes from metropolitan core to oil-deprived periphery of literary history offer a means to "re-couple" our pre-oil energy memories to consider their usefulness for a post-oil world. Reading pre-oil texts from a post-oil perspective becomes particularly instructive. Did people really *walk* "sixty miles each way" on errands and business, as Mr. Earnshaw does in matter-of-fact fashion, near the beginning of Emily Brontë's *Wuthering Heights* (1847)? Will literature after oil become more pedestrian? Certainly post-automobilic narratives of on-foot struggle, such as *The Road* or Joshua Ferris's *The Unnamed* (2010), seem to suggest we re-attune ourselves to an embodied aesthetic with a rich literary history, from Rousseau to Baudelaire, Beckett, and Sebald. Stendhal's famous aphorism from *Le Rouge et Le Noir* (1830), that "a novel is a mirror on a highway walked"—somewhat eclipsed by an age in which the mirror is more likely to reflect a highway burned up by an SUV—comes back into focus here. LeMenager speculates on what might be considered as a "post-petrol style," to challenge the autoerotic, affective concentrations of mass car culture, and asks if, in a world attracted to low growth and reduced output, there might be an "erotics of post-sustainability" on a par close to the "affective intensity" oil living provides (61). An entire corpus of ambulatory fiction awaits this type of analysis, but it is wholly naïve to think of the end, however prolonged, of petroleum as automatically ushering in a new, "older" era of slower movement and localized distance. It might require, as Allan Stoekl has argued, a wholly refurbished theory of energy, involving a redefinition of its utility, necessity, and use value, as well as its physical and philosophical "qualities," to challenge modernity's love of gasoline speed and combustion prowess, its continual pursuit of maximum output, and its captured definitions of energy efficiency and economy. For Stoekl, "we have no choice but (miming Bataille) to elaborate a theory of excess in an era of radical shortage, a practice of human-powered velocity in an era of gas lines" (193). He insists this cannot involve a simple return to a romanticized past, as a "good duality" to carbon-made modernity, without recognising the importance of energy excess and burn as crucial—but nonpolluting— features of human, bodily expenditure.

The extent to which such terms are placed within and against their understanding and operations in the closed global economy of petro-capitalist time and space, presently running out of gas, is crucial. For Stoekl, the solution is to fundamentally rethink animate power, joy, and labor, within a radically re-localized spatiality:

> The radical finitude of fossil fuel—the Nature that refuses to die, even when it gives itself up and runs out (and its running out is its reaffirmation of its singular autonomy) —is the opening of muscle expenditure, the squandering of excited organs. (202)

Such a view, in conjunction with the findings of modern bioenergetics, presents an interesting platform to reconsider the way we re-energize scenes from literary history. Think, for example, of Konstantin Levin's appreciation of a "sea of cheerful common human labor" scything crops in part 3 of *Anna Karenina* (1877). In a novel where the development of steam-driven motive power engages with massive transformations in agro-class development, this renowned scene reminds us of the most primal and fundamental energy form: organic muscular exertion. But it also underlines the long connection between energy "production," resource ownership, and labor exploitation. Similar attention to "alternative" energy sources in the anticipated future-without-oil presents opportunities for historical re-reading. The giant log pile behind Mr. Knightley as he converses with his eventual wife in Jane Austen's *Emma* might have long appeared incidental. Now, in petroleum's deferred wake, it denotes not only an age of wood but also the invested power and prestige in the ownership of stockpiles of energy throughout history.[8] Consider the transformative hydro-active power of a water wheel that runs the nail factory at the commencement of Stendhal's *Le Rouge et Le Noir*. These countless scenes become more than incidental or isolated scenes of the historical entanglement of fuel power, resource-based capitalism, and the class control of extractive production: they become critical fuel for fiction's effective recuperation and recycling of the energy forms made peripheral by the oil age and the cultural forms associated with it. Calvino's narrator's day wasn't, after all, to be about fuel levels, but in the end, in order to move forward, it had to be. However we interpret it, this has to be construed as a problem. If anything, "The Petrol Pump" reminds us that the warning light set in 1973 continues to blink. To properly energize culture in petromodernity's wake requires huge theoretical resolve to jumpstart the practical effort: nothing less than wholesale critical transformation and renewability.

Notes

1. Of course, neo-sustainability arguments pressed into service by corporate and political agencies worldwide maintain that environmental "crisis" is a future-deferred event, however relative the dispute over its temporality and inevitability. A counterargument insists we are already experiencing that crisis in the present.

2. As Bill McKibben notes, in 2012 oil company assets and share values, as well as the financial futures system, relied on approximately $27 trillion priced *un*extracted carbon (2,795 gigatons) to be used eventually, much of it to be burnt.

3. See in particular the international research cluster at www.petrocultures.com.

4. Duncan Clark argues that despite a fall in US emissions, partly due to shale gas fracking, a consequent increase in US coal exports has led to a rise in its carbon extraction and burning. His argument is that carbon measures should automatically be globally based, and on *extraction* rather than national emission rates ("The Rise and Rise").

5. See Huber, "Refined Politics"; and Hitchock, "Everything's Gone Green."

6. Conversely, no one seems to question the seemingly abundant (and presumably "clean") levels of post-fossil energy powering the spaceships and megacities of sf, especially the multiverse energy worlds of Space Opera. These can be safely consigned as "idealist" by petro-realists.

7. This pertains, again, to reappraising cultural perceptions and the acculturation of particular energy forms. Consider, for example, a sport such as Formula One Racing, a pursuit I personally find objectionable on many grounds, not least its contribution of a massive carbon footprint. I recognize, however, its seductive, enthralling aspects: speed, danger, competition, design, and technology—and how the copious *and economic* burning of fossil energy contributes to these as appealing elements to a large amount of people. Does, therefore, the task of theorizing energy not require rethinking what constitutes and defines speed, force, power, competitiveness, etc.? And, following this, how automobility is socially organized, culturally generated, historically contextualized?

8. Emma Woodhouse's name takes on a different hue in a biomass attentive reading!

Works Cited

Berners-Lee, Mike, and Duncan Clark. *The Burning Question*. London: Profile Books, 2013.

Buell, Frederick. "A Short History of Oil Cultures: Or, the Marriage of Catastrophe and Exuberance." *Journal of American Studies* 46, no. 2 (2012): 273–93.

Calvino, Italo. "The Petrol Pump." *Numbers in the Dark and Other Stories*. Translated by Tim Parks. London: Vintage, 1996.

Cassedy Edward, and Peter Z. Grossman. *Introduction to Energy: Resources, Technology and Society*. Cambridge: Cambridge University Press, 1999.

Clark, Duncan. "The Rise and Rise of American Carbon." *Guardian*, August 5, 2013.

Hitchcock, Peter. "Oil in an American Imaginary." *New Formations* 69, no. 4 (2010): 81–97.

———. "Everything's Gone Green: The Environment of BP's Narrative." *Imaginations* 3, no. 2 (2012): 104–14.

Huber, Matthew. "Refined Politics: Petroleum Products, Neoliberalism, and the Ecology of Entrepreneurial Life." *Journal of American Studies* 46, no. 2 (2012): 295–312.

Kapuscinski, Ryszard. *Shah of Shahs*. San Diego: Harcourt Brace Jovanovich, 1985.

LeMenager, Stephanie. "The Aesthetics of Petroleum, after *Oil!*" *American Literary History* 24, no. 1 (2012): 59–86.

Logan, Owen. "Where Pathos Rules: The Resource Curse in Visual Culture." In *Flammable Societies: Studies on the Socio-economics of Oil and Gas*, ed. Owen Logan and John Andrew McNeish. London: Pluto, 2012.

McKibben, Bill. "Global Warming's Terrifying New Math." *Rolling Stone*, July 19, 2012. http://www.rollingstone.com/politics/news/global-warmings-terrifying-new-math -20120719.

Nabokov, Vladimir. *Lolita*. Harmondsworth: Penguin, 2004.

Nixon, Rob. *Slow Violence and the Environmentalism of the Poor*. Cambridge, MA: Harvard University Press, 2011.

Pendakis, Andrew, and Ursula Biemann. "This Is Not a Pipeline: Thoughts on the Politico-aesthetics of Oil." *Imaginations* 3, no. 2 (2012): 6–15.

Smil, Vaclav. *Energy in World History*. Boulder, CO: Westview, 1994.

———. "World History and Energy." In *Encyclopedia of Energy*, vol. 6, ed. J. Vutler Cleveland, 549–61. Amsterdam: Elsevier, 2004.

Stoekl, Allan. *Bataille's Peak: Energy, Religion and Postsustainability*. Minneapolis: University of Minnesota Press, 2007.

Szeman, Imre. "Editors' Column: Literature and Energy Futures." *PMLA* 126, no. 2 (2011): 323–26.

———. "Introduction to Focus: Petrofictions." *American Book Review* 33, no. 3 (March/April 2012): 3.

Szeman, Imre, and Maria Whiteman. "Oil Imag(e)inaries: Critical Realism and the Oil Sands." *Imaginations* 3, no. 2 (2012): 46–67.

Walonen, Michael T. "'The Black and Cruel Demon' and Its Transformations of Space: Towards a Comparative Study of the World Literature of Oil and Place." *Interdisciplinary Literary Studies* 14, no. 1 (2012): 56–78.

Wilson, Sheena, and Andrew Pendakis. "Sight, Site, Cite: Oil in the Field of Vision." *Imaginations* 3, no. 2 (2012): 1–5.

Yaeger, Patricia. "Editor's Column: Literature in the Ages of Wood, Tallow, Coal, Whale-Oil, Gasoline, Atomic Power and Other Energy Sources." *PMLA* 126, no. 2 (2011): 305–10.

Dear Climate: Post-energy Previews

Marina Zurkow, Una Chaudhuri, Fritz Ertl, and Oliver Kellhammer

Dear Climate hacks the aesthetics of public-service instructional posters and the consciousness techniques of mindfulness meditation to construct a semi-serious program of "inner climate change," inviting its viewers and listeners to try on some new approaches to the coming decades of climate chaos and post-energy realities.

Three "movements of mind"—a twelve-step program appropriately accelerated to match the swift pace of climate change—underlie an imagined process of ideological and psychic realignment between inner bewilderment and outer turbulence.

It starts with "meeting" climate change. With "deniers" a dwindling and discredited bunch, this step's getting easier by the minute. Look at the news, look out the window: you know it's arrived. Greet climate change the way you would the stranger who rings the doorbell. Yes? How may I help you? Listen to the answers. They're in the form of hurricanes, super storms, food shortages, species extinctions, floods, droughts, rising sea levels. Instead of watching them on TV and fretting about FEMA, extend them some good old-fashioned courtesy. Salute their strength, acknowledge their reach, inquire about their plans. Mention your own concerns, your grief and mourning, fear and trembling, whining and raging.

From the grist of that mill, advance to step two, and *befriend* climate change. Unleash the ambivalence, expectations, projections, recriminations, resentments, longings, and loyalties that make friendship the wild ride it is. Talk, talk, talk; then sigh, sulk, sneer. Veer wildly between exhausting exasperation and overwhelming affection. Feel what it means to know that something is in your life forever, for good or ill. And that this something has a mind of its own; it's going to do things its own way. Weather used to be the thing they predicted on TV. Your new friend is totally unpredictable, volatile, willful. Seasons used to march along in regular succession, doing their seasonal thing. Your new friend veers wildly from one extreme to another, skipping whole stages, rushing through the gentle transitions— early spring, late autumn—that you loved so much. Now you get sudden summers and abrupt winters. Adjust accordingly. Watch more carefully. React more readily.

Then on to step three, the merging: *become* climate change. Who really knows what that'll look like, but hang around to find out, and jump on any chances to close that yawning gap between yourself and the non-human world. See what parts of socio-spherical life can be adapted for a more atmospherical existence:

flirtation, sex, religion, death rituals, measurement, entertainment, cuisine, word-play, puns . . .

Dear Climate's "movements of mind" attempt to chart a few gentle courses to post-energy life. They look for the little spaces not yet saturated by energy guilt and energy-loss fear. The imagery they generate might come in handy for banners and flags in the fight against energy slavery, though less grandiose uses are more likely and most welcome. We actually prefer the abject, provocative, strange, and lonely: a windswept street, layers of peeling posters in an alleyway or a gallery, a graffiti-walled midnight wheat-paste operation, or air-dropped from drones into the playgrounds of the rich and famous. The posters can be printed from your computer, hung up in your workplace cafeteria, sheet-mulched into your neighborhood carbon-capturing food forest, or slipped into the magazine rack at a freeway filling station. The podcasts could be used for group meditations at weddings, picnics, concerts, sports events, raves, and retreats, or you might enjoy them while lying in your bathtub (empty or full). And a million other things, dreamt up by Dear You.

FOLLOW

THE
CARBON
FOOT
PRINTS.

TRACK
YOUR
CHANGES

REVISE
YOUR
ESTIMATES

ADJUST
YOUR
THERMOSTATS

info, downloads + instructions @ dearclimate.net

GRIEVE

THE
GRID.

DEAR CLIMATE #BF3   info, downloads + instructions @ dearclimate.net

MEET
THE
BEETLES.

GRADE YOUR GRUBS.

DEAR CLIMATE #BC16

info, downloads + instructions @ dearclimate.net

FIND A MOLE MODEL.

DEAR CLIMATE #BC21

info, downloads + instructions @ dearclimate.net

Works Cited

Abbey, Edward. *The Monkey Wrench Gang.* Salt Lake City: Dream Garden, 1985.

Abraham, Itty. *The Making of the Indian Atomic Bomb: Science, Secrecy and the Postcolonial State.* New York: St. Martin's, 1998.

Adams, Jonathan. "Estimates of Total Carbon Storage in Various Important Reservoirs." http://www.esd.ornl.gov/projects/qen/carbon2.html.

Adams, Richard Newbold. *Energy and Structure: A Theory of Social Power.* Austin: University of Texas Press, 1975.

AER Energy Information Administration. "World Net Electricity Generation by Type, Most Recent Annual Estimates, 2006." US Department of Energy. http://www.eia.doe.gov/emeu/international/RecentElectricityGenerationByType.xls.

Agamben, Giorgio. *Homo Sacer: Sovereign Power and Bare Life.* Translated by Daniel Heller-Raozen. Stanford: Stanford University Press, 1998.

———. *The Open: Man and Animal.* Translated by Kevin Attell. Stanford: Stanford University Press, 2004.

———. *State of Exception.* Translated by Kevin Attell. Chicago: University of Chicago Press, 2005.

Agarwal, Anil, and Narain Sunita. *Global Warming in an Unequal World: A Case of Environmental Colonialism.* Delhi: Centre for Science and Environment, 1991.

Agrawal, Arun. *Environmentality: Technologies of Government and the Making of Subjects.* Durham, NC: Duke University Press, 2005.

Aleklett, Kjell, and Colin. J. Campbell. "The Peak and Decline of World Oil and Gas Production." *Minerals and Energy* 18, no. 1 (2003): 5–20.

Alexander, Catherine, and Joshua O. Reno. "From Biopower to Energopolitics in England's Modern Waste Technology." *Anthropological Quarterly* 8, no. 2 (2014): 335–58.

Alianza Mexicana por la Autodeterminación de los Pueblos. "Comunicado de prensa: gobernador de Oaxaca incumplesuscompromisos y desconoce la realidad de lascomunidadesistmeñas." October 2012.

Al-Khafaji, Isam. *Tormented Births: Passages to Modernity in Europe and the Middle East.* London: I. B. Tauris, 2004.

Allison, Ian, et al. *The Copenhagen Diagnosis: Updating the World on the Latest Climate Science.* Sydney: University of New South Wales Climate Change Research Center, 2009.

Altvater, Elmar. *The Poverty of Nations: A Guide to the Debt Crisis from Argentina to Zaire.* Translated by Terry Bond. London: Zed Books, 1991.

Ambrosino, Georges. "La Machine Savante et la vie: Norbert Wiener, *Cybernetics.*" *Critique* 41 (1950): 70–82.

Anderson, Perry. "Scurrying toward Bethlehem." *New Left Review* 10 (2001): 5–31.

Anscombe, E. M. Gertrude. "Modern Moral Philosophy." *Philosophy* 33 (1958): 1–19.

Appadurai, Arjun. *Modernity at Large: Cultural Dimensions of Globalization*. Minneapolis: University of Minnesota Press, 1996.

Apter, Andrew. *The Pan-African Nation: Oil and the Spectacle of Culture in Nigeria*. Chicago: University of Chicago, 2005.

Archer, David. *Global Warming: Understanding the Forecast*. Maiden, MA: Blackwell, 2007.

Arendt, Hannah. *The Origins of Totalitarianism*. New York: Harcourt, 1968.

Aribisala, Karen King. *Kicking Tongues*. Oxford: Heinemann, 1998.

Arrighi, Giovanni. *Adam Smith in Beijing: Lineages of the Twenty-First Century*. London: Verso, 2007.

———. *The Long Twentieth Century: Money, Power, and the Origins of Our Times*. London: Verso, 2006.

Arrow, J. Kenneth, and Gerard Debreu. "Existence of an Equilibrium for a Competitive Economy." *Econometrica* 22, no. 3 (1954): 265–90.

Ashforth, Adam. *Witchcraft, Violence and Democracy in South Africa*. Chicago: University of Chicago Press, 2005.

Asiedu, Elizabeth. *Foreign Direct Investment in Africa*. Helsinki: United Nations University, 2005.

Astill, James, and Paul Brown. "Carbon Dioxide Levels Will Double by 2050, Experts Forecast." *Guardian*, April 5, 2001. http://www.guardian.co.uk/environment/2001/apr/06/usnews.globalwarming.

Attanasi, D. Emil, and Richard F. Meyer. "Natural Bitumen and Extra-Heavy Oil." In *Survey of Energy Resources*, 22nd ed., 123–40. London: World Energy Council, 2010.

Aweto, Albert. "Outline Geography of Urhoboland." N.p., January 2002.

Azar, Christian. "Bury the Chains and the Carbon Dioxide." *Climatic Change* 85 (2007): 473–75.

Badiou, Alain. *Being and Event*. Translated by Oliver Feltham. London: Continuum, 2007.

Bailey, Conner, Peter R. Sinclair, and Mark R. Dubois. "Future Forests: Forecasting Social and Ecological Consequences of Genetic Engineering." *Society and Natural Resources* 17 (2004): 642–45.

Bamberg, James. *The History of British Petroleum*. Vol. 3, *British Petroleum and Global Oil, 1950–1975: The Challenge of Nationalism*. Cambridge: Cambridge University Press, 2000.

Barabas, M. Alicia, and Miguel A. Bartolomé. *Hydraulic Development and Ethnocide: The Mazatecand Chinantec People of Oaxaca, Mexico*. Mexico City: International Work Group for Indigenous Affairs, 1973.

Barcia, Immaculada, and Analía Penchaszadeh. "Ten Insights to Strengthen Responses for Women Human Rights Defenders at Risk." Association for Women's Rights in Development, 2012. http://www.awid.org/sites/default/files/atoms/files/ten_insights_to_strengthen_responses_for_women_human_rights_defenders_at_risk.pdf.

Bardi, Ugo. "Peak Oil's Ancestor: The Peak of British Coal Production in the 1920s." *Newsletter of the Association for the Study of Peak Oil and Gas* 73 (2007): 5–7.

Barker, Holly. "Fighting Back: Justice, the Marshall Islands and Neglected Radiation Communities." In *Life and Death Matters: Human Rights and the Environment at the End of the Millennium*, edited by Barbara Rose Johnston, 290–306. London: Alta Mira, 1997.

Barnes, Sandra. "Global Flows: Terror, Oil and Strategic Philanthropy." *African Studies Review* 48, no. 1 (2005): 1–23.

Barrera, Jorge. "PM Harper Believes Idle No More Movement Creating 'Negative Public Reaction,' Say Confidential Notes." *APTN National Notes*, January 25, 2013. http://aptn .ca/news/2013/01/25/pm-harper-believes-idle-no-more-movement-creating-negative -public-reaction-say-confidential-notes/.

Barrionuevo, Alexei. "Bush Says Lower Oil Prices Won't Blunt New-Fuel Push." *New York Times*, October 13, 2006. http://select.nytimes.com/gst/abstract.html?res=F10910FE3 A540C708DDDA90994DE404482.

Barry, Andrew. *Interdisciplinarity: Reconfigurations of the Social and Natural Sciences.* Edited by Andrew Barry and Georgina Born. London: Routledge, 2013.

———. "Technological Zones." *European Journal of Social Theory* 9, no. 2 (2006): 239–53.

Bataille, Georges. *The Accursed Share: An Essay on General Economy.* Vol. 1, *Consumption.* Translated by Robert Hurley. New York: Zone, 1988.

———. *Oeuvres Complètes.* Vols. 1–12. Paris: Gallimard, 1970.

Bayly, A. Chris. *The Birth of the Modern World, 1780–1914: Global Connections and Comparisons.* Maiden, MA: Blackwell, 2004.

Beaudreau, C. Bernard. *Energy and the Rise and Fall of Political Economy.* Westport, CT: Greenwood, 1999.

Beblawi, Hazem, and Giacomo Luciani. *The Rentier State.* New York: Croom Helm, 1987.

Beer, Gillian. *Darwin's Plots: Evolutionary Narrative in Darwin, George Eliot, and Nineteenth-Century Fiction.* Cambridge: Cambridge University Press, 2000.

Behrends, Andrea, Stephen P. Reyna, and Günther Schlee, eds. *Crude Domination: The Anthropology of Oil.* New York: Berghahn, 2011.

Beinin, Joel, and Zachary Lockman. *Workers on the Nile: Nationalism, Communism, Islam, and the Egyptian Working Class, 1882–1954.* Princeton, NJ: Princeton University Press, 1987.

Bellamy, Brent Ryan, Stephanie LeMenager, and Imre Szeman. "When Energy Is the Focus: Aesthetics, Politics, and Pedagogy: A Conversation." *Postmodern Culture* (forthcoming).

Benjamin, Walter. "Theses on the Philosophy of History." In *Illuminations*, edited by Hannah Arendt, translated by Harry Zohn, 253–64. New York: Schocken, 1969.

Bennett, Jane. *Vibrant Matter: A Political Ecology of Things.* Durham, NC: Duke University Press, 2010.

Benson, Todd. "Africa's Food and Nutrition Security Situation." IFPRI Discussion Paper 37. Washington, DC, 2004.

Berlant, Lauren. "Trauma and Ineloquence." *Cultural Values* 5 (2001): 44.

Berners-Lee, Mike, and Duncan Clark. *The Burning Question.* London: Profile Books, 2013.

Bernstein, Henry. "Considering Africa's Agrarian Questions." *Historical Materialism* 12, no. 4 (2004): 115–44.

Berry, Brian J. L., Edgar C. Conkling, and Michael Ray. *The Global Economy: Resource Use, Locational Choice, and International Trade.* Englewood Cliffs, NJ: Prentice Hall, 1993.

Bishop, Elizabeth. "The Moose." In *The Complete Poems, 1927–1979*, 169–73. New York: Farrar, 1990.

Black, Edwin. *Internal Combustion: How Corporations and Governments Addicted the World to Oil and Derailed the Alternatives*. New York: St. Martin's, 2006.

Blair, M. John. *The Control of Oil*. New York: Pantheon Books, 1976.

Blake, William. "The Divine Image." In *Complete Poetry and Prose of William Blake*, edited by D. V. Erdman, 12–13. New York: Doubleday, 1982.

Blewett, Neal. "The Franchise in the United Kingdom 1885–1918." *Past and Present* 32 (1965): 27–56.

Block, Fred. *The Origins of International Economic Disorder: A Study of United States International Monetary Policy from World War II to the Present*. Berkeley: University of California Press, 1977.

Bloom, Dan. "Can 'Cli-Fi' Help Keep Our Planet Livable?" *Medium*, July 24, 2015. https://medium.com/@clificentral/can-cli-fi-help-keep-our-planet-livable-8b053bd4aa35#.alu099x6d.

Blundell, John. *IEA Turns 50: Celebrating Fisher Meeting Hayek*. Atlas Investor Report, 2005.

Boal, A. Iain, T.J. Clark, Joseph Matthews, and Michael Watts. "Blood for Oil." *Afflicted Powers: Capital and Spectacle in a New Age of War*. New York: Verso, 2005.

Bodansky, Daniel. "May We Engineer the Climate?" *Climatic Change* 33 (1996): 301–21.

Boden, A. Tom, Gregg Marland, and Robert J. Andres. *Global, Regional, and National Fossil-Fuel CO_2 Emissions*. Oak Ridge, TN: Oak Ridge National Laboratory, 2011.

Bogost, Ian. *Alien Phenomenology, or What It's Like to Be a Thing*. Minneapolis: University of Minnesota Press, 2012.

Bois, Y. Alain. *Formless: A User's Guide*. New York: Zone, 1997.

Bolin, Bert. *A History of the Science and Politics of Climate Change: The Role of the Intergovernmental Panel on Climate Change*. Cambridge: Cambridge University Press, 2008.

Borstelmann, Thomas. *Apartheid's Reluctant Uncle: The United States and Southern Africa in the Early Cold War*. New York: Oxford University Press, 1993.

Bowden, Gary. "The Social Construction of Validity in Estimates of U.S. Crude Oil Reserves." *Social Studies of Science* 15, no. 2 (1985): 207–40.

Bowen, Mark. *Censoring Science: Inside the Political Attack on Dr. James Hansen and the Truth of Global Warming*. New York: Dutton, 2008.

Bows Alice, and Anderson Kevin. "Contraction and Convergence: An Assessment of the CCOptions Model." *Climatic Change* 91 (2008): 275–90.

Boyden, Stephen. *Western Civilization in Biological Perspective*. Oxford: Clarendon, 1987.

Boyer, Dominic. "Energopower: An Introduction." *Anthropology Quarterly* 87, no. 2 (2014): 309–33.

———. "On the Ethics and Practice of Contemporary Social Theory: From Crisis Talk to Multiattentional Method." *Dialectical Anthropology* 34, no. 3 (2010): 305–24.

Boyer, Dominic, et al. *The Promise of Infrastructure*. Durham, NC: Duke University Press, Forthcoming.

Boykoff, T. Maxwell. *Who Speaks for the Climate? Making Sense of Media Reporting on Climate Change*. Cambridge: Cambridge University Press, 2011.

Braudel, Fernand. *The Mediterranean and the Mediterranean World in the Age of Philip II*. Translated by Siân Reynolds. London: Collins, 1949.

Briggs, L. Charles, and Mark Nichter. "Biocommunicability and the Biopolitics of Pandemic Threats." *Medical Anthropology* 28, no. 3 (2009): 189–98.

British Petroleum. *BP Statistical Review of World Energy 2007*. 2007.

Bromley, Simon. *American Hegemony and World Oil*. University Park: Pennsylvania State University Press, 1991.

———. "The United States and the Control of World Oil." *Government and Opposition* 40, no. 2 (2005): 225–55.

Bryant, R. Levi. *The Democracy of Objects*. Ann Arbor: Open Humanities, 2011.

Brysse, Kenyn. "Climate Change Prediction: Erring on the Side of Least Drama?" *Global Environmental Change* 23 (2013): 327–37.

Buell, Frederick. "A Short History of Oil Cultures: Or, the Marriage of Catastrophe and Exuberance." *Journal of American Studies* 46, no. 2 (2012): 273–93.

Buell, Lawrence. "Toxic Discourse." *Critical Inquiry* 24 (1998): 639–65.

Bullard, D. Robert. *Dumping in Dixie: Race, Class, and Environmental Quality*. Boulder, CO: Westview, 1990.

Burke, Peter. *The French Historical Revolution: The "Annales" School, 1929–89*. Stanford: Stanford University Press, 1990.

Butler, Judith. *Gender Trouble:* London: Routledge, 1990.

Cabinet Office. *The Torrey Canyon*. London: HMSO, 1967.

Cafaro, Philip. "Thoreau, Leopold, and Carson: Toward an Environmental Virtue Ethics." In *Environmental Virtue Ethics*, edited by Ronald L. Sandler and Philip Cafaro, 31–44. Oxford: Rowman & Littlefield, 2005.

Callon, Michel. "Some Elements of a Sociology of Translation: Domestication of the Scallops and the Fishermen of St. Brieuc Bay." In *Power, Action and Belief: A New Sociology of Knowledge*, edited by John Law, 196–233. London: Routledge, 1986.

Calvin, H. William. *Fever: How to Treat Climate Change*. Chicago: University of Chicago Press, 2008.

Calvino, Italo. "The Petrol Pump." In *Numbers in the Dark and Other Stories*, translated by Tim Parks, 170–75. London: Vintage, 1996.

Campbell, Howard, Leigh Binford, Miguel Bartolomé, and Alicia Barabas, eds. *Zapotec Struggles: Histories, Politics, and Representations from Juchitán, Oaxaca*. Washington, DC: Smithsonian Institution Press, 1993.

Campion-Smith, Bruce. "Idle No More: Spence Urged by Fellow Chiefs to Abandon Her Fast." *Toronto Star*, January 18, 2013. https://www.thestar.com/news/canada/2013/01/18/idle_no_more_spence_urged_by_fellow_chiefs_to_abandon_her_fast.html.

Canning, Kathleen. *Languages of Labor and Gender: Female Factory Work in Germany, 1850–1914*. Ithaca, NY: Cornell University Press, 1996.

Caro, A. Robert. *The Power Broker: Robert Moses and the Fall of New York*. New York: Vintage Books, 1975.

Cassedy, Edward, and Peter Z. Grossman. *Introduction to Energy: Resources, Technology and Society*. Cambridge: Cambridge University Press, 1999.

Castellanos, Bianet. "Don Teo's Expulsion: Property Regimes, Moral Economies, and Ejido Reform." *Journal of Latin American and Caribbean Anthropology* 15, no. 1 (2010): 144–69.

Cawte, Alice. *Atomic Australia, 1944–1990.* Sydney: New South Wales, 2001.

Centre for Strategic and International Studies. "Briefing on the Niger Delta." March 14, 2007. Washington, DC.

Chaca, Roselia. "Presenta Mareña Renovable proyecto eólico en Juchitán." *Noticias: Voz e Imagen,* December 7, 2012. http://www.noticiasnet.mx/portal/oaxaca/general/grupos vulnerables/128457-presenta-marena-renovable-proyecto-eolico-juchitan.

Chakrabarty, Dipesh. "The Climate of History: Four Theses." *Critical Inquiry* 35 (2009): 197–222.

Chernow, Ron. *Titan: The Life of John D. Rockefeller, Sr.* New York: Random House, 1998.

Childe, Gordon. *Man Makes Himself.* London: Watts, 1941.

Choy, Timothy. *Ecologies of Comparison: An Ethnography of Endangerment.* Durham, NC: Duke University Press, 2011.

Church, A. Roy, Quentin Outram, and David. N. Smith. "The Militancy of British Miners, 1893–1986: Interdisciplinary Problems and Perspectives." *Journal of Interdisciplinary History* 22, no. 1 (1991): 49–66.

Cipolla, M. Carlo. *The Economic History of World Population.* Harmondsworth: Penguin, 1978.

Cirincione, Joseph. "Niger Uranium: Still a False Claim." *Carnegie Proliferation Brief* 7, no. 12.

Citino, J. Nathan. "Defending the 'Postwar Petroleum Order': The US, Britain and the 1954 Saudi-Onassis Tanker Deal." *Diplomacy and Statecraft* 11, no. 2 (2000): 137–60.

———. "The Rise of Consumer Society: Postwar American Oil Policies and the Modernization of the Middle East." Paper presented at the 14th International Economic History Congress, Helsinki, 2006.

Clark, Duncan. "The Rise and Rise of American Carbon." *Guardian,* August 5, 2013. https://www.theguardian.com/environment/2013/aug/05/us-emissions-extraction-fracking.

Clark, Nigel. "Aboriginal Cosmopolitanism." *International Journal of Urban and Regional Research* 32, no. 3 (2008): 737–44.

Clark, Pilita. "EU Emissions Trading Faces Crisis." *Financial Times,* January 21, 2013. http://www.ft.com/cms/s/0/42e719c0-63f0-11e2-84d8-00144feab49a.html?ftcamp=published _links%2Frss%2Fworld_europe%2Ffeed%2F%2Fproduct#axzz2uRWfZhwt.

Clark, R. William. *Petrodollar Warfare: Oil, Iraq, and the Future of the Dollar.* Gabriola Island: New Society, 2005.

Clark, Tim. "Derangements of Scale." In *Telemorphosis: Theory in the Era of Climate Change,* vol. 1, edited by Tom Cohen, 148–66. Ann Arbor: University of Michigan Press, 2012.

Clarke, Bruce. *Energy Forms: Allegory and Science in the Era of Classical Thermodynamics.* Ann Arbor: University of Michigan Press, 2001.

Clayton, Susan, and Susan Opotow, eds. *Identity and the Natural Environment: The Psychological Significance of Nature.* Cambridge, MA: MIT Press, 2003.

Coast 2050: Toward a Sustainable Coastal Louisiana. Baton Rouge: Louisiana Department of Natural Resources, 1998.

Cohen, H. Jeffrey. *Cooperation and Community: Economy and Society in Oaxaca.* Austin: University of Texas Press, 1999.

Cohen, Tom. Introduction to *Telemorphosis: Theory in the Era of Climate Change,* vol. 1, 13–42. Ann Arbor: University of Michigan Press, 2012.

Cohen-Joppa, Jack. "Disinformation about Depleted Uranium." *High Beam Research*, November 1, 2004. https://www.highbeam.com/doc/1P3-739697791.html.

Colebrook, Claire. "Not Symbiosis, Not Now: Why Anthropogenic Climate Change Is Not Really Human." *Oxford Literary Review* 34, no. 2 (2012): 198–99.

Collier, Paul, V. L. Elliott, Havard Hegre, Anke Hoeffler, Reynal, Marta Querol, and Nicholas Sambanis. "Breaking the Conflict Trap, Civil War and Development Policy." *World Bank Policy Research Report*. New York, 2003.

Collingwood, G. Robin. *The Idea of History*. New York: Oxford University Press, 1976.

Collins, Sarah, and Tom Kenworthy. *Energy Industry Fights Chemical Disclosure*. Center for American Progress, 2010.

Comaroff, Jean, and John Comaroff. "Ethnography on an Awkward Scale: Postcolonial Anthropology and the Violence of Abstraction." *Ethnography* 4, no. 2 (2003): 147–79.

Connerton, Paul. *How Societies Remember*. Cambridge: Cambridge University Press, 1989.

Connolley, M. William. "Ice Extent in Million Square Kilometers." *Wikimedia Commons*, December 28, 2015. https://commons.wikimedia.org/wiki/File:Seaice-1870-part-2009.png.

Conrad, Peter. *The Art of the City*. New York: Oxford University Press, 1984.

Cooper, Brenda. *Magical Realism in West African Fiction: Seeing with a Third Eye*. New York: Routledge, 1998.

Corbin, David. *Life, Work, and Rebellion in the Coal Fields: The Southern West Virginia Miners, 1880–1922*. Champaign: University of Illinois Press, 1981.

Coronil, Fernando. *The Magical State: Nature, Money, and Modernity in Venezuela*. Chicago: University of Chicago Press, 1997.

Costanza, Robert, et al. "The Value of the World's Ecosystem Services and Natural Capital." *Nature* 387 (1997): 253–60.

Council on Foreign Relations. "More Than Humanitarianism." Task Force Report 56. New York, 2005.

Courson, Elias. "The Burden of Oil: Social Deprivation and Political Militancy in Gbaramatu Clan, Warri Southwest LGA, Delta State, Nigeria." *Niger Delta: Economies of Violence Project*. Berkeley, 2007. http://globetrotter.berkeley.edu/NigerDelta/.

Crate, Susan, and Mark Nuttall, eds. *Anthropology and Climate Change*. Walnut Creek, CA: Left Coast, 2009.

Croce, Benedetto. *The Philosophy of Giambattista Vico*. Translated by R. G. Collingwood. New Brunswick, NJ: H. Latimer, 1913.

Crosby, Alfred W. *The Columbian Exchange: Biological and Cultural Consequences of 1492*. London: Praeger, 2003.

———. "The Past and Present of Environmental History." *American Historical Review* 100, no. 4 (1995): 1177–89.

Crutzen, Paul J. "Albedo Enhancement by Stratospheric Sulfur Injections: A Contribution to Resolve a Policy Dilemma?" *Climatic Change* 77 (2006): 211–19.

———. "Geology of Mankind." *Nature* 415 (2002): 23.

Crutzen, Paul J., and Eugene F. Stroemer. "The Anthropocene." *IGBP [International Geosphere-Biosphere Programme] Newsletter* 41 (2000): 17.

Daly, Herman E. *Steady-State Economics*. 2nd ed. Washington, DC: Island, 1991.

Darier, Eric. "Foucault and the Environment: An Introduction." In *Discourses of the Environment*, edited by Eric Darier, 1–34. Malden, MA: Blackwell, 1999.

Darley, Julian. *High Noon for Natural Gas*. White River Junction, VT: Chelsea Green, 2004.

Daston, Lorraine, and Peter L. Galison. *Objectivity*. Cambridge, MA: Zone Books, 2007.

Davidson, Mark D. "Parallels in Reactionary Argumentation in the US Congressional Debates on the Abolition of Slavery and the Kyoto Protocol." *Climatic Change* 86 (2008): 67–82.

Davis, David B. *Inhuman Bondage: The Rise and Fall of Slavery in the New World*. Oxford: Oxford University Press, 2006.

Davis, Mike. "Living on the Ice Shelf: Humanity's Meltdown." *Tom Dispatch*, June 26, 2008. tomdispatch.com/post/174949.

———. "Los Angeles after the Storm: The Dialectic of Ordinary Disaster." *Antipode* 27 (1995): 221–41.

———. *Planet of Slums*. London: Verso, 2006.

———. "Who Will Build the Ark?" *New Left Review* 61 (2010): 29–46.

Davis, Tracy C. *Stages of Emergency: Cold War Nuclear Civil Defense*. Durham, NC: Duke University Press, 2007.

Dawson, Susan E. "Navajo Uranium Mining Workers and the Effects of Occupational Illnesses: A Case Study." *Human Organization* 51, no. 4 (1992): 389–97.

Debeir, Jean Claude, Jean Paul Deléage, and Daniel Hémery. *In the Servitude of Power: Energy and Civilization through the Ages*. Translated by John Barzman. London: Zed Books, 1991.

Debord, Guy. *Society of the Spectacle*. Translated by Ken Knabb. London: Aldgate, 2006.

Debray, Régis. *Tousazimuts*. Paris: Odile Jacob, 1990.

Deffeyes, Kenneth. *Beyond Oil: The View from Hubbert's Peak*. New York: Hill & Wang, 2005.

de Landa, Manuel. *Intensive Science and Virtual Philosophy*. New York: Continuum, 2002.

de Lauretis, Teresa. *Technologies of Gender: Essays on Theory, Film, and Fiction*. Bloomington: Indiana University Press, 1987.

Deleuze, Gilles. *Difference and Repetition*. Translated by Paul Patton. New York: Columbia University Press, 1994.

Dell'Amore, Christine. "Thousand Walruses Gather on Island as Sea Ice Shrinks." *National Geographic*, October 2, 2013. http://news.nationalgeographic.com/news/2013/10/131002 -walruses-arctic-haulout-science-animals-alaska-global-warming/.

Dennett, Daniel C. *Darwin's Dangerous Idea: Evolution and the Meanings of Life*. New York: Simon & Schuster, 1995.

Dennis, Michael A. "Drilling for Dollars: The Making of US Petroleum Reserve Estimates, 1921–25." *Social Studies of Science* 15, no. 2 (1985): 241–65.

Denord, Francois. "Aux Origines du Néolibéralisme en France: Louis Rougier et le Colloque Walter Lippmann de 1938." *Le Mouvement Social* 195 (2001): 9–34.

Derrida, Jacques. *Acts of Religion*. Edited by Gil Anidjar. New York: Routledge, 2002.

———. "Cogito and the History of Madness." In *Writing and Difference*, translated by Alan Bass, 36–75. Chicago: University of Chicago Press, 1978.

———. "From Restricted to General Economy: A Hegelianism without Reserve." In *Writing and Difference*, translated and edited by Allan Bass, 251–79. Chicago: University of Chicago Press, 1978.

Derrida, Jacques, and Elizabeth Roudinescou. *For What Tomorrow: A Dialogue*. Translated by Jeff Fort. Stanford: Stanford University Press, 2004.

Diamond, Jared. *Collapse: How Societies Chose to Fail or Succeed*. New York: Viking, 2005.

———. *Guns, Germs and Steel*. New York: W. W. Norton, 1999.

Dix, Keith. *What's a Coal Miner to Do? The Mechanization of Coal Mining*. Pittsburgh: University of Pittsburgh Press, 1988.

Dodds, Walter K. *Humanity's Footprint: Momentum, Impact, and Our Global Environment*. New York: Columbia University Press, 2008.

Douthwaite, Richard. "Sharing Out the Rations." *Irish Times*, January 13, 2007.

Dower, John. *War without Mercy: Race and Power in the Pacific War*. New York: Pantheon, 1986.

Dukes, S. Jeffrey. "Burning Buried Sunshine: Human Consumption of Ancient Solar Energy." *Climatic Change* 61, nos. 1–2 (2003): 33–41.

Dworkin, Ronald. *Life's Dominion: An Argument about Abortion, Euthanasia, and Individual Freedom*. New York: Vintage Books, 1993.

Easterly, William. *The White Man's Burden*. London: Penguin, 2006.

Eddington, Arthur. *The Nature of the Physical World*. New York: Macmillan, 1928.

Edwards, Paul K. *Strikes in the United States, 1881–1974*. New York: St. Martin's, 1981.

Edwards, Paul N. *The Closed World: Computers and the Politics of Discourse in Cold War America*. Cambridge, MA: MIT Press, 1996.

———. *A Vast Machine: Computer Models, Climate Data and the Politics of Global Warming*. Cambridge, MA: MIT Press, 2010.

Eley, Geoff. *Forging Democracy: The History of the Left in Europe, 1850–2000*. New York: Oxford University Press, 2002.

Elgin, Duane. *Voluntary Simplicity: Toward a Way of Life That Is Outwardly Simple, Inwardly Rich*. New York: Harper, 2010.

Elkins, Stanley. *Slavery: A Problem in American Institutional and Intellectual Life*. Chicago: University of Chicago Press, 1959.

Emerson, Ralph W. *Essays, First Series*. Boston: Ticknor & Fields, 1863.

Engels, Friedrich. "The Bakunists at Work." In *Revolution in Spain*, edited by Karl Marx and Friedrich Engels, 211–36. London: Lawrence & Wishart, 1939.

Ermarth, Michael. *Wilhelm Dilthey: The Critique of Historical Reason*. Chicago: University of Chicago Press, 1978.

Fabri, Ralph. *Painting Cityscapes*. New York: Watson Guptill, 1975.

Fagge, Roger. *Power, Culture, and Conflict in the Coalfields: West Virginia and South Wales, 1900–1922*. Manchester: Manchester University Press, 1996.

Fahey, Dan. "The Emergence and Decline of the Debate over Depleted Uranium Munitions, 1991–2004." *World Information Service on Energy*, June 20, 2004. http://www.wise-uranium.org/pdf/duemdec.pdf.

Faris, Wendy B. *Ordinary Enchantments: Magical Realism and the Remystification of Narrative*. Nashville: Vanderbilt University Press, 2004.

Fassin, Didier. "The Biopolitics of Otherness: Undocumented Immigrants and Racial Discrimination in the French Public Debate." *Anthropology Today* 17, no. 1 (2001): 3–7.

Ferguson, James. *Global Shadows: Africa in the Neoliberal World Order.* Durham, NC: Duke University Press 2006.

Ferry, Elizabeth Emma, and Mandana E. Limbert, eds. *Timely Assets: The Politics of Resources and Their Temporalities.* Santa Fe: SAR Press, 2008.

Fischer, David. *History of the International Atomic Energy Agency: The First Forty Years.* International Atomic Energy Association, 1997.

Fischett, Mark. "Drowning New Orleans." *Scientific American* 285, no. 4 (2001): 76–85.

Flannery, Tim. *The Weather Makers: The History and Future Impact of Climate Change.* Melbourne: Text Publishing, 2005.

Fleming, Ian. *Goldfinger.* London: Jonathan Cape, 1959.

Fogel, Cathleen. "The Local, the Global, and the Kyoto Protocol." In *Earthly Politics: Local and Global in Environmental Governance,* edited by Sheila Jasanoff and Marybeth Long Martello, 103–25. Cambridge, MA: MIT Press, 2004.

Foot, Philippa. "Morality as a System of Hypothetical Imperatives." *Philosophical Review* 81, no. 3 (1972): 305–16.

Forrestal, James. *The Forrestal Diaries.* Edited by Walter Millis and E. S. Duffield. New York: Viking, 1951.

———. *Papers.* Princeton, NJ: Public Policy Papers Collection, Seeley G. Mudd Manuscript Library, 1941–49.

Foucault, Michel. *Discipline and Punish: The Birth of the Prison.* Translated by Alan Sheridan. New York: Vintage, 1995.

———. *The History of Sexuality.* Vol. 1, *The Will to Knowledge.* London: Penguin, 1978.

———. "'Omneset Singulatim': Toward a Critique of Political Reason." In *Power (The Essential Works of Michel Foucault, 1954–1984, Vol. 3),* edited by James D. Faubion, translated by Robert Hurley, 317–18. New York: New Press, 2000.

———. *Order of Things: An Archaeology of Human Knowledge.* New York: Vintage Books, 1966.

———. "The Risks of Security." In *Power (The Essential Works of Michel Foucault, 1954–1984, Vol. 3),* edited by James Faubion, translated by Robert Hurley, 365–81. New York: New, 2000.

———. *Society Must Be Defended: Lectures at the College De France, 1975–76.* Edited by Mauro Bertani and Alessandro Fontana. Translated by David Macey. New York: Picador, 2003.

Frank, Alison F. *Oil Empire: Visions of Prosperity in Austrian Galicia.* Cambridge, MA: Harvard University Press, 2007.

Friedberg, Susanne. *French Beans and Food Scares.* Oxford: Oxford University Press, 2004.

Frynas, Jędrzej G. "The Oil Boom in Equatorial Guinea." *African Affairs* 103 (2004): 527–46.

Fuller, R. Buckminster. *Utopia or Oblivion: The Prospects for Humanity.* London: Allen Lane, 1969.

Gadamer, Hans Georg. *Truth and Method.* Translated by Joel Weinsheimer and Donald G. Marshall. London: Sheed & Ward, 1975.

Gare, Arran E. *Postmodernism and the Environmental Crisis.* London: Routledge, 1995.

Gary, Ian, and Terry L. Karl. *Bottom of the Barrel: Africa's Oil Boom and the Poor*. Baltimore: Catholic Relief Services, 2003.

Gasset, Jose Ortega Y. *Phenomenology and Art*. Translated by Philip W. Silver. New York: Norton, 1975.

Gatens, Moira. "A Critique of the Sex/Gender Distinction." In *A Reader in Feminist Knowledge*, edited by Sneja Gunew, 139–59. London: Routledge, 1991.

Geyer, Michael. Charles Bright. "World History in a Global Age." *American Historical Review* 100, no. 4 (1995): 1034–60.

Ghamari-Tabrizi, Sharon. *The Worlds of Herman Kahn: The Intuitive Science of Thermonuclear War*. Cambridge, MA: Harvard University Press, 2005.

Ghosh, Amitav. "Petrofiction." *New Republic* 206 (1992): 29–34.

Gibbon, Peter, and Stefano Ponte. *Trading Down*. Philadelphia: Temple University Press, 2005.

Giddens, Anthony. *The Politics of Climate Change*. Cambridge: Polity, 2009.

Gilbert, Scott F., Jan Sapp, and Alfred I. Tauber. "A Symbiotic View of Life: We Have Never Been Individuals." *Quarterly Review of Biology* 87, no. 4 (2012): 325–41.

Gilding, Paul. *The Great Disruption: Why the Climate Crisis Will Bring On the End of Shopping and the Birth of a New World*. New York: Bloomsbury, 2010.

Gillis, Justin. "Poll, Many Link Weather Extremes to Climate Change." *New York Times*, April 17, 2012. http://www.nytimes.com/2012/04/18/science/earth/americans-link-global-warming-to-extreme-weather-poll-says.html.

Gledhill, John. *Neoliberalism, Transnationalization and Rural Poverty: A Case Study of Michoacán, Mexico*. Boulder, CO: Westview, 1995.

Goldberg, Ellis, Erik Wibbels, and Eric Mvukiyehe. "Lessons from Strange Cases: Democracy, Development, and the Resource Curse in the U.S. States." *Comparative Political Studies* 41, nos. 4–5 (2008): 477–514.

Goodrich, Carter. *The Miner's Freedom: A Study of the Working Life in a Changing Industry*. Boston: Marshall Jones, 1925.

Goodrich, Lloyd. *Edward Hopper*. New York: Harry Abrams, 1983.

Goodstein, David. *Out of Gas: The End of the Age of Oil*. New York: W. W. Norton, 2005.

Greenhalgh, Susan, and Edwin A. Winckler. *Governing China's Population: From Leninist to Neoliberal Biopolitics*. Stanford: Stanford University Press, 2005.

Greffrath, Mathias. "Freizeit, Die Siemeinen [Freedom they mean]." *Süddeutsche Zeitung*, June 1998.

Grescoe, Taras. "The Dirty Truth about 'Clean Diesel.' " *New York Times*, January 3, 2016. http://www.nytimes.com/2016/01/03/opinion/sunday/the-dirty-truth-about-clean-diesel.html.

Griswold, Wendy. *Bearing Witness: Readers, Writers, and the Novel in Nigeria*. Princeton, NJ: Princeton University Press, 2000.

Guha, Ramachandra, and Juan Martinez-Alier. *Varieties of Environmentalism: Essays North and South*. London: Earthscan, 1997.

Günel, Gökçe. "Ergos: A New Energy Currency." *Anthropological Quarterly* 87, no. 2 (2014): 359–79.

Gupta, Akhil. "An Anthropology of Electricity from the Global South." *Cultural Anthropology* 30, no. 4 (2015): 555–68.

Gusterson, Hugh. *Nuclear Rites: A Weapons Laboratory at the End of the Cold War*. Berkeley: University of California Press, 1998.

Haberl, Helmut. "The Global Socioeconomic Energetic Metabolism as a Sustainability Problem." *Energy* 31, no. 1 (2006): 87–99.

Hamilton, Clive. *Requiem for a Species: Why We Resist the Truth about Climate Change*. Sydney: Allen & Unwin, 2010.

Hansen, James. "Climate Catastrophe." *New Scientist* 195 (2007): 30–34.

———. "Climate Change and Trace Gases." *Philosophical Transactions of the Royal Society* 365 (2007): 1925–54.

———. "Dangerous Human-Made Interference with Climate: A CISS ModelE Study." *Atmospheric Chemistry and Physics* 7, no. 9 (2007): 2287–312.

Hansen, James, Makiko Sato, Pushker Kharecha, David Beerling, Robert Berner, Valeri Masson-Delmotte, Mark Pagani, Maureen Raymo, Dana L. Royer, and James C. Zachos. "Target Atmospheric CO_2: Where Should Humanity Aim?" *Open Atmospheric Science Journal* 2 (2008): 217–31.

Hansen, James, Makiko Sato, Pushker Kharecha, Gary Russell, David W. Lea, and Mark Siddall. "Climate Change and Trace Gases." *Philosophical Transactions of the Royal Society* 365 (2007): 1925–54.

Haraway, Donna. "Anthropocene, Capitalocene, Chthulucene: Making Kin." *Environmental Humanities* 6 (2015): 159–65.

———. "A Cyborg Manifesto: Science, Technology, and Socialist-Feminism in the Late Twentieth Century." In *Simians, Cyborgs, and Women*, edited by Donna Haraway, 149–81. London: Free Association Books, 1991.

Harman, Graham. *Guerrilla Metaphysics: Phenomenology and the Carpentry of Things*. Chicago: Open Court, 2005.

———. *The Quadruple Object*. Alresford, UK: Zero Books, 2011.

———. *Prince of Networks: Bruno Latour and Metaphysics*. Melbourne: re.press, 2009.

———. *Tool-Being: Heidegger and the Metaphysics of Objects*. Chicago: Open Court, 2002.

Hart, Gillian. *Disabling Globalization*. Berkeley: University of California Press, 2003.

Harvey, David. *The New Imperialism*. New York: Oxford University Press, 2003.

Hastrup, Kirsten, and Martin Skrydstrup, eds. *The Social Life of Climate Change Models: Anticipating Nature*. New York: Routledge, 2013.

Heal, Geoffrey M., and Partha Dasgupta. *Economic Theory and Exhaustible Resources*. Cambridge: Cambridge University Press, 1979.

Hecht, Gabrielle. "Rupture-Talk in the Nuclear Age: Conjugating Colonial Power in Africa." *Social Studies of Science* 32, nos. 5–6 (2002): 691–728.

Heidegger, Martin. *Being and Time*. Translated by Joan Stambaugh. Albany: State University of New York, 1996.

———. "The Origin of the Work of Art." In *Poetry, Language, Thought*, translated by Albert Hofstadter, 17–86. New York: Per Classics, 2001.

———. *Phenomenological Interpretations of Aristotle*. Translated by Richard Rojcwicz. Bloomington: Indiana University Press, 2001.

Heinberg, Richard. *The Party's Over: Oil, War, and the Fate of Industrial Societies*. New York: New Society, 2005.

Helmreich, Jonathan E. *Gathering Rare Ores: The Diplomacy of Uranium Acquisition, 1943–1954*. Princeton, NJ: Princeton University Press, 1986.

Henning, Annette. "Climate Change and Energy Use." *Anthropology Today* 21, no. 3 (2005): 8–12.

Hepburn, Cameron. "Carbon Trading: A Review of the Kyoto Mechanisms." *Annual Review of Environment and Resources* 32 (2007): 375–93.

Heyd, Thomas, ed. *Recognizing the Autonomy of Nature: Theory and Practice*. New York: Columbia University Press, 2005.

Hicks, John. *Value and Capital*. Oxford: Oxford University Press, 1939.

Hill, Thomas. "Ideals of Human Excellence and Preserving the Natural Environment." *Environmental Ethics* 5, no. 3 (1983): 211–24.

Hinde, Robert A. *Why Good Is Good: The Sources of Morality*. London: Routledge, 2002.

Hitchcock, Peter. "Oil in an American Imaginary." *New Formations* 69, no. 4 (2010): 81–97.

Hobsbawm, Eric. *The Age of Empire, 1875–1914*. New York: Vintage, 1989.

Hochroth, Lysa. "The Scientific Imperative: Improductive Expenditure and Energeticism." *Configurations* 3, no. 1 (1995): 47–77.

Hodges, Tony. *Angola: From Afro-Stalinism to Petro-Diamond Capitalism*. Bloomington: Indiana University Press, 2001.

Hoffman, Andrew. *How Culture Shapes the Climate Change Debate*. Stanford: Stanford University Press, 2015.

Holaday, Duncan A. *Control of Radon and Daughters in Uranium Mines and Calculations on Biologic Effects*. Washington, DC: US Government Printing Office, 1957.

Hollier, Denis. "The Dualist Materialism of Georges Bataille." *Yale French Studies* 78 (1990): 124–39.

Howard, Emma. "Rising Numbers of Americans Believe Climate Science, Poll Shows." *Guardian*, October 13, 2015. http://www.theguardian.com/environment/2015/oct/13/rising-numbers-of-american-believe-climate-science-poll-shows.

Howe, Cymene. "Anthropocenic Ecoauthority: The Winds of Oaxaca." *Anthropological Quarterly* 87, no. 2 (2014): 381–404.

———. "Logics of the Wind: Development Desires over Oaxaca." *Anthropology News* 52, no. 5 (2011): 8.

Howe, Cymene, Dominic Boyer, and Edith Barrera. "Los márgenesdel Estado al viento: autonomíay desarrollo de energíasrenovables en el sur de México." Special issue, *Journal of Latin American and Caribbean* (forthcoming).

Hubbert, Marion K. *Nuclear Energy and Fossil Fuels*. Publication no. 95. Exploration and Production Research Division, Shell Development, 1956.

Huber, Matthew. "Oil, Life, and the Fetishism of Geopolitics." *Capitalism Nature Socialism* 22, no. 3 (2011): 32–48.

———. "Refined Politics: Petroleum Products, Neoliberalism, and the Ecology of Entrepreneurial Life." *Journal of American Studies* 46, no. 2 (2012): 295–312.

Huber, Peter W., and Mark P. Mills. *The Bottomless Well: The Twilight of Fuel, the Virtue of Waste, and Why We Will Never Run Out of Energy*. New York: Basic Books, 2005.

Hulme, Mike. "Five Lessons of Climate Change: A Personal Statement." WordPress, March 2008. http://www.mikehulme.org/wp-content/uploads/the-five-lessons-of-climate-change.pdf.

———. *Why We Disagree about Climate Change: Understanding Controversy, Inaction and Opportunity*. Cambridge: Cambridge University Press, 2009.

Human Rights Watch. "Incarcerated America." *Human Rights Watch Backgrounder* (2003). http://www.hrw.org/backgrounder/usa/incarceration.

Huntington, Samuel. *The Clash of Civilizations and the Remaking of the World Order*. New York: Simon & Schuster, 1996.

Illich, Ivan. *Energy and Equity*. London: Calder & Boyars, 1974.

An Inconvenient Truth. Dir. Davis Guggenheim. Paramount, 2006.

Ingold, Tim. *Being Alive: Essays on Movement, Knowledge and Description*. New York: Routledge, 2011.

Inter-American Development Bank (Mexico). "MarenaRenovables Wind Power Project." (MEL1107) Environmental and Social Management Report 21, November 2011.

Intergovernmental Panel on Climate Change. *Fourth Assessment Report*. Geneva: Intergovernmental Panel on Climate Change, 2007.

Intergovernmental Panel on Climate Change. "Summary for Policymakers." In *Climate Change 2007: Impacts, Adaptation and Vulnerability*, ed. M. L. Parry, O. F. Canziani, J. P. Palutikof, P. J. van der Linden, and C. E. Hanson, 7–22. Contribution of Working Group II to the Fourth Assessment Report of the Intergovernmental Panel on Climate Change. Cambridge: Cambridge University Press, 2007.

International Energy Agency. *Coal in World in 2005*. International Energy Agency, 2005.

Jaccard, Mark. *Sustainable Fossil Fuels: The Unusual Suspect in the Quest for Clean and Enduring Energy*. Cambridge: Cambridge University Press, 2006.

Jaggers, Keith, and Tedd R. Gurr. "Tracking Democracy's Third Wave with the Polity III Data." *Journal of Peace Research* 32, no. 4 (1995): 469–82.

Jameson, Fredric. *The Political Unconscious: Narrative as a Socially Symbolic Act*. Ithaca, NY: Cornell University Press 1981.

———. "The Politics of Utopia." *New Left Review* 25 (2004): 35–54.

———. *The Seeds of Time*. New York: Columbia University Press, 1996.

Jamieson, Dale. "Climate Change and Global Environmental Justice." In *Changing the Atmosphere: Expert Knowledge and Global Environmental Governance*, edited by Clark Miller and Paul Edwards, 287–307. Cambridge, MA: MIT Press, 2001.

———. "Energy, Ethics and the Transformation of Nature." In *The Ethics of Global Climate Change*, edited by Denis Arnold, 16–37. London: Cambridge University Press, 2011.

———. *Ethics and the Environment: An Introduction*. Cambridge: Cambridge University Press, 2008.

———. "Ethics, Public Policy, and Global Warming." In *Ethical Adaptation to Climate Change: Human Virtues of the Future*, edited by Allen Thompson and Jeremy Bendik-Keymer, 187–202. Cambridge, MA: MIT Press, 2012.

———. "Ethics, Public Policy, and Global Warming." *Science, Technology, and Human Values* 17, no. 2 (1992): 139–53.

———. *Morality's Progress: Essays on Humans, Other Animals, and the Rest of Nature*. Oxford: Oxford University Press, 2002.

———. "The Question of the Environment." In *Trattato di Biodiritto*, ed. Stefano Canestrari et al., 37–50. Milano: Giuffre Editore, 2010.

———. "When Utilitarians Should Be Virtue Theorists." *Utilitas* 19, no. 2 (2007): 160–83.

Jappe, Anslem. *Guy Debord*. Berkeley: University of California Press, 1999.

Jevons, H. Stanley. *The British Coal Trade*. London: E. P. Dutton, 1915.

Jevons, W. Stanley. *The Coal Question: An Inquiry Concerning the Progress of the Nation and the Probable Exhaustion of our Coal-Mines*. London: Macmillan, 1865.

Johnson, Bob. *Carbon Nation*. Lawrence: University Press of Kansas, 2014.

Jorgensen, Joseph G. *Oil Age Eskimos*. Berkeley: University of California Press, 1990.

Jorgensen, Joseph G., et al. *Native Americans and Energy Development*. Cambridge, MA: Anthropology Resource Center, 1978.

Jowett, Julie. "Fossilised Myths: Fresh Thinking on 'Dirty' Coal." *Guardian Weekly*, March 17–23, 2006, 5.

Kahn, Herman. *On Thermonuclear War*. Princeton, NJ: Princeton University Press, 1960.

Kalyvas, Stathis. "New and Old Civil Wars." *World Politics* 54 (2001): 99–118.

Kander, Astrid, Paolo Malanima, and Paul Warde. *Power to the People: Energy in Europe over the Last Five Centuries*. Princeton, NJ: Princeton University Press, 2014.

Kant, Immanuel. *Critique of Pure Reason*. Translated by Norman Kemp Smith. Boston: St. Martin's, 1965.

———. *Groundwork of the Metaphysics of Morals*. Translated by Allen Wood. New Haven, CT: Yale University Press, 2002.

Kantor, Zack. "How Uber's Autonomous Cars Will Destroy Millions of Jobs and Reshape the US Economy by 2025." *The Personal Blog of Zack Kantor*, January 23, 2015. http://zack kanter.com/2015/01/23/how-ubers-autonomous-cars-will-destroy-10-million-jobs-by -2025/.

Kaplan, Amy. "Homeland Insecurities: Some Reflections on Language and Space." *History Review* 85 (2003): 82–93.

Kapuscinski, Ryszard. *Shah of Shahs*. San Diego: Harcourt Brace Jovanovich, 1985.

Karl, Terry Lynn. *The Paradox of Plenty: Oil Booms and Petro-States*. Berkeley: University of California Press, 1997.

Kashi, Ed. *The Curse of the Black Gold*. Edited by Michael Watts. New York: Powerhouse, 2008.

Katz, Eric. *Nature as Subject: Human Obligation and the Natural Community*. Oxford: Rowman & Littlefield, 1997.

Kerouac, Jack. *On the Road*. New York: Penguin, 1976.

Kerr, Clark, and Abraham J. Siegel. *The Interindustry Propensity to Strike: An International Comparison*. Berkeley: University of California Press, 1955.

Keynes, John Maynard. *The General Theory of Employment, Interest and Money*. London: Macmillan, 1936.

———. "William Stanley Jevons 1835–1882: A Centenary Allocation on His Life and Work as Economist and Statistician." *Journal of the Royal Statistical Society* 99, no. 3 (1936): 516–55.

King, William Lyon MacKenzie. *Industry and Humanity: A Study in the Principles Underlying Industrial Reconstruction*. Boston: Houghton Mifflin, 1918.

Kirby, Maurice W. *The British Coal Mining Industry 1870–1946: A Political and Economic History*. Hamden, CT: Archon Books, 1977.

Klare, Michael. *Blood and Oil: How America's Thirst for Petrol Is Killing Us*. New York: Penguin, 2004.

———. *Resource Wars: The New Landscape of Global Conflict*. New York: Henry Holt, 2001.

———. *Rising Powers, Shrinking Planet: The New Geopolitics of Energy*. New York: Metropolitan Books, 2008.

Klieman, Kairn. "Oil, Politics, and Development in the Formation of a State: The Congolese Petroleum Wars, 1963–68." *International Journal of African Historical Studies* 41, no. 2 (2008): 169–202.

Knauer, Kelly, ed. *Global Warming*. New York: Time Books, 2007.

Knox, Hannah. "Footprints in the City: Models, Materiality, and the Cultural Politics of Climate Change." *Anthropological Quarterly* 87, no. 2 (2014): 405–29.

Kornhauser, Arthur Robert Dubin, and Arthur Ross, eds. *Industrial Conflict*. New York: McGraw-Hill, 1934.

Kothari, Ashish. "The Reality of Climate Injustice." *Hindu*, November 18, 2007. http://www.thehindu.com/todays-paper/tp-features/tp-sundaymagazine/the-reality-of-climate-injustice/article2275734.ece.

Koyré, Alexandre. *From the Closed World to the Infinite Universe*. Baltimore: Johns Hopkins University Press, 1957.

Krauss, Clifford. "Exxon and Russia's Oil Company in Deal for Joint Projects." *New York Times*, April 16, 2012. http://www.nytimes.com/2012/04/17/business/energy-environment/exxon-and-russian-oil-company-agree-to-joint-projects.html?_r=0.

———. "In Global Forecast, China Looms Large as Energy User and Maker of Green Power." *New York Times*, November 9, 2010. http://www.nytimes.com/2010/11/10/business/global/10oil.html.

Krauss, Werner. "The 'Dingpolitik' of Wind Energy in Northern German Landscapes: An Ethnographic Case Study." *Landscape Research* 35, no. 2 (2010): 195–208.

Kruse, John, Judith Kleinfeld, and Robert Travis. "Energy Development on Alaska's North Slope: Effects on the Inupiat Population." *Human Organization* 41, no. 2 (1982): 97–106.

Kunstler, James Howard. *The Long Emergency: Surviving the Converging Catastrophes of the Twenty-First Century*. New York: Atlantic Monthly, 2005.

Lacan, Jacques. *Écrits: A Selection*. Translated by Alan Sheridan. London: Tavistock, 1977.

Larkin, Oliver. *Art and Life in America*. New York: Holt, Rinehart & Winston, 1964.

Latour, Bruno. "Facing Gaia: Six Lectures on the Political Theology of Nature." Gifford Lecture on Natural Religion. University of Edinburgh. February 2013. http://macaulay.cuny.edu/eportfolios/wakefield15/files/2015/01/LATOUR-GIFFORD-SIX-LECTURES_1.pdf.

———. *The Pasteurization of France*. Translated by Alan Sheridan and John Law. Cambridge, MA: Harvard University Press, 1988.

———. *Politics of Nature: How to Bring the Sciences into Society*. Translated by Catherine Porter. Cambridge, MA: Harvard University Press, 2004.

———. *We Have Never Been Modern*. Translated by Catherine Porter. Cambridge, MA: Harvard University Press, 2002.

———. *What Is the Style of Matters of Concern?* Amsterdam: Van Gorcum. 2008.

———. "Why Has Critique Run Out of Steam?" *Critical Inquiry* 30, no. 2 (2004): 225–48.

Law, Robin. *From Slave Trade to "Legitimate Commerce": The Commercial Transition in Nineteenth-Century West Africa.* Cambridge: Cambridge University Press, 1995.

Lear, Jonathan. *Radical Hope: Ethics in the Face of Cultural Devastation.* Cambridge, MA: Harvard University Press, 2006.

LeBlanc, Steven A. *Constant Battles: The Myth of the Peaceful, Noble Savage.* New York: St. Martin's, 2003.

Lee, Katie. *Glen Canyon Betrayed: A Sensuous Elegy.* Flagstaff, AZ: Fretwater, 2006.

LeMenager, Stephanie. "The Aesthetics of Petroleum, after *Oil!*" *American Literary History* 24, no. 1 (2012): 59–86.

———. *Living Oil: Petroleum and Culture in the American Century.* Oxford: Oxford University Press, 2013.

Lenin, Vladimir. *Imperialism: The Highest Stage of Capitalism.* London: Pluto, 1996.

Lifshitz-Goldberg, Yael. "Gone with the Wind? The Potential Tragedy of the Common Wind." *Journal of Environmental Law and Policy* 28, no. 2 (2010): 435–71.

Lindee, Susan. *Suffering Made Real: New York: American Science and the Survivors at Hiroshima.* Chicago: University of Chicago Press, 1994.

Lindfors, Bernth. "Big Shots and Little Shots of the Canon." In *Long Drums and Canons,* 61–75. Trenton, NJ: Africa World, 1995.

———, ed. *Critical Perspectives on Amos Tutuola.* Washington, DC: Three Continents, 1982.

Lindqvist, Sven. *A History of Bombing.* New York: New Press, 2001.

Lingis, Alphonso. *Trust.* Minneapolis: University of Minnesota Press, 2004.

Lippmann, Walter. *The Good Society.* Boston: Little, Brown, 1938.

———. *The Phantom Public.* New York: Harcourt, Brace, 1925.

Logan, Owen. "Where Pathos Rules: The Resource Curse in Visual Culture." In *Flammable Societies: Studies on the Socio-economics of Oil and Gas,* edited by Owen Logan and John Andrew McNeish, 98–130. London: Pluto, 2012.

Lohmann, Larry. "Carbon Trading: A Critical Conversation on Climate Change, Privatisation and Power." Special issue, *Development Dialogue* 48 (2006).

Lord, Barry. "The Culture of a World without Oil." *Medium,* July 27, 2015. https://medium.com/@blord/the-culture-of-a-world-without-oil-130df6e7d63a#.7jdmai22b.

Love, Thomas. "Anthropology and the Fossil Fuel Era." *Anthropology Today* 24, no. 2 (2008): 3–4.

Love, Thomas, and Anna Garwood. "Wind, Sun and Water: Complexities of Alternative Energy Development in Rural Northern Peru." *Rural Society* 20 (2011): 294–307.

Lubeck, Paul, Michael Watts, and Ronnie Lipschutz. "Convergent Interests: US Energy Security and the Securing of Democracy in Nigeria." *International Policy Report.* Washington, DC, 2007.

Luke, Timothy W. "Environmentality as Green Governmentality." In *Discourses of the Environment,* edited by Eric Darier, 121–51. Malden, MA: Blackwell, 1999.

Luxemburg, Rosa. *The Mass Strike, the Political Party, and the Trade Unions.* Detroit: Marxist Educational Society, 1925.

Lynas, Mark. *Six Degrees: Our Future on a Hotter Planet.* Washington, DC: National Geographic, 2008.

Lynn, Martin. *Commerce and Economic Change in West Africa: The Palm Oil Trade in the Nineteenth Century.* Cambridge: Cambridge University Press, 2002.

Macherey, Pierre. *A Theory of Literary Production.* Translated by Geoffrey Wall. London: Routledge, 1978.

Mahdavy, Hossein. "The Patterns and Problems of Economic Development in Rentier States: The Case of Iran." In *Studies in Economic History of the Middle East,* edited by M. A. Cook, 428–67. Oxford: Oxford University Press, 1970.

Makhijani, Arjun, and Stephen I. Schwartz. "Victims of the Bomb." In *Atomic Audit: The Costs and Consequences of U.S. Nuclear Weapons Since 1940,* edited by Stephen I. Schwartz, 395–431. Washington, DC: Brookings Institution, 1998.

Malette, Sebastien. "Foucault for the Next Century: Eco-Governmentality." In *A Foucault for the 21st Century: Governmentality, Biopolitics and Discipline in the New Millennium,* edited by Sam Binkley and Jorge Capetillo, 221–39. Cambridge: Cambridge Scholars, 2009.

Manning, Richard. *Against the Grain: How Agriculture Has Hijacked Civilization.* New York: North Point, 2004.

Marcus, George E. "Ethnography in/of the World System: The Emergence of Multi-Sited Ethnography." *Annual Review of Anthropology* 24 (1995): 95–117.

Marcuse, Herbert. "Nature and Revolution." In *The Essential Marcuse: Selected Writings of Philosopher and Social Critic Herbert Marcuse,* edited by Andrew Feenberg and William Leiss, 233–47. Boston: Beacon, 2007.

Marez, Curtis. "What Is a Disaster?" *American Quarterly* 61 (2009): ix–xi.

Marks, Robert W., ed. *Space, Time, and the New Mathematics.* New York: Bantam Books, 1964.

Martin, Susan. *Palm Oil and Protest: An Economic History of the Ngwa Region, South-Eastern Nigeria, 1800–1980.* Cambridge: Cambridge University Press, 1988.

Martinez, David. "Feast and Famine: A Conversation with Iain Boal on Scarcity and Catastrophe." *Mute Magazine,* February 20, 2006. http://www.metamute.org/editorial/articles/feast-and-famine-conversation-iain-boal-scarcity-catastrophe-and-enclosure.

Marx, Karl. *Capital.* Translated by Ben Fowkes. Harmondsworth: Penguin, 1990.

———. "Critique of the Gotha Program." In *The Marx-Engels Reader,* 2nd ed., edited by Robert C. Tucker, 535–41. New York: W. W. Norton, 1978.

Marx, Karl, and Friedrich Engels. *Selected Works.* Moscow: International, 1969.

Masco, Joseph. "Mutant Ecologies: Radioactive Life in Post-Cold War New Mexico." *Cultural Anthropology* 19, no. 4 (2004): 517–50.

———. *The Nuclear Borderlands: The Manhattan Project in Post–Cold War New Mexico.* Princeton, NJ: Princeton University Press, 2006.

———. "Survival Is Your Business: Engineering Ruins and Affect in Nuclear America." *Cultural Anthropology* 23, no. 2 (2008): 361–98.

Maslin, Mark. *Global Warming: A Very Short Introduction.* Oxford: Oxford University Press, 2004.

Mason, Arthur. "The Rise of Consultant Forecasting in Liberalized Natural Gas Markets." *Public Culture* 19, no. 2 (2007): 367–79.

Mason, Arthur, and Maria Stoilkova. "Corporeality of Consultant Expertise in Arctic Natural Gas Development." *Journal of Northern Studies* 6, no. 2 (2012): 83–96.

Mathews, Andrew. "Statemaking, Knowledge and Ignorance: Translation and Concealment in Mexican Forestry Institutions." *American Anthropologist* 110, no. 4 (2008): 484–94.

Mbembe, Achille. "At the Edge of the World." *Public Culture* 12, no. 1 (2001): 259–84.

McCarthy, Shawn. "Report Slams U.S. Domestic Energy Policy." *Globe and Mail*, October 13, 2006. http://www.theglobeandmail.com/report-on-business/report-slams-us-domestic -energy-policy/article1107489/.

McCloud, Scott. *Understanding Comics: The Invisible Art*. New York: HarperCollins, 1993.

McCright, Aaron M., and Riley E. Dunlap. "Challenging Global Warming as a Social Problem: An Analysis of the Conservative Movement's Counter-claims." *Social Problems* 47 (2000): 499–522.

McCue, Duncan. "Idle No More and Tensions in Thunder Bay." *The Current*, CBC Radio, January 25, 2013. http://www.cbc.ca/player/News/Canada/Audio/ID/2329105939/.

McDonald, James H. "The Neoliberal Project and Governmentality in Rural Mexico: Emergent Farmer Organization in the Michoacán Highlands." *Human Organization* 58, no. 3 (1999): 274–84.

McEnaney, Laura. *Civil Defense Begins at Home: Militarization Meets Everyday Life in the Fifties*. Princeton, NJ: Princeton University Press, 2000.

McGuire, Tom. *History of the Offshore Oil and Gas Industry in Southern Louisiana*. Vol. 2, *Bayou Lafourche: Oral Histories of the Oil and Gas Industry*. Washington, DC: US Department of the Interior, Minerals Management Service, Gulf of Mexico OCS Region, 2008.

McKibben, Bill. *The End of Nature*. New York: Random House, 1989.

———. "Global Warming's Terrifying New Math." *Rolling Stone*, July 19, 2012. http://www .rollingstone.com/politics/news/global-warmings-terrifying-new-math-20120719.

McKie, Robin. "World Will Pass Crucial 2C Global Warming Limit, Experts Warn." *Guardian*, October 10, 2015. http://www.theguardian.com/environment/2015/oct/10/climate -2c-global-warming-target-fail.

McKillop, Andrew. *The Final Energy Crisis*. Edited by Sheila Newman. London: Pluto, 2005.

McKim, Joel. "Of Microperception and Micropolitics: An Interview with Brian Massumi, 15 August 2008." *Inflexions* 3 (2009): 1–19.

McKinzie, D. Richard. "Oral History Interview with Charles P. Kindleberger, Economist with the Office of Strategic Services, 1942–44, 1945; chief, Division German and Austrian Economic Affairs, Department of State, Washington, 1945–48; and Intelligence Officer, 12th U.S. Army group, 1944–45." Independence, MO: Harry S. Truman Library, 1973.

McLuhan, Marshall. *The Gutenberg Galaxy: The Making of Typographic Man*. Toronto: University of Toronto Press, 1962.

———. *Understanding Media: The Extensions of Man*. London: Routledge, 2001.

McNeill, William. *Something New under the Sun: An Environmental History of the Twentieth-Century World*. New York: Norton, 2000.

McNeish, John Andrew, Axel Borchgrevink, and Owen Logan. *Contested Powers: The Politics of Energy and Development in Latin America*. London: Zed Books, 2015.

McNeish, John Andrew, and Owen Logan, eds. *Flammable Societies: Studies on the Socioeconomics of Oil and Gas*. London: Pluto, 2012.

Meadows, Donella H., et al. *The Limits to Growth: A Report for the Club of Rome's Project on the Predicament of Mankind.* New York: Universe Books, 1972.

Meehl, Gerald A., and Thomas F. Stocker. "Global Climate Projections." *Fourth Assessment Report of the Intergovernmental Panel on Climate Change.* Cambridge: Cambridge University Press, 2007.

Meillasoux, Quentin. *After Finitude: An Essay on the Necessity of Contingency.* Translated by Ray Brassier. London: Continuum, 2008.

Metzl, Jonathan. *Prozac on the Couch: Prescribing Gender in the Era of Wonder Drugs.* Durham, NC: Duke University Press, 2003.

Miller, Cecilia. *Giambattista Vico: Imagination and Historical Knowledge.* Basingstoke: St. Martin's, 1993.

Millett, Kate. *Sexual Politics.* New York: Doubleday, 1970.

Mirowski, Philip. *More Heat Than Light: Economics as Social Physics, Physics as Nature's Economics.* Cambridge: Cambridge University Press, 1989.

Mitchell, Timothy. "Carbon Democracy." *Economy and Society* 38, no. 3 (2009): 399–432.

———. "Culture and Economy." In *The Sage Handbook of Cultural Analysis,* edited by Tony Bennett and John Frow, 447–66. Thousand Oaks, CA: Sage, 2008.

———. "Economists and the Economy in the Twentieth Century." In *The Politics of Method in the Human Sciences: Positivism and Its Epistemological Other,* edited by George Steinmetz, 126–41. Durham, NC: Duke University Press, 2005.

———. "Fixing the Economy." *Cultural Studies* 12, no. 1 (1998): 82–101.

———. "McJihad: Islam in the US Global Order." *Social Text* 20, no. 4 (2002): 1–18.

———. *Rule of Experts: Egypt, Techno-politics, Modernity.* Berkeley: University of California Press, 2002.

Mogren, Eric W. *Sands: Uranium Mill Tailings Policy in the Atomic West.* Albuquerque: New Mexico University Press, 2002.

Monbiot, George. *Heat: How to Stop the Planet from Burning.* London: Penguin, 2006.

Moret, Leuren. "A Death Sentence Here and Abroad. Depleted Uranium: Dirty Bombs, Dirty Missiles, Dirty Bullets." Centre for Research on Globalisation. August 21, 2004. http://globalresearch.ca/articles/MOR408A.html.

Morris, Ian. *Foragers, Farmers, and Fossil Fuels: How Human Values Evolve.* Princeton, NJ: Princeton University Press, 2015.

Morrison, James C. "Vico's Principle of Verum Is Factum and the Problem of Historicism." *Journal of the History of Ideas* 39 (1987): 579–95.

Morton, Timothy. *The Ecological Thought.* Cambridge, MA: Harvard University Press, 2012.

———. *Ecology without Nature: Rethinking Environmental Aesthetics.* Cambridge, MA: Harvard University Press, 2007.

Mouawad, Jad. "Estimate Places Natural Gas Reserves 35% Higher." *New York Times,* June 17, 2009. http://www.nytimes.com/2009/06/18/business/energy-environment/18gas .html?_r=0.

Moynihan Report. *The Negro Family: The Case for National Action.* Office of Policy Planning and Research, United States Department of Labor, 1965. http://www.dol.gov/oasam/ programs/history/webid-meynihan.htm.

Mulgan, Tim. *Ethics for a Broken World: Imagining Philosophy after Catastrophe*. Montreal: McGill-Queen's University Press, 2011.

Mumford, Lewis. *Technics and Civilization*. New York: Harcourt, Brace, 1934.

Myers, Norman. "Environmental Refugees: A Growing Phenomenon of the 21st Century." *Biological Sciences* 357 (2002): 609–13. http://www.ncbi.nlm.nih.gov/pmc/articles/PMC 1692964/.

Nabokov, Vladimir. *Lolita*. Harmondsworth: Penguin, 2004.

Nader, Laura. *Energy Choices in a Democratic Society. A Resource Group Study for the Synthesis Panel of the Committee on Nuclear Alternative Energy Systems for the U.S. National Academy of Sciences*. Washington, DC: National Academy of Sciences, 1980.

Nader, Laura, and Stephen Beckerman. "Energy as It Relates to the Quality and Style of Life." *Annual Review of Energy* 3 (1978): 1–28.

Negarestani, Reza. *Cyclonopedia: Complicity with Anonymous Materials*. Melbourne: re.press, 2008.

Negri, Antonio. *Insurgencies: Constituent Power and the Modern State*. Translated by Maurizia Boscagli. Minneapolis: University of Minnesota, 2009.

Neufeld, John. Afterword to *A.D.: New Orleans after the Deluge*, 191–93. New York: Pantheon, 2009.

Neuman, Andrew M. *The Economic Organization of the British Coal Industry*. London: George Routledge, 1934.

Neville, G. Robert. "The Courrières Colliery Disaster, 1906." *Journal of Contemporary History* 13, no. 1 (1978): 33–52.

Nietzsche, Friedrich. *The Gay Science*. Edited by Walter Kaufman. New York: Vintage, 1974.

Nitzan, Jonathan, and Shimshon Bichler. "The Weapondollar-Petrodollar Coalition." In *The Global Political Economy of Israel*, 198–273. London: Pluto, 2002.

Nixon, Rob. "Neoliberalism, Slow Violence, and the Environmental Picaresque." *Modern Fiction Studies* 55 (2009): 443–48.

———. *Slow Violence and the Environmentalism of the Poor*. Cambridge, MA: Harvard University Press, 2011.

Noble, David F. *A World without Women: The Christian Clerical Culture of Western Science*. New York: Knopf, 1992.

Noel, J. M. "Products and By-Products." *BUROTROP Bulletin* 19 (2003).

Nordstrom, Jean Maxwell, et al. *The Northern Cheyenne Tribe and Energy Developments in Southeastern Montana*. Vol. 1, *Social and Cultural Investigations*. Lame Deer: Northern Cheyenne Research Project, 1977.

Nowell, Gregory. *Mercantile States and the World Oil Cartel, 1900–1939*. Ithaca, NY: Cornell University Press, 1994.

Nugent, Daniel, and Ana María Alonso. "Multiple Selective Traditions in Agrarian Reform and Agrarian Struggle: Popular Culture and State Formation in the Ejido of Namiquipa, Chihuahua." In *Everyday Forms of State Formation: Revolution and the Negotiation of Rule in Modern Mexico*, edited by Gilbert M. Joseph and Daniel Nugent, 209–46. Durham, NC: Duke University Press, 1994.

Nye, David E. *Consuming Power: A Social History of American Energies*. Cambridge, MA: MIT Press, 1999.

Oakes, Guy. *The Imaginary War: Civil Defense and American Cold War Culture*. New York: Oxford University Press, 1994.

Obama, Barack. "Text of Obama's Speech on Gulf Oil Spill." *New Haven Register*, June 15, 2010. http://www.nhregister.com/article/NH/20100615/NEWS/306159801.

Oceransky, Sergio. "Wind Conflicts in the Isthmus of Tehuantepec: The Role of Ownership and Decision-Making Models in Indigenous Resistance to Wind Projects in Southern Mexico." *Commoner* 13 (2009): 203–22.

Ochoa, Enrique C. "Neoliberalism, Disorder, and Militarization in Mexico." *Latin American Perspectives* 28, no. 4 (2001): 148–59.

Odum, Howard T., and Elizabeth C. Odum. *A Prosperous Way Down: Principles and Policies*. Boulder: University Press of Colorado, 2001.

Office of the High Commissioner for Human Rights. Fact Sheet No. 14, Contemporary Forms of Slavery. 1991. http://www.ohchr.org/Documents/Publications/FactSheet14en.pdf.

Ogden, Joan. "High Hopes for Hydrogen." *Scientific American* 295, no. 3 (2006): 94–101.

O'Keeffe, Georgia. *Georgia O'Keeffe*. New York: Viking, 1976.

Okonta, Ike. "Nigeria: Chronicle of a Dying State." *Current History* 104 (2005): 203–8.

Okri, Ben. "What the Tapster Saw." In *Stars of the New Curfew*, 183–94. New York: King Penguin, 1989.

O'Neill, John J. "Enter Atomic Power." *Harper's* 181 (1940): 1–10.

O'Neill, John J., Alan Holland, and Andrew Light. *Environmental Values*. New York: Routledge, 2007.

Oosthoek, Jan, and Barry K. Gills. "Humanity at the Crossroads: The Globalization of Environmental Crisis." *Globalizations* 2, no. 3 (2005): 283–91.

Oppenheim, V. H. "Why Oil Prices Go Up: The Past: We Pushed Them." *Foreign Policy* 25 (1976–77): 24–57.

Oreskes, Naomi. "The American Denial of Global Warming." *University of California Television*, December 12, 2007. http://www.uctv.tv/shows/The-American-Denial-of-Global-Warming-Perspectives-on-Ocean-Science-13459.

———. *Climate Change: What It Means for Us, Our Children, and Our Grandchildren*. Edited by Joseph F. C. Dimento and Pamela Doughman. Cambridge, MA: MIT Press, 2007.

———. "Seeing Climate Change." In *Dario Robleto: Survival Does Not Lie in the Heavens*, edited by Gilbert Vicario. Des Moines, IA: Des Moines Art Center, 2011.

———. "The Scientific Consensus on Climate Change." *Science* 306, no. 5702 (2004): 1686.

Oreskes, Naomi, and Erik M. Conway. *The Collapse of Western Civilization: A View from the Future*. New York: Columbia University Press, 2014.

———. *Merchants of Doubt: How a Handful of Scientists Obscured the Truth on Issues from Tobacco to Climate Change*. New York: Bloomsbury, 2010.

Orr, Jackie. *Panic Diaries: A Genealogy of Panic Disorder*. Durham, NC: Duke University Press, 2006.

Osofisan, Femi. *The Oriki of Grasshopper and Other Plays*. Washington, DC: Howard University Press, 1995.

Ostwald, Wilhelm. *Die Energetischen Grundlagen Der Kulturwissenschaften* [The energy basis of the humanities]. Leipzig: Werner Klinkhardt, 1909.

Packer, George. "The Megacity: Decoding the Chaos of Lagos." *New Yorker*, November 13, 2006. http://www.newyorker.com/magazine/2006/11/13/the-megacity.

Painter, David S. "Oil and the Marshall Plan." *Business History Review* 58, no. 3 (1984): 359–83.

Parker, B. Richard. *The October War: A Retrospective*. Gainesville: University Press of Florida, 2001.

Parker, Wendy S. "Predicting Weather and Climate: Uncertainty, Ensembles and Probability." *Studies in History and Philosophy of Modern Physics* 41 (2010): 263–72.

Pasqualetti, Martin J. "Social Barriers to Renewable Energy Landscapes." *Geographical Review* 101, no. 2 (2011): 201–23.

Pateman, Carole. *The Sexual Contract*. Stanford: Stanford University Press, 1988.

Peace, William J. *Leslie A. White: Evolution and Revolution in Anthropology*. Lincoln: University of Nebraska Press, 2004.

Pendakis, Andrew, and Ursula Biemann. "This Is Not a Pipeline: Thoughts on the Politico-aesthetics of Oil." *Imaginations* 3, no. 2 (2012): 6–15.

Perkovich, George. *India's Nuclear Bomb: The Impact on Global Proliferation*. Berkeley: University of California Press, 1990.

Peters, Pauline. "Inequality and Social Conflict over Land." *Journal of Agrarian Change* 4, no. 3 (2004): 269–314.

Pétré-Grenouilleau, Olivier. *Les Traites Négrières: Essai d'Histoire Globale*. Paris: Gallimard, 2004.

Petryna, Adriana. *Life Exposed: Biological Citizens after Chernobyl*. Princeton, NJ: Princeton University Press, 2002.

Pinkus, Karen. *Fuel*. Minneapolis: University of Minnesota Press, 2016.

———. "On Climate, Cars, and Literary Theory." *Technology and Culture* 49, no. 4 (2008): 1002–9.

Podobnik, Bruce. *Global Energy Shifts: Fostering Sustainability in a Turbulent Age*. Philadelphia: Temple University Press, 2006.

Polanyi, Karl. *The Great Transformation*. Boston: Victor Gollancz, 1947.

Pollard, Sidney. *Peaceful Conquest: The Industrialization of Europe, 1760–1970*. Oxford: Oxford University Press, 1981.

Pomeranz, Kenneth. *The Great Divergence: Europe, China, and the Making of the Modern World Economy*. Princeton, NJ: Princeton University Press, 2000.

Porritt, Jonathon. *The World We Made: Alex McKay's Story from 2050*. London: Phaidon, 2013.

Potter, Robert. "Is Atomic Power at Hand?" *Scientific Monthly* 50, no. 6 (1940): 571–74.

Powell, Dana E., and Dáilan J. Long. "Landscapes of Power: Renewable Energy Activism in Diné Bikéyah." In *Indians & Energy: Exploitation and Opportunity in the American Southwest*, edited by Sherry Smith and Brian Frehner, 231–62. Santa Fe: SAR Press, 2010.

Power, Max. *America's Nuclear Wastelands: Politics, Accountability, and Cleanup*. Pullman: Washington State University Press, 2008.

Pratt, Louise Mary. "Planetary Longings: Sitting in the Light of the Great Solar TV." In *World

Writing: Poetics, Ethics, Globalization, edited by Mary Gallagher, 207–23. Toronto: University of Toronto Press, 2008.

Quataert, Donald. *Miners and the State in the Ottoman Empire: The Zonguldak Coalfield, 1822–1920*. New York: Berghahn, 2006.

Quayson, Ato. *Strategic Transformations in Nigerian Writing*. Bloomington: Indiana University Press, 1997.

Rabinow, Paul, and Nikolas Rose. "Biopower Today." *BioSocieties* 1, no. 2 (2006): 195–217.

Rappaport, Roy. "The Flow of Energy in Agricultural Society." In *Biological Anthropology: Readings from Scientific American*, edited by Solomon Katz, 371–87. San Francisco: W. H. Freeman, 1975.

Rawsthorn, Alice. "Innovation for a Better Future." *New York Times*, September 1, 2013. http://www.nytimes.com/2013/09/02/arts/design/Innovation-for-a-Better-Future .html?_r=0.

Reifer, E. Thomas. "Labor, Race & Empire: Transport Workers and Transnational Empires of Trade, Production, and Finance." In *Labor versus Empire: Race, Gender, and Migration*, ed. Gilbert G. Gonzalez et al., 14–32. London: Routledge, 2004.

Renshaw, Edward. "The Substitution of Inanimate Energy for Animal Power." *Journal of Political Economy* 71, no. 3 (1963): 284–92.

Reyna, Stephen, and Andrea Behrends. "The Crazy Curse and Crude Domination: Toward an Anthropology of Oil." *Focal* 52, no. 1 (2008): 3–17.

Rifkin, Jeremy. *The End of Work: The Decline of the Global Workforce and the Dawn of the Post-market Era*. London: Penguin, 2000.

Riles, Annelise. *The Network Inside Out*. Ann Arbor: University of Michigan Press, 2000.

Robbins, Lynn. *The Socioeconomic Impacts of the Proposed Skagit Nuclear Power Plant on the Skagit System Cooperative Tribes*. Bellingham: Lord & Associates, 1980.

Robelius, Fredrik. "Giant Oil Fields—the Highway to Oil: Giant Oil Fields and Their Importance for Future Oil Production." PhD diss., Uppsala University, 2007.

Roberts, D. David. *Benedetto Croce and the Uses of Historicism*. Berkeley: University of California Press, 1987.

Roberts, Paul. *The End of Oil: On the Edge of a Perilous New World*. New York: Mariner Books, 2005.

Robin, Libby, and Will Steffen. "History for the Anthropocene." *History Compass* 5, no. 5 (2007): 1694–719.

Robinson, S. Kim. *Forty Signs of Rain, Fifty Degrees Below, and Sixty Days and Counting*. New York: Spectra, 2005–7.

Rodgers, T. Daniel. *Atlantic Crossings: Social Politics in a Progressive Age*. Cambridge, MA: Belknap, 1998.

Rodrik, Dani. *The New Global Economy and Developing Countries*. London: Open Library, 1999.

Rogers, Douglas. "Energopolitical Russia: Corporation, State and the Rise of Social and Cultural Projects." *Anthropological Quarterly* 87, no. 2 (2014): 431–51.

Ross, Andrew, and Damon Matthews. "Climate Engineering and the Risk of Rapid Climate Change." *Environmental Research Letters* 4, no. 4 (2009): 2–6. http://iopscience.iop .org/article/10.1088/1748-9326/4/4/045103/pdf.

Ross, Deveryn. "Idle No More's Real Challenge." *Winnipeg Free Press*, January 24, 2013. https://www.highbeam.com/doc/1P3-2873117461.html.

Ross, Michael L. "Does Oil Hinder Democracy?" *World Politics*, 53, no. 3 (2001): 325–61.

Rosser, Andrew. "Escaping the Resource Curse: The Case of Indonesia." *Journal of Contemporary Asia* 37, no. 1 (2007): 38–58.

Rossi, Paolo. *The Dark Abyss of Time: The History of the Earth and the History of Nations from Hooke to Vico.* Translated by Lydia G. Cochrane. Chicago: University of Chicago Press, 1979.

Roth, Henry. *Call It Sleep.* New York: Picador, 2005.

Royden, Harrison, ed. *Independent Collier: The Coal Miner as Archetypal Proletarian Reconsidered.* New York: St. Martin's, 1978.

Ruddiman, F. William. "The Anthropogenic Greenhouse Era Began Thousands of Years Ago." *Climatic Change* 61, no. 3 (2003): 261–93.

Russel, Harriet. "An Endangered Species, Oil." In *Sorry, Out of Gas: Architecture's Response to the 1973 Oil Crisis*, edited by Giovanna Borasi and Mirko Zardini, 5–36. Montova: Corraini Edizioni, 2007.

Sachs, Jeffrey D. "The Anthropocene." In *Common Wealth: Economics for a Crowded Planet*, 57–82. New York: Penguin, 2008.

Sachs, Jeffrey D., and Andrew M. Warner. "Natural Resource Abundance and Economic Growth." *Development Discussion Paper No 517a.* Cambridge, MA: Harvard Institute for International Development, 1995. http://www.cid.harvard.edu/ciddata/warner_files/natresf5.pdf.

Sagoff, Mark. "Nature versus the Environment." *Report from the Institute for Philosophy & Public Policy* 11, no. 3 (1991): 5–8.

———. *Price, Principle, and the Environment.* Cambridge: Cambridge University Press, 2004.

Sala-i-Martin, Xavier, and Arvin Subramanian. "Addressing the Natural Resource Curse: An Illustration from Nigeria." *International Monetary Fund* 03.139. Washington, DC, 2003.

Salamé, Ghassan, ed. *Democracy without Democrats? The Renewal of Politics in the Muslim World.* New York: I. B. Tauris, 1994.

Salminen, Antti, and Tere Vadén. *Energy and Experience: An Essay in Nafthology.* Chicago: MCM Prime, 2015.

Samuelson, Paul A. *Foundations of Economic Analysis.* Cambridge, MA: Harvard University Press, 1947.

Sandler, Ronald L. *Character and Environment: A Virtue-Oriented Approach to Environmental Ethics.* New York: Columbia University Press, 2007.

Sandler, Ronald L., and Philip Cafaro, eds. *Environmental Virtue Ethics.* Lanham, MD: Rowman & Littlefield, 2005.

Saro-Wiwa, Ken. "Africa Kills Her Sun." In *Under African Skies: Modern African Stories*, edited by Charles D. Larson, 210–21. New York: Farrar, Straus & Giroux, 1998.

Sartre, Jean P. *Critique of Dialectical Reason.* Vol. 1, *Theory of Practical Ensembles.* London: Verso, 1977.

Sawyer, Suzana, and Terence Gomez, eds. *Crude Chronicles: Indigenous Politics, Multinational Oil, and Neoliberalism in Ecuador.* Durham, NC: Duke University Press, 2004.

———, eds. *The Politics of Resource Extraction: Indigenous Peoples, Corporations and the State.* London: Palgrave Macmillan, 2012.

Scheer, Hermann. *The Solar Economy: Renewable Energy for a Sustainable Global Future.* London: Earthscan, 2004.

Scheinman, Lawrence. *The International Atomic Energy Agency and World Nuclear Order.* Washington, DC: Resources for the Future, 1987.

Schindler, David W., and John P. Smol. "After Rio, Canada Lost Its Way." *Ottawa Citizen,* June 20, 2012. http://www.pressreader.com/canada/ottawa-citizen/20120621/2818314 60814967.

Schreuder, Yda. *The Corporate Greenhouse: Climate Change Policy in a Globalizing World.* London: Zed Books, 2009.

Schumacher, E. F. *Small Is Beautiful: Economics As If People Mattered.* New York: Harper & Row, 1973.

Schwegler, A. Tara. "Take It from the Top (Down)? Rethinking Neoliberalism and Political Hierarchy in Mexico." *American Ethnologist* 35, no. 4 (2008): 682–700.

SENER. "Energíaeólica y la políticaenergéticamexicana." Ing. Alma Santa Rita Feregrino Subdirectora de Energía y MedioAmbiente. Monterrey, México, October 2007.

Serpas, Martha. "Decreation." In *The Dirty Side of the Storm,* 76–80. New York: Norton, 2007.

Shapiro, Mark. "Conning the Climate: Inside the Carbon-Trading Shell Game." *Harper's,* February 2010. http://harpers.org/archive/2010/02/conning-the-climate/.

Sheail, John. "Torrey Canyon: The Political Dimension." *Journal of Contemporary History* 42, no. 3 (2007): 485–504.

Shubin, Neil. "The Disappearance of Species." *Bulletin of the American Academy of Arts and Sciences* 61 (2008): 17–19.

Sieferle, P. Rolf. *The Subterranean Forest: Energy Systems and the Industrial Revolution.* Cambridge: White Horse, 2001.

Silver, Beverly J. *Forces of Labor: Workers' Movements and Globalization since 1870.* Cambridge: Cambridge University Press, 2003.

Simpson, David. "Romanticism, Criticism, and Theory." In *The Cambridge Companion to British Romanticism,* edited by Stuart Curran, 1–24. Cambridge: Cambridge University Press, 1993.

———. *Situatedness; or, Why We Keep Saying Where We're Coming From.* Durham, NC: Duke University Press, 2002.

SIPRI. *Arms Transfers Database.* Stockholm International Peace Research Institute. n.d. https://www.sipri.org/databases/armstransfers.

Skodzinski, Noelle. "Offshoring and the Global Marketplace." *Book Business Magazine,* July 2010. http://www.bookbusinessmag.com/article/offshoring-global-marketplace-12209/10/.

Sloan, John. *Gist of Art.* New York: American Artist Group, 1939.

Smail, Daniel Lord. *On Deep History and the Brain.* Berkeley: University of California Press, 2008.

Smil, Vaclav. *Energy in World History.* Boulder, CO: Westview, 1994.

———. *Energy Transitions: History, Requirements, Prospects.* Westport, CT: Praeger, 2010.

Smith, Bonnie G. "Gender and the Practices of Scientific History: The Seminar and Archival Research in the Nineteenth Century." *American Historical Review* 100 (1995): 1150–76.

Smith, Neil. *The Endgame of Globalization*. New York: Routledge, 2004.

Socolow, Robert H., and Stephen W. Pacala. "A Plan to Keep Carbon in Check." *Scientific American* 295, no. 3 (2006): 50–57.

Solnit, Rebecca. "Diary." *London Review of Books*, August 2010. http://www.lrb.co.uk/v32/n15/rebecca-solnit/diary.

Sorel, Georges. *Reflections on Violence*. Translated by T. E. Hulme. New York: B. W. Huebsch, 1914.

Sörlin, Sverker. "The Changing Nature of Environmental Expertise." *Eurozine*, November 19, 2013. http://www.eurozine.com/articles/2013-11-19-sorlin-en.html.

Stalin, Joseph. *Dialectical and Historical Materialism*. Moscow: International, 1938.

Stambaugh, Joan. *The Finitude of Being*. Albany: State University of New York Press, 1992.

Steffen, Will. "Humans Creating New 'Geological Age.'" *Australian*, March 31, 2008.

Stern, Nicholas. *The Economics of Climate Change: The "Stem Review."* Cambridge: Cambridge University Press, 2007.

Stewart, Kathleen. *Ordinary Affects*. Durham, NC: Duke University Press, 2008.

Stewart, Rory. *Occupational Hazards: My Time Governing in Iraq*. London: Picador, 2006.

Stix, Gary. "A Climate Repair Manual." *Scientific American* 295, no. 3 (2006): 46–49.

Stoekl, Allan. *Bataille's Peak: Energy, Religion, and Postsustainability*. Minneapolis: University of Minnesota, 2007.

Strasser, Johano. *Wenn Der Arbeitsgesellschaft die Arbeit ausgeht* [When the working society runs out of work]. Zürich: Pendo, 1999.

Strauss, Lawrence Guy. "The World at the End of the Last Ice Age." In *Humans at the End of the Ice Age: The Archaeology of the Pleistocene–Holocene Transition*, edited by Lawrence Guy Strauss, Barit Valentin Eriksen, Jon M. Erlandson, and David Y. Yesner, 3–10. New York: Plenum, 1996.

Strauss, Sarah, Thomas Love, and Stephanie Rupp, eds. *Cultures of Energy*. Walnut Creek, CA: Left Coast, 2013.

Strauss, Sarah, and Ben Orlove, eds. *Weather, Climate, Culture*. Oxford: Berg, 2003.

Suny, G. Ronald. "A Journeyman for the Revolution: Stalin and the Labour Movement in Baku, June 1907–May 1908." *Soviet Studies* 23, no. 3 (1972): 373–94.

Suzuki, David, and Amanda McConnell. *The Sacred Balance: Rediscovering Our Place in Nature*. Vancouver: Greystone Books, 1997.

Swans, Elizabeth. *Beyond Terror: Gender, Narrative, Human Rights*. New Brunswick, NJ: Rutgers University Press, 2007.

Szeman, Imre. "Editors' Column: Literature and Energy Futures." *PMLA* 126, no. 2 (2011): 323–26.

———. *Fueling Culture: Politics, History, Energy*. Edited Jennifer Wenzel, Imre Szeman, and Patricia Yaeger. New York: Fordham University Press, 2016.

———. "How to Know about Oil: Energy Epistemologies and Political Futures." *Journal of Canadian Studies / Revue D'Etudes Canadiennes* 47, no. 3 (2013): 145–68.

———. "Introduction to Focus: Petrofictions." *American Book Review* 33, no. 3 (2012): 3.

————. "System Failure: Oil, Futurity and the Anticipation of Disaster." *South Atlantic Quarterly* 106, no. 4 (2007): 805–23.

Szeman, Imre, and Maria Whiteman. "Oil Imag(e)inaries: Critical Realism and the Oil Sands." *Imaginations* 3, no. 2 (2012): 46–67.

Taber, Jane. "Meet Harper's Oil-Sands Muse." *Globe and Mail*, September 10, 2012. http://m .theglobeandmail.com/news/politics/ottawa-notebook/meet-harpers-oil-sands-muse/ article1871340/.

Talim, Valerie. "Rape, Kidnapping Being Investigated as Hate Crime in Thunder Bay." *Indian Country Today Media Network*, January 7, 2013. http://indiancountrytodaymedia network.com/article/rape-kidnapping-being-investigated-hate-crime-thunder-bay -146797.

Tanuro, Daniel. "Marxism, Energy, and Ecology: The Moment of Truth." *Capitalism Nature Socialism* 21, no. 4 (2010): 89–101.

Taylor, Bron. *Dark Green Religion: Nature Spirituality and the Planetary Future*. Berkeley: University of California Press, 2009.

Taylor, Paul. *Respect for Nature: A Theory of Environmental Ethics*. Princeton, NJ: Princeton University Press, 1986.

Thompson, Allen. "Radical Hope for Living Well in a Warming World." *Journal of Agricultural and Environmental Ethics* 23, no. 1 (2010): 43–59.

Thompson, Allen, and Jeremy Bendik-Keymer, eds. *Ethical Adaptation to Climate Change: Human Virtues of the Future*. Cambridge, MA: MIT Press, 2012.

Thompson, E. P. *The Making of the English Working Class*. Harmondsworth: Penguin, 1963.

Thoreau, Henry David. *The Maine Woods*. Edited by Joseph J. Moldenhauer. Princeton, NJ: Princeton University Press, 2004.

Tidwell, Mike. *Bayou Farewell: The Rich Life and Tragic Death of Louisiana's Cajun Coast*. New York: Vintage, 2003.

Tilly, Chris. *Work under Capitalism*. Boulder, CO: Westview, 1998.

Toffel, Michael W., and Arpad Horvath. "Environmental Implications of Wireless Technologies: News Delivery and Business Meetings." *Environmental Science and Technology* 38, no. 11 (2004): 2961–70.

Tomkins, Calvin. "A Sense of Place." *New Yorker*, September 23, 2013, 84.

Torday, Piers. "Why Writing Stories about Climate Change Isn't Fantasy or Sci-fi." *Guardian*, April 21, 2015. http://www.theguardian.com/childrens-books-site/2015/apr/21/climate -change-isnt-fantasy-sci-fi-piers-torday.

Toye, John. *Dilemmas of Development*. Oxford: Basil Blackwell, 1987.

Trainer, Ted F. E. *Consumer Society: Alternatives for Sustainability*. Sydney, Australia: Zed Books, 1995.

Traweek, Sharon. *Beamtimes and Lifetimes: The World of High Energy Physicists*. Cambridge, MA: Harvard University Press, 1988.

Tripp, Charles. *A History of Iraq*. Cambridge: Cambridge University Press, 2007.

Tsing, Anna L. *Friction: An Ethnography of Global Connection*. Princeton, NJ: Princeton University Press, 2004.

Tudge, Colin. *Neanderthals, Bandits, and Farmers: How Agriculture Really Began*. New Haven, CT: Yale University Press, 1999.

Turcotte, Heather M. "Contextualizing Petro-Sexual Politics." *Alternatives: Global, Local, Political* 36, no. 3 (2011): 204.

———. *Petro-sexual Politics: Global Oil, Legitimate Violence, and Transnational Justice.* Charleston: Biblio Bazaar, 2011.

Turner, Jack. *The Abstract Wild.* Tucson: University of Arizona Press, 1996.

Tutuola, Amos. *The Palm-Wine Drinkard.* New York: Grove Weidenfeld, 1984.

UNCTAD. *Review of Maritime Transport 2007.* Geneva: United Nations Commission on Trade and Development, 2007.

UNDP. "The Niger Delta Human Development Report." *United Nations Development Programme.* Abuja, 2006.

United Nations. *United Nations Framework Convention on Climate Change.* 1992. http://un fccc.int/resource/docs/convkp/conveng.pdf.

Updike, John. "Twin Beds in Rome." In *Too Far to Go: The Maples Stories,* 53–64. New York: Knopf Everyman's Pocket Classics, 2009.

US Energy Information Administration. *International Energy Outlook 2011.* Washington, DC: Department of Energy, 2011.

Vattimo, Gianno. *The Transparent Society.* Translated by David Webb. Baltimore: Johns Hopkins University Press, 1994.

Vaughn, Adam. "What Does Canada's Withdrawal from Kyoto Protocol Mean?" *Guardian,* December 13, 2011. http://www.theguardian.com/environment/2011/dec/13/canada-with drawal-kyoto-protocol.

Verchick, M. R. Robert. "Feminist Theory and Environmental Justice." In *New Perspectives on Environmental Justice: Gender, Sexuality, and Activism,* edited by Rachel Stein, 63–77. New Brunswick, NJ: Rutgers University Press, 2004.

Vidal, John. "World Bank: Ditch Fossil Fuel Subsidies to Address Climate Change." *Guardian,* September 21, 2011. http://www.theguardian.com/environment/2011/sep/21/world -bank-fossil-fuel-subsidies.

Vitalis, Robert. *America's Kingdom: Mythmaking on the Saudi Oil Frontier.* Palo Alto, CA: Stanford University Press, 2006.

Vitousek, Peter M., et al. "Human Domination of Earth's Ecosystems." *Science* 277, no. 5325 (1997): 494–99.

WAC Global Services. *Peace and Security in the Niger Delta.* Port Harcourt, 2003.

Walker, Samuel. *Containing the Atom: Nuclear Regulation in a Changing Environment, 1963–1971.* Berkeley: University of California Press, 1992.

Walonen, Michael T. "'The Black and Cruel Demon' and Its Transformations of Space: Towards a Comparative Study of the World Literature of Oil and Place." *Interdisciplinary Literary Studies* 14, no. 1 (2012): 56–78.

Wantchekon, Leonard. "Why Do Resource Dependent Countries Have Authoritarian Governments?" *Journal of African Finance and Economic Development* 5, no. 2 (2002): 17–56.

Ward, Jerry W. *The Katrina Papers: A Journal of Trauma and Recovery.* New Orleans: University of New Orleans Press, 2008.

Watts, Michael, ed. "Petro-Violence: Community, Extraction, and Political Ecology of a Mythic Commodity." In *Violent Environments,* edited by Nancy L. Peluso, 189–212. New York: Cornell University Press, 2001.

————. "Righteous Oil? Human Rights, the Oil Complex and Corporate Social Responsibility." *Annual Review of Environment and Resources* 30 (2005): 373–407.

Weart, Spencer. *The Discovery of Global Warming.* Cambridge, MA: Harvard University Press, 2003.

Weber, Elke U. "Experience-Based and Description-Based Perceptions of Long-Term Risk: Why Global Warming Does Not Scare Us (Yet)." *Climatic Change* 77 (2006): 103–20.

Weinhold, Robert. "A New Pulp Fact? New Research Suggests a Possible Mechanism for Some of the Damage to Fish from Pulp and Paper Mills." *Environmental Science and Technology* 43 (2009): 1242.

Weisman, Alan. *The World without Us.* New York: Harper Perennial, 2007.

Wenzel, Jennifer. "Petro-Magic-Realism: Toward a Political Ecology of Nigerian Literature." *Postcolonial Studies* 9, no. 4 (2006): 449–64.

Wheeler, M. Stephen, and Timothy Beatley, eds. *The Sustainable Urban Development Reader.* New York: Routledge, 2004.

White, Leslie. "Energy and the Evolution of Culture." *American Anthropologist* 45, no. 3 (1943): 335–56.

————. *The Science of Culture: A Study of Man and Civilization.* New York: Farrar, Straus & Giroux, 1949.

Whitehead, John A. "The Partition of Energy by Social Systems: A Possible Anthropological Tool." *American Anthropologist* 89 (1987): 686–700.

Wiggins, David. "Nature, Respect for Nature, and the Human Scale of Values." *Proceedings of the Aristotelian Society* 100, no. 1 (2000): 1–32.

Wilhite, Harold. "Why Energy Needs Anthropology." *Anthropology Today* 21, no. 3 (2005): 1–3.

Williams, Bernard. *Ethics and the Limits of Philosophy.* Cambridge, MA: Harvard University Press, 1985.

Wilson, Edward O. *The Future of Life.* New York: Alfred A. Knopf, 2002.

————. *In Search of Nature.* Washington, DC: Island, 1996.

Wilson, Sheena. "Ethical Oil: The Case for Canada's Oil Sands, Review." *American Book Review* 33, no. 3 (2012): 8–9.

————. Foreword to *Common Wealth: Economics for a Crowded Planet,* by Jeffrey Sachs, xii. New York: Penguin, 2008.

Wilson, Sheena, and Andrew Pendakis. "Sight, Site, Cite: Oil in the Field of Vision." *Imaginations* 3, no. 2 (2012): 1–5.

Winder, George. *The Free Convertibility of Sterling.* London: Batchworth, 1955.

Winther, Tanja. *The Impact of Electricity: Development, Desires and Dilemmas.* Oxford: Berg, 2008.

Wolfe, Cary. *What Is Posthumanism?* Minneapolis: University of Minnesota Press, 2010.

Wollstonecraft, Mary. *A Vindication of the Rights of Woman.* New York: Knopf, 1992.

Wood, Allan. "Kant on Duties Regarding Nonrational Nature." *Proceedings of the Aristotelian Society Supplementary Volume* 72, no. 1 (1998): 189–210.

Wood, Long. "Long-Term World Oil Supply Scenarios: 2000 U.S. Geological Survey World Petroleum Assessment." Accessed September 19, 2006. http://pubs.usgs.gov/dds/dds-060.

Woodhouse, Peter. "African Enclosures: A Default Mode of Development." *World Development* 31, no. 10 (2000): 1705–20.

World Investment Report. United Nations. New York, 2005.

Wright, Oliver. "Britain Meets Gulf Allies over Growing Tensions in Iran." *Independent,* January 24, 2013. http://www.independent.co.uk/news/uk/home-news/britain-meets -gulf-allies-over-growing-tensions-in-iran-8278101.html.

Yaeger, Patricia, Ken Hiltner, Saree Makdisi, Vin Nardizzi, Laurie Shannon, Imre Szeman, and Michael Ziser. "Editor's Column: Literature in the Ages of Wood, Tallow, Coal, Whale Oil, Gasoline, Atomic Power, and Other Energy Sources." *PMLA* 126, no. 2 (2011): 305–26.

Yates, Douglas A. *The Rentier State in Africa: Oil Rent Dependency and Neocolonialism in the Republic of Gabon.* Trenton, NJ: Africa World, 1996.

Yergin, Daniel. "Ensuring Energy Security." *Foreign Affairs* 85, no. 2 (2006): 67–82.

———. *The Prize: The Epic Quest for Oil, Money, and Power.* New York: Free Press, 1991.

Zagorin, Perez. "Vico's Theory of Knowledge: A Critique." *Philosophical Quarterly* 34 (1984): 15–30.

Zalasiewicz, Jan. "Are We Now Living in the Anthropocene?" *GSA Today* 18 (2008): 4–8.

Zalik, Anna. "The Peace of the Graveyard: The Voluntary Principles on Security and Human Rights in the Niger Delta." In *Global Regulation: Managing Crisis after the Imperial Turn,* edited by Kees Van Der Pijl et al., 111–27. London: Macmillan, 2004.

Ziegler, Philip. *The Black Death.* London: Folio Society, 1997.

Ziser, Michael. "Home Again: Peak Oil, Climate Change, and the Aesthetics of Transition." In *Environmental Criticism for the Twenty-first Century,* edited by Stephanie LeMenager, Teresa Shewry, and Ken Hiltner, 181–95. New York: Routledge, 2011.

Zittel, Werner, et al. *Coal: Resources and Future Production.* EWG Paper no. 1/01, July 10, 2007. http://energywatchgroup.org/wp-content/uploads/2014/02/EWG_Report_Coal _10-07-2007ms1.pdf.

Žižek, Slavoj. "Censorship Today: Violence, or Ecology as a New Opium for the Masses, Part 1." http://www.lacan.com/zizecology1.htm.

Zube, H. Ervin, ed. *Landscapes: Selected Writings of J. B. Jackson.* Amherst: University of Massachusetts Press, 1970.

Contributors

MARGARET ATWOOD is the author of more than forty volumes of poetry, non-fiction, and novels. Her newest novel, *MaddAddam* (2013), is the final volume in a series that began with the Man Booker Prize–nominated *Oryx and Crake* (2003). Her nonfiction book *Payback: Debt and the Shadow Side of Wealth* (2008) was adapted for the screen in 2012.

PAOLO BACIGALUPI has been nominated for three Nebula Awards and four Hugo Awards. He won the Theodore Sturgeon Memorial Award for best science fiction short story of the year. His latest novel for adults is *The Water Knife* (2015), a near-future thriller about climate change and drought in the southwestern United States.

LESLIE BATTLER is a widely published poet who won the PRISM international Earle Birney Award (2012) and the University of Calgary Poem of the Season Award (2009). Battler received an MA in English from Concordia University and currently lives in Calgary, where until recently she worked in the petrochemical industry.

URSULA BIEMANN is an artist, writer, and video essayist whose work investigates climate change and the ecologies of oil and water. Biemann is a member of the World of Matter collective project on resource ecologies, and she is a senior researcher at the Zurich University of the Arts.

DOMINIC BOYER is Professor of Anthropology and Director of the Center for Energy and Environmental Research in the Human Sciences at Rice University. He is the author most recently of *The Life Informatic: Newsmaking in the Digital Era* (2013) and part of the editorial collective of the journal *Cultural Anthropology*. Boyer also produces and cohosts the energy humanities podcast series "Cultures of Energy."

ITALO CALVINO was an Italian journalist and writer of short stories and novels. His works include the *Our Ancestors* trilogy (1952–59) and the novels *Invisible Cities* (1972) and *If on a Winter's Night a Traveler* (1979). He was the most translated contemporary Italian writer at the time of his death and a contender for the Nobel Prize for Literature.

WARREN CARIOU is Associate Professor of English at the University of Manitoba and the author of *The Exalted Company of Roadside Martyrs* (1999) and *Lake of the Prairies* (1999). In 2008, Warren was awarded a Canada Research Chair in Narrative, Community and Indigenous Cultures.

DIPESH CHAKRABARTY is currently Lawrence A. Kimpton Distinguished Service Professor in history at the University of Chicago, and he was the recipient of the 2014 Toynbee Prize.

UNA CHAUDHURI is Collegiate Professor and Professor of English and Drama at New York University. With director Fritz Ertl, she has developed a number of theater pieces using a process they call "Research Theatre," and she has worked collaboratively with the artist Marina Zurkow, most recently in a multiplatform project entitled "Dear Climate."

CLAIRE COLEBROOK is Edwin Erle Sparks Professor of English at the Pennsylvania State University. The author of many volumes on literature and philosophy, she has recently completed two books on extinction for Open Humanities Press: *The Death of the Posthuman* (2015) and *Sex after Life* (2015).

STEPHEN COLLIS is Professor of English at Simon Fraser University. His many books of poetry include *The Commons* (2008), *On the Material* (2010), and *To the Barricades* (2013). He has also written two books of criticism, a novel, and a collection of essays on the Occupy movement, *Dispatches from the Occupation* (2012).

ERIK M. CONWAY is a historian of science and technology living in Pasadena, California. He publishes on such diverse topics as aerospace, the history of Mars exploration, and the history of climate science. Recent books (with Naomi Oreskes) include *Merchants of Doubt* (2010) and *The Collapse of Western Civilization* (2014).

AMY DE'ATH is a PhD student at Simon Fraser University and works on the poetics journal *Line*. Her books of poetry include *Erec & Enide* (2010) and *Caribou* (2011). With Fred Wah, she is the editor of a collection of poetry and poetics, *Toward. Some. Air.* (2014).

ADAM DICKINSON is Professor of Poetry at Brock University. His writing focuses on intersections between poetry and science, exploring new ecocritical perspectives and alternative modes of poetic composition. His book *The Polymers* (2013) was a finalist for both the Governor General's Award for Poetry and the Trillium Book Award for Poetry.

Fritz Ertl is head of faculty and curriculum at Playwrights Horizons Theater School in New York. For the past 10 years he has been working on a series of new plays exploring the catastrophic consequences of globalization, including *Carla and Lewis* (aka the ecocide project), by Shonni Enelows.

Pope Francis is the 266th pope of the Roman Catholic Church. The first pope from the Southern Hemisphere, Francis has become known for his adoption of progressive causes, including the mitigation of climate change. In 2015 he published the encyclical *Laudato si'* (2015) on climate change, care for the environment, and sustainable development.

Amitav Ghosh is an author whose work has been translated into more than twenty languages. He has taught in many universities in India and the United States. His novels include *The Circle of Reason* (1986), *The Shadow Lines* (1988), *In an Antique Land* (1992), *The Calcutta Chromosome* (1995), *Dancing in Cambodia* (1998), and *The Hungry Tide* (2004).

Gökçe Günel is a lecturer in Anthropology at Columbia University. She has been a Cultures of Energy Mellon-Sawyer Postdoctoral Fellow at Rice University and an ACLS New Faculty Fellow at Columbia University. Her book *Spaceship in the Desert: Energy, Climate Change and Green Business in Abu Dhabi* is forthcoming.

Gabrielle Hecht is Professor of History at the University of Michigan. Her work addresses themes such as technopolitics, occupational and environmental health, labor, ontological politics, nationalism, and postcoloniality. She is the author of *The Radiance of France: Nuclear Power and National Identity* (2009) and *Being Nuclear: Africans and the Global Uranium Trade* (2012).

Cymene Howe is Associate Professor of Anthropology at Rice University. Her research investigates the overlapping conversations between feminist and queer theory, new materialisms, ontologies, and social movements. She is the author of *Intimate Activism: The Struggle for Sexual Rights in Postrevolutionary Nicaragua* (2013) and *Ecologics: Wind and Power in the Anthropocene* (forthcoming).

Dale Jamieson is Professor of Environmental Studies and Philosophy and Chair of the Environmental Studies Department at New York University. His books include *Ethics and the Environment: An Introduction* (2008) and *Reason in a Dark Time: Why the Struggle to Stop Climate Change Failed—and What It Means for Our Future* (2014).

JULIA KASDORF is an award-winning poet. Her works include *Sleeping Preacher* (1992), *Eve's Striptease* (1998), *The Body and the Book: Writing from a Mennonite Life* (2009), and *Poetry in America* (2011).

OLIVER KELLHAMMER is a land artist, permaculture teacher, activist, and writer. His botanical interventions and public art projects demonstrate nature's surprising ability to recover from damage. Recent writings include "Neo-Eocene," published in *Making the Geologic Now* (ed. Jamie Kruse and Elisabeth Ellsworth, 2012), and "Violent Reactions" in Marina Zurkow's *Petroleum Manga* (2014).

STEPHANIE LEMENAGER is Professor of English at University of Oregon. Her pioneering work in the environmental humanities includes *Living Oil: Petroleum Culture in the American Century* (2014) and her current project, *Weathering* (forthcoming), which focuses on the ecological significance of the humanities in the era of global climate change.

BARRY LORD is internationally known as one of the world's leading museum planners. In *Art & Energy: How Culture Changes* (2014) he highlights the major cultural shifts that have accompanied each energy transition since our mastery of fire to argue that the so-called energy debate is really a conflict of cultures.

GRAEME MACDONALD is Associate Professor of English and Comparative Literary Studies at the University of Warwick, where he is also a member of WreC (Warwick Research Collective), an organization that works on new ways of thinking about world literatures. His research examines global modernity, resource culture, energy and petrofiction, and science fiction and ecocriticism.

JOSEPH MASCO is Associate Professor of Anthropology at University of Chicago. He is the author of *The Nuclear Borderlands: The Manhattan Project in Post–Cold War New Mexico* (2006), which won the 2008 Rachel Carson Prize from the Society for the Social Studies of Science.

JOHN MCGRATH was a playwright and a theorist of radical theater. The founder and Artistic Director of the agitprop theater group 7:84, McGrath took up the cause of Scottish independence, writing on class struggle, exploitation, and economic history in his works.

MARTIN MCQUILLAN is Professor of Literary Theory and Cultural Analysis and Pro-Vice Chancellor (Research) and Dean of the Faculty of Arts and Social Sciences at Kingston University, London. He is Codirector of the London Graduate

School. He is the author of *Deconstruction after 9/11* (2009) and *Roland Barthes; or, The Profession of Cultural Studies* (2011).

TIMOTHY MITCHELL is William B. Ransford Professor of Middle Eastern, South Asian, and African Studies at Columbia University. A political theorist and historian, he is the author of *Colonising Egypt* (1991), *Rule of Experts: Egypt, Techno-Politics, Modernity* (2002), and *Carbon Democracy: Political Power in the Age of Oil* (2011).

TIMOTHY MORTON is Professor and Rita Shea Guffey Chair in English at Rice University. His work explores the intersection of object-oriented thought and ecological studies. He is the author of *Ecology without Nature: Rethinking Environmental Aesthetics* (2007) and *Hyperobjects: Philosophy and Ecology after the End of the World* (2013).

JEAN-FRANÇOIS MOUHOT has been Marie Curie Fellow at Georgetown University and L'école des hautes etudes en sciences sociales. He is the author of *Les Acadiens réfugiés en France: l'Impossible réintégration* (2009) and *Des Esclaves énergétiques: Réflexions sur le changement climatique* (2011).

ABDUL RAHMAN MUNIF was one of the Arab world's best-known novelists. Edward Said called his trilogy *Cities of Salt* (1984) "the only serious work of fiction that tries to show the effect of oil, Americans and the local oligarchy on a Gulf country."

JUDY NATAL is a Chicago-based artist and Professor of Photography and Coordinator of the Graduate Program at Columbia College. Her photography examines the visual narratives landscapes and alterations to those landscapes hold. She is the author of *EarthWords* (2004) and *Neon Boneyard Las Vegas A–Z* (2006).

REZA NEGARESTANI is a philosopher who has contributed extensively to journals and anthologies and lectured at numerous international universities and institutes. He is the author of *Cyclonopedia: Complicity with Anonymous Materials* (2008), one of *Artforum*'s best books of 2009.

PABLO NERUDA was a poet, politician, and winner of the 1971 Nobel Prize for Literature. Gabriel García Márquez proclaimed Neruda "the greatest poet of the 20th century in any language."

DAVID NYE is Professor of American Studies at the Danish Institute of Advanced Study at the University of Southern Denmark. He is the author of *Electrifying*

America: Social Meanings of a New Technology, 1880–1940 (1992), *Consuming Power: A Social History of American Energies* (1999), and *When the Lights Went Out: A History of Blackouts in America* (2010).

NAOMI ORESKES is Professor of the History of Science and Affiliated Professor of Earth and Planetary Sciences at Harvard University. Her 2004 essay "The Scientific Consensus on Climate Change" has been widely cited. She is the author (with Erik M. Conway) of *Merchants of Doubt* (2010) and *The Collapse of Western Civilization* (2014).

ANDREW PENDAKIS is Assistant Professor of Theory and Rhetoric at Brock University. His research focuses on contemporary political culture, with a special interest in the genealogy of centrist reason in the West. He is coeditor of *Contemporary Marxist Theory: An Anthology* (2014) and of *The Bloomsbury Companion to Marx* (forthcoming).

KAREN PINKUS is Professor of Italian and Comparative Literature at Cornell University. She is also a minor graduate field member in Studio Art, a member of the Advisory Board of the Atkinson Center for a Sustainable Future, and a member of the Climate Change Focus Group.

KEN SARO-WIWA was a Nigerian writer, television producer, and environmental activist. Saro-Wiwa led a nonviolent campaign against environmental degradation of the land and waters of Ogoniland by the multinational petroleum industry, and he was also an outspoken critic of the Nigerian government. His execution in 1995 provoked international outrage.

HERMANN SCHEER was a member of the German Parliament, President of the European Association for Renewable Energy (EUROSOLAR), Chairman of the World Council for Renewable Energy (WCRE), and an author. He passed away in 2010.

ROY SCRANTON is a scholar and public intellectual. He has published widely, from *Contemporary Literature* and *Theory & Event* to *Rolling Stone* and the *New York Times*. His essay "Learning How to Die in the Anthropocene" was selected for *The Best American Science and Nature Writing 2014*.

ALLAN STOEKL is Professor of French and Comparative Literature at Penn State University. His recent work has focused on issues of energy use, sustainability, and economy in a literary-cultural and philosophical context. He is the author of *Bataille's Peak: Energy, Religion, and Postsustainability* (2007).

IMRE SZEMAN is Canada Research Chair of Cultural Studies and Professor of English and Film Studies at the University of Alberta. He is coauthor of *After Globalization* (2011) and *After Oil* (2016) and coeditor of *Fueling Culture: 101 Words for Energy and Environment* (2017).

LAURA WATTS is an ethnographer, poet, writer, and Associate Professor at the IT University of Copenhagen. She has worked with the mobile telecoms industry, the renewable energy industry, and the public transport sector. She is currently collaborating in the development of the marine energy industry in the Orkney Islands, Scotland.

MICHAEL WATTS is Professor of Geography at University of California, Berkeley. The author and editor of numerous volumes on political ecology and environmental thought, Watts's work is characterized by a long-standing engagement with the political economy of development and in particular energy and agro-food sectors in Africa.

JENNIFER WENZEL is Associate Professor in the Department of English and Comparative Literature and the Department of Middle Eastern, South Asian, and African Studies at Columbia University. She is author of *Bulletproof: Afterlives of Anticolonial Prophecy in South Africa and Beyond* (2009) and coeditor of *Fueling Culture: 101 Words for Energy and Environment* (2017).

SHEENA WILSON is Assistant Professor of English and Cultural Studies at University of Alberta's Campus Saint-Jean. She is Codirector of the Petrocultures Research Group and Editor in Chief of *Imaginations: Journal of Cross-Cultural Image Studies*. She is also Editor, at University of Alberta Press, of Petrocultures: An Energy Humanities Series.

PATRICIA YAEGER was Henry Simmons Frieze Collegiate Professor of English and Women's Studies at University of Michigan. Her publications include *Honey-Mad Women: Emancipatory Strategies in Women's Writing* (1989), *The Geography of Identity* (1996), and *Dirt and Desire: Reconstructing Southern Women's Writing: 1930–1990* (2000). From 2006 to 2011, she was Editor of *PMLA*.

MARINA ZURKOW is a media artist who has used life science, biomaterials, animation, dinners, and software technologies to foster intimate connections between people and nonhuman agents. Her work spans gallery installations and unconventional public participatory projects. She is a full-time faculty member at ITP / Tisch School of the Arts, New York University.